Microwave Active Circuit Analysis and Design

Microwave Active Circuit Analysis and Design

Clive Poole
Izzat Darwazeh
Department of Electronic and Electrical Engineering
University College London

AMSTERDAM · BOSTON · HEIDELBERG · LONDON
NEW YORK · OXFORD · PARIS · SAN DIEGO
SAN FRANCISCO · SINGAPORE · SYDNEY · TOKYO
Academic Press is an imprint of Elsevier

Academic Press is an imprint of Elsevier
125 London Wall, London, EC2Y 5AS, UK
525 B Street, Suite 1800, San Diego, CA 92101-4495, USA
225 Wyman Street, Waltham, MA 02451, USA
The Boulevard, Langford Lane, Kidlington, Oxford OX5 1GB, UK

Library of Congress Cataloging-in-Publication Data
A catalog record for this book is available from the Library of Congress

British Library Cataloguing in Publication Data
A catalogue record for this book is available from the British Library

For information on all Academic Press publications
visit our website at http://store.elsevier.com/

Printed in the United States of America

ISBN: 978-0-12-407823-9

Working together
to grow libraries in
developing countries

www.elsevier.com • www.bookaid.org

Contents

Preface

This book arose out of our experiences of teaching electronics, communications and microwave engineering over a combined period of 50 years. As a result, we felt there was a need for a concise teaching book that covers the fundamentals but also addresses new development in microwave circuit design. Although the world is not short of text books on microwave circuit design, including some excellent reference works which we cite in this book, many of these are either very broad in scope, covering both active and passive circuits, or else very specialist in nature. We felt that there was room for a new teaching book that is academically rigorous and yet readable and accessible. We also wanted to collect together what we considered to be the foundational elements of the discipline in one text, with a clear and logical flow from one topic to the next, making it ideal for teaching or self-study at undergraduate and postgraduate levels.

The widespread availability of powerful Computer Aided Design (CAD) software has boosted the productivity of the microwave design industry enormously, allowing rapid optimization of complex circuits and eliminating hours of demanding "bench time." The widespread availability of low-cost computer power presents a potential challenge for engineering education in that students may be tempted to cut out the design phase and go straight from concept to simulation. In this book, we have focused on the development of insight and understanding, and have therefore used examples and problems that do not rely on specialized CAD software for their solution.

This book is organized in three parts; covering foundations, analysis, and design.

The foundations part comprises four chapters starting with Chapter 1, which sets out some basic concepts with which all microwave engineers need to be familiar. Chapters 2 and 3 deal with transmission lines while Chapter 4 introduces the Smith Chart, which is a critically important tool for any microwave engineer.

In the second part of the book, we look at various techniques of analysis that are applied to active circuits, specifically immittance parameters and S-parameters, and how they are inter-related. We have included a specific chapter on gain and stability of two-port networks as these concepts are of critical importance when dealing with active devices. Two other chapters in this part cover the analysis (as well as the design) of matching networks, which largely determine the performance of any active element in a circuit. Uniquely, this book includes a chapter detailing three-port network analysis. We considered this material important because the most common active devices used today, namely transistors, are three terminal devices, and the application of three-port analysis techniques allows the microwave engineer extra degrees of freedom in the design process.

The third part of the book first introduces the semiconductor devices used in microwave circuits and how these are integrated in monolithic form. It then goes on to discuss the design of actual system building blocks; amplifiers, oscillators, and mixers. Due to the importance of noise in microwave systems, we have included two chapters dealing with noise in amplifiers and oscillators, and how an amplifier or oscillator designer can take steps to minimize noise in their design.

The book is structured in modular form to facilitate its use in teaching. Each chapter of the first two parts may be considered as the basis of a 2 hour lecture. These chapters (Chapters 1–10) contain worked examples and some tutorial questions to aid learning. The more advanced materials contained in the design part of the book (Chapters 11–17) lend themselves to more advanced teaching and learning. The detailed derivations and worked examples in these chapters will equip readers with the knowledge required to extend their learning with the help of one of the many microwave CAD packages available. The material in these chapters is also suitable for delivery in lecture format, in this case two lectures per chapter. In short, the book can be considered as a teaching book of a seventeen 2 hour lectures.

At the start of each chapter we list the key learning outcomes, in terms of knowledge to be gained and skills to be acquired. At the end of each chapter, we summarize the salient points of the chapter in the form of list of "Takeaways." These listings are not exhaustive; each reader's experience will differ and any reader is therefore encouraged to add their own points to these lists.

In writing this book, we have consulted numerous references and each chapter contains what we consider to be the most significant and useful books and papers on the specific topic of the chapter. We urge the readers of this book to consult these references for greater depth of knowledge and understanding.

In keeping with the teaching focus of the book, we have prepared lecture slides on each chapter. These, as well as some useful MATLAB computer programmes, are available from the companion website.

We hope that this book will be a useful addition to the body of work in this field and that it will inspire the reader to explore this important and fascinating area of electronics. We hope that you will find reading it not only informative but also enjoyable.

Clive Poole and Izzat Darwazeh
Department of Electronic and Electrical Engineering,
University College London,
London 2015

Acknowledgments

The material in this book has been developed from teaching material delivered at various universities. Most of the work on this book, however, was done while the authors were at University College London. We would therefore firstly like to express our gratitude to our students and colleagues over the years for helping us formulate the ideas that have been distilled into this book. We would like to acknowledge, in particular, our UCL colleagues, Dr Ed Romans, Prof Andreas Demosthenous, and Dr Sally Day for their fruitful discussions and advice. We would also like to thank Dr Ryan Grammenos for proofreading the manuscript.

Special thanks are due to Professor Jeremy Everard of University of York for some interesting and enlightening discussions on oscillator phase noise and for permission to use some of his original material in Chapter 16. We also thank Prof Ian Robertson of Leeds University for permission to use some of his MMIC images.

This book was typeset in LaTeX and we would like to thank the numerable members of the LaTeX community whose combined efforts have resulted in a superb medium for the preparation of textbooks such as this one. In particular, we would like to mention Dr Christian Feuersänger for creating the excellent pgfplots package that was used to produce the graphs and diagrams, including most of the Smith Charts, and Massimo A. Redaelli for creating the circuitikz LaTeX package that was used to produce all the schematic diagrams. We also acknowledge the help of Brian Force in creating the MATLAB code that was used to produce the feedback mapping illustrations in Chapter 8.

Clive Poole would like to thank personally some of the people who first inspired him to pursue a career as a microwave engineer, namely Dr David James, formerly of Ferranti Microwave, Poynton, Cheshire, and Ian Clarke and Keith Williams both formerly of Philips Microwave, Hazel Grove, Manchester. He would also like to acknowledge the many thoughtful discussions he has had on microwave engineering (as well as other topics) with Dr Peter Yip who has been an inspirational presence during the many years spent together both in academia and the microwave industry.

Finally, we would like to thank our partners, Lorna Poole and Rachel Darwazeh for their patience and unwavering support throughout the long process of putting this book together.

Clive Poole and Izzat Darwazeh
Department of Electronic and Electrical Engineering,
University College London,
London 2015

List of symbols and abbreviations

The following general conventions will be followed:

1. Upper case letters will be used to represent DC voltages and currents, in the format V_n, I_n.
2. Lower case letters will be used to represent signal voltages and currents, in the format v_n, i_n.
3. Upper case S, in the context of S-parameters, will be used to represent one-port or two-port S-parameters. Any other n-port network where $n > 2$ will be represented by S-parameters in the lower case, that is, s_{ij}.

Specific abbreviations are as follows:

$\angle x$	Phase angle of complex variable x
$\|x\|$	Magnitude of complex variable x
a_i	Incident power wave at ith port of an n-port network
AC	Alternating current
ANA	Automatic Network Analyzer
B	Susceptance (in S) or bandwidth (depending on context)
b_i	Reflected power wave at ith port of an n-port network
B_{ij}	Feedback mapping parameter for the ij plane
BJT	Bipolar junction transistor
β	Feedback fraction or coupling coefficient or transistor common-emitter current gain (depending on context)
C_{ij}	Feedback mapping parameter for the ij plane
C_{fb}	Feedback capacitance
cbg	Common base or common gate configuration
ccd	Common collector or common drain configuration
ces	Common emitter or common source configuration
DC	Direct current
DRO	Dielectric resonator oscillator
DUT	Device under test
Δ	Determinant of two-port S-matrix
Δ_{ij}	Determinant of three-port sub S-matrix
Δf	Bandwidth (depending on context)
E_{ij}	Scattering error parameter
f_o	Resonant frequency
fF	Femto-farad
G	Conductance (in S), or gain (depending on context)
GaAs	Gallium arsenide
GaN	Gallium nitride
GBP	Gain-bandwidth product
GHz	Gigahertz

g_m	Forward transconductance of a transistor
G_{max}	Maximum available gain
Γ_{3smax}	Port-3 termination giving maximum reduced two-port S-parameter
Γ_{in}	Input reflection coefficient
Γ_L	Load reflection coefficient
Γ_{ms}	Source reflection coefficient for maximum gain
Γ_{ml}	Load reflection coefficient for maximum gain
Γ_{out}	Output reflection coefficient
Γ_{rij}	Center of constant resistance mapping circle
γ_{rij}	Radius of constant resistance mapping circle
Γ_s	Source reflection coefficient
Γ_{Tij}	Center of third-port plane constant S_{ij} circle
γ_{Tij}	Radius of third-port plane constant S_{ij} circle
Γ_{xij}	Center of constant reactance mapping circle
γ_{xij}	Radius of constant reactance mapping circle
HBT	Heterojunction bipolar transistor
HEMT	High electron mobility transistor
$Im(x)$	Imaginary part of a complex variable x
K	Rollet's stability factor *or* Kelvin (unit of absolute temperature) depending on context
k_B	Boltzmann constant ($=1.3806488 \times 10^{23}$ J K^{-1})
$\mathscr{L}\{f_m\}$	Phase noise to carrier ratio at frequency offset f_m from carrier
MESFET	Metal semiconductor field effect transistor
MMIC	Monolithic microwave integrated circuits
MSG	Maximum stable gain
ω	Radian frequency
ω_0	Resonant frequency (in radians)
Φ_{3smax}	Port-3 phase giving maximum reduced two-port S-parameter
Q	Stored charge *or* quality factor *or* dissipation matrix *or* transistor number (depending on context)
R	Resistance (in Ω)
$Re(x)$	Real part of complex a variable x
RMS	Root mean square
s_{ij}	Three-port scattering parameter
S_{ij}	Two-port scattering parameter
S_{ija}	Actual scattering parameter
S_{ijm}	Measured scattering parameter
S'_{ij}	Reduced two-port scattering parameter
SiGe	Silicon germanium
T_{ij}	Two-port scattering transfer parameter
THz	Terahertz

VSWR	Voltage standing wave ratio
X	Reactance (in Ω)
X_{fb}	Feedback reactance
Y_{in}	Input admittance
YIG	Yttrium iron garnet
YTO	YIG tuned oscillator
Z_{in}	Input impedance
Z_{o}	System characteristic impedance

About the authors

Clive Poole is a Principal Teaching Fellow and Director of Telecommunications Industry Programmes at University College London (UCL). He has 30 years' experience in the global electronics and telecommunications industries as well as academia. He started his career as a design engineer in several UK microwave companies, designing X-band and Ku-band amplifiers and oscillators for military and telecommunications applications. In the early 1990s, he founded an electronics design consultancy in Hong Kong that developed a number of successful wireless and telecommunications products for Chinese manufacturers. He has run several high technology businesses, including a bespoke paging equipment manufacturer and a large contract manufacturing operation. He was a pioneer in the business of deploying mobile phone networks on ocean going passenger ships. Dr Poole's teaching is focused in the areas of electronic and microwave circuit design, wireless and mobile communications, technology business strategy, and finance. He holds a BSc degree in Electronic Engineering and MSc and PhD degrees in Microwave Engineering from the University of Manchester. He also holds an MBA from the Open University. Dr Poole is a Chartered Engineer and Fellow the Institute of Engineering and Technology (FIET).

Izzat Darwazeh is the Chair of Communications Engineering in UCL and Head of UCL's Communications and Information Systems Group. He is an electrical engineering graduate of the University of Jordan and holds the MSc and PhD degrees from the University of Manchester in the UK. He has been teaching and active in microwave circuit design and communications circuits and systems research since 1991. He has published over 250 scientific papers and is the co-editor of the 1995 IEE book on Analogue Fibre Communications and of the 2008 Elsevier-Newness book on Electrical Engineering. He is also the co-author (with Luis Moura) of the 2005 book on Linear Circuit Analysis and Modelling. He currently teaches mobile and wireless communications and circuit design and his current research interests are in ultra high-speed microwave circuits and in wireless and optical communication systems. In addition to his teaching, Professor Darwazeh acts as a consultant to various engineering firms and government, financial, and legal entities in the UK and worldwide. Professor Darwazeh is a Chartered Engineer and FIET.

Foundations

Introduction

1

CHAPTER OUTLINE

Microwave Active Circuit Analysis and Design. http://dx.doi.org/10.1016/B978-0-12-407823-9.00001-9

INTENDED LEARNING OUTCOMES

- *Knowledge*
  - Understand the characteristics that distinguish high-frequency circuit design from low-frequency circuit design.
  - Be familiar with some basic electromagnetic (EM) theory and understand the importance of Maxwell's equations.
  - Understand some important properties of materials at radio frequencies (RFs) (permittivity and permeability).
  - Understand that parasitic reactances associated with familiar lumped element components become more pronounced at microwave frequencies and will significantly affect their impedance as a function of frequency, and that equivalent circuit models must therefore be used to adequately represent such components.
  - Revise the concept of quality factor, Q, for components and for the components when used in resonant circuits, and specifically its application to microwave resonators.
  - Revise the concept of *maximum power transfer*.
- *Skills*
  - Be able to calculate the inductance of a cylindrical wire.
  - Be able to design a single layer spiral inductor for a given inductance.
  - Be able to calculate the Q of common parallel and series *RLC* circuits.
  - Be able to determine the Q of a generic microwave resonator, based on return loss measurements.

1.1 INTRODUCTION TO MICROWAVE ELECTRONICS
1.1.1 WHAT ARE MICROWAVES?

The term "microwave" commonly refers to the region of the electromagnetic spectrum that extends from 1 to 30 GHz (or 30 to 1 cm wavelengths). Although, strictly speaking, any wavelength below 1 m (i.e., frequencies above 300 MHz) should be considered as being in the "microwave" range, the more general term "RF" (meaning Radio Frequency) is often used to describe frequencies in the hundreds of MHz. Topics covered in this book are also applicable to the millimeter wave and submillimeter wave frequency range which extend from 30 to 300 GHz and 300 to 3 THz, respectively.

The frequency range above 300 MHz can be usefully classified into various bands, each having a letter designation (defined by IEEE Standard 521-1984), as shown in Table 1.1 [1].

Although the application of microwave technology might appear to be a relatively recent development in electronics, the first radio signal transmission and reception to

Table 1.1 Microwave Frequency Bands

Designation	Frequency Range	Free Space Wavelength Range
Ultra high freq. (UHF)	300 MHz to 1 GHz	1 m to 30 cm
L band	1-2 GHz	30.0-15.0 cm
S band	2-4 GHz	15-7.5 cm
C band	4-8 GHz	7.5-3.8 cm
X band	8-12 GHz	3.8-2.5 cm
Ku band	12-18 GHz	2.5-1.7 cm
K band	18-27 GHz	1.7-1.1 cm
Ka band	27-40 GHz	1.1-0.75 cm
Millimeter waves	40-300 GHz	2.7-1 mm
Submillimeter waves	300 GHz to 3 THz	1-0.1 mm

be practically demonstrated (by Heinrich Hertz in 1886) was in fact a microwave, or more precisely a millimeter wave, transmission at around 60 GHz [2].

There is currently a lot of research activity in the submillimeter wave bands, also referred to as "Terahertz" research (although it could be argued that these frequencies should technically be referred to as "subterahertz"), as these sparsely populated bands offer the promise of huge bandwidths [3]. Submillimeter wave engineering occupies a middle ground between microwave electronics and infrared optical technology where coherent transmitter and receiver technology is in its infancy. Although this book addresses microwave circuit design in the GHz frequency ranges, these techniques are being extended to higher and higher frequencies all the time [4,5].

Microwave circuit design is quite different from "conventional" circuit design at lower frequencies because new and unfamiliar circuit elements and design techniques are required, hence the need for specialist text books such as this one. It may also be true to say that circuit design at microwave frequencies is more "difficult" in that designed circuits often do not perform according to expectations and usually require some tuning and alignment before they will do what they were designed to do. Hence the field of microwave circuit design has acquired the reputation of being a bit of a "black art." A lot of this is to do with the fact that, at shorter wavelengths, even a humble piece of connecting wire becomes a *transmission line* with reactive characteristics. The impedance of such a piece of wire can no longer be assumed to be zero when the physical length is comparable to the wavelength of the signal being carried. For circuits with physical dimensions of a few centimeters, for instance, we need to consider something called the *electrical length* of connecting wires when the frequency of operation is more than a few hundred MHz. Electrical length is expressed in terms of a number of wavelengths, such as 0.25λ, 3.75λ, etc., and is simply the physical length divided by the wavelength of the signal in the medium being considered. The essential point is that, as the frequency increases, the physical dimensions of circuits and components need to be taken into account as they have an effect on the electrical properties. This is a concept which sets microwave circuit design apart from circuit design at lower frequencies, where these effects can be safely ignored.

The purpose of this book is to attempt to demystify microwave circuit design and convince the student that it is no more "difficult" than any other area of electronics provided one is armed with the right tools and understanding. We therefore set out by defining a few basic concepts in this chapter.

1.1.2 THE IMPORTANCE OF RADIO FREQUENCY ELECTRONICS

The modern world as we know it would be unimaginable without radio frequency and microwave electronics. Cellular phones, satellite navigation, Wi-Fi, even the humble car locking key-fob all owe their existence to a body of theoretical knowledge about electromagnetic fields and high-frequency electronics that has been built up over the past 150 years or so. The importance of radio frequency technology is only set to increase as we move toward higher and higher operating frequencies for all kinds of electronic devices and systems: microprocessors now have clock frequencies in the GHz frequency range, making them technically "microwave" devices, and radio communications device technology is now approaching the Terahertz (1000 GHz) region [6]. The range of applications of radio frequency electronics is also set to increase dramatically, as everything goes "wireless," mobile networks enter the "fifth generation" [7–9] and we enter the era of the "Internet of Things" [10]. These developments will only require more and more information bandwidth to be made available, which can only be achieved by moving to higher and higher radio frequencies.

Many microwave circuit applications relate to portable, battery-operated equipment, with the modern cellular phone handset being the most obvious example. The microwave design engineer will therefore be faced with a number of challenges and trade-offs, aside from the task of making the microwave circuit behave as intended. Figure 1.1 shows some of these trade-offs.

With battery-operated portable devices physical size and power consumption are the most obvious constraints. Significant advances have been made in both these areas in recent years, as witnessed by the technology and performance embodied in the ubiquitous smartphone, which typically contains several independent radio transceivers, of different frequency bands, modulation schemes, and power levels, all operating simultaneously. None of this would be possible without a high degree of integration of radio frequency electronic circuitry [11]. The progress in this field, starting with the invention of the Integrated Circuit (IC) in 1958 by Jack Kilby [12], through the first Monolithic Microwave Integrated Circuit (MMIC) fabricated by Ray Pengelly and James Turner at Plessey in 1975 [13] right up to today's MMICs that

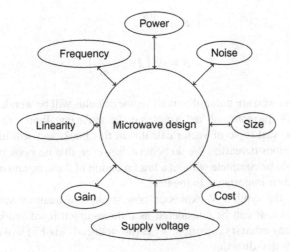

FIGURE 1.1

RF design constraints and trade-offs.

are capable of operating well into the submillimeter wave frequency range [4], and even up to THz frequencies [14], has been truly impressive.

1.1.3 ELECTROMAGNETISM BASICS

Recently we take for granted the idea that electromagnetic waves can be easily produced by a simple electronic circuit, can travel through free space and can be detected by another, remote electronic circuit, and that this arrangement can be used to transmit information over great distances. It is worth remembering, however, that this understanding of the connection between electrical electromagnetic energy is relatively recent. The wireless era only really began in 1886, when Heinrich Hertz experimentally demonstrated the existence of radio waves. He had set out to experimentally verify the existence of a phenomenon that had been predicted theoretically by James Clerk Maxwell 21 years earlier. In a landmark paper entitled "A dynamical theory of the electromagnetic field," published in 1865, Maxwell proposed the following set of equations that relate the vector quantities $\boldsymbol{E}$, $\boldsymbol{B}$, and $\boldsymbol{J}$. These equations are universally known as *Maxwell's equations*:

$$\nabla \cdot \boldsymbol{E} = \frac{\rho}{\varepsilon_0} \tag{1.1.1}$$

$$\nabla \cdot \boldsymbol{B} = 0 \tag{1.1.2}$$

$$\nabla \times E = -\frac{\partial B}{\partial t} \tag{1.1.3}$$

$$\nabla \times B = \mu_0 \left(J + \varepsilon_0 \frac{\partial E}{\partial t} \right) \tag{1.1.4}$$

Those readers who are unfamiliar with vector calculus will be wondering what the strange symbol ∇ (pronounced "del") means in the above equations. We do not intend to delve into the vast topic of vector calculus in this book, and will hardly refer to Maxwell's equations hereafter. We do believe, however, that no book on microwave technology would be complete without a brief mention of these equations, and a brief explanation of their importance to the field.

Simply put, the symbol ∇ expresses how strongly a quantity varies in three-dimensional space. It can be considered as a three-dimensional spatial derivative. If you move in any arbitrary direction, $\nabla \cdot E$ (pronounced "div E") will describe how much E varies in that direction.

By way of illustration, let x, y, z be a system of Cartesian coordinates in three-dimensional space, as shown in Figure 1.2.

Let i, j, k be the corresponding basis of unit vectors. The divergence of a continuously differentiable vector field $E = Ui + Vj + Wk$ is equal to the scalar function:

$$\nabla \cdot E = \frac{\partial U}{\partial x} + \frac{\partial V}{\partial y} + \frac{\partial W}{\partial z} \tag{1.1.5}$$

So $\nabla \cdot E$, being a scalar, simply quantifies the amount of variation there is in the field, E. By contrast, $\nabla \times E$ (pronounced "curl E") measures how much E "curls around," or how much it changes in the perpendicular directions. This function is formally defined, for the vector field $E = Ui + Vj + Wk$, as:

$$\nabla \times E = \left(\frac{\partial E_z}{\partial y} - \frac{\partial E_y}{\partial z} \right) \mathbf{i} + \left(\frac{\partial E_x}{\partial z} - \frac{\partial E_z}{\partial x} \right) \mathbf{j} + \left(\frac{\partial E_y}{\partial x} - \frac{\partial E_x}{\partial y} \right) \mathbf{k} \tag{1.1.6}$$

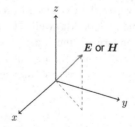

FIGURE 1.2

Three-dimensional coordinate space.

Maxwell's equations are essentially a generalization of previous work by Faraday, Ampere, and Gauss in the area of electric and magnetic fields. Maxwell's starting points were as follows:

- Electric field lines originate on positive charges and terminate on negative charges.
- Magnetic field lines always form closed loops, they do not begin or end anywhere.
- A varying magnetic field induces an electric field (Faraday's law).
- Moving charges or currents give rise to magnetic fields (Ampere's law).

Maxwell built upon this and added some insights of his own to produce the concise set of equations (1.1.1)–(1.1.4) that fully described the interrelationship between the electric field vector E, the magnetic field vector B, the electric charge density ρ (the amount of charge per unit volume), and the electric current density vector J (the amount of electric current flowing through unit area).

The constants used here are physical properties of the medium, namely:

$$\mu_0 = \text{Permeability of free space.}$$

$$\varepsilon_0 = \text{Permittivity of free space.}$$

It is worth spending a little time on these two constants, as they will pop up repeatedly throughout the book. Dielectric permittivity is a property of a medium that determines the strength of the electric field produced by a given electric charge and geometry. Greater values of ε mean that more charge is required to produce the same electric field (for all materials in the liquid or solid phase, ε is greater than the free space value, ε_0). This property of materials is of great importance in microwave engineering, as the physical size of a printed circuit depends on the permittivity of the substrate on which the circuit is fabricated. The use of high permittivity materials is therefore very common at microwave frequencies to facilitate miniaturization.

Greater values of magnetic permeability, μ, in a material mean that a given value of electric current in a nearby conductor produces a stronger magnetic field. Magnetic materials such as iron may have μ values that are thousands of times higher than free space, hence the use of ferrites and other high permeability materials in inductors and transformers as a means of concentrating more magnetic field energy in a small physical area.

Returning now to Maxwell's equations, we can put the four equations of equations (1.1.1)–(1.1.4), in context as follows:

1. Equation (1.1.1) (Gauss's law): Electric charges create electric fields where the strength of the field is determined by the distance from the charge.
2. Equation (1.1.2) (Gauss's law of magnetism): There are no magnetic monopoles—the magnetic field flux through any Gaussian surface sums to zero.
3. Equation (1.1.3) (Faraday's law): A change in magnetic field strength induces a change in electric field strength.
4. Equation (1.1.4) (Ampere's law plus correction): Faraday's law reversed, plus electric current also creates magnetic fields.

By means of these four equations, Maxwell demonstrated that electric and magnetic forces are two complementary aspects of a single phenomenon now known as *electromagnetism*.

In a free space region with no charges ($\rho = 0$) and no currents ($J = 0$), equations (1.1.1)–(1.1.4) reduce to:

$$\nabla \cdot E = 0 \tag{1.1.7}$$

$$\nabla \cdot B = 0 \tag{1.1.8}$$

$$\nabla \times E = -\frac{\partial B}{\partial t} \tag{1.1.9}$$

$$\nabla \times B = \mu_0 \varepsilon_0 \frac{\partial E}{\partial t} \tag{1.1.10}$$

Taking the curl ($\nabla \times$) of the curl equations, and applying a theorem in vector calculus known as the "curl of the curl" identity (i.e., $\nabla \times (\nabla \times A) = \nabla(\nabla \cdot A) - \nabla^2 A$), we obtain the following "wave equations" in three dimensions:

$$\mu_0 \varepsilon_0 \frac{\partial^2 E}{\partial t^2} - \nabla^2 E = 0 \tag{1.1.11}$$

$$\mu_0 \varepsilon_0 \frac{\partial^2 B}{\partial t^2} - \nabla^2 B = 0 \tag{1.1.12}$$

For simplicity, let us consider, say, equation (1.1.11) in one dimension only, so for the x dimension we can write:

$$\frac{\partial^2 E}{\partial x^2} = \mu_0 \varepsilon_0 \frac{\partial^2 E}{\partial t^2} \tag{1.1.13}$$

E and B are mutually perpendicular to each other and the direction of wave propagation, and are in phase with each other. The changing magnetic field creates a changing electric field through Faraday's law (equation 1.1.3). In turn, that electric field creates a changing magnetic field through Maxwell's addition to Ampere's law (equation 1.1.4). It is this perpetual cycle that allows a self-sustaining electromagnetic wave to propagate through space.

Let us now look for a solution to equation (1.1.13) in the form of a sinusoidal wave, with speed v and wavelength λ. Such a wave can be described by the expression:

$$E = E_0 \sin\left(2\pi \frac{x - vt}{\lambda}\right) \tag{1.1.14}$$

Differentiating equation (1.1.14) twice with respect to x and t separately, we get

$$\frac{\partial^2 E}{\partial x^2} = -E_0 \left(\frac{2\pi}{\lambda}\right)^2 \sin\left(2\pi \frac{x - vt}{\lambda}\right) \tag{1.1.15}$$

and

$$\frac{\partial^2 E}{\partial t^2} = -E_o \left(\frac{2\pi v}{\lambda}\right)^2 \sin\left(2\pi \frac{x - vt}{\lambda}\right) \tag{1.1.16}$$

Substituting equations (1.1.15) and (1.1.16) back into the wave equation (1.1.14), we see that we have a solution to equation (1.1.13), provided that:

$$v^2 = \frac{1}{\mu_0 \varepsilon_0} \tag{1.1.17}$$

Using the known values of the physical constants $\mu_0 = 4\pi \times 10^{-7}$ H m^{-1} and $\varepsilon_0 = 8.854187817 \times 10^{-12}$ F m^{-1} we can use equation (1.1.17) to calculate the velocity of electromagnetic wave propagation through a vacuum as:

$$v = 2.99792458\ldots \times 10^8 \text{ m s}^{-1} \tag{1.1.18}$$

Maxwell observed that this velocity happens to be the same as the experimentally measured speed of light, c, and thereby concluded that light is itself an electromagnetic wave. He described his observation thus [15]:

> This velocity is so nearly that of light, that it seems we have strong reason to conclude that light itself (including radiant heat, and other radiations if any) is an electromagnetic disturbance in the form of waves propagated through the electromagnetic field according to electromagnetic laws.

To infer that light itself was a form of electromagnetic wave, and that it differed only in terms of its frequency from the other forms of electromagnetic wave known about at the time, was truly revolutionary. This profound insight can be considered as one of the most significant breakthroughs in the history of physics. It also has some interesting implications, such as the fact that the speed of light in a vacuum is a function only of physical constants μ_0 and ε_0 and does not depend on the speed of the observer. This counter-intuitive consequence of equation (1.1.17) later triggered Einstein to formulate his theory of special relativity.

In materials with relative permittivity ε_r and relative permeability μ_r, the speed of the electromagnetic wave becomes:

$$v_p = \frac{1}{\sqrt{\mu_0 \mu_r \varepsilon_0 \varepsilon_r}} \tag{1.1.19}$$

Which is always less than the speed of light in a vacuum since both μ_r and ε_r are always greater than unity for real materials.

The wave described by equations (1.1.11) and (1.1.12) propagates through space in the positive z-direction and is called a *uniform plane wave*, since it has uniform (constant) properties in a plane perpendicular to the direction of propagation. For the uniform plane wave described by equations (1.1.11) and (1.1.12) the plane of

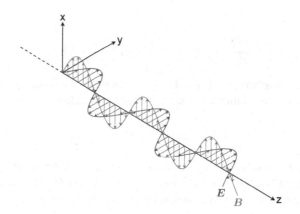

FIGURE 1.3

Plane wave in free space.

uniformity is the xy plane, since the direction of propagation is the z-direction. This is shown in Figure 1.3.

The velocity referred to in equation (1.1.17) called the *phase velocity*. This is not the velocity of any physical entity, but the velocity at which an observer would have to move to see always a constant phase.

It is instructive to consider the ratio of electric and magnetic field magnitudes, which has the units of Ω, that is:

$$Z_0 = \frac{E}{H} = \mu_0 c_0 = \sqrt{\frac{\mu_0}{\varepsilon_0}} = \frac{1}{\varepsilon_0 c_0} \qquad (1.1.20)$$

With the free space values of μ_0 and ε_0 we can calculate the *impedance of free space* as being approximately $376.73031\ldots\Omega$.

This concludes our very brief introduction to Maxwell's equations and plane wave propagation. Readers interested in digging deeper into EM theory are referred to some of the numerous texts on the subject [16–19]. Readers interested in the story of Maxwell's equations are referred to the excellent and very readable book by Forbes and Mahon [20].

1.2 PROPERTIES OF MATERIALS AT MICROWAVE FREQUENCIES

Another contrast between microwave engineering and lower-frequency electronics is that the microwave engineer generally has to have a closer familiarity with the properties of the materials from which the various components are made, as many of these properties change significantly at higher frequencies. The "skin effect," which is a major determinant of the effective resistance of wires at microwave frequencies is just one example of this. Permittivity and permeability, which we briefly introduced in Section 1.1.3, are major determinants of capacitance and inductance, and both are also frequency dependent. In this section, we will delve a bit deeper into permittivity and permeability as we will come across them repeatedly in the following chapters. We will start by looking at the very important property of *resistivity*.

1.2.1 RESISTIVITY

Materials can be broadly classified into *conductors* and *insulators* depending upon how easily electric current can pass through the material. In a conducting material, electrons can be easily dislodged from atoms and can therefore move freely within the material. This means that when an electric potential is applied to a sample of the material then an electric current (which is another way of saying an aggregate movement of electrons) will flow through the sample.

The reciprocal of resistivity is *conductivity*, represented by the symbol σ, which can be used alternatively to describe the same phenomenon. In other words, for a given material:

$$\rho = \frac{1}{\sigma} \tag{1.2.1}$$

Conductivity is related to *electron mobility*, μ_e, in a material as follows:

$$\sigma = nq\mu_e \tag{1.2.2}$$

where n is the number density of electrons in the material and q is the electron charge ($q = 1.60217657 \times 10^{-19}$ coulombs). Electron mobility is a measure of how quickly electrons can move through a material, and is measured in SI unit of mobility, $m^2/(Vs)$, but is often specified in units of $cm^2/(Vs)$. Electron mobility varies widely between different materials and is a major determinant of conductivity (alongside the electron density, n). We will cover mobility in a little more detail in Chapter 11.

By contrast, an insulating material does not permit the free flow of electrons as the electrons are tightly bound to the atoms of the material. The basic property that distinguishes conductors from insulators is *Resistivity*, represented by the symbol ρ, which is a fundamental property of materials and defines to what extent a material will impede the bulk movement of electrons through it. Resistivity is measured in units of Ωm. Generally speaking, conductors will have a resistivity much less that 1 Ωm, whereas insulators have a resistivity greater than about 10^8 Ωm.

There is an exception to the simplistic dichotomy between conductors and insulators in the form of *semiconductors*, which are materials having partial conductivity which can be altered by adding impurities, or *dopants*, to the material. Semiconductor materials are the basis of all modern active electronic devices, and merit their own chapter in this book (Chapter 11).

Resistivities for some typical materials are shown in Table 1.2.

Readers will notice something interesting from Table 1.2 in that gallium arsenide, although listed as a semiconductor, has a much higher resistivity than silicon or

Table 1.2 Resistivity of Typical Materials (Ωm) at 300 K

Material	Resistivity	Classification
Silver	1.59×10^{-8}	Conductor
Copper	1.68×10^{-8}	Conductor
Gold	2.24×10^{-8}	Conductor
Aluminum	2.65×10^{-8}	Conductor
Tungsten	5.65×10^{-8}	Conductor
Iron	9.71×10^{-8}	Conductor
Platinum	10.6×10^{-8}	Conductor
Nichrome (Ni, Fe, Cr alloy)	100×10^{-8}	Conductor
Germanium (intrinsic)	4700	Semiconductor
Silicon (intrinsic)	2.3×10^7	Semiconductor
Gallium arsenide (intrinsic)	10^8	Semiconductor
Glass	10^{11} to 10^{15}	Insulator
Quartz (fused)	7.5×10^{17}	Insulator
PTFE (teflon)	10^{23} to 10^{25}	Insulator

germanium. The fact that it is listed as a semiconductor reflects the fact that its resistivity can be dramatically altered by the addition of impurities. The term "intrinsic" applied to the semiconductors in Table 1.2 means that the resistivity quoted is the resistivity of pure material, without any impurities added. The high resistivity of gallium arsenide, in particular, means that it is sometimes referred to as a "semiinsulating" material. The properties of semiconductors will be covered in more detail in Chapter 11.

It is important not to confuse resistivity with resistance. The resistance of a material sample will depend on the geometry of the particular sample, whereas resistivity is a fundamental property of the material itself.

Consider the resistance of a cylindrical wire made of a conducting material as shown in Figure 1.4. We intuitively understand that the resistance will increase with the length of the wire, l, and that thinner wires have higher resistance than thicker wires. We can therefore infer that:

$$R \propto \frac{l}{A}$$

(1.2.3)

where $A = \pi r^2$ is the cross-sectional area of the wire of radius r. It turns out that the constant of proportionality in equation (1.2.3) is the *resistivity*. We can then write equation (1.2.3) as:

$$R = \rho \frac{l}{A}$$

(1.2.4)

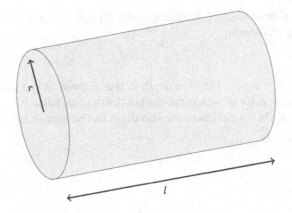

FIGURE 1.4

Resistance of a cylindrical wire.

1.2.2 THE SKIN EFFECT

There are a number of peculiar physical effects which alter the behavior of common materials and components as the frequency of operation increases. One of the most important is the *skin effect* which describes the fact that alternating current tends to accumulate near the surface of a solid conductor at higher frequencies. This effectively limits the cross-sectional area of the conductor with a corresponding increase in the resistance of that conductor above what would be expected at DC.

Consider the cross section of a cylindrical conductor shown in Figure 1.5. As the frequency increases, the current density in a conductor, J, decreases exponentially

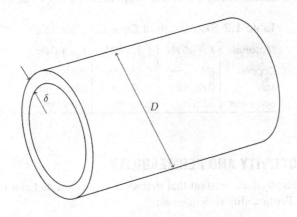

FIGURE 1.5

The skin effect.

from its value at the surface J_S according to the depth d from the surface according to the following relationship:

$$J = J_S \, e^{-d/\delta} \tag{1.2.5}$$

where δ is the skin depth. The skin depth is thus defined as the depth below the surface of the conductor at which the current density has fallen by a factor of $1/e$ (about 0.37) of J_S. In normal cases, the skin depth can be approximated by [21]:

$$\delta = \sqrt{\frac{2\rho}{\omega \mu_r \mu_0}} \tag{1.2.6}$$

or alternatively:

$$\delta = \frac{1}{\sqrt{\pi f \mu_r \mu_0 \sigma}} \tag{1.2.7}$$

where $\rho = 1/\sigma$ is the resistivity of the conductor and ω is the angular frequency.

A long cylindrical conductor such as the wire shown in Figure 1.5, having a diameter D large compared to the skin depth, δ, will have a resistance approximately that of a hollow tube with wall thickness δ carrying direct current. Using a material of resistivity ρ we then find the AC resistance of a wire of length l to be:

$$R \approx \frac{\rho l}{\pi (D - \delta)\delta} \tag{1.2.8}$$

In case $\delta \ll D$, equation (1.2.8) can be approximated by:

$$R \approx \frac{l\rho}{\pi D \delta} \tag{1.2.9}$$

Table 1.3 lists the skin depth of some common materials at various frequencies [22]. Note the significant difference in skin depth between copper and iron.

Table 1.3 Skin Depth of Common Materials

Material	$f = 60$ Hz	$f = 1$ MHz	$f = 1$ GHz
Copper	8.61 mm	0.067 mm	2.11 μm
Iron	0.65 mm	5.03 μm	0.016 μm
Sea water	32.5 m	0.25 m	7.96 mm

1.2.3 PERMITTIVITY AND PERMEABILITY

Permeability is a physical constant that defines how much a material responds to a magnetic field. Permeability is defined as:

$$\mu = \frac{B}{H} \tag{1.2.10}$$

where B is the magnitude of the flux density, and H is the magnitude of the magnetic field strength. The units of permeability are Henries/meter. The permeability of a vacuum is denoted by μ_0 and has the value $4\pi \times 10^{-7}$ (approximately $1.25663706 \times 10^{-6}$). Most materials have permeability very close to that of a vacuum. Materials that contain iron, chrome, or nickel, however, will have a higher relative permeability (μ_r). Relative permeabilities of typical materials are shown in Table 1.4.

The materials in Table 1.4 are also classified in terms of their response to an externally applied magnetic field [19]. Some materials exhibit an induced magnetic field in a direction opposite to an externally applied magnetic field. These are called *diamagnetic*. Diamagnetic materials are repelled by an externally applied magnetic field. Some other materials are attracted by an externally applied magnetic field and form internal, induced magnetic fields in the direction of the applied field. These materials are called *paramagnetic*. Ferromagnetic materials are those that are able to retain magnetization in the absence of an external magnetic field. These are the material of which permanent magnets are made.

Permeability is an important determinant of skin depth, since the higher the relative permeability, the less an electromagnetic wave will penetrate into the material.

Permittivity refers to the ability of a material to polarize in response to an externally applied electric field and thereby reduce the total electric field inside the material. In other words, permittivity is a measure of a material's ability to transmit (or "permit") an electric field within it. As with permeability, the permittivity generally depends on the frequency of the applied field. This frequency dependence reflects the fact that a material's polarization does not respond instantaneously to an

Table 1.4 Relative Permeabilities of Typical Materials

Material	Type	Relative Permeability (μ_r)
Silver	Diamagnetic	0.99998
Lead	Diamagnetic	0.999983
Copper	Diamagnetic	0.999991
Water	Diamagnetic	0.999991
Vacuum	Nonmagnetic	1
Air	Paramagnetic	1.0000004
Aluminum	Paramagnetic	1.00002
Ferrite (nickel zinc)	Ferromagnetic	16-640
Cobalt	Ferromagnetic	250
Nickel	Ferromagnetic	600
Ferrite (manganese zinc)	Ferromagnetic	640
Mild steel (0.2% C)	Ferromagnetic	2000
Iron (0.2% impurity)	Ferromagnetic	5000
Silicon iron	Ferromagnetic	7000

Table 1.5 Relative Permittivity (Dielectric Constant) of Typical Materials

Material	Dielectric Constant (ε_r)	Loss Tangent[a] (tan δ_e)
Vacuum	1.0	0
Teflon	2.1	0.0003
Nylon	2.4	0.0083
Sandy soil (dry)	2.55	0.0062
Silicon dioxide	3.9	0.001
Thermoset polyester	4.0	0.0050
Paper	3-4	0.0125-0.0333
Concrete (dry)	4.5	0.0111
Glass	4-7	0.0050
Soda lime glass	6.0	0.02
Alumina	9.0	0.0006
RT/duroid 5870 (microstrip substrate)	2.33	0.0009

[a]Loss tangent is measured at low frequency (<1 MHz).

applied field. There is a certain delay between the application of the applied field and the response, which can be represented by a phase difference at a given frequency.

Table 1.5 shows the relative permittivity (ε_r) of some common materials. The table also lists something called the *loss tangent* that will be explained in the next section.

1.2.4 LOSSES IN DIELECTRIC AND MAGNETIC MATERIALS AT MICROWAVE FREQUENCIES

When a time-varying electric field is applied to a material, the polarization dipoles inside the material will flip back and forth in response to the field. The finite mass of the charge carriers has two important consequences. First, work has to be done to move them, which means that some of the applied energy will be "lost" in the material. Second, it takes a finite time for these dipoles to move, which means that the polarization vector will lag behind the applied electric field [23]. The result of this phase lag is that permittivity should be more correctly defined as a complex number, that is:

$$\varepsilon = \varepsilon' - j\varepsilon'' \tag{1.2.11}$$

$$= \varepsilon' - j\frac{\sigma}{\omega} \tag{1.2.12}$$

Both the real and imaginary parts of equation (1.2.11) are frequency dependent [18], but at lower frequencies the imaginary part of ε' is small and is usually ignored.

The real part of equation (1.2.11), ε', is a measure of how much energy from an external electric field is stored in the material. The imaginary part of permittivity, ε'', is called the *loss factor* and is a measure of how dissipative or lossy a material is. The loss factor includes the effects of both dielectric loss and conductivity, σ.

Dielectric losses are more commonly expressed in terms of the *loss tangent*, which is defined as the ratio of the imaginary and real parts of the complex permittivity. So from equation (1.2.11) we have:

$$\tan \delta_\varepsilon = \frac{\varepsilon''}{\varepsilon'} = \frac{\sigma}{\omega \varepsilon'} \tag{1.2.13}$$

For very low loss materials it is customary to make the approximation $\tan \delta_\varepsilon \approx \delta_\varepsilon$. Loss tangents for some typical materials are shown in Table 1.5.

The power dissipated per unit volume in the material when an external electric field is applied is given by [18]:

$$P_e = \frac{\text{Power}}{\text{Volume}} = \frac{\text{Current}}{\text{Area}} \cdot \frac{\text{Voltage}}{\text{Length}} \tag{1.2.14}$$

$$= J|E| = \sigma |E|^2 \tag{1.2.15}$$

$$= \omega \varepsilon' \tan \delta_\varepsilon |E|^2 \quad (\text{Wm}^{-3}) \tag{1.2.16}$$

In stating equation (1.2.14), we have made the assumption that the power is uniform throughout the volume and that thermal equilibrium has been achieved. This is not always the case, but equation (1.2.14) does provide a useful approximation for many practical purposes.

A similar argument applies to magnetic materials where there are also losses and delays due to the energy required, and time taken, for the magnetic dipoles inside a material to respond to changes in the applied magnetic field. This means that the permeability, μ, is also a complex quantity, and can be written as:

$$\mu = \mu' + j\mu'' \tag{1.2.17}$$

The ratio of the imaginary to the real part of equation (1.2.17) is called the magnetic *loss tangent*, which is defined as:

$$\tan \delta_\mu = \frac{\mu''}{\mu'} \tag{1.2.18}$$

In a similar way as we described the power dissipated in a material due to dielectric losses, we can describe the power that is absorbed per unit volume (W m^{-3}) of the sample due to magnetic losses as follows:

$$P_\mu = \omega \mu'' |H|^2 \tag{1.2.19}$$

$$= \omega \mu' \tan \delta_\mu |H|^2 \quad (\text{W m}^{-3}) \tag{1.2.20}$$

Table 1.6 Loss Tangent Versus Frequency Common Dielectric Materials

Material	Loss Tangent (tan δ)			
	$f = 1$ kHz	$f = 1$ MHz	$f = 100$ MHz	$f = 3$ GHz
Alumina	0.0006	0.00033	0.0003	0.001
Barium titanate	0.00044	0.0002	NA	0.0023
Silicon dioxide	0.00075	0.0001	0.0002	0.00006
Polyethylene	<0.0002	<0.0002	0.0002	0.00031
Teflon	0.0003	<0.0002	<0.0002	0.00015
Sodium chloride	0.0001	<0.0002	NA	0.0005
Water (distilled)	NA	<0.0400	<0.0050	0.1570

From equations (1.2.14) and (1.2.19) we can see that the power absorbed in a material when an external alternating electric or magnetic field is applied, varies linearly with frequency, permeability, and the loss tangent and with the square of the magnitude of applied field (electric or magnetic). The frequency dependence of the loss tangent for some common dielectric materials is shown in Table 1.6.

1.3 BEHAVIOR OF REAL COMPONENTS AT MICROWAVE FREQUENCIES

Readers who are familiar with lower-frequency electronic design will probably be accustomed to considering passive components, such as resistors, capacitors, and inductors as having a single electrical "personality" (e.g., resistance) that is frequency independent. As the frequency of interest approaches the microwave region, however, the behavior of real passive components departs from this ideal. At microwave frequencies passive components need to be considered as complex impedance elements that may be capacitive in one region of the frequency spectrum and inductive in another. This implies that these components have one or more self-resonant frequencies where the reactance transitions from positive to negative and vice versa. In this section, we will briefly outline the high-frequency behavior of the most important passive components, starting with the humble interconnecting wire.

1.3.1 WIRE

Readers may be accustomed to thinking of interconnecting wire as a zero resistance, ideal, point-to-point connection whose electrical characteristics can be ignored in the design. This assumption becomes less valid as the frequency increases.

First, when current is flowing through a wire, a magnetic field is induced around the wire. If the magnetic field is forced to expand and contract by changes in the current, a voltage will be induced in the wire that will tend to oppose the change in current flow (Faraday's law). This effect manifests itself as a *self-inductance*.

A good estimate of the self-inductance of a cylindrical wire can be obtained from the following empirical formula [24]:

$$L = 0.002l \left[2.3 \log_{10} \left(\frac{4l}{d} \right) - 0.75 \right] \mu H \qquad (1.3.1)$$

where l is the length of wire in cm and d is the diameter of the wire in cm. According to this formula, 5 cm of wire of 1 mm diameter will have an inductance of 50 nH, which translates to the rather appreciable reactance of 314 Ω at 1 GHz.

Second, the resistance of a wire interconnection will increase with frequency due to the skin effect discussed previously. Finally, depending on the physical length of the wire in relation to the wavelength of interest, it may need to be considered as a *transmission line* (to be elaborated on in Chapter 2), that can, in theory, assume any reactive impedance, including capacitance. These phenomena can completely disrupt the operation of a circuit and produce quite unexpected results if not taken into account during the design process.

1.3.2 RESISTORS

If we now think about a conventional leaded resistor in the context of the above discussion, we can immediately see that the high-frequency equivalent circuit of the resistor must also include some inductance, due to the leads. In addition, because the ends of the resistor are generally at different potentials, we would also expect to see some parasitic capacitance. We can therefore intuitively draw an equivalent circuit model of such a resistor, as shown in Figure 1.6 [25].

The inductors, L, in Figure 1.6 are the lead inductances, C_b is the inter-lead capacitance and C_a is the parasitic capacitance. The effective impedance of the circuit in Figure 1.6 is as follows:

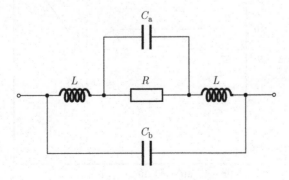

FIGURE 1.6

High-frequency equivalent circuit of a typical resistor.

$$Z_{R_{\text{equiv}}} = \cfrac{1}{j\omega C_b + \cfrac{1}{j2\omega L + \cfrac{1}{G + j\omega C_a}}} \tag{1.3.2}$$

where the conductance $G = 1/R$.

The inter-lead capacitance, C_b, is usually much smaller than the inherent parasitic capacitance, C_a, and can often be ignored, in which case equation (1.3.2) simplifies to:

$$Z_{R_{\text{equiv}}} = j2\omega L + \frac{1}{G + j\omega C_a} \tag{1.3.3}$$

The frequency dependence of this equivalent circuit shows some surprising behavior at high frequencies as shown in Figure 1.7, where we have picked the following typical values for the circuit in Figure 1.6:

$$R = 1 \text{ k}\Omega$$

$$L = 60 \text{ nH}$$

$$C = 4 \text{ pF}$$

Although the resistor maintains its nominal value of 1 kΩ up to about 20 MHz, above that the parasitic capacitances start to have an increasingly significant effect, causing the component to have a negative reactance that decreases with frequency. The magnitude of the impedance decreases until the self-resonant frequency is reached at about 500 MHz. Above that frequency the parasitic inductance starts to dominate and the total impedance increases with frequency. We therefore conclude

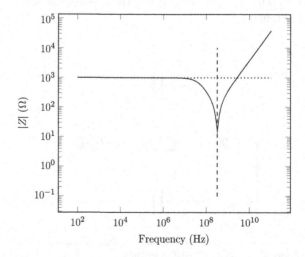

FIGURE 1.7

|Z| versus frequency for a typical resistor (nominal value: 1 kΩ).

that great care must be taken in using "conventional" resistors, even at frequencies of just a few hundred MHz. Readers may be struck by how low this self-resonant frequency is for what is essentially a very typical leaded resistor.

Since the inductive components in Figure 1.6 are mainly due to the connecting leads, the simplest way to increase the self-resonant frequency of a resistor is to dispense with the leads altogether. Consequently, surface mount device (SMD) resistors tend to be preferred in high-frequency applications, aside from their miniaturization advantages.

SMD resistors come in various sizes. The smaller the size, the better the high-frequency performance, but the less the power handling capacity. The standard notation for the physical size of chip components is length-width in 10 mil increments. For example, chip resistors are available in 1206, 0805, 0603, 0402, and 0201 dimensions where a 0402 would be 40×20 mils (1 mm $\times$ 0.5 mm).

Representative values of parasitic elements for a 1206 chip resistor are $L = 1.2$ nH and $C = 0.03$ pF, which pushes the resonance "dip" shown in Figure 1.7 out to around 20 GHz.

1.3.3 CAPACITORS

Capacitors are very widely used in microwave circuits as DC blocking and decoupling elements, as well as reactive elements in filters, tuners, and matching networks.

Ceramic is the most common dielectric for microwave capacitors due to its low loss at high frequencies. Some ceramics, however, have large variation of capacitance with temperature (or temperature coefficient, typically in ppm/°C), so care must be exercised in selecting ceramic capacitors for filters and tuning elements.

As with leaded resistors, leaded capacitors can have significant self-inductance at relatively low microwave frequencies.

The loss in the dielectric, as discussed in Section 1.2.4, results in an *effective series resistance* or ESR which needs to be taken into account. A typical equivalent circuit for a high frequency capacitor that takes these parasitic effects into account is shown in Figure 1.8.

The effective impedance of the circuit in Figure 1.8 is given by:

$$Z_{C_{equiv}} = R_s + j\omega L + \frac{1}{G_c + j\omega C} \tag{1.3.4}$$

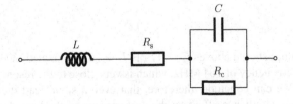

FIGURE 1.8

High-frequency equivalent circuit of a typical capacitor.

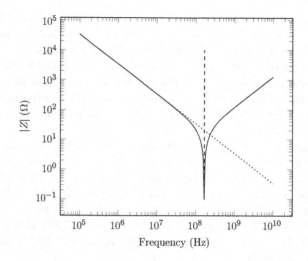

FIGURE 1.9

|Z| versus frequency for a typical capacitor (nominal value: 47 pF).

where $G_c = 1/R_c$.

Figure 1.9 shows the magnitude of the capacitor impedance versus frequency, according to the equivalent circuit of Figure 1.8 for a typical 47 pF capacitor with the following parasitic component values:

$$R_s = 0.1 \ \Omega$$
$$R_c = 100 \ \text{k}\Omega$$
$$L = 20 \ \text{nH}$$

Once again, we see a self-resonant phenomenon at around 200 MHz. We can simplify the equivalent circuit of Figure 1.8 by combining the resistors R_s and R_c, which reduces Figure 1.8 to a simple series RLC network. We can estimate the self-resonant frequency simply by:

$$\omega_0 = \frac{1}{\sqrt{LC}} \tag{1.3.5}$$

With the values shown above ($C = 47$ pF, $L = 20$ nH), equation (1.3.5) gives a series resonant frequency of 164 MHz, which is very close to the resonant dip shown in Figure 1.9. We can conclude, therefore, that even a small lead inductance, like 20 nH, will have an adverse effect on the performance of the leaded capacitor. We therefore tend to use SMD "chip" capacitors almost exclusively above a few hundred MHz. The self-inductance of a typical chip capacitor will be 1 nH or less.

Electrolytic and tantalum capacitors have particularly high self-inductances and relatively high dielectric losses and should therefore be avoided entirely in the signal path of any microwave circuit.

1.3.4 INDUCTORS

Inductors are widely used in RF and microwave circuits as either reactive elements in tuning and matching networks or as "chokes" in biasing networks.

Because an inductor will usually consist of a wire coil of some kind, the main parasitic elements in a real inductor are the resistance of the wire and the aggregate effect of the inter-winding capacitance. An equivalent circuit of a typical spiral inductor is shown in Figure 1.10.

By inspection the impedance of the equivalent circuit in Figure 1.10 is:

$$Z_{L_{equiv}} = \frac{R_s + j\omega L}{1 + j\omega C_s(R_s + j\omega L)} \tag{1.3.6}$$

Because there is a capacitive parasitic element in Figure 1.10, the inductor will also exhibit self-resonance and the capacitive parasitics will become dominant at high frequencies. This causes the apparent inductance to become quite frequency dependent as we approach the resonant frequency, as shown in the frequency response of $|Z|$ in Figure 1.11 for a typical chip inductor with the following parasitic component values:

$$R_s = 0.05 \ \Omega$$

$$L = 25 \ \text{nH}$$

$$C_s = 0.5 \ \text{pF}$$

In this case, the self-resonance occurs at around 1.5 GHz.

Christopher Bowick's book [24] contains a lot of useful information on constructing different types of practical RF inductors from wire, such as single-layer air-core inductors and toroidal inductors. We will repeat here a useful design formula for

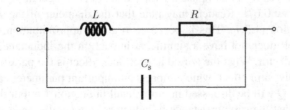

FIGURE 1.10

High-frequency equivalent circuit of a typical inductor.

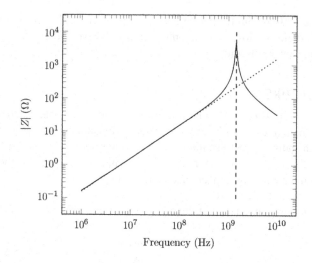

FIGURE 1.11

$|Z|$ versus frequency for a typical inductor (nominal value: 25 nH).

single-layer air-core inductors that every radio engineer will find useful, as these are probably the most common form of inductor:

$$L = \frac{0.394r^2N^2}{9r + 10l} \tag{1.3.7}$$

where

$$L = \text{Inductance in } \mu\text{H}$$

$$l = \text{Coil length in cm}$$

$$r = \text{Coil radius in cm}$$

$$N = \text{Number of turns}$$

The formula (1.3.7) is an approximation that starts to become less accurate as l decreases below $0.67r$. Readers may note that the diameter of the wire does not appear in equation (1.3.7). This is because, to a first approximation, the diameter of the wire itself does not have a significant impact on the inductance of a single-layer wound inductor. What the wire diameter does affect is the parasitic resistance, resulting from the skin effect, which impacts an important parameter called the "Q," of the inductor. Q will be discussed in more detail in Section 1.6. For the time being let us just say that parasitic resistance in inductors is generally undesirable, so it is always a good idea to use the largest diameter wire possible in any given application. The diameter of wire that can be used will be limited by the length l of the inductor, which is specified in equation (1.3.7). As a rule of thumb, notwithstanding other considerations, optimum Q is obtained when $l = 2r$ [24].

Example 1.1 (Inductor design).

Problem. Design a 35-nH single-layer air-core inductor using a 6-mm diameter core former.

Solution. We will set the length l equal to the diameter to maximize the Q of the inductor. Therefore, we set $l = 6$ mm.

Using equation (1.3.7) we therefore have:

$$L = \frac{0.394 \times (0.3)^2 \times N^2}{9 \times 0.3 + 10 \times 0.6} = 35 \times 10^{-3}$$

Which gives:

$$N = \sqrt{\frac{8.7 \times 35 \times 10^{-3}}{0.0355}}$$

$$\approx 3 \text{ turns}$$

The thickness of wire used should be chosen to be as thick as possible consistent with getting three turns within a length of 6 mm.

1.3.5 SURFACE MOUNT DEVICES

In the majority of microwave applications, except for specialist applications, such as high power amplifiers, the components encountered will be in Surface Mount Device (SMD) packaging. SMD technology allows electronics to be miniaturized, and for this reason SMD technology has largely replaced (leaded or "through-hole") component technology across the modern electronics industry. For the microwave engineer, SMD components have the added advantage of much lower parasitics than leaded components due, primarily, to the absence of lead inductances. On the downside, SMD components, because of their small size, generally have a smaller power rating than leaded components and this may limit their use in some designs.

The physical dimensions of two-terminal SMD components is identified by a four-digit size code. The first two digits in the size code refer to the length (L) from termination-to-termination. The second two digits refer to the width (W) of the termination. It is important to be aware that this four-digit size code can represent either millimeters or inches. The fact that the same code, for example, "1210" can represent two different sizes, one being 0.12×0.10 inches and the other being 1.2×1.0 mm, can lead to confusion. It is therefore necessary, when specifying a component, to verify whether the size code is in metric or inches. Some of the most common SMD two-terminal component sizes are shown in Table 1.7, with the typical power rating if the component is a resistor.

The thickness of the component is not included in the four-digit size code.

Table 1.7 SMD Two-Terminal Component Sizes

Identifier		Dimensions		
Imperial	Metric	Imperial (in)	Metric (mm)	Power Rating (W) (Resistors)
0402	1005*	0.04 × 0.02	1.0 × 0.5	0.1
0504	1210*	0.05 × 0.04	1.2 × 1.0	0.1
0603	1508	0.06 × 0.03	1.5 × 0.8	0.1
0805	2012	0.08 × 0.05	2.0 × 1.2	0.125
1005*	2512	0.10 × 0.05	2.5 × 1.2	0.125
1206	3216	0.12 × 0.06	3.2 × 1.6	0.25
1210*	3225	0.12 × 0.10	3.2 × 2.5	0.5
1812	4532	0.18 × 0.12	4.5 × 3.2	0.75
2512	6332	0.25 × 0.13	6.4 × 3.2	1.0

Note: In the table, cases where the same component identifier is used for two different component sizes (imperial vs metric) are indicated by *.

1.4 THE IMPORTANCE OF IMPEDANCE MATCHING
1.4.1 MAXIMUM POWER TRANSFER

A lot of microwave design involves moving signal power most efficiently from one place to another. The maximum transfer of signal power implies that losses in the signal path should be minimized. This requires that the source and load impedances be *matched*. In this section, we will explain what this means and how matching can be achieved. We will elaborate on the actual design techniques for impedance matching networks in Chapters 9 and 10.

As a starting point, it is worth briefly reviewing the DC maximum power transfer theorem which will be familiar to the reader from basic circuit theory.

Consider the DC power being delivered to the resistive load, R, in Figure 1.12 when it is connected to a battery with internal resistance r:

$$P_L = I^2 R = \left(\frac{V}{R+r}\right)^2 R \qquad (1.4.1)$$

If we consider how load power, P_L, changes with load resistance, R, we can write:

$$\frac{dP_L}{dR} = V^2 \left[\frac{(R+r)^2 - 2R(R+r)}{(R+r)^4}\right] \qquad (1.4.2)$$

Load power is maximized when the above derivative equals 0, that is when:

$$(R+r)^2 = 2R(R+r) \qquad (1.4.3)$$

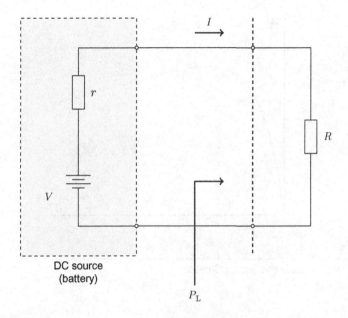

FIGURE 1.12

Maximum power transfer: DC case.

Solving equation (1.4.3) yields the familiar condition for maximum power transfer in the DC case, namely:

$$R = r \qquad (1.4.4)$$

Figure 1.13 shows the relationship between P_L and R, and shows that maximum power is transferred when $R = r$.

When operating at microwave frequencies we need to take into account the reactive elements of both load and source. We therefore have the situation shown in Figure 1.14(a) for complex source and load impedances, Z_S and Z_L, respectively.

We proceed by matching the resistive and reactive elements of Z_S and Z_L independently. Matching of the resistive elements requires that equation (1.4.4) be satisfied, as in the DC case. In addition, the reactive elements of source and load need to cancel each other out (i.e., be of equal magnitude but opposite sign). We therefore have the following requirement for maximum power transfer in the RF case:

$$R_S + jX_S = R_L - jX_L \qquad (1.4.5)$$

or more simply:

$$Z_S = Z_L^* \qquad (1.4.6)$$

where * indicates the complex conjugate.

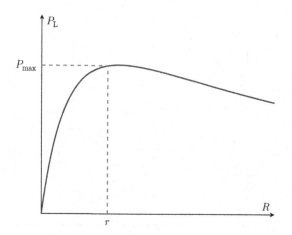

FIGURE 1.13

Load power versus load resistance: DC case.

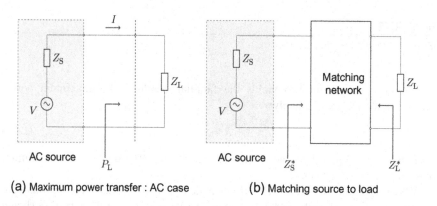

(a) Maximum power transfer : AC case   (b) Matching source to load

FIGURE 1.14

Impedance matching. (a) Maximum power transfer: AC case. (b) Matching source to load.

There are many design situations where we do not have control over the values of either the source or the load impedances. For example, we may need to connect a coaxial cable, having a certain characteristic impedance (representing Z_S in Figure 1.14(a)) to an antenna which may have a different impedance (representing Z_L in Figure 1.14(a)). The purpose of impedance matching is to address such general cases where condition (1.4.6) is not satisfied for a given source/load combination. In such general cases, a *matching network* needs to be inserted between the source and load as shown in Figure 1.14(b). The purpose of the matching network is to satisfy condition (1.4.6) at the source and load simultaneously, thereby allowing maximum power transfer from source to load. This means that, with reference to Figure 1.14(b), the input impedance of the matching network needs to be Z_S^* and the output impedance of the matching network needs to be Z_L^*. Maximum power transfer from source to load in Figure 1.14(b), of course, also implies that the matching network itself should be lossless.

    The design of matching networks to satisfy these conditions will be covered in detail in Chapters 9 and 10.

1.5 COMMON MICROWAVE METRICS

There are a number of metrics that are commonly used at microwave frequencies but do not appear at lower frequencies. Two examples are *reflection coefficient* and *standing wave ratio* (SWR). Both of these quantities relate to the behavior of currents and voltages on *transmission lines*. They originally gained importance because of their relative ease of measurement at very high frequencies, when compared with the traditional electrical quantities of voltage and current.

1.5.1 REFLECTION COEFFICIENT

We will formally define the reflection coefficient, represented by the symbol Γ, in terms of the ratio of forward and reflected traveling voltage waves in Chapter 2. Mathematically, it as a bilinear transformation that translates complex impedance in the impedance plane to a dimensionless complex number in the reflection coefficient plane. Because of the bilinear nature of this transformation, the reflection coefficient of a given impedance is unique to that impedance and vice versa.

    If we consider a "normalized" arbitrary impedance, $z = r + jx$, (i.e., an ohmic impedance $Z = R + jX$ divided by the "system characteristic impedance, Z_o"), we can define the reflection coefficient as [17]:

$$\Gamma = \frac{z-1}{z+1} = \frac{(r-1)+jx}{(r+1)+jx} \tag{1.5.1}$$

and conversely:

$$z = \frac{1+\Gamma}{1-\Gamma} \tag{1.5.2}$$

where z is the normalized impedance, that is, $z = (R+jX)/Z_o$.

    We can similarly define the reflection coefficient where the impedance concerned is expressed as a normalized admittance, that is, $y = g + jb$, where $y = 1/z$:

$$\Gamma = \frac{1-y}{1+y} = \frac{(1-g)-jb}{(1+g)+jb} \tag{1.5.3}$$

and conversely:

$$y = \frac{1-\Gamma}{1+\Gamma} \tag{1.5.4}$$

Note above a convention we will use throughout this book, namely that lower case letters (z and y) will represent normalized quantities and upper case letters (Z and Y) will represent unnormalized quantities.

For all passive impedances, the real part of the impedance, r, will be non-negative. According to equation (1.5.1), the magnitude of the reflection coefficient for a passive impedance can therefore never be greater than unity. Note that for a short-circuited load impedance ($z = 0$ or $y = \infty$) we have $\Gamma = 1 \angle 180°$, while an open-circuited load ($z = \infty$ or $y = 0$) produces $\Gamma = 1 \angle 0°$. In view of the fact that the magnitude of Γ can never exceed unity for passive loads, it follows that all possible passive impedances whose values reflect a range extending from a short-circuit to an open-circuit map to coordinates lying within, or directly on, the unit circle centered at the origin of the reflection plane, as shown in Figure 1.15, where the reflection plane is taken as a coordinate system whose horizontal axis is the real part of Γ and whose vertical axis is the imaginary part of Γ. This is the basis of the *Smith Chart*, which will be described in detail in Chapter 4.

1.5.2 STANDING WAVE RATIO (SWR)

Unlike the reflection coefficient, SWR is a scalar quantity and does not provide any phase information. This means that there isn't a one-to-one correspondence between SWR and reflection coefficient or load impedance, since there can be more than one possible value of complex load giving the same value of SWR. SWR was used historically because it can be easily measured using a simple device called a *slotted line*. A slotted line consists of an air-dielectric coaxial transmission line with a slot

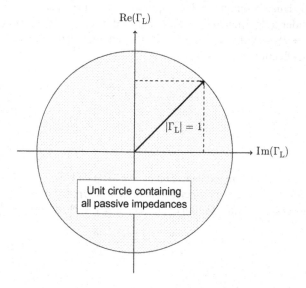

FIGURE 1.15

Cartesian coordinate system on the reflection coefficient (Γ) plane.

in the outer conductor allowing a voltage probe to be inserted. The probe is moved along the line to find the points of maximum and minimum standing wave voltage amplitude, V_{max} and V_{min}. The ratio of these two measurements gives the *voltage standing wave ratio* (VSWR) which is defined as follows:

$$VSWR = \frac{|V_{min}|}{|V_{max}|} \tag{1.5.5}$$

The relationship between VSWR and reflection coefficient is given by [17]:

$$VSWR = \frac{1 + |\Gamma|}{1 - |\Gamma|} \tag{1.5.6}$$

1.5.3 RETURN LOSS

Return loss is the loss of power in the signal returned/reflected by a *discontinuity*. A discontinuity typically occurs where there is a mismatch between two impedances in a transmission line system. Return loss is usually expressed in decibels (dB) as a log of the ratio of two powers, as follows:

$$RL_{dB} = 10 \log_{10} \frac{P_i}{P_r} \tag{1.5.7}$$

The relationship between reflection coefficient and return loss can be found by applying equation (1.5.1) to equation (1.5.7), which gives:

$$RL_{dB} = -10 \log |\Gamma|^2 = -20 \log |\Gamma| \tag{1.5.8}$$

The reader might expect a quantity called "return loss," expressed in dB, to be a positive number. In the case of return loss, however, the common practice is to express it as a negative number. An increasingly negative return loss value therefore means "more loss."

Return loss is a measure of the quality of the match. A good match means the return loss is high, meaning less power is "returned" to the source from the discontinuity. The relationship between these various quantities for some important matching cases is shown in Table 1.8.

Table 1.8 Reflection Coefficient, VSWR, and Return Loss for Various Matching Cases

Description	Load Impedance	$|\Gamma|$	% of Power Reflected	VSWR	Return Loss
Perfect match	Z_0	0	0	1.00	$-\infty$
Fairly good match	$\simeq Z_0$	0.03	0.1	1.06	−31 dB
Poor match	e.g., $2 \times Z_0$ or $0.5 \times Z_0$	0.33	11	2.00	−9.5 dB
Very poor match	e.g., $10 \times Z_0$ or $0.1 \times Z_0$	0.82	67	10.00	−1.7 dB
Open circuit	∞	1.00	100	∞	0
Short circuit	0	1.00	100	∞	0

Due to the extremely wide range of power levels in microwave systems, ranging from picowatts (10^{-12} W) to gigawatts (10^9 W), it is more convenient to represent power on a logarithmic scale in units of "dBm," being defined as the ratio of the power level, in milliwatts, to one milliwatt, that is:

$$P(\text{dBm}) = 10 \log_{10} \left(\frac{P(\text{mW})}{1 \text{ mW}} \right) \quad (1.5.9)$$

One milliwatt is chosen as the reference level simply because many of the systems of interest are operating with powers of this order of magnitude. Any powers quoted in dBm are assumed to be referenced relative to a 50 Ω system impedance. If this is not the case then it needs to be stated.

Typical values of power and their equivalents in dBm are listed in Table 1.9. Note that a doubling or halving of power level corresponds to a change of +3 and −3 dBm, respectively.

Table 1.9 dBm Versus Milliwatts

Power	Power (dBM)
1 μW	−30
5 μW	−23
10 μW	−20
0.1 mW	−10
0.5 mW	−3
1 mW	0
2 mW	3
10 mW	10
100 mW	20
200 mW	23
1 W	30
1 kW	60

1.6 QUALITY FACTOR, Q

1.6.1 THE MEANING OF Q

Energy loss occurs in all real passive components, including reactive components such as capacitors and inductors. In Section 1.3, we showed that all real-world passive components should be represented as resonant circuits at microwave frequencies, where energy is being exchanged back and forth between inductive and capacitive energy storage elements of the equivalent circuit and some of the energy is being "lost" in parasitic resistances. Given the presence of real-world losses, we can define a measure of *Quality*, or "*Q*" for any such equivalent circuit as follows:

$$Q = \frac{\text{Energy stored}}{\text{Average power dissipated}} \tag{1.6.1}$$

Higher *Q* indicates a lower rate of power loss in the circuit relative to the energy stored.

The stored energy is the sum of energies stored in all the lossless reactive elements (inductors and capacitors), whereas the energy dissipated is the sum of the energies lost in all the resistive elements per cycle. In other words, a circuit containing only ideal reactive elements would have an infinite *Q*. Real-world components and circuits all exhibit some electrical losses and therefore have a finite *Q*. In the case of individual reactive components such as capacitors and inductors, the higher the *Q* the closer the component approaches the ideal.

Figures 1.16 and 1.17 show how *Q* is calculated for various combinations of a reactive component with series and parallel resistances, respectively. Consider the

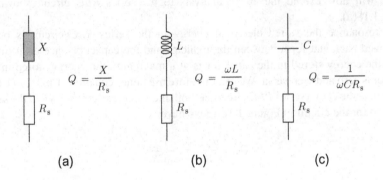

(a)   (b)   (c)

FIGURE 1.16

Q of series combinations.

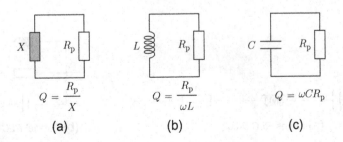

(a)   (b)   (c)

FIGURE 1.17

Q of parallel combinations.

series LR circuit shown in Figure 1.16(b). The peak energy stored in the inductor is $\frac{1}{2}LI^2$, whereas the energy dissipated in the resistor in one cycle, T, is equal to $\frac{1}{2}I^2R_sT = \frac{1}{2}I^2R_s(1/f)$. Hence, for the circuit of Figure 1.16(b):

$$Q_s = 2\pi \frac{\frac{1}{2}LI^2_{max}}{\frac{1}{2}I^2_{max}R\,(1/f)} = \frac{\omega L}{R} \tag{1.6.2}$$

where ω is the angular frequency ($\omega = 2\pi f$). If we now similarly consider the energy stored in the capacitor in the series RC circuit of Figure 1.16(c), we can write an equivalent expression to equation (1.6.2) as:

$$Q_s = 2\pi \frac{\frac{1}{2}I^2_{max}/\omega^2C}{\frac{1}{2}I^2_{max}R_s\,(1/f)} = \frac{1}{\omega CR_s} \tag{1.6.3}$$

This is generalized for an arbitrary reactance, X, in Figure 1.16(a). It is left as an exercise for the reader to derive equivalent relationships for the Q of the parallel circuits in Figure 1.17.

We will now extend the above analysis to the series RLC circuit shown in Figure 1.18(a).

At resonance the stored electrical energy in the series RLC circuit is being exchanged back and forth between the inductor and the capacitor once each cycle. When the energy stored in the capacitor is at a maximum, the energy stored in the inductor is zero and vice versa. We can therefore use either equation (1.6.2) or (1.6.3) to calculate the Q of a series RLC circuit, as both these expressions will yield the same result, so for the circuit in Figure 1.18(a) we have

$$Q_s = \frac{\omega_0 L}{R} = \frac{1}{\omega_0 CR} \tag{1.6.4}$$

At the resonant frequency, ω_0, the net reactance of a series RLC circuit is zero, so the impedance of the circuit in Figure 1.18(a) at resonance is simply R, which is the lowest value of impedance obtainable for this circuit at any frequency.

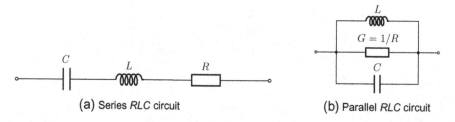

(a) Series RLC circuit   (b) Parallel RLC circuit

FIGURE 1.18

RLC circuits. (a) Series RLC circuit. (b) Parallel RLC circuit.

The parallel *RLC* circuit of Figure 1.18(b) can be considered as the dual of Figure 1.18(a). In other words, the net susceptance is zero at resonance so the admittance is simply $1/R$, which is the lowest value of admittance obtainable for this circuit at any frequency.

In the case of the parallel *RLC* circuit of Figure 1.18(b), we can carry out a similar analysis in terms of admittances. We therefore have the following expressions for Q:

$$Q_p = \frac{R}{\omega_0 L} = \omega_0 CR \tag{1.6.5}$$

This all makes sense if we consider the fact that infinite Q in the case of Figure 1.18(a) requires $R = 0$, whereas infinite Q in the case of Figure 1.18(b) requires $R = \infty$.

We now consider, for example, the impedance of the series resonant circuit shown in Figure 1.18(a), which can be written as follows:

$$Z = R + j\omega L + \frac{1}{j\omega C} \tag{1.6.6}$$

We can eliminate either L or C from equation (1.6.6) by noting that the resonant frequency, ω_0, relates these quantities via $\omega_0 = 1/\sqrt{LC}$. We can therefore rewrite equation (1.6.6) as:

$$Z = R + j\frac{\omega}{\omega_0^2 C} + \frac{1}{j\omega C} = R + j\frac{1}{\omega_0 C}\left(\frac{\omega}{\omega_0} - \frac{\omega_0}{\omega}\right) \tag{1.6.7}$$

Substituting the expression for the Q of the series circuit, as defined by equation (1.6.4), into equation (1.6.7) gives:

$$Z = R\left[1 + jQ_S\left(\frac{\omega}{\omega_0} - \frac{\omega_0}{\omega}\right)\right] \tag{1.6.8}$$

Greater clarity can be achieved by introducing a *fractional detuning* parameter, "δ," which represents a per-unit deviation from the resonant frequency and is defined as follows:

$$\delta = \frac{\omega - \omega_0}{\omega_0} \tag{1.6.9}$$

Using this definition of δ, we can redefine the quantity inside the brackets of equation (1.6.8) in terms of δ as follows [26]:

$$\left(\frac{\omega}{\omega_0} - \frac{\omega_0}{\omega}\right) = 1 + \delta - \frac{1}{1+\delta} = \delta\left(\frac{2+\delta}{1+\delta}\right) \tag{1.6.10}$$

Equation (1.6.8) now becomes:

$$Z = R\left[1 + j\delta Q_S\left(\frac{2+\delta}{1+\delta}\right)\right] \tag{1.6.11}$$

The admittance of Figure 1.18(a) is thus given by:

$$Y = \frac{1}{Z} = \frac{1}{R} \cdot \frac{1}{\left[1 + j\delta Q_S \left(\dfrac{2+\delta}{1+\delta}\right)\right]} \tag{1.6.12}$$

Noting that the admittance of Figure 1.18(a) at resonance is $Y_o = 1/R$, we can write:

$$\frac{Y}{Y_o} = \frac{1}{1 + j2\delta Q_S \left(\dfrac{2+\delta}{1+\delta}\right)} \tag{1.6.13}$$

For very small frequency deviations around the resonant frequency, that is, for $\delta \ll 1$, we can approximate equation (1.6.13) by [27]:

$$\frac{Y}{Y_o} \approx \frac{1}{1 + j2\delta Q_S} \tag{1.6.14}$$

A graph of Y/Y_o versus frequency is shown in Figure 1.19. The *half-power* points are indicated in Figure 1.19, being the frequencies, labeled ω_1 and ω_2, either side of ω_o, where Y/Y_o falls to a value of 0.707 of its value at resonance. The distance between ω_1 and ω_2 is called the *half-power* bandwidth of the circuit, or $\Delta\omega$. These parameters are related to Q by the following, which also serves as another working definition of Q for a resonant circuit [25]:

$$Q = \frac{\omega_o}{\Delta\omega} = \frac{f_o}{\Delta f} \tag{1.6.15}$$

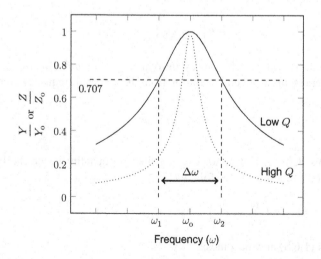

FIGURE 1.19

Resonant frequency and Q.

The value of Q calculated using equation (1.6.15) for any given circuit will, of course, be identical to that calculated using any of the other expressions for Q such as equation (1.6.4) or (1.6.5).

For every series combination in Figure 1.16, at a given frequency, there is an equivalent parallel combination, and vice versa. The ability to convert between series and parallel representations is important in network design. This is accomplished as follows:

Consider the circuit of Figure 1.16(a). We can write the admittance as:

$$Y = \frac{1}{R_s + jX} \tag{1.6.16}$$

$$= \frac{R_s - jX}{R_s^2 + X^2} \tag{1.6.17}$$

$$= \frac{R_s}{R_s^2 + X^2} - j\frac{X}{R_s^2 + X^2} \tag{1.6.18}$$

Which can be rewritten as:

$$Y = \frac{1/R_s}{R_s\left[1 + \left(\dfrac{X}{R_s}\right)^2\right]} + \frac{1/X}{j\left[\left(\dfrac{R_s}{X}\right)^2 + 1\right]} \tag{1.6.19}$$

In terms of the Q of the series circuit of Figure 1.16(a), Q_s, we can write:

$$Y = \frac{1}{R_s(1 + Q_s^2)} + \frac{1}{jX\left[\dfrac{1}{Q_s^2} + 1\right]} \tag{1.6.20}$$

So, from equation (1.6.20) we can see that the series combination of Figure 1.16(a) is equivalent to the parallel combination in Figure 1.17(a), where the parallel elements (conductance and susceptance) are:

$$G = \frac{1}{R_s(1 + Q_s^2)} \tag{1.6.21}$$

$$B = \frac{-Q_s}{X(1 + Q_s^2)} \tag{1.6.22}$$

Since the conductance, G, in equation (1.6.21) is the reciprocal of the equivalent parallel resistance, R_p, we can state the following relationship between R_s and R_p:

$$R_p = R_s(1 + Q_s^2) \tag{1.6.23}$$

Which implies that R_p will always be greater than R_s for any $Q_s > 0$, as one would expect. Rearranging equation (1.6.23), we obtain the following useful formula for the Q of a resonant circuit in terms of series and parallel resistances that will pop up repeatedly in different forms throughout this book and especially in Chapter 9, where we introduce matching network design:

$$Q_s = \sqrt{\left(\frac{R_p}{R_s}\right) - 1}$$

(1.6.24)

1.6.2 LOADED *Q* AND EXTERNAL *Q*

So far we have analyzed *RLC* circuits in isolation, but in the real world these circuits are usually embedded in a larger system. At the very least, our *RLC* circuit must be connected to a source and a load, both of which will contain their own resistive elements. When a resonant circuit is connected to the outside world, the total losses, for the purpose of calculating Q, will have to include losses in the source and load resistances. A typical situation where a parallel *RLC* resonant circuit is connected to a source, R_S and load, R_L, is shown in Figure 1.20.

The loaded Q may be determined by considering the overall Q of the entire circuit with external lossy elements taken into account, in other words, considering only the energy stored in the capacitor:

$$Q_L = \frac{\text{Total susceptance}}{\text{Total conductance}} = \frac{\omega_0 C}{G_{\text{total}}} = \frac{\omega_0 C}{G_S + G + G_L}$$

(1.6.25)

where $G_s = 1/R_s$, $G = 1/R$, and $G_L = 1/R_L$.

We could use an equivalent expression by considering only the energy stored in the inductor and the results will be the same. We can immediately see, from equation (1.6.25), that any finite values of G_S and G_L will have the effect of reducing the overall value of Q. This concurs with our intuitive understanding that the loaded Q of a passively loaded circuit can never be higher than its unloaded Q, since the addition of external passive circuit elements can only serve to add losses.

Another way of considering loaded Q is to introduce the concept of *external Q*, which represents the effect of all the elements that are not intrinsic to the "unloaded" resonant circuit. Unloaded Q_u and external Q_e are combined together to give the

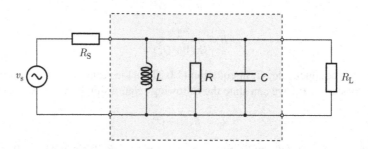

FIGURE 1.20

Parallel *RLC* circuit with source and load.

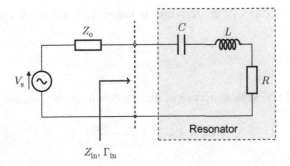

FIGURE 1.21

One-port *RLC* resonator.

loaded Q_L of the circuit according to the following general equation [28]:

$$\frac{1}{Q_L} = \frac{1}{Q_u} + \frac{1}{Q_e} \tag{1.6.26}$$

By comparing equations (1.6.26) and (1.6.25) we can see that the external Q of the circuit in Figure 1.20 is given by:

$$Q_e = \frac{\omega_0 C}{G_s + G_L} \tag{1.6.27}$$

Readers may notice the similarity in form between equation (1.6.26) and the equation for the total resistance of two resistors in parallel. In the same way that the total resistance of two resistors in parallel will always be less than the resistance of either of the individual resistors alone, the effect of adding external Q_e will always be to reduce the overall Q_L of any given resonant circuit.

1.6.3 Q OF A ONE-PORT RESONATOR

We will now combine our understanding of Q and reflection coefficient to look at the case of an *RLC* circuit applied as a one-port *resonator*. A resonator is basically any passive microwave device that is designed to resonate at a specific frequency. Resonators find application in oscillators and filters where they are used for frequency control and we will introduce a number of different types of resonator, such as cavity resonators, YIG resonators, and dielectric resonators in Chapter 15. The electrical properties of any resonator can be modeled, more or less accurately, by a simple series or parallel *RLC* circuit like those in Figure 1.18, with appropriately chosen component values. Consider the series *RLC* equivalent of a one-port resonator shown in Figure 1.21. In this case, we assume that the resonator is being driven by a variable frequency AC source, having an internal resistance which we will refer to as "Z_0" for reasons that will become apparent in Chapter 2.

The input impedance of the resonator in Figure 1.21 at a single frequency, ω, is given by:

$$Z_{\text{in}} = R + j\omega L - j\frac{1}{\omega C} \tag{1.6.28}$$

Which can be re-written in terms of the resonant frequency, ω_0, as:

$$Z_{\text{in}} = R + j\hat{Z}_0 \left(\frac{\omega}{\omega_0} - \frac{\omega_0}{\omega} \right) \tag{1.6.29}$$

where $\hat{Z}_0$ is a parameter we will refer to as the *characteristic impedance* of the resonator defined by:

$$\hat{Z}_0 = \sqrt{\frac{L}{C}} \tag{1.6.30}$$

and ω_0 is the nominal resonant frequency defined by:

$$\omega_0 = \frac{1}{\sqrt{LC}} \tag{1.6.31}$$

We have written Z_{in} in the form of equation (1.6.29) in order to emphasize the fact that Z_{in} reduces to simply R when $\omega = \omega_0$. The input reflection coefficient of the resonator of Figure 1.21 at the frequency, ω, can be calculated from equation (1.5.1) as:

$$\Gamma_{\text{in}} = \frac{Z_{\text{in}} - Z_0}{Z_{\text{in}} + Z_0} \tag{1.6.32}$$

Applying equation (1.6.29) we get:

$$\Gamma_{\text{in}} = \frac{R - Z_0 + j\hat{Z}_0 \left(\dfrac{\omega}{\omega_0} - \dfrac{\omega_0}{\omega} \right)}{R + Z_0 + j\hat{Z}_0 \left(\dfrac{\omega}{\omega_0} - \dfrac{\omega_0}{\omega} \right)} \tag{1.6.33}$$

$$\Gamma_{\text{in}} = \frac{\left(\dfrac{R - Z_0}{\hat{Z}_0} \right) + j \left(\dfrac{\omega}{\omega_0} - \dfrac{\omega_0}{\omega} \right)}{\left(\dfrac{R + Z_0}{\hat{Z}_0} \right) + j \left(\dfrac{\omega}{\omega_0} - \dfrac{\omega_0}{\omega} \right)} \tag{1.6.34}$$

The various Q factors for the circuit in Figure 1.21 are defined as:
Unloaded Q:

$$Q_{\text{u}} = \frac{\omega_0 L}{R} = \frac{1}{\sqrt{LC}} \cdot \frac{L}{R} = \sqrt{\frac{L}{C}} \cdot \frac{1}{R} = \frac{\hat{Z}_0}{R} \tag{1.6.35}$$

External Q:

$$Q_e = \frac{\omega_0 L}{Z_0} = \frac{1}{\sqrt{LC}} \cdot \frac{L}{Z_0} = \sqrt{\frac{L}{C}} \cdot \frac{1}{Z_0} = \frac{\hat{Z}_0}{Z_0} \qquad (1.6.36)$$

Loaded Q:

$$Q_L = \frac{\omega_0 L}{(R + Z_0)} = \frac{Q_u}{(1 + Z_0/R)} = \frac{R Q_u}{(R + Z_0)} = \frac{\hat{Z}_0}{(R + Z_0)} \qquad (1.6.37)$$

By combining equations (1.6.35)–(1.6.37) with equation (1.6.26) we can now write equation (1.6.34) in terms of the various Q factors as follows:

$$\Gamma_{in} = \frac{\left(\dfrac{1}{Q_u} - \dfrac{1}{Q_e}\right) + j\left(\dfrac{\omega}{\omega_0} - \dfrac{\omega_0}{\omega}\right)}{\left(\dfrac{1}{Q_u} + \dfrac{1}{Q_e}\right) + j\left(\dfrac{\omega}{\omega_0} - \dfrac{\omega_0}{\omega}\right)} \qquad (1.6.38)$$

Equation (1.6.38) is a universal expression for the input reflection coefficient of any microwave resonator in terms of the resonant frequency and the two Q factors Q_u and Q_e. The utility of equation (1.6.38) lies in the fact that these parameters can usually be more easily measured than the fundamental components R, L, and C, especially in the case of metal cavities and dielectric resonators, where R, L, and C do not exist as discrete physical entities.

From equation (1.6.38), the squared magnitude of the input reflection coefficient is given by:

$$|\Gamma_{in}(\omega)|^2 = \frac{\left(\dfrac{1}{Q_u} - \dfrac{1}{Q_e}\right)^2 + \left(\dfrac{\omega}{\omega_0} - \dfrac{\omega_0}{\omega}\right)^2}{\left(\dfrac{1}{Q_u} + \dfrac{1}{Q_e}\right)^2 + \left(\dfrac{\omega}{\omega_0} - \dfrac{\omega_0}{\omega}\right)^2} \qquad (1.6.39)$$

Which gives the input return loss as a function of ω, as:

$$RL_{dB}(\omega) = 10\log_{10}\left[\frac{\left(\dfrac{1}{Q_u} - \dfrac{1}{Q_e}\right)^2 + \left(\dfrac{\omega}{\omega_0} - \dfrac{\omega_0}{\omega}\right)^2}{\left(\dfrac{1}{Q_u} + \dfrac{1}{Q_e}\right)^2 + \left(\dfrac{\omega}{\omega_0} - \dfrac{\omega_0}{\omega}\right)^2}\right] \qquad (1.6.40)$$

The return loss at resonance can be found by setting $\omega = \omega_0$ in equation (1.6.40), giving:

$$RL_{dB}(\omega_0) = 10\log_{10}\left[\frac{Q_e - Q_u}{Q_e + Q_u}\right] \qquad (1.6.41)$$

We can define a specific frequency, ω_L, which satisfies the following:

$$\left(\frac{\omega_L}{\omega_0} - \frac{\omega_0}{\omega_L}\right)^2 = \frac{1}{Q_L^2} = \left(\frac{1}{Q_u} + \frac{1}{Q_e}\right)^2 \tag{1.6.42}$$

At the frequency ω_L the reflection coefficient therefore becomes:

$$|\Gamma_{in}(\omega_L)|^2 = \frac{\left(\dfrac{1}{Q_u} - \dfrac{1}{Q_e}\right)^2 + \left(\dfrac{1}{Q_u} + \dfrac{1}{Q_e}\right)^2}{\left(\dfrac{1}{Q_u} + \dfrac{1}{Q_e}\right)^2 + \left(\dfrac{1}{Q_u} + \dfrac{1}{Q_e}\right)^2}$$

$$= \frac{1}{2} + \frac{\left(\dfrac{1}{Q_u} - \dfrac{1}{Q_e}\right)^2}{2\left(\dfrac{1}{Q_u} + \dfrac{1}{Q_e}\right)^2}$$

$$= \frac{1}{2} + \frac{1}{2}|\Gamma_{in}(\omega_0)|^2 \tag{1.6.43}$$

Based on the above analysis, we can outline a methodology for measuring the various Qs of any resonator without knowledge of the equivalent circuit component values and based solely on reflection coefficient measurements, as follows:

1. Measure the return loss of the resonator at resonance and thereby calculate $|\Gamma_{in}(\omega_0)|$.
2. Determine $|\Gamma_{in}(\omega_L)|$ from equation (1.6.43).
3. Plot a curve of $|\Gamma_{in}|$ (or return loss) versus frequency and determine the frequencies for which $|\Gamma_{in}| = |\Gamma_{in}(\omega_L)|$. Note that there will be two such frequencies, $\pm\omega_L$, one either side of ω_0.
4. Calculate Q_L given ω_L from equation (1.6.42) as follows:

$$Q_L = \frac{\omega_0 \omega_L}{\omega_L^2 - \omega_0^2} \tag{1.6.44}$$

5. Calculate Q_e from:

$$Q_e = \frac{2Q_L}{1 \pm |\Gamma_{in}(\omega_0)|^2} \tag{1.6.45}$$

The choice of sign in equation (1.6.45) depends on the phase of $\Gamma_{in}(\omega_0)$. If $R < Z_0$ then the negative sign is used. If $R > Z_0$ then the positive sign is used.
6. Calculate Q_u from equation (1.6.26).

Let us consider the example of a series RLC resonator with the following component values:

$$Z_0 = 50 \ \Omega$$

$$R = 2 \ \Omega$$

$$L = 3.3 \text{ nH}$$

$$C = 4.7 \text{ pF}$$

These values give the following derived parameters:

$$f_0 = 1.28 \text{ GHz}$$

$$Q_u = 13.2$$

$$Q_L = 0.51$$

$$Q_e = 0.53$$

We can use equation (1.6.39) to plot $|\Gamma_{in}|$ versus frequency for this resonator, as shown in Figure 1.22(a). It is also useful to plot the return loss, as defined by equation (1.6.40) versus frequency as shown in Figure 1.22(b). This is another name for $|\Gamma_{in}|^2$ measured in dB and can be thought of as a measure of the proportion of incident power that is reflected back from the input of the resonator.

At frequencies far above or below the resonant frequency in Figure 1.22, $|\Gamma_{in}| = 1$ and the return loss is zero dB. This means that all the incident power is reflected back from the resonator. As the resonant frequency is approached from either direction, $|\Gamma_{in}|$ decreases and return loss becomes increasingly negative.

At the resonant frequency of 1.28 GHz, the drop in value of $|\Gamma_{in}|$ is only about 8%, from 1.0 to 0.923. This translates to a return loss of about -0.7 dB. In other words, most of the power incident on the resonator is still being reflected back to the source, even at resonance.

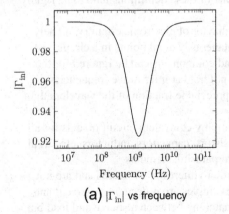

(a) $|\Gamma_{in}|$ vs frequency

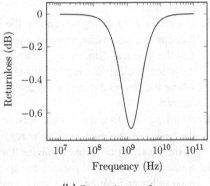

(b) Return loss vs frequency

FIGURE 1.22

$|\Gamma_{in}|$ and return loss versus frequency for the one-port *RLC* resonator of Figure 1.21. (a) $|\Gamma_{in}|$ vs frequency. (b) Return loss vs frequency.

We can see the relationship between the loss in the resonator (which is entirely due to the resistor, R) and the return loss at resonance by setting $\omega = \omega_0$ in equation (1.6.33), which results in:

$$\Gamma_{in_0} = \frac{R - Z_0}{R + Z_0} \tag{1.6.46}$$

This implies that as R approaches zero, Γ_{in_0} approaches unity, which means that 100% of incident power will be reflected at any frequency, even ω_0. This makes sense if we consider that if $R = 0$, there can be no power dissipated in the resonator. Similarly, as R approaches infinity, Γ_{in_0} also approaches unity, which also makes sense as the resonator will always appear as an open circuit to the source. So what value of R gives us the lowest value of Γ_{in_0}? From equation (1.6.46) we can determine that setting $R = Z_0$ gives $\Gamma_{in_0} = 0$. This means that all the power will be absorbed in the resonator at the resonant frequency, ω_0.

An interesting discussion of the measurement of unloaded, loaded, and external Q of one and two-port resonators based in reflection coefficient magnitudes is given by Bray and Roy [29].

1.7 TAKEAWAYS

1. "Microwave" frequencies are formally defined as those above 300 MHz (wavelengths less than 1 m), but informally we consider microwave frequencies as being above about 1 GHz. Frequencies above about 30 GHz are referred to as "millimeter wave" and those above 300 GHz are referred to as "submillimeter wave." It must be said, however, that none of these definitions tend to be rigidly adhered to.

2. Some implicit assumptions about the behavior of electronic circuitry, namely that currents and voltages appear instantaneously at all points in a circuit and that the physical dimensions of wires and components can be ignored for the purposes of electric circuit analysis, do not hold at microwave frequencies, where the physical dimensions are an appreciable fraction of the wavelength of interest.

3. At microwave frequencies we need to employ equivalent circuit models for all the basic components, such as wires, resistors, capacitors, inductors, etc., as all of these exhibit complex behavior due to parasitic elements.

4. Some commonly used metrics are specific to microwave systems and are not used in lower-frequency systems. Two examples are *reflection coefficient* and *SWR*. Both can indicate the quality of matching between a source and load but, whereas reflection coefficient is a complex number that can be used to specify the load impedance precisely, SWR is a scalar quantity that does not contain any phase information. SWR alone cannot, therefore, be used to specify a load impedance.

5. The "dB" is a relative measurement used to represent the ratio of two powers, such as power gain or power loss (attenuation), assuming a common reference impedance. Absolute power levels are sometimes referenced to 1 mW and expressed in "dBm."

6. The Quality factor, or "Q" factor characterizes the rate of energy loss in a given circuit relative to the amount of energy stored. Higher Q indicates a lower rate of energy loss relative to the stored energy, or higher "quality."

7. Loaded Q is defined as the Q of a given circuit when it is connected to external circuitry such as load resistances, etc. This is in contrast to the unloaded Q of the circuit alone, in isolation. Because external circuitry can only add (not subtract) losses, the loaded Q will always be less than the unloaded Q.

8. There is a relationship between loaded Q and bandwidth: the higher the Q the narrower the bandwidth of a given circuit.

9. A "resonator" may be fabricated out of discrete electronic components (resistors, capacitors, and inductors) or as a physical structure, such as a metal cavity or dielectric sample.

TUTORIAL PROBLEMS

Problem 1.1 (Resistance and resistivity) Calculate the resistance of a 50 cm length of copper wire of cylindrical cross section with a diameter of 1 mm at 100 MHz, 1 GHz, and 10 GHz, given that the resistivity of copper is 1.68×10^{-8} Ωm.

Problem 1.2 (Wire resistance at RF) Derive a simple formula giving the ratio of DC resistance to RF resistance for a wire of circular cross section.

Problem 1.3 (Resistance and inductance of a wire) Calculate the self-inductance of the wire in Problem 1.1. What would be the inductance of the same piece of wire if it is now formed into a cylindrical coil of inner diameter 5 mm and length 10 cm? What is the unloaded Q of the wire at 1 GHz in each case?

Problem 1.4 (Impedance and admittance) Show that, although a given impedance, $Z = R + jX$, can be represented as an equivalent admittance $Y = 1/Z$, where $Y = G + jB$, it is not correct to say that $G = 1/R$ and $B = 1/X$.

Problem 1.5 (Return loss) A load, that is known to be purely resistive, is connected to a microwave signal generator having an output impedance of 50 Ω. The return loss is measured to be 18 dB. Determine the value of the load resistance.

Problem 1.6 (Maximum power transfer) Find the value of R_L in Figure 1.23 that results in maximum power being delivered to R_L and calculate the value of this maximum power. When R_L is adjusted for maximum power transfer, what percentage of the power emanating from the source is actually delivered to R_L.

Problem 1.7 (Insertion loss) The shunt resistor, R_x, inserted between source and load in Problem 1.6 may be characterized by an *insertion loss*, which is defined as the difference between the power delivered to the load, in dB, without R_x present and the power delivered to the load, in dB, with R_x present.

If we set $Z_s = Z_L = Z_o$, show that the insertion loss arising from the presence of any generalized shunt admittance, $Y_x = G_x + jB_x$, in place of R_x in Figure 1.23 is given by:

$$\alpha_L(\text{dB}) = 10\log_{10}\left[\left(1 + \frac{G_x Z_o}{2}\right)^2 + \left(\frac{B_x Z_o}{2}\right)^2\right]$$

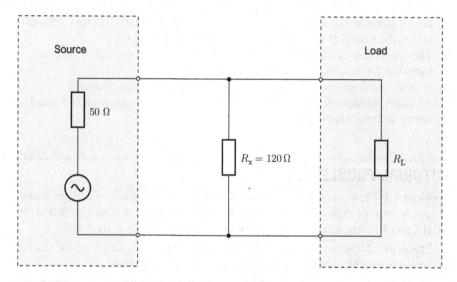

FIGURE 1.23

Maximum power transfer example.

Hence, calculate insertion loss of the circuit in Figure 1.23 at 1 GHz if the 120 Ω resistor in Figure 1.23 is replaced by a series combination of a 100 Ω resistor a 33 nH inductor.

Problem 1.8 (Q of a resonator) Calculate Q_L, Q_u, and Q_e for the following microwave resonator:

- resonant frequency, f_o, = 13 GHz
- return loss at resonance = -22 dB
- $f_L = \pm 0.2$ GHz (referenced to f_o).

REFERENCES

[1] IEEE Std 521-1984, IEEE Standard Letter Designations for Radar-Frequency Bands, 1984, http://dx.doi.org/10.1109/IEEESTD.1984.81588.

[2] J. Wiltse, History of millimeter and submillimeter waves, IEEE Trans. Microw. Theory Tech. 32 (9) (1984) 1118-1127, ISSN 0018-9480, http://dx.doi.org/10.1109/TMTT.1984.1132823.

[3] H.-J. Song, T. Nagatsuma, Present and future of terahertz communications, IEEE Trans. Terahertz Sci. Technol. 1 (1) (2011) 256-263, ISSN 2156-342X, http://dx.doi.org/10.1109/TTHZ.2011.2159552.

[4] K. Eriksson, I. Darwazeh, H. Zirath, InP DHBT wideband amplifiers with up to 235 GHz bandwidth, in: 2014 IEEE MTT-S International Microwave Symposium (IMS), 2014, pp. 1-4, http://dx.doi.org/10.1109/MWSYM.2014.6848436.

[5] C. Poole, I. Darwazeh, H. Zirath, K. Eriksson, D. Kuylenstierna, S. Lai, Design and characterization of a negative resistance Common Emitter InP Double Heterojunction Bipolar Transistor subcircuit for millimeter wave and submillimeter wave applications, in: 2014 44th European Microwave Conference (EuMC), 2014, pp. 933-936, http://dx.doi.org/10.1109/EuMC.2014.6986589.

[6] I. Hosako, N. Sekine, M. Patrashin, S. Saito, K. Fukunaga, Y. Kasai, P. Baron, T. Seta, J. Mendrok, S. Ochiai, H. Yasuda, At the dawn of a new era in terahertz technology, Proc. IEEE 95 (8) (2007) 1611-1623, ISSN 0018-9219, http://dx.doi.org/10.1109/JPROC.2007.898844.

[7] J. Andrews, S. Buzzi, W. Choi, S. Hanly, A. Lozano, A. Soong, J. Zhang, What will 5G be?, IEEE J. Sel. Areas Commun. 32 (6) (2014) 1065-1082, ISSN 0733-8716, http://dx.doi.org/10.1109/JSAC.2014.2328098.

[8] F. Boccardi, R. Heath, A. Lozano, T. Marzetta, P. Popovski, Five disruptive technology directions for 5G, IEEE Commun. Mag. 52 (2) (2014) 74-80, ISSN 0163-6804, http://dx.doi.org/10.1109/MCOM.2014.6736746.

[9] T. Rappaport, S. Sun, R. Mayzus, H. Zhao, Y. Azar, K. Wang, G. Wong, J. Schulz, M. Samimi, F. Gutierrez, Millimeter wave mobile communications for 5G cellular: it will work!, IEEE Access 1 (2013) 335-349, ISSN 2169-3536, http://dx.doi.org/10.1109/ACCESS.2013.2260813.

[10] ITU, The Internet of Things, ITU Internet Reports, 2005.

[11] H. Howe, Microwave integrated circuits: an historical perspective, IEEE Trans. Microw. Theory Tech. 32 (9) (1984) 991-996, ISSN 0018-9480, http://dx.doi.org/10.1109/TMTT.1984.1132812.

[12] J. Kilby, Invention of the integrated circuit, IEEE Trans. Electron Devices 23 (7) (1976) 648-654, ISSN 0018-9383, http://dx.doi.org/10.1109/T-ED.1976.18467.

[13] R. Pengelly, J. Turner, Monolithic broadband GaAs F.E.T. amplifiers, Electron. Lett. 12 (10) (1976) 251-252, ISSN 0013-5194, http://dx.doi.org/10.1049/el:19760193.

[14] L. Samoska, An overview of solid-state integrated circuit amplifiers in the submillimeter-wave and THz regime, IEEE Trans. Terahertz Sci. Technol. 1 (1) (2011) 9-24, ISSN 2156-342X, http://dx.doi.org/10.1109/TTHZ.2011.2159558.

[15] M. Shamos, Great Experiments in Physics: Firsthand Accounts from Galileo to Einstein, Dover Publications, 2012, ISBN 9780486139623.

[16] W. Hayt, J. Buck, Engineering Electromagnetics, McGraw-Hill Higher Education, 2012, ISBN 9780071089012.

[17] S. Ramo, J. Whinnery, T. Van Duzer, Fields and Waves in Communication Electronics, Wiley, 1994, ISBN 9780471585510.

[18] J. Kraus, Electromagnetics With Applications, McGraw-Hill College, 1998, ISBN 9780072356632.

[19] A. Kip, Fundamentals of Electricity and Magnetism. International Student edition, McGraw-Hill, 1981.

[20] N. Forbes, B. Mahon, Faraday, Maxwell, and the Electromagnetic Field: How Two Men Revolutionized Physics, Prometheus Books, 2014, ISBN 9781616149420.

[21] D. Pozar, Microwave Engineering, second ed., John Wiley and Sons Inc., New York, USA, 1998.

[22] Z. Popović, B. Popović, Introductory Electromagnetics, Prentice Hall, 2000, ISBN 9780201326789.

[23] K. Fenske, D. Misra, Dielectric materials at microwave frequencies, Appl. Microw. Wirel. 12 (12) (2000) 92-100.

[24] C. Bowick, RF Circuit Design, Newnes Elsevier, Burlington, MA, USA, 2008.

[25] R. Ludwig, G. Bogdanov, RF Circuit Design, second ed., Pearson Education Inc., Upper Saddle River, NJ, USA, 2009.

[26] H. Skilling, Electrical Engineering Circuits, Wiley, 1963, ISBN 9780471794363.

[27] A. Fitzgerald, D. Higginbotham, A. Grabel, Basic Electrical Engineering: Circuits Electronics Machines Controls. International Student Edition, McGraw-Hill Series in Electrical Engineering: Networks and Systems, McGraw-Hill Book Company, 1985.

[28] R. Collin, Foundations for Microwave Engineering, second ed., John Wiley and Sons Inc., New York, USA, 2005, ISBN 9788126515288.

[29] J. Bray, L. Roy, Measuring the unloaded, loaded, and external quality factors of one- and two-port resonators using scattering-parameter magnitudes at fractional power levels, in: IEE Proceedings Microwaves, Antennas and Propagation, vol. 151, ISSN 1350-2417, 2004, pp. 345-350, http://dx.doi.org/10.1049/ip-map:20040521.

Transmission line theory

2

INTENDED LEARNING OUTCOMES

- *Knowledge*
  - Understand that electrical energy travels at a finite speed in any medium, and the implications of this.
  - Understand the behavior of lossy versus lossless transmission lines.
  - Understand power flows on a transmission line and the effect of discontinuities.

- *Skills*
  - Be able to determine the location of a discontinuity in a transmission line using time domain refractometry.
  - Be able to apply the telegrapher's equations in a design context.
  - Be able to calculate the reflection coefficient, standing wave ratio of a transmission line of known characteristic impedance with an arbitrary load.
  - Be able to calculate the input impedance of a transmission line of arbitrary physical length, and terminating impedance.
  - Be able to determine the impedance of a load given only the voltage standing wave ratio (VSWR) and the location of voltage maxima and minima on a line.

2.1 INTRODUCTION

To someone with a superficial knowledge of electrical technology it seems obvious that when two physically separated points are connected using a length of conducting wire, and assuming the resistance of the wire can be ignored, the voltage at the remote end of the wire will be the same as that at the source. In other words, the physical length of the wire and physical location of the source and destination, again ignoring resistive effects, are immaterial to understanding of how the circuit works.

This common sense understanding of electric circuit behavior is all well and good as long as the voltage changes in the circuit occur in a time interval which significantly exceeds the time it takes for the signal to travel the length of the wire from source to destination. The limitations of this common sense view were dramatically demonstrated, however, in August 1858, when the first transatlantic telegraph cable, a cross section of which is shown in Figure 2.1, was activated [1]. Although the project presented enormous engineering challenges, in terms of manufacturing and cable laying, in electrical terms the circuit was believed to be quite simple, being essentially a pair of wires 3000 km long, with a DC voltage source at one end and a mirror galvanometer at the other end to detect changes in current on the line. The unexpectedly poor performance of this hugely expensive project, in terms of the speed at which Morse Code data pulses could be transmitted across the Atlantic (the first message took over 17 h to transmit, at an average data rate of one Morse character every 2 min) exposed the limitations of the contemporary understanding of the behavior of very long cables.

After a few unsuccessful "brute force" attempts at increasing the data rate by, for example, increasing the drive voltage at the source end, it was realized that the time had come to apply some brain power to the problem. Some of the greatest scientists of the time, such as James Clerk Maxwell, Lord Kelvin, and Oliver Heaviside set about constructing a rigorous mathematical analysis of the behavior of electrical impulses on long cables [3,4]. In 1880, Heaviside published the first paper that described his analysis of propagation in cables using the *telegrapher's equations* [5], which we will elaborate on in Section 2.4 of this chapter. Thus arose the first conceptual

FIGURE 2.1

A section of the first transatlantic telegraph cable, c1858 (reproduced by kind permission of the History of Diving Museum [2]).

understanding of a *transmission line*, and a set of mathematical tools for transmission line analysis that we still use to this day.

So we can see that, in the case of a submarine cable several thousand kilometers long, the physical length of the wire cannot be ignored. The deciding factor, however, in determining whether transmission line analysis is appropriate is not the physical length of the line, *per se*, but the relationship between the physical length of the line and the frequency (or wavelength) of the electrical signals on the line. In other words, the transmission line theory originally developed to analyze very long cables at relatively low frequencies also applies to much shorter transmission lines at much higher frequencies. The key parameter here is what is known as the *electrical length* of the line, which is defined as the ratio of the physical length of the line to the wavelength of the signal of interest as it appears on the line. The speed of propagation, v, in any given transmission line will be fixed. The wavelength of a signal of frequency, f, in such a line is therefore given by:

$$\lambda = \frac{v}{f} \tag{2.1.1}$$

where λ is in m, f is in Hz, and v is in m s^{-1}.

We have purposely referred to the speed of propagation in the transmission line, as this will depend on the dielectric constant of the insulating material used to fabricate the line and will generally be less than the free space speed, since [6]:

$$v = \frac{v_0}{\sqrt{\varepsilon_r}} \tag{2.1.2}$$

where v_0 is the speed of propagation in free space and ε_r is the dielectric constant.

We can say that, irrespective of the frequency of operation, any line with an electrical length in the order of one wavelength must be treated as a transmission line [7]. We can summarize the characteristics of a transmission line as follows:

(i) The physical length of the line is comparable to the wavelength at the frequency of interest.

(ii) Instantaneous values of voltage (or current) are not the same at all points on the line, but vary with physical position along the line.

(iii) Voltage (or current) must be treated as "waves" that travel down the line at a finite speed. Such "waves" will be reflected from discontinuities in the line such as loads, junctions, and interconnections, causing reflected waves to travel back down the line toward the source.

(iv) These forward and reverse traveling waves exist simultaneously on the same line. The instantaneous voltage or current at any point on the line will be comprised of the sum of the forward and reverse traveling waves.

The above characteristics result in a behavior that is quite counter-intuitive for students familiar with low-frequency circuits. In this chapter, we will elaborate on these characteristics and set out the tools for the analysis of transmission line problems.

2.2 PROPAGATION AND REFLECTION ON A TRANSMISSION LINE

Let us consider a simple lossless transmission line, which could be simply a pair of parallel wires, terminated in a resistive load and connected to a DC source, such as a battery having a finite internal resistance, R_S. We will add one minor refinement to this simple picture in the form of a changeover switch at the source, as shown in Figure 2.2, which enables us to disconnect the battery by moving the switch from position A to position B, but ensures that the line continues to see the same source impedance, R_S, whether the battery is connected or not.

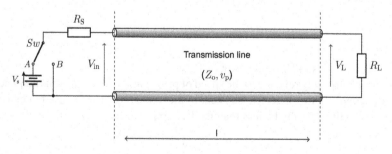

FIGURE 2.2

Simple transmission line system, DC driven.

We can conduct a simple thought experiment by considering the switch is initially in position B and is then instantaneously moved to position A at time we shall define as $t = 0$. The casual observer, applying basic circuit theory, might conclude that the voltage at the input of the line, V_{in}, is determined by the potential divider formed by the source resistance R_s and the load resistance R_L as follows:

$$V_{in} = V_s \left(\frac{R_L}{R_s + R_L} \right)$$
(2.2.1)

This is indeed the case once the circuit has reached a *steady state*; that is, after the switch has been in position A for sometime. What we are interested in, however, is the value of V_{in} *at the exact instant* the switch is moved to position A, that is at time $t = 0$. It is crucial to bear in mind that electrical energy propagating in any medium will travel at a finite speed. Therefore, at time $t = 0$, the source has no way of "knowing" the nature of the termination at the other end of the line. At $t = 0$, the line may well be infinitely long as far as the determination of V_{in} is concerned. We do know, however, that a finite current does start to flow into the line at $t = 0$, in order for the circuit to reach the steady-state condition. The fact that a finite current is flowing at $t = 0$ suggests that the source is "seeing" a finite impedance at the input of the line, and a corresponding finite voltage, V_{in}, is being developed across it. Since we are talking about $t = 0$, that is before the current has had time to traverse the full physical length of the line, we must conclude that this mystery line impedance, which we shall call the *characteristic impedance*, is an inherent property of this particular line and is independent of both the physical length of the line and the value of the load.

If we assign Z_0 to this characteristic impedance, we can now write the voltage, V_{in}, at time $t = 0$ as follows:

$$V_{in} = V_s \left(\frac{Z_0}{R_s + Z_0} \right)$$
(2.2.2)

The concept of characteristic impedance will be elaborated on later in this chapter, but for now the reader is asked to accept that this is the impedance of an infinitely long transmission line of specific physical characteristics. This can be understood by considering, once again, the fact that at time $t = 0$ any transmission line of nonzero length will appear to be infinitely long, as far as the source is concerned.

So, as time progresses, from $t = 0$ onward, a voltage wave, of amplitude V_{in}, will propagate down the line toward the load. We will refer to this forward-propagating voltage wave as the *incident* voltage, V_i. The corresponding incident current, I_i, propagating down the line in a direction away from the source is given by:

$$I_i = \frac{V_i}{Z_0}$$
(2.2.3)

Let us take our thought experiment one stage further by assuming that the switch, having been moved to position A at time $t = 0$, is moved back to position B, a short time later at a time $t = \Delta t$. We choose Δt to be much less than the time, τ, it takes for electrical energy to propagate down the full length of the line from source to load, that is:

$$\Delta t \ll \tau \qquad (2.2.4)$$

where

$$\tau = \frac{l}{v_p} \qquad (2.2.5)$$

where v_p is the velocity (or speed) of propagation on the line.

After the switch has been moved back to the B position, we now have a rectangular pulse of electrical energy, of voltage amplitude V_i and width Δt, propagating down the line at a velocity v_p, toward the load, in the transmission line system shown in Figure 2.3.

What happens when this energy pulse arrives at the load? In the general case, the pulse will encounter a step change in impedance (an impedance *discontinuity*) at the point of termination, that is $R_L \neq Z_0$. Kirchhoff's laws and Ohm's law must always be satisfied at any point in space and time. We therefore have the following boundary conditions at the instant the voltage pulse arrives at the load:

$$V_L = I_L R_L \qquad (2.2.6)$$

and

$$V_i = I_i Z_0 \qquad (2.2.7)$$

but, at the instant the pulse arrives at the load, $V_L = V_i$, we therefore have, from equations (2.2.6) and (2.2.7):

$$I_L R_L = I_i Z_0 \qquad (2.2.8)$$

We know, however, that in the general case $R_L \neq Z_0$, so equation (2.2.8) implies that $I_L \neq I_i$. How, then, can Kirchhoff's current law be satisfied at the load? In order

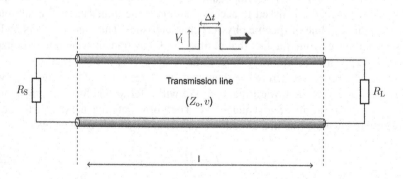

FIGURE 2.3

Single pulse on a transmission line.

to account for the difference between I_L and I_i at the load, we introduce a new term, I_r, which relates to I_L and I_i as follows:

$$I_L = I_i - I_r \tag{2.2.9}$$

If we assume that no electromagnetic energy is radiated from the load, the only place for this balancing current, I_r, to flow is back down the line toward the source. We therefore deduce that I_r (and the corresponding voltage, $V_r = Z_0 I_r$) constitutes a *reflected* energy pulse that travels back down the line toward the source.

Applying Kirchhoff's voltage law at the load now gives the following:

$$V_L = V_i + V_r \tag{2.2.10}$$

Combining equation (2.2.10) with equations (2.2.6), (2.2.7), and (2.2.9) results in:

$$\frac{V_i + V_r}{R_L} = \frac{V_i}{Z_0} - \frac{V_r}{Z_0} \tag{2.2.11}$$

A simple rearrangement of equation (2.2.11) results in:

$$V_i \left[\frac{1}{Z_0} - \frac{1}{R_L} \right] = V_r \left[\frac{1}{Z_0} + \frac{1}{R_L} \right] \tag{2.2.12}$$

Of particular interest is the ratio of reflected voltage amplitude to incident voltage amplitude, which can be found from equation (2.2.12) as:

$$\frac{V_r}{V_i} = \frac{R_L - Z_0}{R_L + Z_0} = \Gamma_L \tag{2.2.13}$$

This ratio is referred to as the *load voltage reflection coefficient* or simply the *load reflection coefficient*, and is given the symbol Γ_L. From equation (2.2.13), we can observe that if $R_L = Z_0$ then $\Gamma_L = 0$. In other words, if the load is equal to the characteristic impedance of the line then there will be no reflected energy and all the incident energy will be absorbed in the load. Under this condition, we say that the line is *matched*. As far as the source is concerned, a matched line, of any length, is indistinguishable from a line of infinite length.

We can further deduce from equation (2.2.13) that if $R_L < Z_0$ then Γ_L will be negative, whereas if $R_L > Z_0$ then Γ_L will be positive. This means that the polarity of the reflected pulse from a load having an impedance less than Z_0 will be of the opposite polarity to the incident pulse, whereas the polarity of the reflected pulse from a load having an impedance greater than Z_0 will be the same polarity as the incident. In extremis, the maximum value of load impedance is infinity (i.e., an open circuit) and the minimum value of load impedance is zero (i.e., a short circuit). The reflection characteristics of a line with extreme load values is summarized in Table 2.1 and shown in Figure 2.4.

Table 2.1 Reflection Coefficient Extremes

Load	Γ_L	Reflected Pulse
Open circuit ($Z_L = \infty$)	1	Same amplitude, same polarity as the incident pulse
Matched ($Z_L = Z_0$)	0	No reflected pulse
Short circuit ($Z_L = 0$)	−1	Same amplitude but opposite polarity to incident pulse

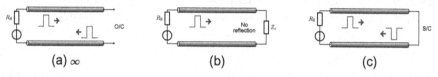

(a) ∞ (b)   (c)

FIGURE 2.4

Incident and reflected pulses with extreme load values. (a) Open circuit ($Z_L = \infty$). (b) Matched ($Z_L = Z_0$). (c) Short circuit ($Z_L = 0$).

Returning to our thought experiment, we now observe a reflected voltage pulse, of width Δt and voltage amplitude that we will now refer to as "V_{r1}" traveling back down the line from the load toward the source. Employing equation (2.2.13), the voltage amplitude of this reflected pulse is given by:

$$V_{r1} = \Gamma_L V_i \tag{2.2.14}$$

When this reflected pulse arrives back at the source, at a time $t = 2\tau$ later, some of the pulse energy will be absorbed in R_S, and the remainder of the energy will be reflected back down the line toward the load again, this time in the form of another voltage pulse of width Δt and voltage amplitude which we will refer to as V_{r2}, as follows:

$$V_{r2} = \Gamma_L \Gamma_S V_i \tag{2.2.15}$$

where Γ_S is the *source voltage reflection coefficient*, which is defined, with respect to Z_0, similarly to equation (2.2.13), as:

$$\Gamma_S = \frac{R_S - Z_0}{R_S + Z_0} \tag{2.2.16}$$

This secondary pulse will be reflected again at the load resulting in another reflected voltage pulse, of further reduced amplitude, traveling back toward the source again. This process continues indefinitely as long as R_S and R_L remain connected to the line as in Figure 2.3. Provided both source and load terminations are purely resistive (i.e., $0 < R_S < \infty$ and $0 < R_L < \infty$), the magnitudes of Γ_S and Γ_L will always be less than or equal to unity, each successive pulse will be of smaller amplitude than the one before, ultimately becoming vanishingly small (but never zero). We can understand this intuitively if we consider that, with each pass, some of the energy in the pulse is being absorbed in R_S and R_L. Furthermore, because both $R_L < Z_0$ and $R_S < Z_0$ the polarity of each successive voltage pulse is reversed with respect to the previous one.

If we were to place a suitable voltage measuring instrument exactly half way along a line terminated with $R_S < Z_0$ and $R_L < Z_0$, we would observe the picture shown in Figure 2.5.

It is instructive to now consider the sequence of events that occurs when we connect the DC source in Figure 2.2 to a transmission line terminated with an open circuit. Let us assume we connect the source to the line at time $t = 0$ and leave it permanently connected. Since there is an open circuit at the end of the line, there are two things we can say for certain:

1. At the exact instant the switch is closed, at $t = 0$, the voltage across the input of the transmission line, V_{in}, will be given by equation (2.2.2), which implies (for finite values of Z_0 and R_S) that $V_{in} \neq V_S$.
2. In the steady-state condition, that is, sometime after the switch has been closed, the voltage across the input of the transmission line, V_{in}, and indeed at all points along the line, must be equal to V_S.

How does the voltage change as we transition from state (1) to state (2) above? Consider the *transient analysis* shown in Figure 2.6 [8].

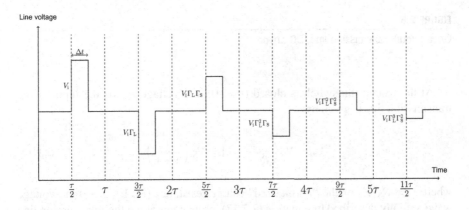

FIGURE 2.5

Voltage pulse and reflections, as seen at a point exactly half way along the transmission line with $R_L < Z_0$ and $R_S < Z_0$.

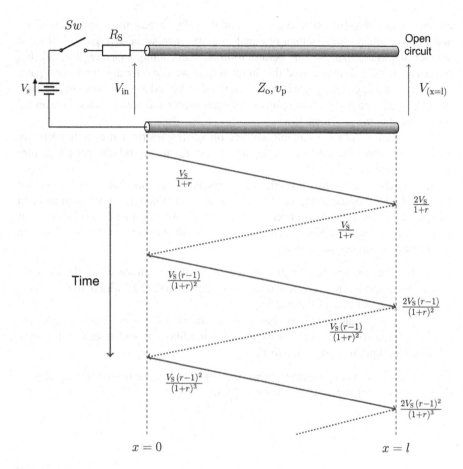

FIGURE 2.6

Open-circuit transmission line, DC driven.

At the instant the switch is closed ($t = 0$), the voltage across the input of the transmission line is given by:

$$V_{in} = V_S \left(\frac{Z_o}{R_S + Z_o} \right) = V_S \left(\frac{1}{1+r} \right) \qquad (2.2.17)$$

where $r = R_S/Z_o$ or the "normalized" source resistance [8]. For $t > 0$ the voltage wave, initially described by equation (2.2.17), propagates along the transmission line in a finite time and arrives at the far end of the line a fixed time, $t = \tau$, later. Since the line is terminated in an open circuit ($\Gamma_L = 1$), this voltage is totally reflected back toward the source as shown in Figure 2.6.

When the reflected voltage propagating back down the line reaches the source, it sees a reflection coefficient, Γ_S, given by:

$$\Gamma_S = \frac{R_S - Z_0}{R_S + Z_0} = \frac{r - 1}{r + 1} \tag{2.2.18}$$

Hence, a fraction of this voltage is reflected back down the line toward the open circuit with a value given by:

$$\left(\frac{V_S}{1 + r}\right)\Gamma_S = \frac{V_S}{1 + r}\left(\frac{r - 1}{r + 1}\right) = V_S \frac{r - 1}{(r + 1)^2} \tag{2.2.19}$$

which in turn gets reflected at the open circuit and so on, as shown in Figure 2.6. The total voltage at $x = l$ builds up as t increases. The final voltage, at $t = \infty$, is given by the addition of all these partial voltages, that is to say:

$$V_{(x=l)} = 2V_S\left(\frac{1}{r + 1} + \frac{r - 1}{(r + 1)^2} + \frac{(r - 1)^2}{(r + 1)^3} + \cdots\right) \tag{2.2.20}$$

or

$$V_{(x=l)} = \left(\frac{2V_S}{r + 1}\right)\sum_{k=0}^{\infty}\left(\frac{r - 1}{r + 1}\right)^k \tag{2.2.21}$$

Equation (2.2.21) contains a sum of an infinite geometric series, which can be replaced by its equivalent, resulting in:

$$V_{(x=l)} = \left(\frac{2V_S}{r + 1}\right)\left(\frac{r + 1}{2}\right) \tag{2.2.22}$$

$$= V_S \tag{2.2.23}$$

In other words, the voltage at the open end of the transmission line in Figure 2.6 will eventually come to equal the source voltage, V_S, although this will not happen instantaneously. What is interesting is how the output voltage gets to V_S and what happens in the first few moments after the switch is closed, which depends on the value of r.

Figure 2.7 shows the voltage at the load end of the transmission line ($x = l$) versus time for three values of the ratio r, namely $r = 4$, $r = 1$, and $r = 0.25$ [8]. We observe that in all three cases the voltage at the load tends to V_S, as we would expect. Note that in Figure 2.7(b) the voltage at the load equals V_S for any $t > \tau$, since $r = 1$ means that $R_S = Z_0$. When the first reflection arrives back at the source, it is completely absorbed by the source resistance and there are no subsequent reflections.

Example 2.1 (Transmission line: Time Domain Reflectometry). A time domain reflectometer (TDR) is a piece of test equipment which is used to locate transmission line discontinuities, such as breaks or short circuits. The TDR works by injecting a pulse at one end of the line and showing the various reflections on an oscilloscope type display (i.e., voltage amplitude vs time). Reflections shown on the display represent discontinuities due to abrupt changes in the impedance along a transmission line. The magnitude of the impedance discontinuity determines the magnitude of the

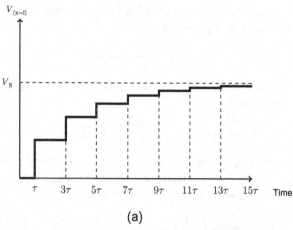

(a)

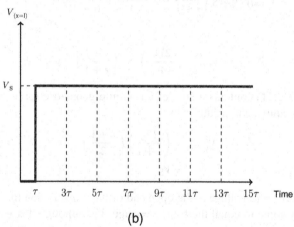

(b)

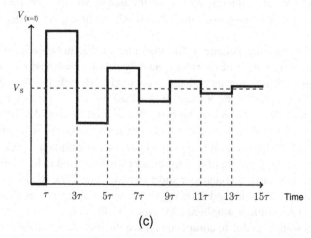

(c)

FIGURE 2.7

Load voltage development on an open-circuit terminated line for various values of
normalized source resistance, *r*. (a) *r* = 4. (b) *r* = 1. (c) *r* = 0.25.

reflection and the time delay between transmitted and reflected pulses represents the distance to the discontinuity.

Problem. A serious fault (i.e., either a short circuit or an open circuit) is suspected to lie at some point in a long coaxial cable. Using a TDR, a square pulse of width 1 μs is transmitted down the cable. The leading edge of the reflected pulse arrives back at the source 3.2 μs after the pulse is applied to the line. Determine how far down the cable the short circuit is located, given that the velocity of propagation in the cable is 10^8 m s^{-1}.

Given that the returned pulse is the same polarity as the transmitted pulse, determine the nature of the fault (short or open circuit).

Solution. The distance traveled by the pulse is $2d$, where d is the distance from the source to the fault. We can therefore write:

$$2d = v\tau$$

where v is the velocity of propagation and τ is the round trip time. Therefore,

$$2d = 10^8 \times 3.2 \times 10^{-6} = 320 \text{ m}$$

The fault is therefore located 160 m from the source end of the cable.

The fact that the returned pulse is the same polarity as the transmitted pulse indicates that the fault is an open circuit, as shown in Figure 2.8.

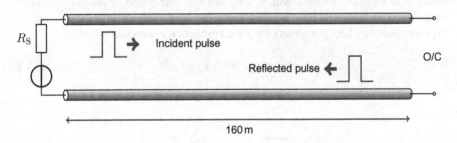

FIGURE 2.8

TDR example.

2.3 SINUSOIDAL STEADY-STATE CONDITIONS: STANDING WAVES

In the previous section, we established that the terminating impedance of a transmission line differs from the characteristic impedance of the line then electrical energy will reflected from the termination and that this reflected energy will travel back down the line toward the source.

So far we have only considered discrete energy pulses on lines. We will now extend this analysis to consider the more significant case (from an RF perspective) of steady-state sinusoidal excitation. If a mismatched line is continuously driven by a sinusoidal source, we can envisage an incident sinusoidal voltage wave, V^-, traveling from the source to the load and the discontinuity at the mismatched load giving rise to a continuous reflected sinusoidal voltage wave, V^+, traveling in the reverse direction from load to source. In steady-state conditions, both these waves will co-exist on the line, as shown in Figure 2.9.

At any particular point on the line, the instantaneous voltage will be the sum of the instantaneous values of the forward traveling wave, V^-, and the reverse traveling wave, V^+. These waves will interfere with each other, and the vector sum of the two traveling waves will create a "standing wave" pattern of voltage on the line that remains static with time. V^- and V^+ interfere with each other to produce voltage maxima (where they add in phase) and minima (where they add 180° out of phase).

In order to capture both magnitude and phase of these traveling waves, we will represent them in their complex exponential form. The voltage at any given point on a lossless transmission line, $V(x)$, can therefore be written as the sum of two complex traveling waves, that is:

$$V(x) = V^+ e^{-j\beta x} + V^- e^{j\beta x} \qquad (2.3.1)$$

where β is a *phase constant* which describes how the phase of the wave changes with distance x along the line. We shall be saying a lot more about β shortly. We can express equation (2.3.1) in terms of the load reflection coefficient as follows:

$$V(x) = V^+ e^{-j\beta x} + \Gamma_L V^+ e^{j\beta x} \qquad (2.3.2)$$

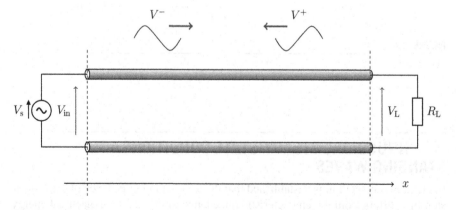

FIGURE 2.9

Transmission line with sinusoidal excitation.

or

$$V(x) = V^+ e^{-j\beta x} \left[1 + \Gamma_L e^{j2\beta x} \right] \tag{2.3.3}$$

We are primarily interested in the magnitude of the resultant voltage, so equation (2.3.3) gives us:

$$|V| = |V^+||1 + \Gamma_L e^{j2\beta x}| \tag{2.3.4}$$

But we are interested in the voltage at a distance l back from the load. So we substitute $x = -l$ in equation (2.3.4) to get:

$$|V| = |V^+| \cdot |1 + \Gamma_L e^{-j2\beta l}| \tag{2.3.5}$$

Since Γ_L is in general complex, we can write it as:

$$\Gamma_L = |\Gamma_L| e^{j\theta_L} \tag{2.3.6}$$

where θ_L represents the phase difference between V^+ and V^- at the load. Equation (2.3.5) can now be written as:

$$|V| = |V^+| \cdot |1 + |\Gamma_L| e^{j(\theta_L - 2\beta l)}| \tag{2.3.7}$$

By inspection of equation (2.3.7) we see that the maximum voltage on the line occurs when the phase term, $e^{j(\theta_L - 2\beta l)} = 1$. Thus we have:

$$|V|_{\max} = |V^+|(1 + |\Gamma_L|) \tag{2.3.8}$$

Similarly, the minimum voltage on the line occurs when the phase term, $e^{j(\theta_L - 2\beta l)} = -1$. Thus we have:

$$|V|_{\min} = |V^+|(1 - |\Gamma_L|) \tag{2.3.9}$$

Further inspection of equation (2.3.7) as a function of l will reveal that maxima occur when:

$$\beta l - \frac{\theta_L}{2} = n\pi \tag{2.3.10}$$

and minima occur when:

$$\beta l - \frac{\theta_L}{2} = n\pi + \frac{\pi}{2} \tag{2.3.11}$$

where $n = 0, 1, 2, 3, \ldots$.

In other words, the distance between a given voltage maximum and the adjacent voltage minimum will always be $\pi/2$ in terms of phase or $\lambda/2$ in terms of electrical length, as shown for an arbitrary load impedance (not a short or open circuit) in Figure 2.10.

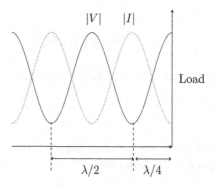

FIGURE 2.10

Voltage maxima and minima on a transmission line.

An important ratio is that of $|V_{\max}|$ to $|V_{\min}|$, which is referred to as the *VSWR*, that is:

$$\text{VSWR} = \frac{|V_{\max}|}{|V_{\min}|} \tag{2.3.12}$$

We can also define a *current (I) Standing Wave Ratio*, ISWR, as being the ratio of $|I_{\max}|$ to $|I_{\min}|$. The value of the ISWR is the same as the VSWR, although the current maxima occur at the location of voltage minima and vice versa.

The relationship between VSWR and reflection coefficient can be simply derived from equations (2.3.8) and (2.3.9) as:

$$\text{VSWR} = \frac{1 + |\Gamma|}{1 - |\Gamma|} \tag{2.3.13}$$

and, conversely

$$|\Gamma| = \frac{\text{VSWR} - 1}{\text{VSWR} + 1} \tag{2.3.14}$$

VSWR is an important and commonly used parameter in RF engineering because it gives a clear indication of the degree of match/mismatch on a transmission line. Its popularity is also due to its ease of measurement. Since VSWR is a scalar quantity, however, it does not, by itself, give any indication of the type of load on the line (i.e., capacitive, inductive) only the degree of match or mismatch. In order to determine the complete nature of a complex load we need phase information in addition to scalar information provided by the VSWR. This phase information can be obtained from the location of voltage minima on the line, as shown in Example 2.2.

Example 2.2 (Transmission line: determination of an arbitrary load impedance). Back in the days before computerized microwave test equipment, the slotted line was an essential tool in any microwave lab. The slotted line consists of an air dielectric coaxial transmission line plus a sliding probe usually containing a Schottky diode detector connected to a simple voltmeter. The slotted line enables the magnitude of the standing wave to be measured at various points along the line. As well as the rather obvious application in the measurement of VSWR by detecting the voltage maxima and minima, the slotted line can also be used to measure impedance as follows.

Problem. Consider an unknown load connected to an air dielectric coaxial slotted line of 50 Ω characteristic impedance and operating at 1.5 GHz. The slotted line is used to measure the location and magnitudes of the voltage minima and maxima and hence the VSWR is calculated as 2.0. The first voltage minimum is located at $\lambda/4$ from the load. Calculate the load impedance, assuming the slotted line is lossless.

When a different load is connected to the slotted line the VSWR remains the same but the first voltage minimum is seen to move 0.05λ toward the load. Determine the value of this second load.

Solution. We first calculate the magnitude of the reflection coefficient from equation (2.3.14) as follows:

$$|\Gamma| = \frac{2 - 1}{2 + 1} \approx 0.334$$

We know that the first voltage minimum occurs when $e^{j(\theta_L - 2\beta l)} = -1$ where θ_L is the phase angle of the load reflection coefficient, that is:

$$\Gamma_L = |\Gamma_L| e^{j\theta_L}$$

The phase angle of the load reflection coefficient is then given by:

$$\theta_L = \pi + 2\beta l_{min}$$

where l_{min} is the distance from the load to the first minimum. For the first load $l_{min} = \lambda/4$. We can therefore state that:

$$\theta_{L1} = \pi + \pi = 2\pi$$

So the load reflection coefficient is:

$$\Gamma_{L1} = |\Gamma_{L1}| \times (\cos(\theta_{L1}) + j\sin(\theta_{L1}))$$
$$\Gamma_{L1} = 0.334 \times (1 + j0) \tag{2.3.15}$$

We can now have all the information we need to calculate the load using equation (2.7.5):

$$Z_{L1} = 50 \times \frac{(1 + 0.334)}{(1 - 0.334)}$$
$$Z_{L1} = 100 \ \Omega \tag{2.3.16}$$

For the second load, we have:

$$\theta_{L2} = \pi + 0.4\pi = 1.4\pi$$

So the load reflection coefficient is:

$$\Gamma_{L2} = |\Gamma_{L2}| \times (\cos(\theta_{L2}) + j\sin(\theta_{L2}))$$

$$\Gamma_{L2} = 0.334 \times (-0.3090 - j0.9511) = -0.1032 - j0.3177 \qquad (2.3.17)$$

Again, we now calculate the load using equation (2.7.5):

$$Z_{L2} = 50 \times \frac{(1 + (-0.1032 - j0.3177))}{(1 - (-0.1032 - j0.3177))}$$

$$Z_{L2} = 33.70 - j24.10 \ \Omega \qquad (2.3.18)$$

2.4 PRIMARY LINE CONSTANTS

We can gain a deeper understanding of how transmission lines behave with steady-state sinusoidal excitation by modeling the transmission line as an infinite series of two-port network elements, comprised of elementary lumped components. Each elementary two-port represents an infinitesimally short segment of the transmission line as shown in Figure 2.11. The reader will recall, from the previous sections, that all voltages and currents on such a line are functions of physical location, x. Because we are talking about steady-state sinusoidal excitation, these voltages and currents are also functions of time, so at any particular point on the line we must refer to $v(x, t)$ and $i(x, t)$ (we are using lower case letters to represent AC small signal quantities).

The components R, L, C, and G are referred to as *primary line constants* and are defined in Table 2.2.

To aid further analysis, the elementary circuit shown in Figure 2.11 can be redrawn in simplified form in Figure 2.12.

The line voltage $V(x)$ and the current $I(x)$ in Figure 2.12 can be expressed in the frequency domain using the so-called *Telegrapher's equations*:

$$\frac{\partial V(x)}{\partial x} = -(R + j\omega L)I(x) \qquad (2.4.1)$$

$$\frac{\partial I(x)}{\partial x} = -(G + j\omega C)V(x) \qquad (2.4.2)$$

where $I(x)$ and $V(x)$ describe the current and voltage along the transmission line, as a function of position x. The functions $I(x)$ and $V(x)$ are complex, where the magnitude and phase of the complex functions describe the magnitude and phase of the sinusoidal time function $e^{j\omega t}$. Only those functions of $I(x)$ and $V(x)$ that satisfy the telegrapher's equations can exist on a transmission line.

So, we now need to find functions $I(x)$ and $V(x)$ that satisfy both telegrapher's equations, equations (2.4.1) and (2.4.2). We will first combine the telegrapher's

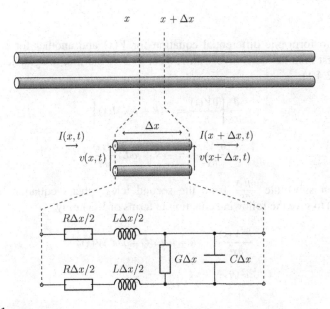

FIGURE 2.11

Distributed element model of a transmission line.

Table 2.2 Primary Line Constants

Constant	Meaning	Units
R	Ohmic resistance of the conductors	Ohms per meter
L	Inductance associated with the magnetic field around the conductors	Henries per meter
C	Capacitance associated with the electric field in the dielectric medium separating the conductors	Farads per meter
G	Conductivity of the dielectric material separating the conductors	Siemens per meter

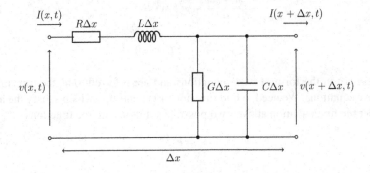

FIGURE 2.12

Transmission line incremental section: lumped element equivalent circuit.

equations to form one differential equation for $V(x)$ and another for $I(x)$. To do this, we first take the derivative with respect to x of the first telegrapher equation, equation (2.4.1):

$$\frac{\partial}{\partial x}\left\{\frac{\partial V(x)}{\partial x} = -(R+j\omega L)I(x)\right\} \tag{2.4.3}$$

$$\frac{\partial^2 V(x)}{\partial x^2} = -(R+j\omega L)\frac{\partial I(x)}{\partial x} \tag{2.4.4}$$

We then substitute $\frac{\partial I(x)}{\partial x}$ from the second telegrapher's equation into equation (2.4.4) to give the following equation in terms of $V(x)$ only:

$$\frac{\partial^2 V(x)}{\partial x^2} = (R+j\omega L)(G+j\omega C)V(x) \tag{2.4.5}$$

$$\frac{\partial^2 V(x)}{\partial x^2} = \gamma^2 V(x) \tag{2.4.6}$$

where γ is the *propagation constant* defined by:

$$\gamma = \sqrt{(R+j\omega L)(G+j\omega C)} = \alpha + j\beta \tag{2.4.7}$$

where α is called the *attenuation* (or loss) factor, which, by convention, is measured in units of Nepers (Np) per meter. The Neper is a logarithmic unit of power ratio, like the decibel, except that it uses the natural logarithm, rather than base 10 logarithm. Since 1 Np corresponds to a power ratio of e^2, 1 Np is equivalent to [7]:

$$1\ \text{Np} = 10\ \log(e^2) = 8.686\ \text{dB} \tag{2.4.8}$$

The β in equation (2.4.7) happens to be the *phase* (or velocity) factor referred to previously, which is measured in units of radians per second. We can now restate both telegrapher's equations as follows:

$$\frac{\partial^2 V(x)}{\partial x^2} - \gamma^2 V(x) = 0 \tag{2.4.9}$$

$$\frac{\partial^2 I(x)}{\partial x^2} - \gamma^2 I(x) = 0 \tag{2.4.10}$$

These are in the form of wave equations and are referred to as the transmission line wave equations. We need to find functions $V(x)$ and $I(x)$ which satisfy the above. Consider the first equation above: two possible solutions are the functions

$$V(x) = e^{-\gamma x} \tag{2.4.11}$$

and

$$V(x) = e^{+\gamma x} \tag{2.4.12}$$

Which represent sinusoidal waves traveling in the negative x-direction and the positive x-direction, respectively.

Since the transmission line wave equation is a linear differential equation, a weighted superposition of the two solutions is also a solution. So the general solution to these wave equations is:

$$V(x) = V^+ e^{-\gamma x} + V^- e^{\gamma x} \qquad (2.4.13)$$

$$I(x) = I^+ e^{-\gamma x} + I^- e^{\gamma x} \qquad (2.4.14)$$

The two terms in each solution describe the two waves propagating in opposite directions on the transmission line, as discussed previously. The reader may note that equation (2.4.13) is actually a generalization of equation (2.3.1) where β has been replaced by $\gamma = \alpha + j\beta$. By convention, the first term in equation (2.4.13) represents a wave propagating in the positive ($+x$) direction. The second term represents a wave propagating in the opposite direction ($-x$).

2.5 THE LOSSLESS TRANSMISSION LINE

An interesting set of results is obtained if we make the simplifying assumption that the line is lossless; that is, it is constructed of perfect conductors ($R = 0$) and perfect insulators ($G = 0$). If we set $R = 0$ and $G = 0$ in equation (2.4.7), we get $\alpha = 0$, so the propagation constant reduces to:

$$\gamma = j\beta = j\omega\sqrt{LC} \qquad (2.5.1)$$

Replacing equation (2.5.1) in equations (2.4.9) and (2.4.10) gives us the second-order steady-state telegrapher's equations for a lossless line as follows:

$$\frac{\partial^2 V(x)}{\partial x^2} - \omega^2 LC V(x) = 0 \qquad (2.5.2)$$

$$\frac{\partial^2 I(x)}{\partial x^2} - \omega^2 LC I(x) = 0 \qquad (2.5.3)$$

We note that equations (2.5.2) and (2.5.3) are in the form of wave equations with a *phase velocity* which is given by:

$$\upsilon_p = \frac{1}{\sqrt{LC}} \qquad (2.5.4)$$

Given equation (2.5.1) we can also see that:

$$\upsilon_p = \frac{\omega}{\beta} \qquad (2.5.5)$$

Phase velocity has been defined as the velocity at which an observer would have to travel down the line in order to observe a constant phase for the wave propagating on the line [9]. In general, the phase velocity, υ_p, is the speed of propagation in free space (c) scaled by the ratio of guided wavelength to the free space wavelength as follows:

$$v_p = \frac{\lambda_g}{\lambda_o} c \qquad (2.5.6)$$

Finally we observe, from the fact that $\beta = \omega\sqrt{LC}$, that a signal traveling down a lossless transmission line experiences a phase shift directly proportional to its frequency.

2.6 DERIVATION OF THE CHARACTERISTIC IMPEDANCE

We introduced the concept of *characteristic impedance* earlier in this chapter, as the "mystery" impedance seen by the source when looking into an infinitely long transmission line. We will now proceed to derive an equation for the characteristic impedance of a line in terms of the primary line parameters introduced in the previous section.

Consider a sinusoidal voltage source connected to an infinitely long transmission line which can be modeled as an infinite series of the unit elements shown in Figure 2.12. In Figure 2.13, we have separated out the first of these infinitesimal elements to aid the analysis to follow.

Consider the instant, $t = 0$, when the switch in Figure 2.13 is closed. In order for any current to flow into the line, the source must see a finite impedance, which we now know to be the characteristic impedance of the line, Z_0. It is worth remembering that, even though we are using the theoretical abstraction of an infinitely long line, which does not exist in practice, any line, whatever its physical length, will appear to the source as if it were infinitely long at the instant the switch is closed.

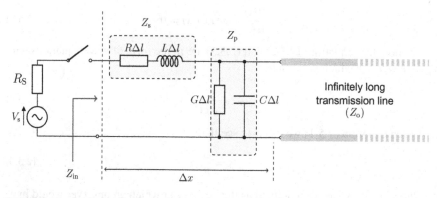

FIGURE 2.13

Characteristic impedance of an infinitely long transmission line.

Considering the circuit of Figure 2.13, we can define the impedance seen by the source at the instant $t = 0$ as:

$$Z_{in} = Z_s + \frac{Z_p Z_o}{Z_p + Z_o}$$ (2.6.1)

where Z_s and Z_p are the unit series and parallel elements in Figure 2.13 defined by:

$$Z_p = \frac{1}{(G + j\omega C)\Delta x}$$ (2.6.2)

$$Z_s = (R + j\omega L)\Delta x$$ (2.6.3)

At the instant $t = 0$ the line may as well be infinitely long as far as the source is concerned. Therefore, at $t = 0$, $Z_{in} = Z_o$ and we have, from equation (2.6.1):

$$Z_o = Z_s + \frac{Z_p Z_o}{Z_p + Z_o}$$ (2.6.4)

Rearranging equation (2.6.4) gives:

$$Z_o^2 - Z_o Z_s - Z_p Z_s = 0$$ (2.6.5)

So we now have Z_o in terms of Z_p and Z_s:

$$Z_o = \frac{Z_s \pm \sqrt{Z_s^2 + 4Z_p Z_s}}{2}$$ (2.6.6)

Which, in terms of the primary line parameters, becomes:

$$Z_o = \frac{(R + j\omega L)\Delta x}{2} \pm \frac{1}{2}\sqrt{[(R + j\omega L)\Delta x]^2 + 4\frac{(R + j\omega L)}{(G + j\omega C)}}$$ (2.6.7)

Since Δx is arbitrarily small, we can reduce it to zero, at which point equation (2.6.7) becomes:

$$Z_o = \sqrt{\frac{(R + j\omega L)}{(G + j\omega C)}}$$ (2.6.8)

Another way of arriving at equation (2.6.8) is to consider the instantaneous voltage and current relationships on the line as defined by equations (2.4.1) and (2.4.4). Substituting equation (2.4.4) into equation (2.4.1) and differentiating we get the following:

$$I(x) = \frac{\gamma}{(R + j\omega L)}V_o^+ e^{+\gamma x} + V_o^- e^{+\gamma x}$$ (2.6.9)

Since the ratio of voltage and current is always an impedance, we can define the *characteristic impedance* of the line as follows:

$$Z_0 = \frac{(R + j\omega L)}{\gamma} = \sqrt{\frac{(R + j\omega L)}{(G + j\omega C)}} \qquad (2.6.10)$$

At this point, we can make some interesting observations about characteristic impedance. First, we can see from equation (2.6.8) that, if the line is lossy, that is $R \neq 0$ and/or $G \neq 0$, then the characteristic impedance will be complex. On the other hand, if the line is lossless then equation (2.6.8) reduces to

$$Z_0 = \sqrt{\frac{L}{C}} \qquad (2.6.11)$$

which means that the characteristic impedance of a lossless transmission line is a real number. We should pause for a moment and consider the profound implications of equation (2.6.11). Here we have a lossless circuit, comprising purely reactive elements, that gives rise to an input impedance that appears purely real. It looks like a real resistance, in the sense that current and voltage are in phase, but it is not a true "resistance," as normally understood, since no power is being dissipated in a lossless line (by definition).

2.7 TRANSMISSION LINES WITH ARBITRARY TERMINATIONS

We will now further investigate the behavior of a transmission line, of characteristic impedance Z_0, terminated with an arbitrary load, Z_L, under conditions of steady-state sinusoidal excitation, as shown in Figure 2.14. We will first deal with the general case of a lossy line, having a complex propagation constant $\gamma = \alpha + j\beta$. The specific case of a lossless line will be covered in the next section.

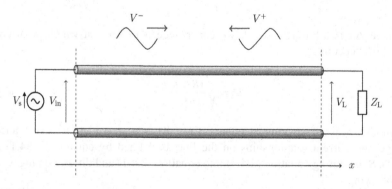

FIGURE 2.14

Transmission line with complex load.

Under these steady-state conditions we will observe a standing wave of voltage and current on the line, that is, voltage and current are functions of position, x, only. The expressions for voltage and current on the line are given by equations (2.4.13) and (2.4.14):

$$V(x) = V^+ e^{-\gamma x} + V^- e^{\gamma x} \qquad (2.4.13)$$

$$I(x) = I^+ e^{-\gamma x} - I^- e^{\gamma x} \qquad (2.4.14)$$

where V^+ and V^- are the amplitudes of the forward and reverse propagating voltage waves on the line, and I^+ and I^- are the amplitudes of the corresponding current waves. At the load, that is at $x = 0$, we have:

$$V_L = V^+ + V^- \qquad (2.7.1)$$

$$I_L = I^+ - I^- \qquad (2.7.2)$$

But $V^+ = Z_0 I^+$ and $V^- = Z_0 I^-$, hence equation (2.7.2) can be written as:

$$I_L = \frac{1}{Z_0}(V^+ - V^-) \qquad (2.7.3)$$

Combining equations (2.7.1) and (2.7.3), we have, at the load:

$$\frac{V_L}{I_L} = Z_0 \frac{(V^+ + V^-)}{(V^+ - V^-)} \qquad (2.7.4)$$

At this point, we note that $V_L/I_L = Z_L$ and $V^-/V^+ = \Gamma_L$, the voltage reflection coefficient at the load. We can therefore rewrite equation (2.7.4) as:

$$Z_L = Z_0 \frac{(1 + \Gamma_L)}{(1 - \Gamma_L)} \qquad (2.7.5)$$

which, when re-arranged, is consistent with equation (2.2.13), that is:

$$\Gamma_L = \frac{(Z_L - Z_0)}{(Z_L + Z_0)} \qquad (2.7.6)$$

This gives us the voltage reflection coefficient at only one location on the line, namely at the load where we have defined $x = 0$. What we now wish to do is generalize this to obtain an expression for the reflection coefficient at any point on the line. We can write the voltage reflection coefficient at any arbitrary distance, $x = -l$, back from the load, toward the source, as the ratio of reverse and forward propagating voltage waves, that is:

$$\Gamma(l) = \frac{V^- e^{-\gamma l}}{V^+ e^{\gamma l}} = \frac{V^-}{V^+} e^{-2\gamma l} \qquad (2.7.7)$$

We have already established that $V^-/V^+ = \Gamma_L$, the reflection coefficient of the load, so we can write:

$$\Gamma(l) = \Gamma_L e^{-2\gamma l} \tag{2.7.8}$$

which is the reflection coefficient looking into the line at a distance, l, from the load.

2.7.1 INPUT IMPEDANCE AT AN ARBITRARY POINT ON A TERMINATED LINE

Let us define $Z_{in}(l)$ as the impedance looking into a terminated line, toward the load, at location $x = -l$. We will now refer to the normalized input impedance at this point which we shall denote by $z_{in}(l)$, and define as follows:

$$z_{in}(l) = \frac{Z_{in}(l)}{Z_0} = \frac{V(l)}{I(l)Z_0} \tag{2.7.9}$$

Applying equations (2.4.13) and (2.4.14) to equation (2.7.9) gives:

$$z_{in}(l) = \frac{V^+ e^{\gamma l} + V^- e^{-\gamma l}}{V^+ e^{\gamma l} - V^- e^{-\gamma l}} \tag{2.7.10}$$

Dividing through by V^+ and applying equation (2.7.7), results in:

$$z_{in}(l) = \frac{e^{\gamma l} + \Gamma_L e^{-\gamma l}}{e^{\gamma l} - \Gamma_L e^{-\gamma l}} \tag{2.7.11}$$

Which we can simplify as:

$$z_{in}(l) = \frac{1 + \Gamma_L e^{-2\gamma l}}{1 - \Gamma_L e^{-2\gamma l}} \tag{2.7.12}$$

Replacing Γ_L with equation (2.7.5) and noting that $e^x = \cosh(x) + \sinh(x)$, we can rewrite equation (2.7.12) as:

$$z_{in}(l) = \frac{Z_L \cosh(\gamma l) + Z_0 \sinh(\gamma l)}{Z_0 \cosh(\gamma l) + Z_L \sinh(\gamma l)} \tag{2.7.13}$$

Which can be further rewritten as:

$$z_{in}(l) = \frac{Z_L + Z_0 \tanh(\gamma l)}{Z_0 + Z_L \tanh(\gamma l)} \tag{2.7.14}$$

In the special case of a lossless line, where $\alpha = 0$ and thus $\gamma = j\beta$, equation (2.7.14) reduces to:

$$z_{in}(l) = \frac{Z_L + jZ_0 \tan(\beta l)}{Z_0 + jZ_L \tan(\beta l)} \tag{2.7.15}$$

It is left as an exercise for the reader to show that the impedance, $z_{in}(l)$, given by equation (2.7.15) is a periodic function of position along the line, l, with period $\lambda/2$.

Example 2.3 (Transmission line: reflection coefficient and impedance).

Problem. A 50 Ω transmission line is connected to a load consisting of a 25 Ω resistor in series with a 5 pF capacitor.

1. Find the reflection coefficient, Γ_L, at the load for a 100 MHz signal.
2. Find the impedance, Z_{in}, at the input of the transmission line, given that the line is 0.125λ long.

Solution (i). The load impedance is calculated as:

$$Z_L = R_L - \frac{j}{\omega C}$$

$$= 25 - \frac{j}{2\pi \times 10^8 \times 5 \times 10^{-12}} \qquad (2.7.16)$$

$$= 25 - j318.31 \ \Omega \qquad (2.7.17)$$

The normalized load impedance is:

$$z_L = \frac{25 - j318.31}{50} = 0.5 + j6.37$$

The reflection coefficient at the load is calculated as:

$$\Gamma_L = \frac{z_L - 1}{z_L + 1}$$

$$= \frac{0.5 + j6.37 - 1}{0.5 + j6.37 + 1} \qquad (2.7.18)$$

$$= 0.930 + j0.297 \qquad (2.7.19)$$

$$= \underline{0.976\angle 17.7^\circ} \qquad (2.7.20)$$

Solution (ii). In this case, $\beta l = 2\pi \times 0.125 = \pi/4$. We can employ equation (2.7.15) to determine the input impedance of the line at 0.125λ from the load:

$$Z_{in}(l = 0.125\lambda) = Z_0 \frac{z_L + j\tan(\pi/4)}{1 + jz_L \tan(\pi/4)}$$

$$= Z_0 \frac{z_L + j}{1 + jz_L} \qquad (2.7.21)$$

$$= 50 \times \frac{0.5 + j6.37 + j}{1 + j(0.5 + j6.37)} \qquad (2.7.22)$$

$$= \underline{1.719 - j68.46} \qquad (2.7.23)$$

2.7.2 SPECIAL CASES: SHORT-CIRCUIT LINE

When the line is terminated with a short circuit we have $Z_L = 0$. Therefore, equation (2.7.15) simply reduces to:

$$Z_{in} = jZ_0 \tan \beta l \qquad (2.7.24)$$

The normalized input impedance versus electrical length for an open-circuit transmission line is shown in Figure 2.15.

Figure 2.16 shows an enlarged portion of Figure 2.15 around the origin. We can see that, if the electrical length of the line is short enough (say, less than 0.1λ), then equation (2.7.24) is an approximately linear function of β, and thus Z_{in} is approximately proportional to frequency. This means that such short sections of short-circuit terminated line can be used to substitute for an inductor. This is a common technique in the design of Monolithic Microwave Integrated Circuits (MMIC), as a short section of transmission line takes up less space on a semiconductor substrate than, say, a spiral inductor.

The reactance of a short-circuit line can also present a capacitive reactance, depending on the length. By way of practical illustration, Figure 2.17 shows a photomicrograph of a submillimeter wave negative device comprising a singe common emitter heterojunction bipolar transistor (HBT) with capacitive series feedback, implemented in MMIC form in InP material [10].

Using a capacitor as a series feedback element in this circuit would have required a bypass choke to be placed in parallel with the capacitor to carry the DC bias current. The inductance of the bypass choke would have to be large enough to make its

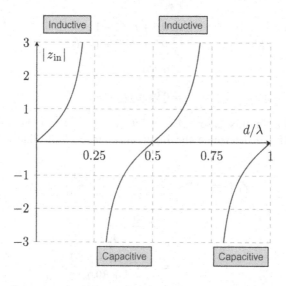

FIGURE 2.15

Normalized input impedance versus electrical length for a short-circuit transmission line.

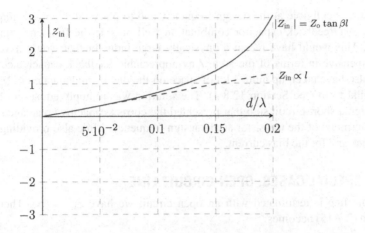

FIGURE 2.16

Use of a short section of short-circuit terminated transmission line as an inductor.

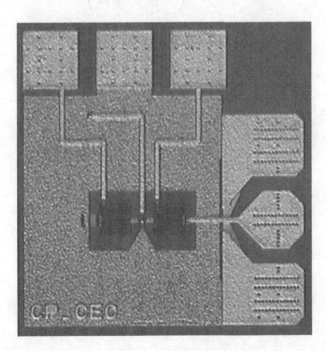

FIGURE 2.17

Submillimeter wave negative resistance MMIC device using an open-circuit transmission line to provide capacitive series feedback to a common emitter HBT [10] (die size 0.38 mm × 0.38 mm)

susceptance negligible at the design frequency and to place the resonant frequency of the choke/feedback capacitor combination well outside the frequency range of interest. This would have meant a physically large inductor that would have been quite expensive in terms of die area. Any appreciable feedback capacitance value would also have taken up quite a lot of space on the die, if implemented in MIM or interdigital form (see Section 12.8.4). The solution was to apply equation (2.7.24) and design a short-circuit stub that presented the correct capacitive reactance to the emitter terminal of the transistor at the design frequency, while also providing a DC path to ground for the bias current.

2.7.3 SPECIAL CASES: OPEN-CIRCUIT LINE

When the line is terminated with an open circuit we have $Z_L = \infty$. Therefore, equation (2.7.15) becomes:

$$Z_{in} = \lim_{Z_L \to \infty} \left(Z_0 \frac{1 + j\frac{Z_0}{Z_L} \tan \beta l}{\frac{Z_0}{Z_L} + j \tan \beta l} \right) \tag{2.7.25}$$

$$Z_{in} = \frac{Z_0}{j \tan \beta l} \tag{2.7.26}$$

$$Z_{in} = -jZ_0 \cot \beta l \tag{2.7.27}$$

The normalized input impedance versus electrical length for a open-circuit transmission line is shown in Figure 2.18.

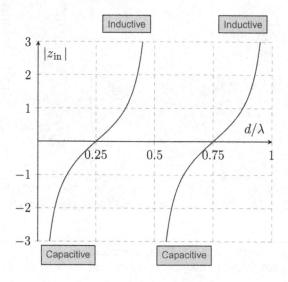

FIGURE 2.18

Normalized input impedance versus electrical length for an open-circuit transmission line.

2.7.4 SPECIAL CASES: MATCHED LINE

When the line is terminated with a load equal to the characteristic impedance ($Z_L = Z_0$). The input impedance of such a line, from equation (2.7.15), is simply:

$$Z_{in} = Z_0 \qquad\qquad (2.7.28)$$

In other words, a line terminated in the characteristic impedance will appear to the source as if it was an infinitely long line of characteristic impedance Z_0.

2.8 THE EFFECT OF LINE LOSSES

2.8.1 CHARACTERISTIC IMPEDANCE OF LOSSY LINES

All the analysis we have covered so far has assumed that the transmission line is lossless, which is a good enough approximation in most practical RF circuits. It is instructive to now consider the effect of losses in the line, as the effect of losses may need to be taken into account in some practical design scenarios, for example, where the line is of appreciable physical length.

A lossy line is one that contains appreciable series resistance and/or shunt conductance. Different frequencies travel at different speeds on lossy lines, hence, any signal containing multiple frequency components, such as a rectangular pulse, will not remain rectangular but will become distorted in shape as it travels down the line. The presence of losses changes the velocity of propagation and causes the wave to be attenuated as it travels. Therefore, any lossy line of sufficient length will appear to be well matched to a source of the same characteristic impedance, irrespective of the actual load impedance.

We can extend the analysis covered in the previous sections by reintroducing the attenuation constant, α, and the propagation constant given by equation (2.4.7). The reflection coefficient looking into the line at a distance, x, back from the load Γ_L is given by equation (2.7.8). In the case of a lossless line, $\alpha = 0$ and so equation (2.7.8) reduces to:

$$\Gamma(x) = \Gamma_L e^{-2\beta x} \qquad\qquad (2.8.1)$$

In the case of a lossy line, we cannot ignore α, so equation (2.7.8) should be written as:

$$\Gamma(x) = \Gamma_L e^{-2(\alpha + j\beta)x}$$

$$= \Gamma_L e^{-2\alpha x} e^{-j2\beta x} \qquad\qquad (2.8.2)$$

Comparing equations (2.8.2) and (2.8.1) we can see that the introduction of the attenuation constant, α, has the effect of reducing the magnitude of $\Gamma(x)$ exponentially as x increases. In the limit as $x \to \infty$ we find that the magnitude of the reflection coefficient looking into the line tends to zero. In other words, any sufficiently long length of lossy line will appear to be matched, *irrespective of the value of the load.* This makes sense when we consider that, in a lossy line, energy in both the forward and reflected waves is being continuously dissipated as they travel along the line. Any reflections from an arbitrary load will be reduced exponentially with physical length, thereby making the load appear better matched than it would be if the line was lossless.

Another implication of line loss can be understood by considering equation (2.6.8). If either $R \neq 0$ or $G \neq 0$ then the characteristic impedance will be complex. One implication of this is that a purely resistive source can never be perfectly matched to a lossy line, irrespective of the electrical length.

2.8.2 DISPERSION

In general, the phase velocity, v_p, is a function of wavelength (and thus frequency). In other words, in the case of a signal containing many frequency components, which is any signal other than a single frequency sinusoid, the various frequency components will travel down the line at different speeds. The result of this is that the signal will become "dispersed," that is the various components of the signal will arrive at the load at different times. This can be catastrophic for signals that are composed of many frequency components, such as square pulses (i.e. digital signals) which end up being "smeared out" in time as they travel down the line. If the line is long enough, square pulses will end up being so distorted that they will be unrecoverable at the receiver. This implies that the only way to prevent dispersion is to use a lossless line, so that v_p is independent of frequency. This suggests that dispersion is unavoidable, since all real-world lines have losses.

But there is another way. In the course of his work on the transatlantic cable problem outlined in the introduction to this chapter, Oliver Heaviside showed that a transmission line would be dispersionless if the line parameters exhibited the following ratio [11]:

$$\frac{R}{L} = \frac{G}{C} \tag{2.8.3}$$

This is called the *Heaviside condition*, which can be verified by substituting equation (2.8.3) into equation (2.4.7):

$$\gamma = \sqrt{LC(R/L + j\omega)(G/C + j\omega)}$$

$$= \sqrt{LC(R/L + j\omega)(R/L + j\omega)}$$

$$= (R/L + j\omega)\sqrt{LC}$$

$$= R\sqrt{\frac{C}{L}} + j\omega\sqrt{LC} \tag{2.8.4}$$

Given that $\gamma = \alpha + j\beta$ we can write equation (2.8.4) as:

$$\alpha = R\sqrt{\frac{C}{L}} \tag{2.8.5}$$

and

$$\beta = \omega\sqrt{LC} \tag{2.8.6}$$

We note that equation (2.8.6) is the same as equation (2.5.1), meaning that we have achieved the same result as for a lossless line. We also have the following phase velocity:

$$v_p = \frac{\omega}{\beta} = \frac{1}{\sqrt{LC}} \tag{2.8.7}$$

Since the above phase velocity is independent of frequency, we have created a lossy line that behaves like a lossless line, provided that equation (2.8.3) is satisfied. Thus Oliver Heaviside showed that we can eliminate dispersion from a lossy transmission line by adding series inductors periodically along the line so as to satisfy equation (2.8.3).

2.9 POWER CONSIDERATIONS

Since the ultimate purpose of any transmission line system is to deliver *power* from one place to another, we complete this chapter by looking at how power flows in a transmission line system.

If a lossless line is matched, that is $Z_L = Z_0$, then all of the power contained in the incident wave will be delivered to the load, that is, no power will be reflected back from the load. If the line is not matched, that is, $Z_L \neq Z_0$, then some of the power incident at the load will be reflected back down the line toward the source. In general, the power dissipated in a load is given by:

$$P_L = \frac{1}{2}\mathrm{Re}(V_L I_L^*) \tag{2.9.1}$$

where V_L and I_L represent complex load voltage and current, respectively, and (*) represents the complex conjugate. Equation (2.9.1) can be expressed in terms of forward and reflected traveling waves on the line arriving at the load as follows:

$$P_L = \frac{1}{2}\mathrm{Re}[(V^+ + V^-)(I^+ - I^-)^*] \tag{2.9.2}$$

$$P_L = \frac{1}{2}\mathrm{Re}\left[(V^+ + V^-)\left(\frac{(V^+ - V^-)^*}{Z_0}\right)\right] \tag{2.9.3}$$

$$P_L = \frac{1}{2}\mathrm{Re}\left[\frac{|V^+|^2}{Z_0}(1 + \Gamma_L)(1 - \Gamma_L)^*\right] \tag{2.9.4}$$

Which simplifies to:

$$P_L = \frac{1}{2}\frac{|V^+|^2}{Z_0}(1 - |\Gamma_L|^2) \tag{2.9.5}$$

The physical interpretation of equation (2.9.5) is that some portion of the incident power will be absorbed in the load and the remainder will be reflected back toward the source. Equation (2.9.5) will apply, in general, to any discontinuity at any point in a transmission line. In such situations, some portion of the incident power will be reflected back toward the source, and some will be transmitted onward down the line. The respective powers will be determined by the magnitude of the reflection coefficient at the discontinuity, Γ_x, as shown in Figure 2.19.

Where the reflection coefficient at the discontinuity is defined by:

$$\Gamma_x = \frac{Z_x - Z_0}{Z_x + Z_0} \tag{2.9.6}$$

When considering power flows, another parameter that is often used to describe the degree of match or mismatch on a transmission line is *return loss*, which is basically the ratio of incident to reflected power, expressed in dB:

$$RL_{dB} = 10\log_{10}\left[\frac{P_i}{P_r}\right] \tag{2.9.7}$$

In terms of the reflection coefficient, Γ, at the discontinuity in question (which could be a load or antenna etc.), we have:

$$RL_{dB} = 10\log_{10}\left[\frac{1}{|\Gamma|^2}\right] = -20\log_{10}|\Gamma| \tag{2.9.8}$$

The higher the return loss the better the match, which can be interpreted as more of the incident power being "lost" in the load, so less power being reflected back toward the source.

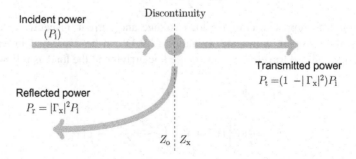

FIGURE 2.19

Generalized picture of power reflection and transmission at a transmission line discontinuity.

Table 2.3 Reflection Coefficient, VSWR, and Return Loss Extremes

Load Type	Z_L	Γ_L	VSWR	Return Loss
Open circuit	∞	1	∞	0 dB
Matched	Z_0	0	1	∞
Short circuit	0	-1	∞	0 dB

It is useful to now extend Table 2.1 to include VSWR and return loss as shown in Table 2.3.

2.10 TAKEAWAYS

1. Because electromagnetic energy travels at a finite speed, any set of interconnecting wires with a physical length comparable to the wavelength of interest need to be treated as a *transmission line.*
2. When both an incident and reflected wave are simultaneously present on a transmission line, a standing wave is said to be present. This means that a stationary pattern of voltage maxima and minima is present. The ratio of the maximum voltage to the minimum voltage is called the VSWR.
3. Every transmission line will have a *characteristic impedance*, which is a function only of the physical geometry of the line and material from which it is made.
4. The line is said to be *matched* to a load termination when the load impedance equals the characteristic impedance. A matched line will appear to be infinitely long as far as the source is concerned.
5. The characteristic impedance of a lossless line is purely real, whereas the characteristic impedance of a lossy line is complex.
6. An ideal lossless line transmits signals without distortion, whereas, in a lossy line, the velocity of propagation is a function of frequency, which results in *dispersion.*
7. The speed at which voltage maxima propagate on the line is called the phase velocity v_p.
8. Traveling waves on a line will be reflected at impedance discontinuities. If the line continues beyond the discontinuity, a portion of the wave is transmitted onward. The reflected and transmitted waves are related by the *reflection coefficient* at the discontinuity.
9. The positions of the voltage maxima are determined by the phase angle of the load reflection coefficient. The spacing between each pair of adjacent maxima is $\lambda/2$ and the distance between a maximum and the adjacent minimum is $\lambda/4$.
10. The impedance $Z(x)$ at any point on a line is a periodic function of position along the line, with period $\lambda/2$.

TUTORIAL PROBLEMS

Problem 2.1 (Transmission line terminology). Define the following terms in the context of transmission line propagation:

1. characteristic impedance;
2. velocity factor;
3. reflection coefficient; and
4. return loss.

Problem 2.2 (Pulses on transmission lines). Reproduce Figure 2.5 for the following scenarios:

1. $R_L > Z_o$, $R_S < Z_o$
2. $R_L > Z_o$, $R_S > Z_o$
3. $R_L < Z_o$, $R_S > Z_o$

Problem 2.3 (Voltage maxima on a transmission line). Show that the input impedance looking into a lossless line at the location where the magnitude of the voltage is a maximum is purely real.

Problem 2.4 (VSWR and reflection coefficient). Measurements on a 50 Ω transmission line produced a maximum voltage $V_{max} = 10\,\text{mV}$ and a minimum voltage $V_{min} = 2\,\text{mV}$ at an operating frequency $f = 1\,\text{GHz}$. The first voltage minimum from the load was measured at 25.5 mm. If the transmission line has a $\varepsilon_r = 2$, determine

1. the VSWR;
2. the wavelength in the medium;
3. the magnitude of the reflection coefficient; and
4. the load impedance.

Problem 2.5 (Time domain refractometry). We want to determine the nature and location of a failure in a 430-m long coaxial cable by time domain refractometry. We

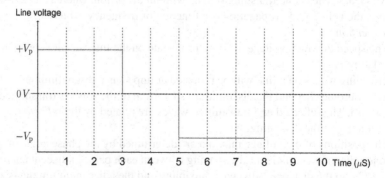

FIGURE 2.20

Waveform seen at the input of a faulty cable.

decide to send a 5 μs pulse down the cable and monitor the reflection. Figure 2.20 shows the waveform monitored at the input of the cable. Determine the nature and location of the fault, given that the cable has an inductance per meter of 250 nH and a capacitance per meter of 100 pF. Line losses can be ignored. (This problem has been adapted from Darwazeh and Moura [8].)

Problem 2.6 (Transmission line input impedance). Calculate the input impedance of a 50 Ω transmission line having an electrical length 6.7λ and terminated in a load impedance $Z_L = 35 + j35$ Ω. At certain points on the line the impedance looking into the line, toward the load, will be purely real. Find the location of these points.

Problem 2.7 (Reflection coefficient). Calculate the complex reflection coefficient and return loss (dB) for reflections at a load impedance $75 + j13$ Ω connected to a 50 Ω transmission line having velocity factor of 0.6.

Problem 2.8 (Power flow in a transmission line). A transmitter is connected to an antenna by 50 Ω coaxial cable 100 m long. The cable has a matched line loss of 0.6 dB/100 m. The antenna impedance is $25 + 45j$ Ω. The transmitter delivers 35 W of power into the line. Calculate

1. the amount of power delivered to the load;
2. the power lost in the line; and
3. the VSWR at the antenna.

Problem 2.9 (Power flow in a transmission line). A transmission line with characteristic impedance $Z_0 = 50$ Ω is terminated with a load impedance $Z_L = 75 + j25$ Ω. What percentage of the incident power is reflected back into the line?

Problem 2.10 (Power delivered to a load). A lossless transmission line of characteristic impedance $Z_0 = 100$ Ω operates with a standing wave ratio of 5 and a maximum voltage on the line of 1.5 V (rms). Calculate the power being delivered to the load.

REFERENCES

[1] G. Cookson, The Cable—The Wire that Changed the World, Tempus Publishing Limited, Gloucestershire, UK, 2003, ISBN 0752423665.
[2] The History of Diving Museum, 2011, "Not So Basic" Cable, http://historyofdivingmuseum.blogspot.co.uk/2011/07/not-so-basic-cable.html.
[3] N. Forbes, B. Mahon, Faraday, Maxwell, and the Electromagnetic Field: How Two Men Revolutionized Physics, Prometheus Books, 2014, ISBN 9781616149420.
[4] I. Yavetz, From Obscurity to Enigma: The Work of Oliver Heaviside, 1872-1889. Modern Birkhäuser Classics, Springer, Basel AG, 2011, ISBN 9783034801768.
[5] O. Heaviside, On induction between parallel wires, J. Soc. Telegraph Eng. 9 (34) (1880) 427-458, http://dx.doi.org/10.1049/jste-1.1880.0047.

[6] J. Kraus, Electromagnetics with Applications, McGraw-Hill College, Boston, 1998, ISBN 9780072356632.

[7] D. Pozar, Microwave Engineering, second ed., John Wiley and Sons Inc., New York, USA, 1998.

[8] I. Darwazeh, L. Moura, Introduction to Linear Circuit Analysis and Modelling, Newnes, 2005, ISBN 0750659327.

[9] R. Collin, Foundations for Microwave Engineering, second ed., John Wiley and Sons Inc., New York, USA, 2005, ISBN 9788126515288.

[10] C. Poole, I. Darwazeh, H. Zirath, K. Eriksson, D. Kuylenstierna, S. Lai, Design and characterization of a negative resistance Common Emitter InP Double Heterojunction Bipolar Transistor subcircuit for millimeter wave and submillimeter wave applications, in: 2014 44th European Microwave Conference (EuMC), 2014, pp. 933-936, http://dx.doi.org/10.1109/EuMC.2014.6986589.

[11] O. Heaviside, Electromagnetic Theory, Cosimo, Incorporated, 2008, ISBN 9781605206172.

Practical transmission lines

3

CHAPTER OUTLINE

INTENDED LEARNING OUTCOMES

- *Knowledge*
  - Be familiar with various types of practical transmission lines, their construction, properties, and applications.
  - Be familiar with various modes of electromagnetic propagation in practical transmission line structures.

- Understand the strengths and weaknesses of a variety of practical transmission line structures.
- Be familiar with several common microstrip discontinuities and their effects.
- *Skills*
  - Be able to calculate the cut-off frequency of rectangular and circular waveguides of specified dimensions.
  - Be able to calculate the characteristic impedance of a co-axial cable of specified dimensions.
  - Be able to calculate the characteristic impedance of a microstrip line of specified dimensions (analysis) and design a microstrip line of specified characteristic impedance (synthesis).
  - Be able to model various common microstrip discontinuities as equivalent circuits or two-port networks.

3.1 INTRODUCTION

In Chapter 2, the concept of a transmission line was introduced, along with some theoretical models of lossy and lossless transmission lines. In this chapter, we describe how practical transmission lines are constructed, analyze their electrical behavior, and compare the advantages and disadvantages of the various types.

3.2 WAVEGUIDE

Waveguide is the exception to all the other types of practical transmission line to be discussed in this chapter, which all have two separate conductors. Waveguide, by contrast, has only one conducting surface consisting of a hollow metal tube filled with a dielectric material, usually air. Signals are transmitted in a waveguide in the form of electromagnetic energy that propagates within the dielectric.

The first waveguide was demonstrated experimentally in 1936 independently by George C. Southworth of the AT&T Company and W.L. Barrow of MIT, although Oliver Heaviside, Lord Rayleigh (John William Strutt), and others had postulated such a form of transmission line in the late nineteenth century [1].

The rectangular cross-section waveguide shown in Figure 3.1 is the most common form of waveguide construction. Analysis of waveguide is a major area of study and involves heavy doses of electromagnetic field theory, which we do not intend to delve into in this short section. Our purpose is simply to provide the microwave engineer with a basic working knowledge of waveguide and why it may be the right choice of transmission line in certain circumstances. Readers wanting to further explore the theoretical background of waveguide are referred to some of the classic books which cover this in much more detail, such as Pozar [2], Collin [3], or Ramo and Whinnery [4].

Waveguide propagation is analyzed by solving Maxwell's equations with boundary conditions determined by the dimensions and properties of the metal walls and the dielectric medium. These equations have multiple solutions, or *modes*, which define how the electric and magnetic fields can exist within the structure at various frequencies. There are basically two modes of propagation in waveguide, namely:

1. TE (transverse electric modes), where there is no electric field in the direction of propagation.
2. TM (transverse magnetic modes), where there is no magnetic field in the direction of propagation.

Since waveguide has only one conducting surface it cannot support another mode of propagation called *transverse electro-magnetic* (TEM), which does exist in some of the other types of transmission line we will cover later in this chapter.

Waveguide modes are referred to by the nomenclature $TE_{m,n}$ for electric field modes and $TM_{m,n}$ for magnetic field modes. The "T" refers to the fact that the electric field is *transverse* to the direction of propagation and m and n refer to the number of half wave-lengths ($\lambda/2$) across the a- and b-axes in Figure 3.1, respectively.

Figure 3.2 is a diagramatic representation of the strength of electric field across the cross section of a rectangular waveguide. In the case of the dominant $TE_{1,0}$ mode, the electric field wave just fits exactly into the waveguide along the a-axis.

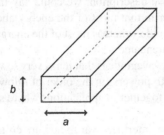

FIGURE 3.1

Rectangular waveguide.

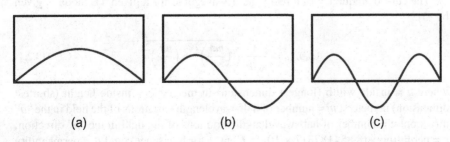

FIGURE 3.2

TE modes for a rectangular waveguide. (a) $TE_{1,0}$ mode. (b) $TE_{2,0}$ mode. (c) $TE_{3,0}$ mode.

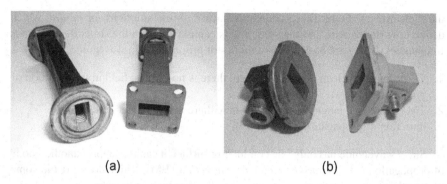

(a)   (b)

FIGURE 3.3

Examples of rectangular waveguide sections and waveguide to co-axial transitions (reproduced by kind permission of Department of Electronic and Electrical Engineering, University College London).

From this we may infer that signals with a frequency lower than (i.e., a wavelength longer than) that shown in Figure 3.2(a) cannot propagate within this waveguide. We therefore refer to this frequency as the *cut-off frequency*, and the $TE_{1,0}$ mode as the *dominant* mode.

On the basis of the above description, we could say that waveguide essentially behaves as a high pass filter in that most of the energy above the cut-off frequency will pass through the waveguide, whereas most of the energy that is below the cut-off frequency will be heavily attenuated.

The main advantages of waveguide is that it has very low loss (low value of α) and can also handle very high RF powers (in the order of kilowatts). It is therefore often used to connect subsystems together in applications where signal power is either very low or very high.

It is often necessary to interface waveguide to co-axial cable. This can be accomplished by means of a *waveguide transition*, such as those shown in Figure 3.3(b). Such transitions are purchased as separate components that have been carefully designed to ensure as little signal loss as possible in transitioning between one medium and another.

The cut-off frequency of a rectangular waveguide for a given TE mode is given by [5]:

$$(\omega_c)_{m,n} = \frac{1}{\sqrt{\mu\varepsilon}}\sqrt{\left(\frac{m\pi}{a}\right)^2 + \left(\frac{n\pi}{b}\right)^2} \qquad (3.2.1)$$

where a = inside width (longest dimension) in meters; b = inside height (shortest dimension) in meters; m = number of half-wavelength variations of the field in the "a" direction; n = number of half-wavelength variations of the field in the "b" direction; ε = permittivity ($8.854187817 \times 10^{-12}\ F \cdot m^{-1}$ for free space); and μ = permeability ($4\pi \times 10^{-7}\ H \cdot m^{-1}$ for free space).

For an air-filled rectangular waveguide (the most common type) we can simplify equation (3.2.1) to:

$$(f_c)_{m,n} = \frac{c}{2}\sqrt{\left(\frac{m}{a}\right)^2 + \left(\frac{n}{b}\right)^2} \tag{3.2.2}$$

where c is the speed of light in air ($\approx 3 \times 10^8$ m s^{-1}).

Example 3.1 (Rectangular waveguide cut-off).

Problem. Determine the cut-off frequency for TE$_{1,0}$ (the mode with the lowest cut-off frequency) in WR284 S-band waveguide which has the following dimensions:

width (a dimension) = 2.840 in (7.214 cm)
height (b dimension) = 1.340 in (3.404 cm)

Solution. Applying equation (3.2.1) we have:

$$(f_c)_{1,0} = \frac{c}{2}\sqrt{\left(\frac{m}{a}\right)^2 + \left(\frac{n}{b}\right)^2}$$

$$= \frac{3 \times 10^8}{2}\sqrt{\left(\frac{100}{7.214}\right)^2 + \left(\frac{0}{3.404}\right)^2}$$

$$= 2.078 \text{ GHz}$$

So far we have only considered rectangular waveguide, but there is also another important waveguide geometry, namely the *circular* waveguide, which essentially resembles a cylindrical metal pipe, as shown in Figure 3.4. Applications where an

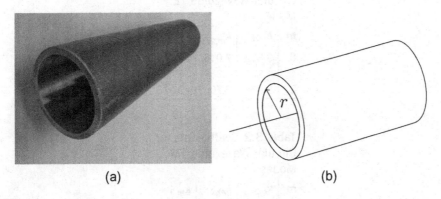

(a)   (b)

FIGURE 3.4

Circular waveguide (photo reproduced by kind permission of Department of Electronic and Electrical Engineering, University College London).

antenna is required to rotate, for example a radar installation, will require a rotary joint in the waveguide at some point. This is only really practical when circular cross-section waveguide is used.

The lower cut-off wavelength for a particular $TE_{m,n}$ mode in circular waveguide of internal radius "r," as defined in Figure 3.4(b), is determined by [5]:

$$\lambda_c = \frac{2\pi r}{k'_{m,n}} \tag{3.2.3}$$

where $k'_{m,n}$ are the roots of the electric wave equation for circular waveguide, as listed in Table 3.1.

The lower cut-off wavelength for a particular $TM_{m,n}$ mode in circular waveguide of radius "r" is determined by:

$$\lambda_c = \frac{2\pi r}{k_{m,n}} \tag{3.2.4}$$

where $k_{m,n}$ are the roots of the magnetic wave equation for circular waveguide, as listed in Table 3.2.

One of the pioneers in the field of circular waveguide was Professor Harold Barlow, who investigated their use as a means of transporting microwave energy over long distance for telecommunications purposes [6]. A PhD student of Professor Barlow, during his tenure at University College London, was a certain Charles Kao, who investigated the effect of filling a circular waveguide with a solid dielectric. Around 1966, he showed that, as the electromagnetic radiation is almost entirely contained within the dielectric, the metal walls could be removed altogether. Thus was born the concept of the *optical fiber* as an efficient communications medium

Table 3.1 Coefficients for Circular Waveguide TE Modes

m	$k_{m,1}$	$k'_{m,2}$	$k'_{m,3}$
0	3.832	7.016	10.174
1	1.841	5.331	8.536
2	3.054	6.706	9.970

Table 3.2 Coefficients for Circular Waveguide TM Modes

m	$k_{m,1}$	$k_{m,2}$	$k_{m,3}$
0	2.405	5.520	8.654
1	3.832	7.016	10.174
2	5.135	8.417	11.620

which was to go on to revolutionize long distance telecommunications, as well as earning Charles Kao a Nobel Prize in 2009 [7].

Example 3.2 (Circular waveguide design).

Problem. You are asked to design an air filled circular waveguide having $TE_{1,1}$ as the dominant mode of propagation. The cut-off frequency needs to be 3 GHz. Calculate the required radius of the waveguide.

Solution. The required cut-off frequency is 3 GHz, so the cut-off wavelength can be calculated from:

$$\lambda_c = \frac{c}{f_c} = \frac{3 \times 10^8}{3 \times 10^9} = 10 \text{ cm}$$

Applying equation (3.2.3) and Table 3.1 we then have

$$r = \frac{10 \times 1.841}{2\pi} = 2.93 \text{ cm}$$

3.3 CO-AXIAL CABLE

Co-axial cable, or "co-ax" is a very common form of RF transmission line that consists of a solid inner conductor together with a tubular outer conductor, the space between them being filled with a dielectric material, as shown in Figure 3.5. The term *co-axial* comes from the inner conductor and the outer shield sharing a geometric axis. Co-axial cable was invented by English engineer and mathematician Oliver Heaviside, who patented the design in 1880 [8].

The insulating material separating the inner and outer conductors may be a solid dielectric, such as polyethylene (PE), polypropylene (PP), fluorinated ethylene propylene (FEP), or polytetrafluoroethylene (PTFE). Alternatively, larger diameter co-axial cables may use air as a dielectric, in which case solid dielectric spacers

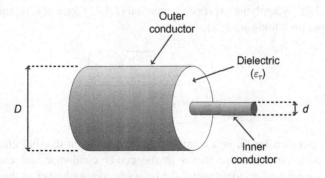

FIGURE 3.5

Co-axial cable construction.

need to be included to provide physical separation between the two conductors. The outer conductor may be of solid or mesh construction, the latter offering increased flexibility.

RF co-axial cable should not be confused with the numerous types of *shielded cable* used to carry lower frequency signals, although these may be superficially similar in appearance. In the case of RF co-axial transmission line, the dimensions must be precisely controlled to achieve a specific conductor spacing, so as to maintain a constant characteristic impedance. By contrast, the conductor spacing in low-frequency shielded cable is of no significance.

For the co-axial cable of Figure 3.5 the shunt capacitance per unit length, in farads per meter is given by [3]:

$$C = \frac{2\pi\varepsilon}{\ln(D/d)} = \frac{2\pi\varepsilon_0\varepsilon_r}{\ln(D/d)} \tag{3.3.1}$$

where D is the outer conductor internal diameter and d is the inner conductor diameter. Series inductance per unit length, in henrys per meter is given by [3]:

$$L = \frac{\mu}{2\pi}\ln\left(\frac{D}{d}\right) = \frac{\mu_0\mu_r}{2\pi}\ln\left(\frac{D}{d}\right) \tag{3.3.2}$$

At low frequencies, the resistance per unit length is the sum of the DC resistance of the inner conductor and the outer conductor. At higher frequencies, the skin effect increases the effective resistance by confining the conduction to a thin layer of each conductor. The shunt conductance is usually very small because the types of insulators in use today all have a very low loss tangent.

Characteristic impedance of co-axial cable

Neglecting resistance per unit length (a reasonable assumption for most practical cables), the characteristic impedance is determined from the capacitance per unit length, given by equation (3.3.1), and the inductance per unit length, given by equation (3.3.2), by applying expression equation (2.5.11) for a generic transmission line as follows (assuming $\mu_r = 1$):

$$Z_0 = \sqrt{\frac{L}{C}} = \frac{1}{2\pi}\sqrt{\frac{\mu_0\mu_r}{\varepsilon_0\varepsilon_r}}\ln\left(\frac{D}{d}\right) \tag{3.3.3}$$

$$\boxed{\approx \frac{138\,\Omega}{\sqrt{\varepsilon_r}}\log_{10}\left(\frac{D}{d}\right)}$$

The loss per unit length is a combination of the loss in the dielectric material filling the cable, and resistive losses in the center conductor and outer shield. These losses are frequency dependent; the losses becoming higher as the frequency increases. Skin effect losses in the conductors can be reduced by increasing the diameter of the cable. Consequently, very low loss co-axial cable tends to be of large diameter and often has air as a dielectric (with solid dielectric spacers).

The velocity of propagation inside the cable depends on the dielectric constant and relative permeability (which is usually 1), that is:

$$v = \frac{1}{\sqrt{\varepsilon\mu}} = \frac{c}{\sqrt{\varepsilon_r\mu_r}} \qquad (3.3.4)$$

The co-axial cable could be considered as a circular waveguide with the addition of a center conductor. This means that energy will propagate in the cable in distinct modes. The dominant mode (the mode with the lowest cut-off frequency) is the TEM mode, which propagates all the way down to DC (i.e., it has a cut-off frequency of zero). The mode with the next lowest cut-off is the TE_{11} mode. This mode has one "wave" (two reversals of polarity) in going around the circumference of the cable. To a good approximation, the condition for the TE_{11} mode to propagate is that the wavelength in the dielectric is no longer than the average circumference of the insulator; that is that the frequency is at least:

$$f_c \approx \frac{1}{\pi\left(\dfrac{D+d}{2}\right)\sqrt{\mu\varepsilon}} = \frac{c_0}{\pi\left(\dfrac{D+d}{2}\right)\sqrt{\mu_r\varepsilon_r}} \qquad (3.3.5)$$

Hence, the cable is single-mode from DC up to this frequency, and might in practice be used up to 90% of this frequency [9].

Assuming the dielectric properties of the material inside the cable do not vary appreciably over the operating range of the cable, the characteristic impedance given in equation (3.3.3) is frequency independent above about five times the *shield cut-off frequency*. The shield cut-off frequency is the frequency above which the energy in the cable is partially propagating in the form of both waveguide modes and TEM modes, all traveling at different velocities. Below the shield cut-off frequency, the energy propagates through the cable as a TEM wave with no electric or magnetic field component in the direction of propagation.

Co-axial cables are widely used to carry RF energy from A to B, sometimes over considerable distances. In the case of flexible co-axial cables, the inner conductor is usually made from multi-stranded wire and the outer conductor is made from wire braid. At microwave frequencies, we often come across "rigid" or "semi-rigid" co-axial cables where both inner and outer conductors are fabricated as solid metal cylinders, common materials being tinned or silver plated copper, copper clad steel, and copper clad aluminum. The advantage of rigid and semi-rigid cables is that they have lower losses than flexible co-ax cables and are generally used in applications where flexibility is not so important, such as fixed interconnections between subsystems inside pieces of equipment. Some sections of semi-rigid co-ax, with typical RF connectors, are shown in Figure 3.6.

Finally, it is worth noting that for low-power situations (e.g., cable TV), co-axial transmission lines are optimized for low loss, which works out to be a characteristic impedance of about 75 Ω (for co-axial transmission lines with air dielectric). For RF and microwave communication and radar applications, where higher powers are more

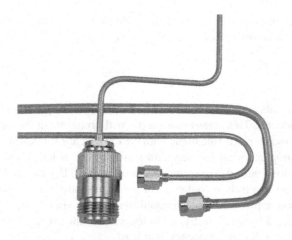

FIGURE 3.6

A selection of semi-rigid co-axial cables, with RF connectors (reproduced by kind permission of L-com Global Connectivity, 45 Beechwood Dr. N. Andover, MA, USA (www.l-com.com)).

often encountered, co-axial transmission lines are designed to have a characteristic impedance of 50 Ω, a compromise between maximum power handling (occurring at 30 Ω) and minimum loss [10].

3.4 TWISTED PAIR

Arguably the simplest possible transmission line, in terms of ease of construction, is two parallel insulated wires placed side by side. An added refinement is to twist the two wires together to form a *twisted pair*. The twisting serves three purposes, first, electromagnetic interference from external sources is reduced as the conductors will present alternating profiles to any external interference source. Second, the twisting maintains an almost constant distance between conductors, thereby ensuring a more stable characteristic impedance and third, twisting the conductors together provides additional mechanical rigidity. Twisted pair cables are very widely used in telephony and networking applications to carry signals up to the low GHz range. They may also be used for internal interconnection inside equipment operating at microwave frequencies.

A typical twisted pair line and its cross section are shown in Figure 3.7.

The characteristic impedance of a twisted pair is a function of the size and spacing of the conductors and the type and thickness of dielectric material used to insulate the wires. The fact that the conductors are not held rigidly in place, however, means that the characteristic impedance at a particular point can change if the spacing between the conductors is changed by, for example, bending of the cable.

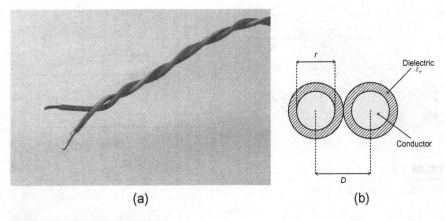

(a)   (b)

FIGURE 3.7

Twisted pair cable and cross section.

The characteristic impedance of the twisted pair shown in Figure 3.7(b) can be calculated as follows [11].

$$Z_0 = \frac{120}{\sqrt{\varepsilon_r}} \cdot \ln\left[\frac{D}{r}\right] \tag{3.4.1}$$

where D is the distance between the two conductors (center to center); r is the radius of the conductors; and ε_r is the effective dielectric constant of the insulating material.

The major influence on characteristic impedance and other secondary coefficients is the capacitance. This is largely determined by the dielectric constant of the wire insulation. In most twisted pair cable applications, the characteristic impedance is between 100 and 150 Ω.

3.5 MICROSTRIP

The practical transmission lines discussed so far are mainly used for carrying signals from one place to another with minimal losses. When it comes to fabricating microwave circuits in printed circuit board (PCB) form; however, we need a type of transmission line that allows active and passive components to be integrated easily with the transmission line medium to build complete circuits. In other words, we need a transmission line that can be fabricated in *planar* form.

Microstrip [12,13] is the most common form of PCB transmission line, due to its compatibility with standard PCB fabrication techniques and ease of access to components during both assembly and test [14].

A typical microstrip transmission line consists of a narrow strip of conducting material, of width *w*, separated from a ground plane by a dielectric substrate of

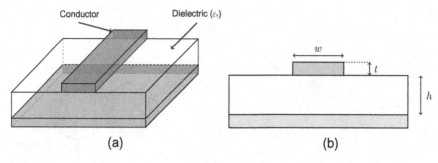

Conductor   Dielectric (ε_r)

(a)   (b)

FIGURE 3.8

Microstrip transmission line.

thickness h, as shown in three-dimensional form in Figure 3.8(a) and in cross-section form in Figure 3.8(b). Although the upper conductive strip is shown as having a finite thickness, t, in Figure 3.8; this thickness is assumed to be negligible for the purposes of most analysis.

Some typical dielectric substrates are *RT/Duroid* (a trademark of Rogers corporation), which is available with several values of dielectric constant ($\varepsilon_r = 2.23$, $\varepsilon_r = 6$, and $\varepsilon_r = 10.5$), quartz ($\varepsilon_r = 3.7$), and aluminum oxide or "alumina" ($\varepsilon_r = 9$). The higher the dielectric constant the shorter the wavelength in the dielectric and, therefore, the smaller the transmission line components will be at a given frequency of operation.

When fabricating microstrip lines in Monolithic Microwave Integrated Circuit (MMIC) form, the underlying MMIC substrate, such as gallium arsenide (GaAs), or indium phosphide (InP) forms the substrate material. Indium phosphide has a dielectric constant of 12.5. In the case of gallium arsenide, the dielectric constant is 12.9. Gallium arsenide is also *piezoelectric* which means that acoustic modes can propagate in the substrate and can couple to the electromagnetic waves on the conductors. This may have to be taken into account in some circumstances, particularly when lines are very long.

The electromagnetic wave carried by a microstrip line exists partly in the dielectric substrate, and partly in the free space above it, making an exact mathematical analysis of this type of transmission line challenging. Since the dielectric constant of the substrate will generally be different (and greater) than that of the air above it, we can say that the wave is traveling in an *inhomogeneous* medium. In consequence, the propagation velocity will be somewhere between the speed of the electromagnetic wave in the substrate, and the speed of the electromagnetic wave in air. The inhomogeneity of the dielectric in microstrip is commonly modeled by means of an *effective dielectric constant*, ε_{eff}, this being the dielectric constant of an equivalent homogeneous medium (i.e., one having the same propagation velocity as the inhomogeneous microstrip configuration). In order to understand the range of ε_{eff}, consider the cases of a very wide microstrip line versus a very narrow microstrip line. In the case of a very wide microstrip line, nearly all of the electric field will be concentrated between the line and the ground plane, and the whole structure will resemble a parallel plate capacitor. We can thus write:

$$\varepsilon_{eff} \approx \varepsilon_r \qquad (3.5.1)$$

At the other extreme, with a very narrow microstrip line, the electric field will be roughly equally distributed between the air and the substrate dielectric. We can therefore write, intuitively:

$$\varepsilon_{eff} \approx \frac{1}{2}(\varepsilon_r + 1) \qquad (3.5.2)$$

We can thus approximate the range of ε_{eff} as:

$$\frac{1}{2}(\varepsilon_r + 1) \leq \varepsilon_{eff} \leq \varepsilon_r \qquad (3.5.3)$$

Most analytical models include closed form expressions for ε_{eff}, examples of which are given in equations (3.5.11) and (3.5.13). The phase velocity of an electromagnetic wave in microstrip will be lower than the phase velocity in free space and the wavelength will be shorter than that in free space both by a factor of $1/\sqrt{\varepsilon_{eff}}$ as follows:

$$v_p = \frac{c_0}{\sqrt{\varepsilon_{eff}}} \qquad (3.5.4)$$

and

$$\lambda_g = \frac{v_p}{f} = \frac{\lambda_0}{\sqrt{\varepsilon_{eff}}} \qquad (3.5.5)$$

where c_0 is the speed of light in free space (i.e., $\approx 3 \times 10^8$ m s^{-1}). It is therefore good to have a high dielectric constant substrate as this means that circuits can be made physically smaller for a given frequency (wavelength).

At nonzero frequencies, both the E and H fields will have longitudinal components (a hybrid mode) [15,16]. The longitudinal components are small, however, and so the dominant mode is referred to as *quasi-TEM*.

Further consequences of an inhomogeneous medium include:

(i) *The line is dispersive*: With increasing frequency, the effective dielectric constant gradually approaches that of the substrate, so that the phase velocity gradually decreases [16,17]. This is true even with a nondispersive substrate material (the substrate dielectric constant will usually fall with increasing frequency).

(ii) *The characteristic impedance of the line is frequency dependant*: Even with a nondispersive substrate material the characteristic impedance of non-TEM modes is not uniquely defined, and depending on the precise definition used, the impedance of microstrip either rises, falls, or falls then rises with increasing frequency [18]. The low-frequency limit of the characteristic impedance is referred to as the *quasi-static* characteristic impedance, and is the same for all definitions of characteristic impedance.

The challenges associated with analyzing microstrip mathematically have resulted in a large number of published formulas for analysis and synthesis of microstrip lines which are often the result of a combination of mathematical analysis and empirical modeling. These formulas can be rather daunting and usually have to be employed with care, as they are often only accurate over only a limited range of parameters. We will present the most commonly used formulas here without going into any detail about their derivation.

Synthesis formulas for microstrip

By *Synthesis* we mean the process of determining the physical dimensions of the line W and h and ε_{eff} starting with a desired value of Z_0 and a given ε_r. In other words, we wish to design a microstrip transmission line with a specified characteristic impedance.

Seminal work on the characteristic impedance of microstrip lines was done by Schneider [19], Wheeler [20], and Hammerstad [21]. A commonly used formula for synthesis of a microstrip line with a given characteristic impedance for narrow strips, that is, $Z_0 > (44 - 2\varepsilon_r)\ \Omega$, is as follows [22]:

$$\frac{W}{h} = \left[\frac{e^H}{8} - \frac{1}{4e^H} \right]^{-1} \tag{3.5.6}$$

where the parameter H is defined as:

$$H = \frac{Z_0\sqrt{2(\varepsilon_r + 1)}}{119.9} + \frac{1}{2}\left(\frac{\varepsilon_r - 1}{\varepsilon_r + 1}\right)\left[\ln\left(\frac{\pi}{2}\right) + \frac{1}{\varepsilon_r}\ln\left(\frac{4}{\pi}\right)\right] \tag{3.5.7}$$

For wide strips, that is, $Z_0 < (44 - 2\varepsilon_r)\ \Omega$, the synthesis formula becomes [22]:

$$\frac{W}{h} = \frac{\pi}{2}\left[(d_e - 1) - \ln(2d_e - 1)\right] + \frac{\varepsilon_r - 1}{\pi\varepsilon_r}\left[\ln(d_e - 1) + 0.293 - \frac{0.517}{\varepsilon_r}\right] \tag{3.5.8}$$

where the parameter d_e is given by:

$$d_e = \frac{59.95\pi^2}{Z_0\sqrt{\varepsilon_r}} \tag{3.5.9}$$

Analysis formulas for microstrip

By *Analysis* we mean the process of determining Z_0 given a specific value of W, h and ε_r. In other words, we are given a microstrip transmission line of specific dimensions and we need to determine the characteristic impedance.

Early work on obtaining an approximation for the characteristic impedance of a "wide" microstrip (i.e., $W/h > 3.3$) was again carried out by Wheeler [20] and Schneider [19]. Hammerstad showed that Schneider and Wheeler's equations were prone to quite large errors for certain values of characteristic impedance [21]. Hammerstad proposed a more accurate equation for characteristic impedance which tends to be more widely used these days.

Hammerstad claims that the accuracy of the following expressions fall within $\pm 1\%$ of Wheeler's numerical results. The following expressions cover the range of practical dimensions for microstrip, that is, $0.05 \leq W/h \leq 20$ and $\varepsilon_r \leq 16$ [21].

For $W/h < 1$ we have:

$$Z_0 = \frac{60}{\sqrt{\varepsilon_{\text{eff}}}} \cdot \ln\left(\frac{8h}{W} + \frac{W}{4h}\right) \tag{3.5.10}$$

where

$$\varepsilon_{\text{eff}} = \frac{\varepsilon_r + 1}{2} + \frac{\varepsilon_r - 1}{2}\left[\left(1 + \frac{12h}{W}\right)^{-1/2} + 0.04\left(1 - \frac{W}{h}\right)^2\right] \tag{3.5.11}$$

For $W/h \geq 1$ we have:

$$Z_0 = \frac{\pi}{\sqrt{\varepsilon_{\text{eff}}}} \cdot \frac{120}{\frac{W}{h} + 1.393 + 0.667\ln\left(\frac{W}{h} + 1.444\right)} \tag{3.5.12}$$

where

$$\varepsilon_{\text{eff}} = \frac{\varepsilon_r + 1}{2} + \frac{\varepsilon_r - 1}{2}\left(1 + 12\frac{h}{W}\right)^{-1/2} \tag{3.5.13}$$

Example 3.3 (Microstrip design).

Problem. Design a 50 Ω microstrip transmission line using FR4 substrate material ($\varepsilon_r = 4.6$) of thickness 1 mm.

Solution. We note that $Z_0 > (44 - 2\varepsilon_r)$ Ω, so we use equations (3.5.6) and (3.5.7) to calculate W/h. We start by calculating the parameter H using equation (3.5.7) as follows:

$$H = \frac{Z_0\sqrt{2(\varepsilon_r + 1)}}{119.9} + \frac{1}{2}\left(\frac{\varepsilon_r - 1}{\varepsilon_r + 1}\right)\left[\ln\left(\frac{\pi}{2}\right) + \frac{1}{\varepsilon_r}\ln\left(\frac{4}{\pi}\right)\right] \tag{3.5.14}$$

$$= \frac{50 \times \sqrt{2 \times (4.6 + 1)}}{119.9} + \frac{1}{2}\left(\frac{4.6 - 1}{4.6 + 1}\right)\left[\ln\left(\frac{\pi}{2}\right) + \frac{1}{4.6}\ln\left(\frac{4}{\pi}\right)\right] \tag{3.5.15}$$

$$= 1.39560 + \frac{1}{2} \times 0.64286 \times \left[0.451583 + \frac{1}{4.6} \times 0.241564\right] \tag{3.5.16}$$

$$= 1.5576 \tag{3.5.17}$$

We now calculate W/h using equation (3.5.6):

$$\frac{W}{h} = \left[\frac{e^H}{8} - \frac{1}{4e^H}\right]^{-1}$$

$$= \left[\frac{e^{1.5576}}{8} - \frac{1}{4e^{1.5576}}\right]^{-1}$$

$$= 1.8492$$

The width of the 50 Ω line in this case will therefore be 1.8492 mm.

We can verify the above result by means of equations (3.5.12) and (3.5.13), since $W/h > 1$. We first calculate the effective dielectric constant using equation (3.5.13) as follows:

$$\varepsilon_{\text{eff}} = \frac{\varepsilon_r + 1}{2} + \frac{\varepsilon_r - 1}{2}\left(1 + 12\frac{h}{W}\right)^{-1/2}$$

$$= \frac{4.6 + 1}{2} + \frac{4.6 - 1}{2}\left(1 + \frac{12}{1.8492}\right)^{-1/2}$$

$$= 2.8 + 1.8 \times 0.3654$$

$$= 3.4577$$

Note that the effective dielectric constant of microstrip, in this case, is significantly lower than the dielectric constant of the substrate material itself (3.4577 as compared to 4.6). This illustrates the fact that microstrip is an inhomogeneous medium where the electric and magnetic field energy is distributed between the substrate material below the line and the air above it.

Using the calculated value of ε_{eff} we can now calculate the actual value of the microstrip line using equation (3.5.12):

$$Z_0 = \frac{\pi}{\sqrt{\varepsilon_{\text{eff}}}} \cdot \frac{120}{\dfrac{W}{h} + 1.393 + 0.667\ln\left(\dfrac{W}{h} + 1.444\right)}$$

$$= \frac{\pi}{1.8595} \cdot \frac{120}{1.8492 + 1.393 + 0.667\ln(1.8492 + 1.444)}$$

$$= \frac{\pi}{1.8595} \cdot \frac{120}{4.03717}$$

$$= 50.218\ \Omega$$

So, the calculated value of Z_0 is within 4.3% of the design value. The reader may wish to verify, by similar calculations, that this error increases for very high or very low characteristic impedance lines, that is as the ratio W/h moves further away from unity.

3.6 **MICROSTRIP DISCONTINUITIES**

Once we start to design actual circuits in the microstrip medium, we will inevitably introduce *discontinuities*. Any practical circuit implemented in microstrip must contain a number of bends, gaps, and junctions if we are to connect the various components together. Generally speaking, such discontinuities give rise to *parasitic* capacitances and inductances which are typically quite small (often <0.1 pF and <0.1 nH). The reactance of these parasitics can, however, become significant at high microwave and millimeter wave frequencies, so they often need to be accounted for. For ease of analysis, we normally model these discontinuities by an equivalent circuit consisting of lumped parasitic capacitances and inductances. In this section, we will look at the most common types of microstrip discontinuity, as follows:

1. open-circuit "edge" effects;
2. series gaps;
3. bends and curves; and
4. step width changes.

A large body of theoretical and experimental know-how has been built up over the past few decades relating to the electrical properties of the above discontinuity types. This topic is worth spending a little time on, not simply because discontinuities are unavoidable in most circuit designs, but because an understanding of discontinuities can actually be used deliberately in the design process to achieve a specific result. For example, a gap in a microstrip line can be used as a DC block, in place of a discrete capacitor.

3.6.1 **EDGE EFFECTS IN MICROSTRIP**

We frequently come across microstrip lines that are designed to end abruptly in an open circuit. This typically happens when our design includes a *short-circuit stub*, as will be explained in Chapter 10. Due to the electromagnetic field patterns at the end of an open microstrip line, we cannot simply assume that the line ends abruptly with the length we have designed. The "edge effect" can be modeled as a parasitic capacitance connected across the end of the line, but more usefully as a simple addition to the physical length of the line, resulting in the *effective length* being slightly longer than the designed length.

There is a lot of literature on microstrip end effects. A comprehensive analysis of open end effects in microstrip has been carried out by Silvester and Benedek [23], Kirschning et al. [24], and Hammerstad [25].

Silvester and Benedek [23] have modeled the open end by a small length extension, Δl, which is defined by:

$$\frac{\Delta l}{h} = 0.412 \frac{\varepsilon_{\text{eff}} + 0.3}{\varepsilon_{\text{eff}} - 0.258} = \frac{W/h + 0.262}{W/h + 0.813} \tag{3.6.1}$$

Equation (3.6.1) should suffice for most practical purposes where the edge effect needs to be taken into account, but in case greater accuracy is required, Hammerstad [25] offers an alternative equation to which he claims to be more accurate for $W/h < 20$ as follows:

$$\frac{\Delta l}{h} = 0.102 \frac{W/h + 0.106}{W/h + 0.264} \left[1.166 + \frac{\varepsilon_r + 1}{\varepsilon_r} \left(0.9 + \ln\left(\frac{W}{h} + 2.475\right) \right) \right] \qquad (3.6.2)$$

Hammerstad claims that the numerical error in equation (3.6.2) is less than 1.7% for $W/h < 20$.

3.6.2 MICROSTRIP GAPS AND DC BLOCKS

It is often necessary to provide a DC break in a microstrip line, while allowing the RF signal to pass through with the minimum amount of attenuation. Such DC blocks are typically required between the various stages of a multi-stage amplifier, in order to isolate the DC bias levels of one stage from the next. The DC blocking function is often carried out using a chip capacitor connected across a gap in the microstrip line. It is possible, however, to use the gap itself as a DC block, which eliminates the need for an additional component.

A representation of a microstrip gap between two microstrip lines of different widths is shown in Figure 3.9(a). This gap may be approximately modeled by the equivalent π-network of Figure 3.9(b).

The capacitor values of the equivalent π-network in Figure 3.9(b), in "pF," are given by Kirschning [26]:

$$C_S = 500 \cdot h \cdot \exp\left(-1.86 \cdot \frac{s}{h}\right) Q_1 \left[1 + 4.19 \left(1 - \exp\left[-0.785 \cdot \sqrt{\frac{h}{W_1}} \cdot \frac{W_2}{W_1} \right] \right) \right] \qquad (3.6.3)$$

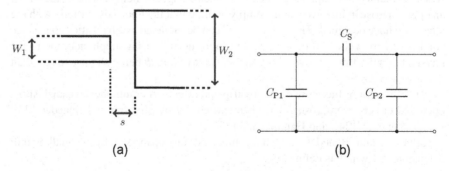

(a) (b)

FIGURE 3.9

Microstrip discontinuity and its electrical equivalent circuit.

$$C_{P1} = C_1 \cdot \frac{Q_2 + Q_3}{Q_2 + 1} \tag{3.6.4}$$

$$C_{P2} = C_2 \cdot \frac{Q_2 + Q_4}{Q_2 + 1} \tag{3.6.5}$$

where s is the width of the gap, h is the height of the substrate, and W_1 and W_2 are the line widths indicated in Figure 3.9(a). C_1 and C_2 are the open end capacitances of the respective microstrip lines in Figure 3.9(a) according to either equation (3.6.1) or (3.6.2), depending on the degree of accuracy required. The coefficients Q_n are defined by Kirschning as follows:

$$Q_1 = 0.04598 \cdot \left(0.03 + \left(\frac{W_1}{h}\right)^{Q_5}\right) \cdot (0.272 + 0.07 \cdot \varepsilon_r) \tag{3.6.6}$$

$$Q_2 = 0.107 \cdot \left(\frac{W_1}{h} + 9\right) \cdot \left(\frac{s}{h}\right)^{3.23} + 2.09 \left(\frac{s}{h}\right)^{1.05} \cdot \frac{1.5 + 0.3 \cdot W_1/h}{1 + 0.6 \cdot W_1/h} \tag{3.6.7}$$

$$Q_3 = \exp\left(-0.5978 \cdot \left(\frac{W_2}{W_1}\right)^{1.35}\right) - 0.55 \tag{3.6.8}$$

$$Q_4 = \exp\left(-0.5978 \cdot \left(\frac{W_1}{W_2}\right)^{1.35}\right) - 0.55 \tag{3.6.9}$$

The numerical error of the capacitive admittances is less than 0.1 mS provided that the following conditions are met:

$$0.1 \leq W_1/h \leq 3$$

$$0.1 \leq W_2/h \leq 3$$

$$1 \leq W_2/W_1 \leq 3$$

$$6 \leq \varepsilon_r \leq 13$$

$$0.2 \leq s/h \leq \infty$$

$$0.2 \text{ GHz} \leq f \leq 18 \text{ GHz}$$

Once the capacitor values in the equivalent circuit of Figure 3.9(b) have been calculated using the above formulas, the Y-parameters for the two-port equivalent circuit can be calculated as follows:

$$Y = \begin{bmatrix} j\omega(C_{P1} + C_S) & -j\omega C_S \\ -j\omega C_S & j\omega(C_{P2} + C_S) \end{bmatrix} \tag{3.6.10}$$

These Y-parameters can be converted to S-parameters using the relationships in Appendix A. This rather long-winded process allows us to treat the DC block as a two-port network for the purpose of further analysis. Needless to say, most of the above calculations will nowadays be carried out within a CAD environment, but they are presented here to give an insight into the complexity of modeling even quite simple discontinuities in microstrip.

Interested readers are referred to a useful paper on DC blocks by LaCombe and Cohen [27], which contains design information based mainly on experimental results. Synthesis equations for microstrip DC blocks are given in another paper by Kajfez [28]. Another useful reference on this topic is the short paper by Borgaonkar and Rao [29].

3.6.3 MICROSTRIP BENDS AND CURVES

Any practical microstrip circuit will contain a number of bends or curves, and these will also introduce parasitic effects. A simple 90° microstrip bend is shown in Figure 3.10(a). Such a bend can be modeled by the equivalent circuit of Figure 3.10(b).

According to Kirschning [26], the values of the equivalent circuit components in Figure 3.10(b) are as follows:

$$C = W \cdot \left((10.35 \cdot \varepsilon_r + 2.5) \cdot \frac{W}{h} + (2.6 \cdot \varepsilon_r + 5.64) \right) \text{ pF} \tag{3.6.11}$$

$$L = 220 \cdot h \cdot \left(1 - 1.35 \cdot \exp\left(-0.18 \cdot \left(\frac{W}{h} \right)^{1.39} \right) \right) \text{ nH} \tag{3.6.12}$$

The parasitic effects of the simple 90° bend in Figure 3.10(a), especially the parasitic capacitor, C, in Figure 3.10(b), can be reduced by *mitering*. The resultant shape is shown in Figure 3.11(a). For a 50% mitered bend, the values of L and C are as follows [26].

$$C = W \cdot \left((3.93 \cdot \varepsilon_r + 0.62) \cdot \frac{W}{h} + (7.6 \cdot \varepsilon_r + 3.80) \right) \text{ pF} \tag{3.6.13}$$

$$L = 440 \cdot h \cdot \left(1 - 1.062 \cdot \exp\left(-0.177 \cdot \left(\frac{W}{h} \right)^{0.947} \right) \right) \text{ nH} \tag{3.6.14}$$

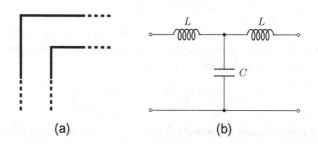

(a)   (b)

FIGURE 3.10

Microstrip right-angle bend and its electrical equivalent circuit.

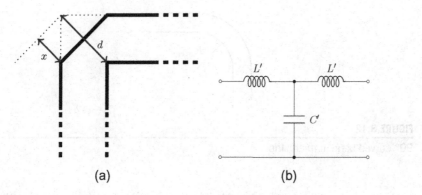

FIGURE 3.11

Microstrip mitered bend and its electrical equivalent circuit.

With W being width of the microstrip line and h thickness of the substrate. These formulas are valid for $W/h = 0.2$ to 6.0 and for $\varepsilon_r = 2.36$ to 10.4 and up to 14 GHz. The precision is approximately 0.3%.

The equivalent circuit of the mitered bend in Figure 3.11(b) is essentially the same as Figure 3.10(b) but with different component values, most notably a value of C' that is considerably lower than C in Figure 3.10(b).

The optimum miter for a wide range of microstrip geometries has been determined experimentally by Douville and James [30], who show that, subject to $W/h \geq 0.25$ and $\varepsilon_r \leq 25$, a good fit for the optimum percentage miter is given by:

$$M = 100\frac{x}{d}\% = \left(52 + 65e^{-\frac{27}{20}\frac{w}{h}}\right) \tag{3.6.15}$$

where W is the width of the line and x and d are defined in Figure 3.11.

For the purposes of incorporating these bends into our circuit design, the two-port Z-matrices for the equivalent circuits in Figures 3.10(b) and 3.11(b) can be calculated as per equation (3.6.16), where L and C represent the respective series inductances and shunt capacitance in either figure. The resultant Z-matrix can be converted to S-parameters using the relationships in Appendix A.

$$Z = \begin{bmatrix} j\omega L + \dfrac{1}{j\omega C} & \dfrac{1}{j\omega C} \\ \dfrac{1}{j\omega C} & j\omega L + \dfrac{1}{j\omega C} \end{bmatrix} \tag{3.6.16}$$

An alternative to using a "bend" to change the direction of a microstrip line is to use a "curve" [31], as shown in Figure 3.12.

When the curving radius is larger than twice the width of the line, the main parasitic effect is a change in the effective line length. The effective length of the curve ($3 < R/W < 7$) can be estimated by assuming the effective radius to be:

$$R_{\text{eff}} = R_{\text{inner}} + 0.3W \tag{3.6.17}$$

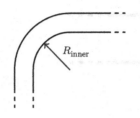

FIGURE 3.12

90° curved bend in microstrip.

For both the curved and mitered bends, the electrical length is somewhat shorter than the physical path length of the microstrip line.

3.6.4 MICROSTRIP STEP WIDTH CHANGE

A step change in microstrip characteristic impedance is quite often required in circuit design and is physically implemented as a step change in the width of the microstrip. This has a number of applications, such as the quarter wave transformer, which is introduced in Chapter 11. A microstrip filter, for example, may employ many resonators of different characteristic impedances, implemented as a cascade of microstrip sections of differing widths.

Figure 3.13 shows a step change in impedance from high to low (in moving from left to right).

In addition to the desired effect of changing the line impedance, a step width change will introduce a *discontinuity capacitance* at the end of the lower impedance line that will have the effect of increasing its electrical length.

The method of compensating for excess capacitance in a step width change is similar to that used to compensate for that in an open ended line, and is based on an expression for the length correction, l_S, required for the lower impedance line, W_2, proposed by Edwards [22] as follows:

$$\frac{l_S}{h} = 0.412 \frac{\varepsilon_{\text{eff}} + 0.3}{\varepsilon_{\text{eff}} - 0.258} = \frac{W/h + 0.262}{W/h + 0.813} \left[1 - \frac{W_1}{W_2} \right] \qquad (3.6.18)$$

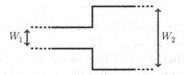

FIGURE 3.13

Microstrip step discontinuity.

Applying (3.6.1), this can be written as:

$$\frac{l_S}{h} = \frac{\Delta l}{h}\left[1 - \frac{W_1}{W_2}\right]$$

(3.6.19)

where $\Delta l/h$ is the value of length correction required in an open-ended line, obtained from equation (3.6.1). If more accuracy is required then equation (3.6.2) may be used to calculate the $\Delta l/h$ term in equation (3.6.19).

3.7 STRIPLINE

Stripline is a form of printed circuit transmission line where the signal trace is sandwiched between upper and lower ground planes, as shown in three-dimensional form in Figure 3.14(a) and in cross-section form in Figure 3.14(b). There are a number of advantages to such an arrangement, most important of which being that the electromagnetic radiation is entirely enclosed within a homogeneous dielectric, thus minimizing emissions and providing natural shielding against incoming spurious signals. It is worth mentioning, also, that although the two "ground" planes represent AC grounds, they can be at different DC potentials, making for convenient distribution of DC power.

All the electromagnetic energy propagating in a stripline is contained between the top and bottom ground planes and is therefore propagating within a homogeneous dielectric, unlike the case with microstrip. The effective permittivity of striplines is therefore simply equal to the relative permittivity of the dielectric substrate. As a result, stripline is nondispersive, unlike microstrip.

The characteristic impedance of the stripline in Figure 3.14 has traditionally been modeled by replacing it with an equivalent cylindrical wire of diameter d, where d is defined in terms of t and w by [32]:

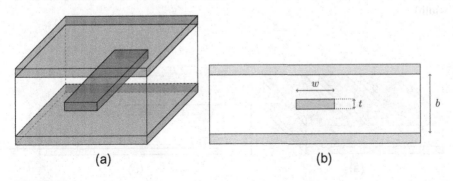

(a)   (b)

FIGURE 3.14

Stripline transmission line.

$$d = \left[1 + \frac{t}{w}\left(1 + \ln\left(\frac{4\pi w}{t}\right) + 0.51\pi\left(\frac{t}{w}\right)^2\right)\right]$$ (3.7.1)

The characteristic impedance of the stripline can now be calculated by employing equation (3.7.1) in the conventional equation for a cylindrical center conductor [32], as follows:

$$Z_0 = \frac{60}{\sqrt{\varepsilon}} \ln\left(\frac{4b}{\pi d}\right)$$ (3.7.2)

where b is the separation between the ground planes in Figure 3.14(b). Note that, again by contrast with microstrip, the thickness of the center conductor, t, cannot be ignored in the above analysis.

3.8 COPLANAR WAVEGUIDE

Coplanar waveguide is a printed transmission line construction that was first proposed by C.P. Wen in 1969 [33] and has become increasingly popular as it performs better than microstrip at higher frequencies, yet can still be fabricated using conventional PCB technology. Coplanar waveguide transmission lines can also be easily fabricated on MMIC.

A coplanar waveguide is a printed circuit structure whereby the upper metalization consists of a central metallic strip flanked by two narrow slits with ground plane on either side. This upper structure is separated from the ground plane by a dielectric substrate, as shown in three-dimensional form in Figure 3.15(a) and in cross-section form in Figure 3.15(b).

The important dimensions of a coplanar waveguide are the central strip width W and the width of the slots s. The structure is normally symmetrical along a vertical plane running in the middle of the central strip (i.e., both slots are the same width).

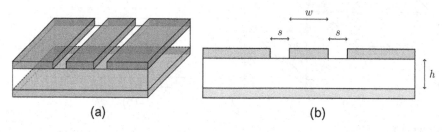

(a)   (b)

FIGURE 3.15

Coplanar waveguide transmission line.

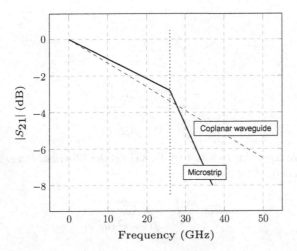

FIGURE 3.16

Comparison of insertion loss for 6.3 cm lengths of microstrip and coplanar waveguide. (© Rogers Corporation, Inc. 2012. Reproduced with Permission, Courtesy of Rogers Corporation).

Coplanar waveguide is slightly more complex to design and fabricate than microstrip, takes up more board space, and is a little more tricky in terms of component placement, so why bother?

The reason the extra complexity is justified can be seen by referring to Figure 3.16, which shows the insertion loss ($|S_{21}|$ in dB) for standard microstrip and coplanar waveguide, with the same dielectric material.

Although the insertion loss of microstrip increases markedly for frequencies above about 27 GHz, the insertion loss of coplanar waveguide increases monotonically with frequency. The result is that at frequencies above about 29 GHz, coplanar waveguide has an increasingly lower insertion loss than an equivalent microstrip.

The characteristic impedance of a coplanar waveguide is primarily determined by the separation of metallic traces and not by substrate thickness. In fact some coplanar waveguide implementations may dispense with the ground plane altogether.

According to Visser [34], the characteristic impedance of the coplanar waveguide in Figure 3.15(b), in terms of the dimensions w, s, and h are calculated as follows [35]:

$$Z_0 = \frac{30\pi}{\sqrt{\varepsilon_e}} \frac{K(k')}{K(k)} \tag{3.8.1}$$

where

$$\varepsilon_e = 1 + \frac{\varepsilon_r - 1}{2} \left[\frac{K(k')K(k_1)}{K(k)K(k'_1)} \right] \tag{3.8.2}$$

and

$$k = 1 + \frac{W}{W + 2S'} \tag{3.8.3}$$

$$k_1 = \frac{\sinh\left(\dfrac{\pi W}{4H}\right)}{\sinh\left(\dfrac{\pi(W + 2S)}{4H}\right)} \tag{3.8.4}$$

In the above equations (3.8.1)–(3.8.4), $K(k)$ is the *Elliptical Integral of the First Kind*, which is defined by [36]:

$$K(k) = \int_0^\phi \frac{d\theta}{\sqrt{(1 - k^2 \sin^2 \theta)}} \tag{3.8.5}$$

and

$$k_n' = \sqrt{(1 - k_n^2)} \tag{3.8.6}$$

Visser [34] offers a simplification to the above whereby the ratio of complete elliptic functions in equation (3.8.1) may be approximated by the following [34]:

$$\frac{K(k)}{K(k')} \approx \begin{cases} \dfrac{\dfrac{1}{2\pi} \ln\left[2\dfrac{\sqrt{1+k} + \sqrt[4]{4k}}{\sqrt{1+k} - \sqrt[4]{4k}}\right]}{2\pi} \\ \dfrac{2\pi}{\ln\left[2\dfrac{\sqrt{1+k} + \sqrt[4]{4k}}{\sqrt{1+k} - \sqrt[4]{4k}}\right]} \end{cases} \tag{3.8.7}$$

Coplanar waveguide lends itself well to MMIC implementation and is often employed in this format. An example of an MMIC using coplanar waveguide transmission lines is shown in Figure 3.17.

3.9 TAKEAWAYS

1. The most popular forms of practical transmission lines are waveguide, co-axial, microstrip, stripline, and coplanar waveguide. Each of these has their own strengths and weaknesses, depending on the application.
2. Electromagnetic energy propagates in waveguide in the form of distinct *modes*, which define how the electric and magnetic fields can exist within the structure at various frequencies.
3. There are basically two modes of propagation in rectangular waveguide, namely "TE" modes (transverse electric), where there is no electric field in the direction

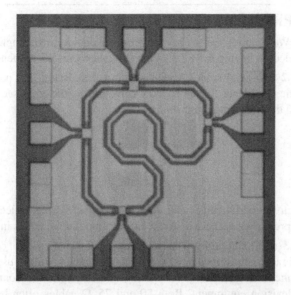

FIGURE 3.17

A Ku-band branch line coupler MMIC employing coplanar waveguide (reproduced by kind permission of Ian Robertson, Leeds University) [37].

of propagation and TM modes (transverse magnetic), where there is no magnetic field in the direction of propagation.

4. A consequence of the existence of modes in waveguide means that there exists a *cut-off* frequency below which energy will not propagate.

5. Microstrip is a popular *planar* transmission line structure that can be easily fabricated using standard PCB technology, and lends itself readily to the construction of complex microwave circuits containing a mixture of lumped and distributed components.

6. Because the electromagnetic energy in microstrip propagates in TEM mode, with part of the field in the dielectric and part in air, mathematical analysis of microstrip lines is quite complex. Closed form equations have been published, however, that allow analysis and synthesis of microstrip lines.

7. Most practical microstrip circuits will contain *discontinuities*, which can usually be modeled as parasitic lumped components or as virtual extensions to the line length, as in the case of an open-circuit line.

8. Other printed circuit transmission line structures exist, such as stripline and coplanar waveguide, which have advantages and disadvantages depending on the application. For example, coplanar waveguide has lower losses than microstrip at higher frequencies.

TUTORIAL PROBLEMS

Problem 3.1 (Waveguide). Let us say that the waveguide in Example 3.1 is filled with a solid dielectric having $\varepsilon_r = 2.3$. What is the new cut-off frequency?

Problem 3.2 (Co-axial cable). Calculate the transmission line parameters and the characteristic impedance for the co-axial cable having the cross-section shown in Figure 3.18 and the following dimensions:

$$a = 1 \text{ mm}$$
$$b = 1.5 \text{ mm}$$
$$t = 0.5 \text{ mm}$$

The conductor material (inner and outer) is copper and the dielectric is PTFE. The necessary properties of these material can be found in the tables in Chapter 1.

Problem 3.3 (Co-axial cable). The two most popular categories of co-axial cable, in terms of characteristic impedance, are the 50 Ω cables widely used in RF and microwave applications and the 75 Ω cables more often used to connect antennas to domestic television equipment. Both 50 and 75 Ω cables often have the same diameter of outer conductor, so as to allow the use of a standard BNC connector in both cases.

Given the above information, determine the ratio between the diameter of the center conductor of a 50 Ω co-axial cable and a 75 Ω co-axial cable.

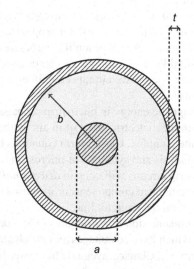

FIGURE 3.18

Co-axial cross section (for Problem 3.2).

Problem 3.4 (Microstrip design). Design a microstrip line having a characteristic impedance of 70 Ω using alumina material ($\varepsilon_r = 9.0$) with a thickness of 0.3 mm.

Problem 3.5 (Microstrip calculations). Write a computer programme, in a language of your choice, to plot the characteristic impedance of a microstrip line versus the W/h ratio and a given value of substrate ε_r over the range $0.1 < W/h < 4.0$.

Problem 3.6 (Microstrip bends). Determine the percentage reduction in parasitic capacitance obtained by employing a 50% mitered right-angle bend in the microstrip line designed in Problem 3.4.

Problem 3.7 (Microstrip discontinuities). If the 70 Ω microstrip line in Problem 3.4 is subject to a step change to a lower impedance of 35 Ω at some point along its length, calculate the effective length correction required to the lower impedance line section.

REFERENCES

[1] K. Packard, The origin of waveguides: a case of multiple rediscovery, IEEE Trans. Microw. Theory Tech. 32 (9) (1984) 961-969, ISSN 0018-9480, http://dx.doi.org/10.1109/TMTT.1984.1132809.

[2] D. Pozar, Microwave Engineering, second ed., John Wiley and Sons Inc., New York, USA, 1998.

[3] R. Collin, Foundations for Microwave Engineering, second ed., John Wiley and Sons Inc., New York, USA, 2005, ISBN 9788126515288.

[4] S. Ramo, J. Whinnery, Fields and Waves in Modern Radio, second ed., John Wiley and Sons, Inc., New York, 1962.

[5] J. Kraus, Electromagnetics with Applications, McGraw-Hill College, Boston, 1998, ISBN 9780072356632.

[6] H. Barlow, H. Effemey, Propagation characteristics of low-loss tubular waveguides, Proc. IEE B Radio Electron. Eng. 104 (15) (1957) 254-260, http://dx.doi.org/10.1049/pi-b-1.1957.0148.

[7] H. Griffiths, K. Tong, Y. Yang, Charles K. Kao and other telecommunication pioneers, IEEE Commun. Mag. 48 (3) (2010) S20-S27, ISSN 0163-6804, http://dx.doi.org/10.1109/MCOM.2010.5434374.

[8] P. Nahin, Oliver Heaviside: The Life, Work, and Times of an Electrical Genius of the Victorian Age, Oliver Heaviside, Johns Hopkins University Press, Baltimore, MD, 2002, ISBN 9780801869099.

[9] G. Kizer, Microwave Communication, Iowa State University Press, Ames, 1990, ISBN 9780813800264.

[10] M. Golio, The RF and Microwave Handbook, in: Electrical Engineering Handbook. Taylor & Francis, 2000, ISBN 9781420036763.

[11] E. da Silva, High Frequency and Microwave Engineering, Butterworth-Heinemann, Oxford, 2001.

[12] A. Fuller, Microwaves: An Introduction to Microwave Theory and Techniques, Pergamon International Library, Pergamon Press, New York, 1979.

[13] E. Fooks, R. Zakarevičius, Microwave Engineering Using Microstrip Circuits, Prentice Hall, New York, 1990, ISBN 9780136916505.

[14] A. Horn, J. Reynolds, J. Rautio, Conductor profile effects on the propagation constant of microstrip transmission lines, in: 2010 IEEE MTT-S International Microwave Symposium Digest (MTT), ISSN 0149-645X, 2010, pp. 1-1, http://dx.doi.org/10.1109/MWSYM.2010.5517477.

[15] G. Zysman, D. Varon, Wave propagation in microstrip transmission lines, in: 1969 G-MTT International Microwave Symposium, 1969, pp. 3-9, http://dx.doi.org/10.1109/GMTT.1969.1122648.

[16] E. Denlinger, A frequency dependent solution for microstrip transmission lines, IEEE Trans. Microw. Theory Tech. 19 (1) (1971) 30-39, ISSN 0018-9480, http://dx.doi.org/10.1109/TMTT.1971.1127442.

[17] H. Cory, Dispersion characteristics of microstrip lines, IEEE Trans. Microw. Theory Tech. 29 (1) (1981) 59-61, ISSN 0018-9480, http://dx.doi.org/10.1109/TMTT.1981.1130287.

[18] B. Bianco, L. Panini, M. Parodi, S. Ridella, Some considerations about the frequency dependence of the characteristic impedance of uniform microstrips, IEEE Trans. Microw. Theory Tech. 26 (3) (1978) 182-185, ISSN 0018-9480, http://dx.doi.org/10.1109/TMTT.1978.1129341.

[19] M. Schneider, Microstrip lines for microwave integrated circuits, Bell Syst. Tech. J. 48 (1969) 1421-1444.

[20] H. Wheeler, Transmission-line properties of parallel strips separated by a dielectric sheet, IEEE Trans. Microw. Theory Tech. 13 (2) (1965) 172-185, ISSN 0018-9480, http://dx.doi.org/10.1109/TMTT.1965.1125962.

[21] E. Hammerstad, Equations for microstrip circuit design, in: 5th European Microwave Conference, Hamburg, 1975, pp. 268-272, http://dx.doi.org/10.1109/EUMA.1975.332206.

[22] T. Edwards, Foundations for Microstrip Circuit Design, John Wiley & Sons Canada, Limited, 1992, ISBN 9780471930624.

[23] P. Silvester, P. Benedek, Equivalent capacitances of microstrip open circuits, IEEE Trans. Microw. Theory Tech. 20 (8) (1972) 511-516, ISSN 0018-9480, http://dx.doi.org/10.1109/TMTT.1972.1127798.

[24] M. Kirschning, R. Jansen, N. Koster, Accurate model for open end effect of microstrip lines, Electron. Lett. 17 (3) (1981) 123-125, ISSN 0013-5194, http://dx.doi.org/10.1049/el:19810088.

[25] E. Hammerstad, Computer-aided design of microstrip couplers with accurate discontinuity models, in: 1981 IEEE MTT-S International Microwave Symposium Digest, ISSN 0149-645X, 1981, pp. 54-56, http://dx.doi.org/10.1109/MWSYM.1981.1129818.

[26] M. Kirschning, R. Jansen, N. Koster, Measurement and computer-aided modeling of microstrip discontinuities by an improved resonator method, in: 1983 IEEE MTT-S International Microwave Symposium Digest, ISSN 0149-645X, 1983, pp. 495-497, http://dx.doi.org/10.1109/MWSYM.1983.1130959.

[27] D. Lacombe, J. Cohen, Octave-band microstrip DC blocks (short papers), IEEE Trans. Microw. Theory Tech. 20 (8) (1972) 555-556, ISSN 0018-9480, http://dx.doi.org/10.1109/TMTT.1972.1127808.

[28] D. Kajfez, B. Vidula, Design equations for symmetric microstrip DC blocks, IEEE Trans. Microw. Theory Tech. 28 (9) (1980) 974-981, ISSN 0018-9480, http://dx.doi.org/10.1109/TMTT.1980.1130205.

[29] S. Borgaonkar, S. Rao, Analysis and design of DC blocks, Electron. Lett. 17 (2) (1981) 101-103, ISSN 0013-5194, http://dx.doi.org/10.1049/el:19810073.

[30] R. Douville, D. James, Experimental study of symmetric microstrip bends and their compensation, IEEE Trans. Microw. Theory Tech. 26 (3) (1978) 175-182, ISSN 0018-9480, http://dx.doi.org/10.1109/TMTT.1978.1129340.

[31] A. Weisshaar, S. Luo, M. Thorburn, V. Tripathi, M. Goldfarb, J. Lee, E. Reese, Modeling of radial microstrip bends, in: IEEE MTT-S International Microwave Symposium Digest, 1990, vol. 3, 1990, pp. 1051-1054, http://dx.doi.org/10.1109/MWSYM.1990.99760.

[32] H. Howe, Stripline Circuit Design, Modern Frontiers in Applied Science, Artech House, Dedham, 1974.

[33] C. Wen, Coplanar waveguide: a surface strip transmission line suitable for nonreciprocal gyromagnetic device applications, IEEE Trans. Microw. Theory Tech. 17 (12) (1969) 1087-1090, ISSN 0018-9480, http://dx.doi.org/10.1109/TMTT.1969.1127105.

[34] H. Visser, Antenna Theory and Applications, Wiley, Chichester, 2012, ISBN 9781119945215.

[35] G. Ghione, C. Naldi, Analytical formulas for coplanar lines in hybrid and monolithic MICs, Electron. Lett. 20 (4) (1984) 179-181, ISSN 0013-5194, http://dx.doi.org/10.1049/el:19840120.

[36] K. Stroud, D. Booth, Advanced Engineering Mathematics, Palgrave Macmillan Limited, Basingstoke, 2011, ISBN 9780230275485.

[37] I. Robertson, S. Lucyszyn, Institution of Electrical Engineers, RFIC and MMIC Design and Technology, IEE Circuits, Devices and Systems Series, Institution of Engineering and Technology, 2001, ISBN 9780852967867.

The Smith Chart

CHAPTER OUTLINE

INTENDED LEARNING OUTCOMES

- *Knowledge*
  - Understand the derivation of the Smith Chart based on a *bilinear transformation*.
  - Understand the benefits of the Smith Chart as a circuit design tool.
  - Be aware of the significance of various regions on the Smith Chart in terms of resistance/reactance categories.
  - Understand the meaning and application of constant Q contours on the Smith Chart.
  - Be familiar with variants of the Smith Chart, such as the Compressed Smith Chart, and their applications.

- *Skills*
  - Be able to plot an impedance or admittance on the Smith Chart and convert to reflection coefficient (and vice versa).
  - Be able to use the Smith Chart to determine the input resistance of a transmission line at a certain distance from a load.
  - Be able to draw constant Q contours on the Smith Chart and apply them in circuit design.

4.1 INTRODUCTION TO THE SMITH CHART

In the days before electronic calculators, engineers resorted to a variety of graphical calculation devices, or *nomograms*, to solve complex engineering problems quickly and with an acceptable degree of accuracy (typically up to three decimal places). The Smith Chart is the best known example of such a nomogram and will be familiar to all microwave engineers. The Smith Chart is named after Phillip H. Smith who invented it in 1939 while he was working for the Bell Telephone Laboratories [1]. Although the Smith Chart has been largely superseded in its role as a calculator by modern computer and calculator software, the chart still provides a very helpful way of visualizing the relationship between immittances and reflection coefficients and is routinely used to display data related to the performance of microwave devices and circuits. This explains its enduring popularity, decades after its original conception.

4.2 SMITH CHART DERIVATION

As outlined in Chapter 2, both the complex impedance and reflection coefficient looking into a transmission line will vary as we move away from the load toward the generator along a lossless transmission line. Although the variation in impedance along the line is a little complicated (as defined by equation (2.6.15)), the variation in reflection coefficient is quite simple, in that only the angle of the reflection coefficient changes whereas the magnitude remains fixed and equal to the magnitude of the load reflection coefficient, as illustrated by equation (2.6.8). This means that our path away from the load toward the generator is described by a circle of fixed radius on the reflection coefficient plane, that was introduced in Section 1.5. A graphical means of determining the impedance at any point on the transmission line could be achieved by drawing an overlay of contours of constant resistance and constant reactance on the reflection coefficient plane of Figure 1.15. This is essentially how a Smith Chart is created.

The Smith Chart can be considered as a graphical representation of another equation that was introduced in Chapter 1, namely equation (1.5.1), that relates impedance and reflection coefficient, thus:

$$\Gamma = \frac{Z - Z_0}{Z + Z_0} \tag{1.5.1}$$

Mathematically speaking, equation (1.5.1) is a *bilinear transformation* (or Möbius transformation) [2], that is to say it is of the form:

$$w = \frac{Az + B}{Cz + D} \tag{4.2.1}$$

where w and z are complex variables and A, B, C, and D are complex constants satisfying [3]:

$$\Delta = AD - BC \neq 0 \tag{4.2.2}$$

The bilinear transformation of equation (4.2.1) is a one-to-one mapping, meaning that every point in the z plane corresponds to a unique point in the w plane, and vice versa.

By presenting contours of complex Z (resistance and reactance) on the complex Γ plane, the Smith Chart allows us to convert readily between impedance (or admittance) and reflection coefficient, and to determine the impedance (or admittance) at any point on a transmission line connecting a given source to a given load.

We have established that a circle on the Smith Chart, centered at the origin, represents a constant value of reflection coefficient magnitude. From equation (1.5.6), it should be clear that there is a one-to-one correspondence between reflection coefficient magnitude and VSWR. For this reason, such circles drawn on the Smith Chart centered at the origin are referred to as *constant VSWR circles*.

Although low-cost computing power abounds these days, allowing transmission line problems to be quickly solved numerically, it is still important for any RF/microwave engineer to be conversant with the use of a Smith Chart to solve simple transmission line problems. The Smith Chart allows the engineer to instantly visualize a variety of RF situations, such as changes in device parameters with frequency, and for this reason it continues to be very commonly used. The Smith Chart has become such a standard means of representing parameters at microwave frequencies that computer programs and test equipment will often present their output data in Smith Chart format.

We will start by defining the reflection coefficient in terms of the load impedance Z_L as follows:

$$\Gamma_L = \text{Re}(\Gamma_L) + j\text{Im}(\Gamma_L) = \frac{Z_L - Z_o}{Z_L + Z_o} \tag{4.2.3}$$

The normalized impedance, z_L, is defined as the load impedance, Z_L, divided by the system characteristic impedance, Z_o. Therefore,

$$\Gamma_L = \frac{(Z_L - Z_o)/Z_o}{(Z_L + Z_o)/Z_o} = \frac{z_L - 1}{z_L + 1} = r + jx \tag{4.2.4}$$

We first reverse equation (4.2.4) to give the normalized resistance and reactance in terms of reflection coefficient:

$$z_L = r + jx = \frac{1 + \Gamma_L}{1 - \Gamma_L} = \frac{1 + \text{Re}(\Gamma_L) + j\text{Im}(\Gamma_L)}{1 - \text{Re}(\Gamma_L) - j\text{Im}(\Gamma_L)} \tag{4.2.5}$$

Therefore, we can write the normalized resistance as:

$$r = \frac{1 - \text{Re}(\Gamma_L)^2 - \text{Im}(\Gamma_L)^2}{1 - 2\text{Re}(\Gamma_L) + \text{Re}(\Gamma_L)^2 + \text{Im}(\Gamma_L)^2} \tag{4.2.6}$$

or in a more compact form:

$$r = \frac{1 - |\Gamma_L|^2}{1 - 2\text{Re}(\Gamma_L) + |\Gamma_L|^2} \tag{4.2.7}$$

and the normalized reactance as:

$$x = \frac{2\text{Im}(\Gamma_L)}{1 - 2\text{Re}(\Gamma_L) + \text{Re}(\Gamma_L)^2 + \text{Im}(\Gamma_L)^2} \tag{4.2.8}$$

or in a more compact form:

$$x = \frac{2\text{Im}(\Gamma_L)}{1 - 2\text{Re}(\Gamma_L) + |\Gamma_L|^2} \tag{4.2.9}$$

4.2.1 DERIVATION OF CONSTANT RESISTANCE CIRCLES

Taking equation (4.2.6) first, our task is to rearrange this equation for normalized resistance into a parametric equation of the form:

$$(x - a)^2 + (y - b)^2 = R^2 \tag{4.2.10}$$

which represents a circle in the complex (x, y) plane with center at $[a, b]$ and radius R. This rearrangement proceeds through the following steps:

$$r + r\,\text{Re}(\Gamma_L)^2 - 2r\,\text{Re}(\Gamma_L) + r\,\text{Im}(\Gamma_L)^2 = 1 - \text{Re}(\Gamma_L)^2 - \text{Im}(\Gamma_L)^2 \tag{4.2.11}$$

$$\text{Re}(\Gamma_L)^2 + r\,\text{Re}(\Gamma_L)^2 + 2r\,\text{Re}(\Gamma_L) + r\,\text{Im}(\Gamma_L)^2 + \text{Im}(\Gamma_L)^2 = 1 - r$$

$$(1 + r)\text{Re}(\Gamma_L)^2 + 2r\,\text{Re}(\Gamma_L) + (1 + r)\text{Im}(\Gamma_L)^2 = 1 - r$$

$$\text{Re}(\Gamma_L)^2 - \frac{2r}{(1 + r)}\text{Re}(\Gamma_L) + \text{Im}(\Gamma_L)^2 = \frac{1 - r}{1 + r}$$

We now take equation (4.2.11) and add $\frac{r^2}{(1+r)^2}$ to both sides to give:

$$\text{Re}(\Gamma_L)^2 - \frac{2r}{(1 + r)}\text{Re}(\Gamma_L) + \frac{r^2}{(1 + r)^2} + \text{Im}(\Gamma_L)^2 - \frac{r^2}{(1 + r)^2} = \frac{1 - r}{1 + r} \tag{4.2.12}$$

Rearranging equation (4.2.12) gives

$$\left(\text{Re}(\Gamma_L) - \frac{r}{1+r}\right)^2 + \text{Im}(\Gamma_L)^2 = \frac{1-r}{1+r} + \frac{r^2}{(1+r)^2} = \frac{1}{(1+r)^2} \qquad (4.2.13)$$

Finally, therefore, we have:

$$\left(\text{Re}(\Gamma_L) - \frac{r}{1+r}\right)^2 + \text{Im}(\Gamma_L)^2 = \left(\frac{1}{1+r}\right)^2 \qquad (4.2.14)$$

Note that the variable x, (normalized reactance) is entirely absent from equation (4.2.14), which is in the form of equation (4.2.10), that is, it describes a circle in the complex Γ plane with a center at:

$$\text{Center}_r = \frac{r}{(r+1)} + j0 \qquad (4.2.15)$$

and having a radius of:

$$\text{Radius}_r = \left|\frac{1}{(r+1)}\right| \qquad (4.2.16)$$

These circles are called the *constant resistance circles* and are shown in Figure 4.1.

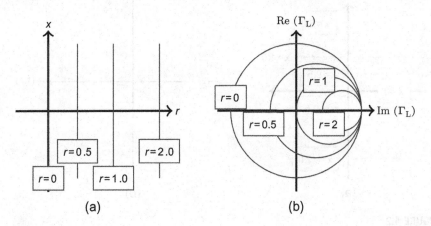

(a) (b)

FIGURE 4.1

Constant resistance contours. (a) Constant resistance contours in the normalized Z plane. (b) Constant resistance contours in the Γ plane.

4.2.2 DERIVATION OF CONSTANT REACTANCE CIRCLES

We now develop the equations for constant reactance circles in a similar way, starting with equation (4.2.8) and rearranging as follows:

$$x + x\,\text{Re}(\Gamma_L)^2 - 2x\,\text{Re}(\Gamma_L) + x\,\text{Im}(\Gamma_L)^2 = 2\text{Im}(\Gamma_L) \tag{4.2.17}$$

$$1 + \text{Re}(\Gamma_L)^2 - 2\text{Re}(\Gamma_L) + \text{Im}(\Gamma_L)^2 = \frac{2\text{Im}\Gamma_L}{x}$$

$$\text{Re}(\Gamma_L)^2 + 2\text{Re}(\Gamma_L) + 1 + \text{Im}(\Gamma_L)^2 - \frac{2}{x}\text{Im}(\Gamma_L) = 0$$

$$\text{Re}(\Gamma_L)^2 - 2\text{Re}(\Gamma_L) + 1 + \text{Im}(\Gamma_L)^2 - \frac{2}{x}\text{Im}(\Gamma_L) + \frac{1}{x^2} - \frac{1}{x^2} = 0$$

Finally, therefore, we have:

$$(\text{Re}(\Gamma_L) - 1)^2 + \left(\text{Im}(\Gamma_L) - \frac{1}{x}\right)^2 = \frac{1}{x^2} \tag{4.2.18}$$

Once again we have, in equation (4.2.18) a parametric equation in the form of equation (4.2.10) in the complex Γ plane, describing a circle centered at:

$$\text{Center}_x = 1 + j\frac{1}{x} \tag{4.2.19}$$

and having a radius of:

$$\text{Radius}_x = \left|\frac{1}{x}\right| \tag{4.2.20}$$

These circles are called the *constant reactance circles* and are shown in Figure 4.2.

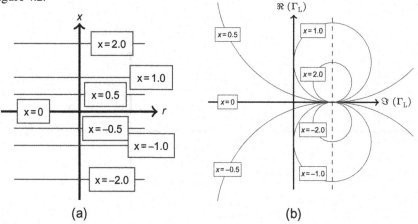

(a)   (b)

FIGURE 4.2

Constant reactance contours. (a) Constant reactance contours in the normalized Z plane. (b) Constant reactance contours in the Γ plane.

4.2.3 BUILDING THE COMPLETE SMITH CHART

We can now proceed to construct the complete Smith Chart, by combining a set of constant resistance circles and another set of constant reactance circles together on the same axis, as shown in Figure 4.3.

The standard Smith Chart is limited to passive impedances only, so the boundary of the Smith Chart is set at $|\Gamma| = 1$. A standard published Smith Chart, such as the one shown in Figure 4.4, includes two scales around the perimeter, one indicating the angle of reflection coefficient (in degrees) and the other indicating electrical lengths in fractions of a wavelength.

Although both wavelength scales cover the same range $(0\text{-}0.50\lambda)$, they are opposite in direction. The outer scale is calibrated clockwise and it represents wavelengths toward the generator. The inner scale is calibrated counter-clockwise and it represents wavelengths toward the load. These two scales are complementary at all points. Thus, 0.09λ on the outer scale corresponds to $(0.50 - 0.09) = 0.41\lambda$

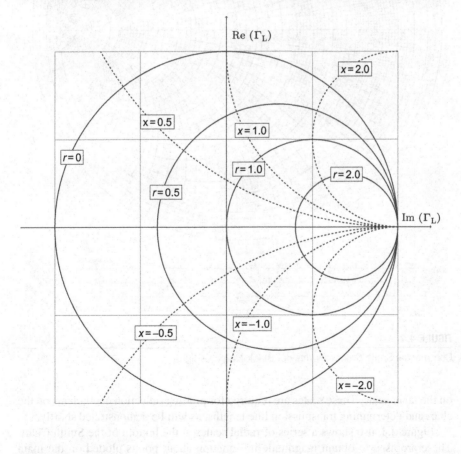

FIGURE 4.3

Combined constant resistance and reactance contours in the Γ plane.

The complete Smith Chart
Black magic design

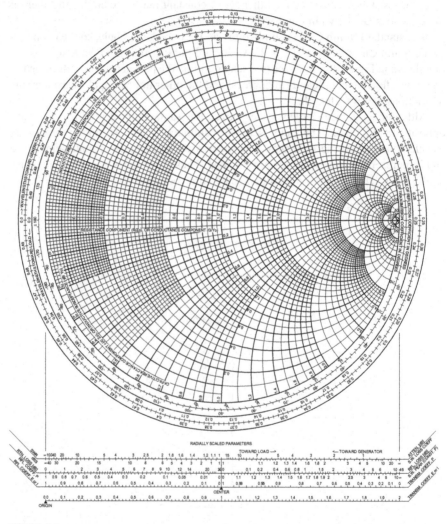

FIGURE 4.4

Commercial Smith Chart (courtesy of Black Magic Design).

on the inner scale. These scales are required for plotting reflection coefficients on the chart and determining transmission line lengths, as will be demonstrated shortly.

Figure 4.4 also shows a series of radial scales at the bottom of the Smith Chart. These are used to obtain magnitude information about points plotted on the main chart. The number of radial scales presented may vary depending on the particular brand of Smith Chart being used. The radial scales in Figure 4.4 will be described as follows, starting at the top left and moving counter clockwise around the group:

1. "SWR" refers to the standing wave ratio (also voltage standing wave ratio, VSWR). This ratio is defined as:

$$\text{SWR} = \frac{1 + |\Gamma|}{1 - |\Gamma|}$$

The range of VSWR values is between unity, at the center of the Smith Chart (perfect match) and infinity, at the boundary of the chart (perfect mismatch).

2. "dBs" refers to the SWR in decibels, which is defined as:

$$\text{VSWR}_{\text{dB}} = 20 \log_{10} \left(\frac{1 + |\Gamma|}{1 - |\Gamma|} \right)$$

The range of VSWR_{dB} values is between zero, at the center of the Smith Chart (perfect match) and infinity, at the boundary of the chart (perfect mismatch).

3. "RTN. LOSS [dB]" refers to the return loss in decibels, defined as:

$$L_r = 10 \log_{10} \left(|\Gamma|^2 \right) = 20 \log_{10} (|\Gamma|)$$

At the center of the Smith Chart (perfect match) nothing is reflected and the return loss is infinite. At the boundary of the chart we have full reflection, since $|\Gamma| = 1$. Therefore, the return loss is equal to 0 dB.

4. "REF. COEFF. P" refers to the power reflection coefficient, which is defined as:

$$\Gamma_{\text{pwr}} = |\Gamma|^2$$

There is no reflected power for the perfectly matched case (i.e., the center of the Smith Chart), but 100% reflected power at the boundary of the chart where $\Gamma_{\text{pwr}} = 1$.

5. "REF. COEFF. E or I" shows the voltage (or current) reflection coefficient, which is defined as the ratio of reflected voltage/current to incident voltage at a given point on the line. The radial scale represents only the magnitude of Γ, which is defined as follows:

$$|\Gamma| = \frac{|V_{\text{ref}}|}{|V_{\text{inc}}|} = \frac{|I_{\text{ref}}|}{|I_{\text{inc}}|}$$

Since we are only concerned with the magnitude of Γ, the scale can refer to both voltage or current, as we have omitted the signs. The reflection coefficient is zero at the Smith Chart center and unity at the boundary of the chart.

6. "TRANS. COEFF. E or I" refers to the voltage (or current) transmission coefficient, which is defined as:

$$T = 1 - \Gamma$$

7. "TRANS. COEFF. P" refers to the *power transmission coefficient*, which is defined as the transmitted power as a function of mismatch, that is:

$$T_{\text{pwr}} = 1 - \Gamma_{\text{pwr}} = 1 - |\Gamma|^2$$

Thus, in the center of the Smith Chart (i.e., $|\Gamma| = 0$), all the power is transmitted. At the boundary of the chart we have total reflection, that is no power is being transmitted. For intermediate values, such as $|\Gamma| = 0.5$ we see that 75% of the incident power is transmitted.

8. "RFL. LOSS [dB]" refers to the *reflection loss*, which is defined as:

$$L_{\text{r}} = 10 \log_{10} \left(1 - |\Gamma|^2\right)$$

Reflection loss relates the amount of power lost in transmission and is not to be confused with the return loss, which refers to the amount of incident power reflected. Figure 2.19 in Section 2.8 clarifies the difference between these two quantities. The scale shows reflection loss in dB.

9. "ATTN. [dB]" use of this scale assumes that we are measuring an attenuator, or lossy transmission line, which itself is terminated by an open- or short-circuit load ($\Gamma_L = 1\angle 0°$ or $\Gamma_L = 1\angle 180°$). The wave will travel twice through the attenuator (forward and backward). The value of this attenuator can be between zero, at the center, and some very high number corresponding to the matched case. The lower scale of this ruler displays the same situation but in terms of VSWR.

We can divide the Smith Chart into a number of distinct regions, as shown in Figure 4.5 and described as follows [4]:

1. The upper half of the chart, above the horizontal axis, represents all impedances with a positive reactive part (i.e., inductive impedances).
2. The lower half of the chart represents all impedances with a negative reactive part (i.e., capacitive impedances).
3. The horizontal axis represents all pure resistances.
4. The outer perimeter of the Smith Chart represents all purely reactive impedances (i.e., zero resistance). These are pure inductances on the upper semicircle and pure capacitances on the lower semicircle.
5. The rightmost point on the horizontal axis represents infinite impedance (a perfect open circuit).
6. The leftmost point on the horizontal axis represents zero impedance (a perfect short circuit).

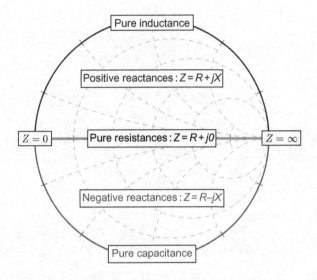

FIGURE 4.5

Smith Chart regions.

4.3 USING THE SMITH CHART
4.3.1 CONVERSION BETWEEN IMMITTANCE AND REFLECTION COEFFICIENT

One of the most basic functions of the Smith Chart is conversion from immittance to reflection coefficient representation and vice versa. The immittance in question can be either impedance or admittance, as the Smith Chart can be used in either mode. When dealing with impedances, the Smith Chart circles are interpreted as normalized resistance and reactance. When dealing with admittances, the Smith Chart circles are interpreted as normalized conductance and susceptance.

Figure 4.6 shows the technique by means of the example of a load $Z_L = 25 - j50\ \Omega$. All immittances plotted on the Smith Chart must be normalized, so we normalize the load impedance as:

$$z_L = \frac{(25 - j50)}{50} = 0.5 - j$$

We plot this value on the Smith Chart at the intersection of the $r = 0.5$ and $x = -1$ normalized resistance and reactance circles, shown as point z_L in Figure 4.6. We draw a constant SWR circle through this point, centered at the origin, and draw a line through this point and the origin that intersects the constant SWR circle at another point 180° away at the point y_L. The radius of the SWR circle, corresponding to the magnitude of the reflection coefficient, can be determined by looking at the scale labeled "RFL. COEFF, E or I" at the bottom of the standard Smith Chart. This

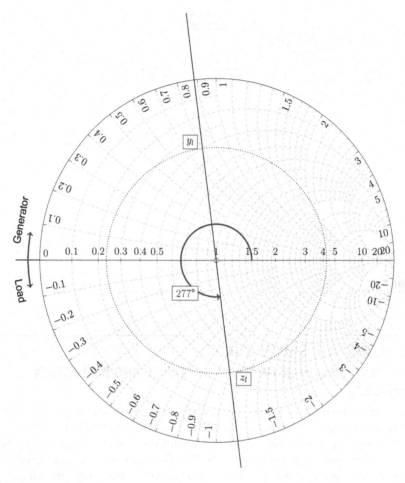

FIGURE 4.6

Immitance/reflection coefficient conversion.

gives us a reflection coefficient magnitude of 0.62. We determine the angle of the reflection coefficient by looking at where the radial passing through z_L intersects the "ANGLE OF REFLECTION COEFFICIENT" scale around the outside of the Smith Chart. We thus determine the reflection coefficient of the load as: $\Gamma = 0.62 \angle 277°$.

The point y_L represents the normalized admittance of the load. We can determine its value by reading off the constant conductance and constant susceptance circles that this point lies on. This gives us a normalized admittance of:

$$y_L = 0.4 + j0.8$$

We can denormalize this to get the ohmic load admittance Y_L as:

$$Y_L = \frac{(0.4 + j0.8)}{50} = 0.008 + j0.016 \text{ S}$$

4.3.2 IMPEDANCE AT ANY POINT ON A TRANSMISSION LINE

Circles of various radii on the Smith Chart, with centers at the origin, represent a constant SWR, which is equivalent to a constant magnitude of reflection coefficient. Figure 4.7 shows this for VSWR = 3.0, VSWR = 8.0, and VSWR = 1.5. The radius of any such circle determines the magnitude of the reflection coefficient, which lies in the range 0-1 in the case of a standard (not compressed) Smith Chart. This encompasses all possible passive values of impedance (or admittance).

Any point on one of these circles, therefore, represents a point on a lossless transmission line at some distance from the load, since, as we travel away from the

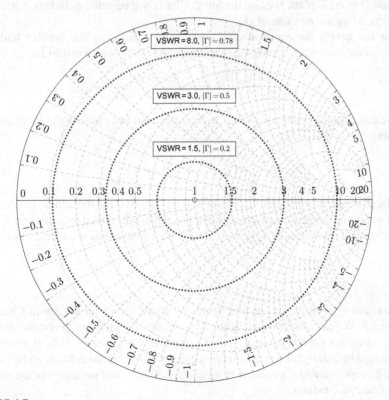

FIGURE 4.7

Constant VSWR circles.

load on a lossless line, the reflection coefficient magnitude remains constant but the angle of the reflection coefficient changes.

4.3.3 CONSTANT Q CONTOURS ON THE SMITH CHART

We introduced the important concept of network Q in Section 1.6. We will now show how Q can be interpreted on the Smith Chart, a useful technique in matching network design. The Q of a normalized series impedance, $z = r + jx$ is defined as:

$$Q = \frac{|x|}{r} \tag{4.3.1}$$

From equation (4.3.1) we can infer that a constant value of Q would describe a straight line described by $x = Qr$ in the impedance planes shown in Figures 4.1(a) and 4.2(a), where Q defines the slope of the line. We have already established that the Smith Chart is a bilateral transformation that transforms impedance plane circles into Γ plane circles. We also note that a straight line is just a special case of a circle having an infinite radius. We would therefore expect that, in general, contours of constant Q in the Γ plane (i.e., on the Smith Chart) will describe a circle or a straight line, depending on the value of Q.

We can derive the equation of a constant Q contour on the Smith Chart by applying equations (4.2.7) and (4.2.9) to equation (4.3.1), which results in:

$$Q = \frac{2|\text{Im}(\Gamma_L)|}{1 - |\Gamma_L|^2} = \frac{2|\text{Im}(\Gamma_L)|}{1 - \text{Re}(\Gamma_L)^2 - \text{Im}(\Gamma_L)^2} \tag{4.3.2}$$

We can rearrange equation (4.3.2) to be in the form of a circle in the Γ_L plane, as follows:

$$\text{Im}(\Gamma_L)^2 + \left(\text{Re}(\Gamma_L) \pm \frac{1}{Q}\right)^2 = 1 \pm \frac{1}{Q^2} \tag{4.3.3}$$

The plus sign in equation (4.3.3) is used if x is positive and the minus sign is used if x is negative. The centers of the constant Q contours described by equation (4.3.3) are located at coordinates $[\mp 1/Q, 0]$ in the Γ_L plane and the radii are given by:

$$\sqrt{1 + \frac{1}{Q^2}} \tag{4.3.4}$$

Constant Q contours for several values of Q are shown on the Smith Chart in Figure 4.8. We can observe that, as the Q increases, the constant Q contour moves further away from the horizontal ($x = 0$) axis. At the extremes, $Q = 0$ would be represented by a straight line contour lying exactly on the horizontal axis, and $Q = \infty$ would be represented by a contour lying exactly on the outer perimeter of the Smith Chart (the $r = 0$ radius).

Constant Q contours are very useful in matching network design, as discussed in Chapters 9 and 10.

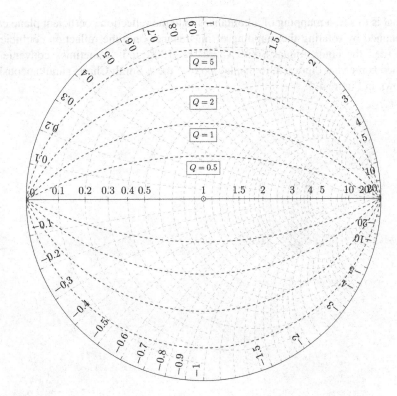

FIGURE 4.8

Constant Q contours on the Smith Chart.

4.4 SMITH CHART VARIANTS
4.4.1 COMBINED IMPEDANCE-ADMITTANCE SMITH CHART

The Smith Chart can be used to represent both impedance and admittances simultaneously by overlaying another reflection coefficient plane in a different color. For normalized admittances we have:

$$\Gamma_y = \frac{y-1}{y+1} \tag{4.4.1}$$

where y is the normalized admittance in question, normalized to $Y_0 = 1/Z_0$. It can easily be demonstrated that for any given impedance/admittance:

$$\Gamma_y = -\Gamma_z = \Gamma_z e^{j\pi} \tag{4.4.2}$$

That is to say, a mapping of admittances onto the reflection coefficient plane can be obtained by rotating the mapping of impedances onto the reflection coefficient plane (i.e., the conventional Smith Chart) by 180°. It is sometimes convenient, although somewhat cluttered, to display both of these Smith Charts simultaneously, as shown in Figure 4.9.

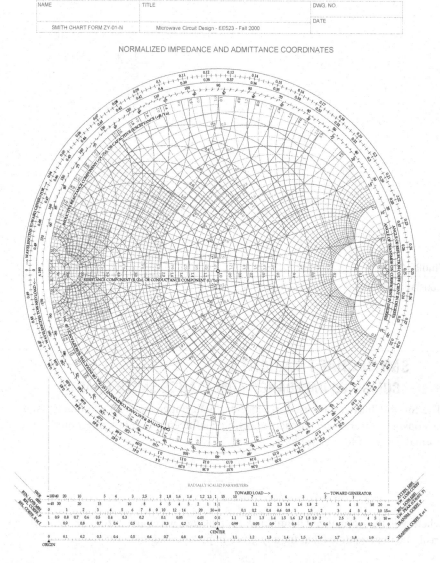

FIGURE 4.9

Combined (impedance/admittance) Smith Chart.

4.4.2 COMPRESSED SMITH CHART

The Smith Chart, as it has been presented up to this point, is a plot of reflection coefficient for magnitudes either equal to or less than 1, thereby encompassing all real, positive values of resistance. In some cases, where active microwave devices are involved, we need to represent values of negative resistance on the Smith Chart, which implies a reflection coefficient magnitude greater than unity. As explained in Chapter 15, negative resistance is simply a way of representing the fact that the power wave reflected from a given termination is greater in magnitude than the incident power wave. In such cases, the reflection coefficient magnitude is greater than unity and we refer to "return gain" at the termination, rather than return loss.

Negative resistance values plotted on a Smith Chart lie outside the $|\Gamma| = 1$ boundary of the conventional Smith Chart. In order to represent negative resistances we need to compress the conventional Smith Chart to be a subset of a larger chart, which typically has a radius of $|\Gamma| = 3.16$, this value being chosen to represent 10 dB return gain. An example of the use of a compressed Smith Chart to plot the negative resistance behavior of an MHG9000 GaAs MESFET, in terms of shunt feedback pole locations (as defined in Chapter 8) on the Γ_3 plane from 2 to 18 GHz is shown in Figure 4.10. From 6 to 10 GHz, the pole lies inside the $\Gamma_3 = 1$ boundary of the Smith Chart in Figure 4.10, indicating that negative resistance can be generated in this device using passive shunt feedback. The use of a compressed Smith Chart therefore allows the designer to visualize device parameters over the complete frequency range, where both positive and negative resistance behavior may be exhibited. This is useful in oscillator design, where designers routinely have to work with negative resistance values [5].

4.5 TAKEAWAYS

1. The Smith Chart is an example of a nomogram (graphical calculator) that was invented by Phillip H. Smith in 1939.
2. The Smith Chart is a conformal mapping of contours of complex Z (resistance and reactance) on the complex Γ plane.
3. The main applications of the Smith Chart are in solving problems involving transmission lines and matching circuits, but it is also widely used to display measurement data simultaneously in impedance and reflection coefficient format.
4. The Smith Chart is a quick and convenient way of converting impedance or admittance to reflection coefficient and vice versa.
5. The Smith Chart is a quick and convenient way of determining the input resistance of a transmission line at a certain distance from a complex load.
6. Constant Q contours on the Smith Chart are a useful tool in the design of matching networks.

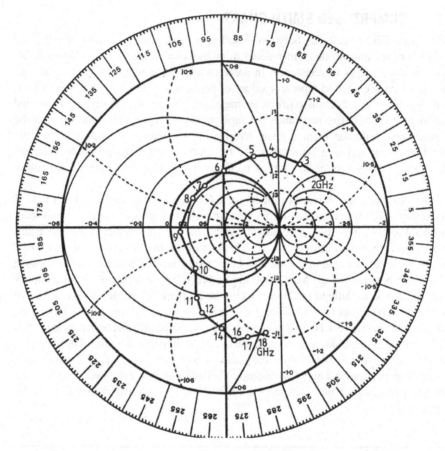

FIGURE 4.10

Compressed Smith Chart used to plot negative resistance behavior of an MHG9000 MESFET.

TUTORIAL PROBLEMS

Problem 4.1 (Smith Chart 1). Locate the following impedance and admittance points on the Smith Chart:

1. $Z_1 = 15 + j25\ \Omega$
2. $Y_2 = 0.2 - j4 \times 10^{-2}\ \text{S}$
3. $Z_3 = j70\ \Omega$
4. $Y_4 = 0.1 + j0.5\ \text{S}$
5. $Z_5 = 25 + j100\ \Omega$
6. $Y_6 = j4 \times 10^{-3}\ \text{S}$

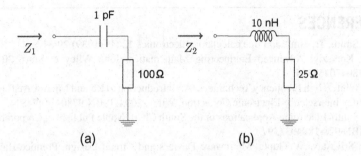

FIGURE 4.11

Circuits for Problem 4.2.

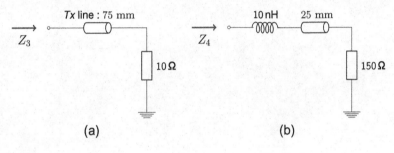

FIGURE 4.12

Circuits for Problem 4.3.

Problem 4.2 (Smith Chart 2). Assuming an operating frequency of 1 GHz, locate the load impedances, Z_1 and Z_2 shown in Figure 4.11 on the Smith Chart:

Problem 4.3 (Smith Chart 3). Assuming an operating frequency of 1 GHz, locate the load impedances, Z_3 and Z_4 shown in Figure 4.12 on the Smith Chart:

Problem 4.4 (Smith Chart 4). A 50 Ω transmission line is terminated by a 40 Ω resistor in series with a single capacitor of 1 pF. There are two such points on the line where the input impedance is purely resistive, one having an input resistance smaller than 50 Ω and one having an input resistance larger than 50 Ω. Determine the input resistance of the line at each of these points and the distance from the load to each point. Assume an operating frequency of 2 GHz.

Problem 4.5 (Smith Chart 5). Use a Smith Chart to estimate the Q of a single *LC* matching network used to match a load $Z_L = 6 + j0$ Ω to a 50 Ω line at 2.5 GHz. Again using the Smith Chart, determine the VSWR looking into the matching network at 2, 2.5, and 3 GHz.

REFERENCES

[1] P. Smith, Transmission line calculator, Electronics 12 (1) (1939) 29-31.
[2] E. Kreyszig, Advanced Engineering Mathematics, John Wiley & Sons, 2010, ISBN 9780470458365.
[3] J. White, High Frequency Techniques: An Introduction to RF and Microwave Engineering, Wiley Interscience Electronic Collection, Wiley, 2004, ISBN 9780471474814.
[4] P. Smith, Electronic Applications of the Smith Chart, Noble Publishing Corporation, 2000, ISBN 978-1884932397.
[5] G. Srivastava, V. Gupta, Microwave Devices and Circuit Design, Prentice-Hall of India Pvt Ltd, New Delhi, 2006.

Microwave circuit analysis

Immittance parameters

CHAPTER OUTLINE

INTENDED LEARNING OUTCOMES

- *Knowledge*
  - Be familiar with the most common types of immittance parameter representations (Z, Y, h, and $ABCD$-parameters), their respective strengths, weaknesses, and applications.
  - Be aware of the use of Miller's theorem to simplify two-port feedback problems.

- *Skills*
  - Be able to calculate the input immittance of a two-port network with an arbitrary load termination, and the output immittance of a two-port network with an arbitrary source termination.
  - Be able to convert a given dataset from one immittance parameter representation to another.
  - Be able to calculate the immittance parameters of a two-port device with shunt or series feedback.
  - Be able to apply immittance parameters to determine whether a given two-port is "active," "passive," or "lossless."

5.1 INTRODUCTION

One widely used approach to network analysis and design is to treat the particular component or subcircuit of interest as a *two-port* network. Under this regime, the component or subcircuit is treated as a black box where only the input and output ports are accessible (hence the name "two-port"). In the absence of any information about the internal composition of the black box, we can fully characterize it in terms of relationships between measured signal currents and voltages at the external terminals, assuming the circuitry inside the box is linear with no independent current or voltage sources.

Figure 5.1 illustrates this by showing the signal voltages and currents we can measure at the terminals of a two-port. The direction of the voltages and current arrows in Figure 5.1 should be carefully noted; that is, currents are always defined as positive when entering the two-port and negative when leaving the two-port.

The two-port approach is based on a number of assumptions. First, it is assumed that the relationships between the terminal voltages and currents is *linear*. This is generally not the case when active devices, such as diodes and transistors, are involved, but may be approximately true if the range of current and voltage change is small. We refer to this restriction as *small-signal* operation. Another assumption implicit in the two-port approach is that the black box does not contain any *independent*

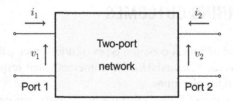

FIGURE 5.1

Two-port network with port definitions.

current or voltage sources. In other words, it can be completely described in terms of relationships between terminal voltages v_1 and v_2 and currents i_1 and i_2, two of which are *independent* variables and the other two being *dependent* variables.

In this chapter, we explain the most commonly used two-port parameters and their applications. When the parameter is a ratio of voltage to current, the parameter has the dimensions of ohms; that is, it is an *impedance* parameter. When the parameter is a ratio of current to voltage, the parameter has the dimensions of siemens; that is, it is an *admittance* parameter. Ratios of voltage to voltage and current to current are dimensionless. All these are collectively referred to as *immittance parameters*.

The most commonly used two-port immittance parameter types are the impedance (*Z*) parameters, admittance (*Y*) parameters, hybrid (*h*) parameters, and *ABCD* (or "chain") parameters. These each have specific applications, which we will discuss in detail in this chapter.

5.1.1 THE ADMITTANCE OR *Y*-PARAMETERS

We will start by defining the admittance parameters, or *Y*-parameters, in which case terminal signal voltages, v_1 and v_2, are the independent variables and signal currents, i_1 and i_2, are the dependent variables. We can represent a linear two-port as an equivalent circuit containing two admittance elements, y_{11} and y_{22} and two current dependent current sources with coefficients y_{12} and y_{21}, as shown in Figure 5.2. From Figure 5.2 we can write down the terminal voltage and current relationships as follows:

$$i_1 = Y_{11}v_1 + Y_{12}v_2 \tag{5.1.1}$$
$$i_2 = Y_{21}v_1 + Y_{22}v_2 \tag{5.1.2}$$

or in matrix form:

$$\begin{bmatrix} i_1 \\ i_2 \end{bmatrix} = \begin{bmatrix} Y_{11} & Y_{12} \\ Y_{21} & Y_{22} \end{bmatrix} \begin{bmatrix} v_1 \\ v_2 \end{bmatrix} \tag{5.1.3}$$

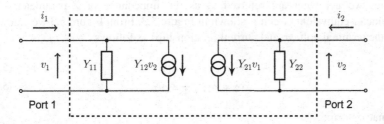

FIGURE 5.2

Two-port *Y*-parameter equivalent circuit.

Table 5.1 *Y*-Parameter Definitions

Y_{11}:	Input admittance with output short circuited
Y_{12}:	Reverse transconductance with input short circuited
Y_{21}:	Forward transconductance with output short circuited
Y_{22}:	Output admittance with input short circuited

In general, the *Y*-parameters are complex parameters, representing both amplitude and phase relationships between the terminal voltages and currents. By convention the *Y*-parameters are depicted in Cartesian form, that is $Y_{ij} = G_{ij} + jB_{ij}$, where G_{ij} is conductance and B_{ij} is susceptance.

A method of measuring the *Y*-parameters can be deduced from Figure 5.2. If we apply a short circuit at the output terminals, that is, $v_2 = 0$, and apply a unit voltage generator across the input terminals, then the input and output port currents i_1 and i_2, both defined as having positive direction toward the two-port, are equal to y_{11} and y_{21}, respectively. Similarly, short circuiting the input terminals and applying a unit voltage to the output terminals provides Y_{12} and Y_{22} from i_1 and i_2, respectively. The measurement of *Y*-parameters is summarized in Table 5.1.

Referring to Figure 5.2, the *Y*-parameters are defined as follows:

$$\left. \begin{array}{l} Y_{11} = \left. \dfrac{i_1}{v_1} \right|_{v_2=0} \\[2mm] Y_{12} = \left. \dfrac{i_1}{v_2} \right|_{v_1=0} \\[2mm] Y_{21} = \left. \dfrac{i_2}{v_1} \right|_{v_2=0} \\[2mm] Y_{22} = \left. \dfrac{i_2}{v_2} \right|_{v_1=0} \end{array} \right\} \tag{5.1.4}$$

5.1.2 THE IMPEDANCE OR *Z*-PARAMETERS

If we choose i_1 and i_2 as the independent variables and v_1 and v_2 as the dependent variables we get what are referred to as the impedance or Z-parameters. The Z-parameter equivalent circuit is shown in Figure 5.3. From Figure 5.3 we can write down the terminal voltage and current relationships as follows:

$$v_1 = Z_{11}i_1 + Z_{12}i_2 \tag{5.1.5}$$

$$v_2 = Z_{21}i_1 + Z_{22}i_2 \tag{5.1.6}$$

or in matrix form:

$$\begin{bmatrix} v_1 \\ v_2 \end{bmatrix} = \begin{bmatrix} Z_{11} & Z_{12} \\ Z_{21} & Z_{22} \end{bmatrix} \begin{bmatrix} i_1 \\ i_2 \end{bmatrix} \tag{5.1.7}$$

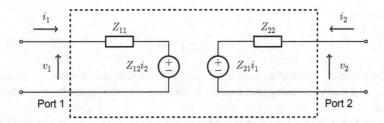

FIGURE 5.3

Two-port Z-parameter equivalent circuit.

Table 5.2 Z-Parameter Definitions

Z_{11}:	Input impedance with output open circuited
Z_{12}:	Reverse transimpedance with input open circuited
Z_{21}:	Forward transimpedance with output open circuited
Z_{22}:	Output impedance with input open circuited

In general, the Z-parameters are complex parameters, representing both amplitude and phase relationships between the terminal voltages and currents. By convention, the Z-parameters are depicted in Cartesian form, that is, $Z_{ij} = R_{ij} + jX_{ij}$, where R_{ij} is resistance and X_{ij} is reactance.

Z-parameters are measured by measuring terminal signal voltages and currents with open-circuit terminations applied to the appropriate port, according to Table 5.2.

Referring to Figure 5.3, the Z-parameters are defined as follows:

$$\left. \begin{aligned} Z_{11} &= \left. \frac{v_1}{i_1} \right|_{i_2=0} \\[2mm] Z_{12} &= \left. \frac{v_1}{i_2} \right|_{i_1=0} \\[2mm] Z_{21} &= \left. \frac{v_2}{i_1} \right|_{i_2=0} \\[2mm] Z_{22} &= \left. \frac{v_2}{i_2} \right|_{i_1=0} \end{aligned} \right\} \tag{5.1.8}$$

From equation (5.1.8) we can see that Z-parameters are *open-circuit parameters*, since the measurement of Z-parameters is performed with open circuits applied to the ports. When applying Y- and Z-parameters at microwave frequencies, it is common practice to *normalize* them to the system characteristic impedance. By convention, lower case letters are used to represent normalized quantities. The respective normalized parameters are defined by:

$$z_{ij} = \frac{Z_{ij}}{Z_o}$$

$$y_{ij} = Z_o Y_{ij}$$

If the internal circuit of a passive two-port is known, it may be possible to derive the Y- or Z-parameters using traditional circuit theory, as illustrated by the following example.

Example 5.1 (Calculation of Y- and Z-parameters for a passive T-network).

Problem. Calculate the two-port Y- and Z-parameters for the passive T-network shown in Figure 5.4.

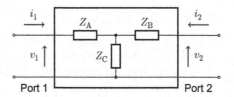

FIGURE 5.4

T-network.

Solution. With a T-network, the Z-parameters are easier to determine than the Y-parameters. Considering open-circuit terminal voltages and currents in Figure 5.4, and applying the definitions in equation (5.1.8) we can write:

$$Z_{11} = Z_A + Z_C$$

$$Z_{12} = Z_C$$

$$Z_{21} = Z_C$$

$$Z_{22} = Z_B + Z_C$$

To calculate the Y-parameters, we derive terminal and transfer impedances from Figure 5.4 under short-circuit conditions at the ports. Now applying the definitions in equation (5.1.4) to Figure 5.4, the reader might like to verify the following:

$$Y_{11} = \frac{Z_B + Z_C}{Z_A Z_B + Z_A Z_C + Z_B Z_C}$$

$$Y_{12} = \frac{-Z_C}{Z_A Z_B + Z_A Z_C + Z_B Z_C}$$

$$Y_{21} = \frac{-Z_C}{Z_A Z_B + Z_A Z_C + Z_B Z_C}$$

$$Y_{22} = \frac{Z_A + Z_C}{Z_A Z_B + Z_A Z_C + Z_B Z_C}$$

5.1.3 THE HYBRID OR *h*-PARAMETERS

If we choose v_2 and i_1 as the independent variables and v_1 and i_2 as the dependent variables we get what are referred to as the "hybrid" or *h*-parameters. The network equations for the equivalent circuit shown in Figure 5.5 are:

$$v_1 = i_1 h_{11} + v_2 h_{12} \tag{5.1.9}$$

$$i_2 = i_1 h_{21} + v_2 h_{22} \tag{5.1.10}$$

or in matrix form:

$$\begin{bmatrix} v_1 \\ i_2 \end{bmatrix} = \begin{bmatrix} h_{11} & h_{12} \\ h_{21} & h_{22} \end{bmatrix} \begin{bmatrix} i_1 \\ v_2 \end{bmatrix} \tag{5.1.11}$$

By convention, *h*-parameters are always represented as lower case, even though they are not usually normalized to the system characteristic impedance.

The *h*-parameters are ideally suited to the characterization of bipolar transistors, which, in the common emitter configuration, have a low input impedance (base-emitter junction) and high output impedance (collector-emitter junction). For this reason, the *h*-parameters are widely used for transistor circuit design at lower frequencies.

h-parameters are measured as shown in Table 5.3.

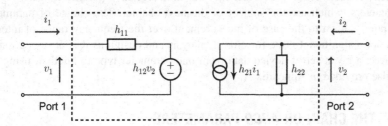

FIGURE 5.5

Two-port *h*-parameter equivalent circuit.

.

Table 5.3 *h*-Parameter Definitions

h_{11}:	Input impedance with output short circuited
h_{12}:	Reverse voltage transfer ratio with input open circuited
h_{21}:	Forward current transfer ratio with output short circuited
h_{22}:	Output admittance with input open circuited

$$h_{11} = \left.\frac{v_1}{i_1}\right|_{v_2=0}$$

$$h_{12} = \left.\frac{v_1}{v_2}\right|_{i_1=0}$$

$$h_{21} = \left.\frac{i_2}{i_1}\right|_{v_2=0}$$

$$h_{22} = \left.\frac{i_2}{v_2}\right|_{i_1=0}$$

(5.1.12)

The *h*-parameters are called *hybrid*, because measurement of *h*-parameters is performed with a combination of open and short circuits applied to the ports. Specifically h_{11} and h_{21} are measured with the output short circuited, whereas h_{12} and h_{22} are measured with the input open circuited. From equation (5.1.12) we can see that h_{11} has the dimensions of impedance, h_{12}, and h_{21} are dimensionless and h_{22} has the dimensions of admittance.

The *h*-parameters are the preferred parameter set for bipolar junction transistor characterization at low to medium frequencies, since the parameter h_{21} (the forward current transfer ratio) is consistent with BJT operation. Manufacturers routinely supply *h*-parameters in transistor data sheets, or at least the parameter h_{21} in the form of "h_{fe}," which is easily understood as the forward current gain of the device in common emitter configuration. As the frequency of interest increases, it becomes more and more impractical to measure *h*-parameters. At a few hundred MHz, *h*-parameters give way to *Y*-parameters in manufacturer's data sheets, since *Y*-parameters can be measured solely under short-circuit conditions. As the operating frequency range extends yet further into the GHz region, *Y*-parameters give way to *S*-parameters, which are the subject of the next chapter. The choice of parameter set is purely down to the ease of measurement over the frequency range of interest and is not a significant issue for circuit analysis and design as it is always possible to convert a given set of device data from one parameter type to another, using the formulae provided in Appendix A.

5.1.4 THE CHAIN OR *ABCD*-PARAMETERS

Finally we introduce the rather unimaginatively named *ABCD*-parameters, which are defined by taking v_1 and i_1 as the independent variables and v_2 and i_2 as the dependent variables. The network equations in this case are:

$$v_1 = Av_2 - Bi_2 \tag{5.1.13}$$

$$i_1 = Cv_2 - Di_2 \tag{5.1.14}$$

or in matrix form:

$$\begin{bmatrix} v_1 \\ i_1 \end{bmatrix} = \begin{bmatrix} A & B \\ C & D \end{bmatrix} \begin{bmatrix} v_2 \\ -i_2 \end{bmatrix} \tag{5.1.15}$$

The *ABCD*-parameters are measured as follows:

$$A = \left.\frac{v_2}{v_1}\right|_{i_1=0}$$

$$B = \left.\frac{v_2}{i_1}\right|_{v_1=0}$$

$$C = \left.\frac{-i_2}{v_1}\right|_{i_1=0}$$

$$D = \left.\frac{-i_2}{i_1}\right|_{v_1=0}$$

(5.1.16)

The *ABCD*-matrix is not as widely used for circuit analysis and design as the other immittance matrices, and the reader may ask why the world needs yet another set of immittance parameters. The primary application of the *ABCD*-matrix is the calculation of the overall immittance parameters for a cascade of two or more two-port networks as shown in Figure 5.6.

For two-port "a" in Figure 5.6 we can write the following (note that all superscripts are only for identification and do not represent "to the power of"):

$$\begin{bmatrix} v_1^a \\ i_1^a \end{bmatrix} = \begin{bmatrix} A^a & B^a \\ C^a & D^a \end{bmatrix} \begin{bmatrix} v_2^a \\ -i_2^a \end{bmatrix}$$

(5.1.17)

Similarly, for two-port "b" we have:

$$\begin{bmatrix} v_1^b \\ i_1^b \end{bmatrix} = \begin{bmatrix} A^b & B^b \\ C^b & D^b \end{bmatrix} \begin{bmatrix} v_2^b \\ -i_2^b \end{bmatrix}$$

(5.1.18)

But from Figure 5.6 we can see that:

$$\begin{bmatrix} v_2^a \\ i_2^a \end{bmatrix} = \begin{bmatrix} v_1^b \\ -i_1^b \end{bmatrix}$$

(5.1.19)

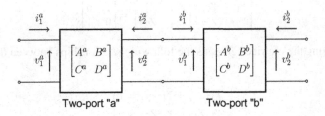

Two-port "a"   Two-port "b"

FIGURE 5.6

Cascaded two-ports represented by *ABCD*-parameters.

Therefore, we can combine equations (5.1.17) and (5.1.18) to obtain:

$$
\begin{bmatrix} v_1^a \\ i_1^a \end{bmatrix} = \begin{bmatrix} A^a & B^a \\ C^a & D^a \end{bmatrix} \begin{bmatrix} A^b & B^b \\ C^b & D^b \end{bmatrix} \begin{bmatrix} v_2^b \\ -i_2^b \end{bmatrix} \tag{5.1.20}
$$

or

$$
\begin{bmatrix} v_1^a \\ i_1^a \end{bmatrix} = \begin{bmatrix} A^{ab} & B^{ab} \\ C^{ab} & D^{ab} \end{bmatrix} \begin{bmatrix} v_2^b \\ -i_2^b \end{bmatrix} \tag{5.1.21}
$$

where

$$
\begin{bmatrix} A^{ab} & B^{ab} \\ C^{ab} & D^{ab} \end{bmatrix} = \begin{bmatrix} A^a & B^a \\ C^a & D^a \end{bmatrix} \begin{bmatrix} A^b & B^b \\ C^b & D^b \end{bmatrix}
$$
$$
= \begin{bmatrix} (A^a A^b + B^a C^b) & (A^a B^b + B^a D^b) \\ (C^a A^b + D^a C^b) & (C^a B^b + D^a D^b) \end{bmatrix} \tag{5.1.22}
$$

So, we can determine the *ABCD*-matrix of a cascade of two two-ports simply by multiplying their *ABCD*-matrices together. It should be stressed that the *ABCD*-matrix is only defined for two-port networks.

5.2 CONVERSION BETWEEN IMMITTANCE PARAMETERS

Since all the immittance parameters so far discussed are ratios of terminal voltages and/or currents, we would expect to be able to derive closed form equations that would allow us to translate one set of parameters into another. Taking the *Y*- and *Z*-parameters as an example, given equations (5.1.3) and (5.1.7) we can write:

$$
\begin{bmatrix} i_1 \\ i_2 \end{bmatrix} = \begin{bmatrix} Y_{11} & Y_{12} \\ Y_{21} & Y_{22} \end{bmatrix} \begin{bmatrix} Z_{11} & Z_{12} \\ Z_{21} & Z_{22} \end{bmatrix} \begin{bmatrix} i_1 \\ i_2 \end{bmatrix} \tag{5.2.1}
$$

Since the current vectors are equal, we can write:

$$
\begin{bmatrix} Y_{11} & Y_{12} \\ Y_{21} & Y_{22} \end{bmatrix} \begin{bmatrix} Z_{11} & Z_{12} \\ Z_{21} & Z_{22} \end{bmatrix} = \begin{bmatrix} 1 & 0 \\ 0 & 1 \end{bmatrix} \tag{5.2.2}
$$

Multiplying the matrices gives us the following relationships between the various elements:

$$
\left. \begin{aligned}
Y_{11}Z_{11} + Y_{12}Z_{21} &= 1 \\
Y_{11}Z_{12} + Y_{12}Z_{22} &= 0 \\
Y_{21}Z_{11} + Y_{22}Z_{21} &= 0 \\
Y_{21}Z_{12} + Y_{22}Z_{22} &= 1
\end{aligned} \right\} \tag{5.2.3}
$$

We can solve the simultaneous equations equation (5.2.3) to obtain the Y-parameters in terms of Z-parameters or vice versa, as follows:

$$\begin{bmatrix} Y_{11} & Y_{12} \\ Y_{21} & Y_{22} \end{bmatrix} = \begin{bmatrix} \left(\dfrac{Z_{22}}{Z_{11}Z_{22} - Z_{12}Z_{21}} \right) & \left(\dfrac{-Z_{12}}{Z_{11}Z_{22} - Z_{12}Z_{21}} \right) \\ \left(\dfrac{-Z_{21}}{Z_{11}Z_{22} - Z_{12}Z_{21}} \right) & \left(\dfrac{Z_{11}}{Z_{11}Z_{22} - Z_{12}Z_{21}} \right) \end{bmatrix} \quad (5.2.4)$$

and

$$\begin{bmatrix} Z_{11} & Z_{12} \\ Z_{21} & Z_{22} \end{bmatrix} = \begin{bmatrix} \left(\dfrac{Y_{22}}{Y_{11}Y_{22} - Y_{12}Y_{21}} \right) & \left(\dfrac{-Y_{12}}{Y_{11}Y_{22} - Y_{12}Y_{21}} \right) \\ \left(\dfrac{-Y_{21}}{Y_{11}Y_{22} - Y_{12}Y_{21}} \right) & \left(\dfrac{Y_{11}}{Y_{11}Y_{22} - Y_{12}Y_{21}} \right) \end{bmatrix} \quad (5.2.5)$$

We can carry out similar analysis for the h- and $ABCD$-parameters. The resulting relationships are shown in the conversion table in Appendix A.1.

5.3 INPUT AND OUTPUT IMPEDANCE OF A TWO-PORT IN TERMS OF IMMITTANCE PARAMETERS

The input impedance (or admittance) of a linear two-port is generally a function of the value of load termination. Likewise, the output impedance (or admittance) will generally be a function of the value of source termination. We use the word "generally" here because there is a special case where this rule does not apply. This is the case where the device is *unilateral*, as will be discussed and defined in Section 5.4.2.

The general case is shown in Figure 5.7 where a two-port characterized by a Z-matrix, $[Z]$ (or admittance matrix, $[Y]$) is connected to a source, Z_S (or Y_S) and a load Z_L (or Y_L).

Let us choose to describe the two-port of Figure 5.7 in terms of Y-parameters. Examining Figure 5.7 we can write:

$$-i_2 = Y_L v_2 \quad (5.3.1)$$

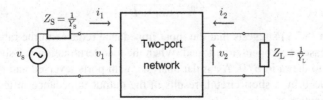

FIGURE 5.7

Two-port network with load and source terminations.

Combining equations (5.1.1) and (5.3.1) we have:

$$\frac{v_2}{v_1} = \frac{-Y_{21}}{Y_{22} + Y_L} \tag{5.3.2}$$

Substituting for v_2 in equation (5.1.1) we have:

$$i_1 = Y_{11}v_1 - \frac{Y_{21}Y_{12}}{Y_{22} + Y_L}v_1 \tag{5.3.3}$$

Hence, we can write the input admittance, Y_{in}, as:

$$Y_{in} = \frac{i_1}{v_1} = Y_{11} - \frac{Y_{21}Y_{12}}{Y_{22} + Y_L} \tag{5.3.4}$$

The reader may like to undertake a similar analysis to prove that the output admittance, Y_{out}, is given by:

$$Y_{out} = Y_{22} - \frac{Y_{12}Y_{21}}{Y_{11} + Y_S} \tag{5.3.5}$$

(Note that the voltage source v_S can be treated as a short circuit for the purpose of determining Y_{out}.)

If we prefer to work with Z-parameters, we could follow a similar line of analysis to the above which would result in the following:

$$Z_{in} = Z_{11} - \frac{Z_{21}Z_{12}}{Z_{22} + Z_L} \tag{5.3.6}$$

$$Z_{out} = Z_{22} - \frac{Z_{12}Z_{21}}{Z_{11} + Z_S} \tag{5.3.7}$$

The input and output impedances in terms of $ABCD$-parameters can be determined by applying equations (5.1.13) and (5.1.14) to the two-port in Figure 5.7. This gives the input impedance as:

$$Z_{in} = \frac{v_1}{i_1} = \frac{Av_2 - Bi_2}{Cv_2 - Di_2} \tag{5.3.8}$$

But since $v_2 = -Z_L i_2$ we have:

$$Z_{in} = \frac{AZ_L + B}{CZ_L + D} \tag{5.3.9}$$

Equation (5.3.9) suggests that the input impedance reduces to the ratio A/C for the special case of an open-circuit load. When the load termination is a short circuit, Z_{in} reduces to the ratio B/D. A similar analysis, with ports reversed and the voltage source replaced by a short circuit, results in the output impedance in terms of the $ABCD$-parameters as:

$$Z_{out} = \frac{DZ_S + B}{CZ_S + A} \tag{5.3.10}$$

5.4 CLASSIFICATION OF IMMITTANCE MATRICES

5.4.1 ACTIVITY AND PASSIVITY

Two-ports may be categorized as either *active* or *passive* [1]. An active two-port is one that can be used to provide amplification of a signal or can be made to oscillate. Passive two-ports do not exhibit amplification or oscillation. Any circuit that comprises only resistors, capacitors, and inductors will be passive. An active circuit must contain transistors or some other type of active device. Having said this, a circuit containing active devices can still be passive if the total average small-signal power entering all the ports is zero or positive. If this sum is zero the circuit is referred to as *lossless*. If the sum is positive the circuit is *lossy*.

Consider the two-port of Figure 5.1 on page 144. The total power entering the two-port is given by:

$$P_{\text{tot}} = \text{Re}(V_1 I_1^*) + \text{Re}(V_2 I_2^*) = \text{Re}(V_1^* I_1) + \text{Re}(V_2^* I_2) \tag{5.4.1}$$

where the asterisk in superscript denotes conjugate and $\text{Re}(x)$ denotes the real part of the complex quantity, x.

Equation (5.4.1) can be rewritten as:

$$P_{\text{tot}} = \frac{1}{2}(V_1 I_1^* + V_1^* I_1) + \frac{1}{2}(V_2 I_2^* + V_2^* I_2) \tag{5.4.2}$$

Equation (5.4.2) can be rearranged to give:

$$P_{\text{tot}} = \frac{1}{2}(V_1^* I_1 + V_2^* I_2) + \frac{1}{2}(I_1^* V_1 + I_2^* V_2) \tag{5.4.3}$$

which can be written in matrix form as:

$$P_{\text{tot}} = \frac{1}{2}[V^*]^T[I] + \frac{1}{2}[I^*]^T[V] \tag{5.4.4}$$

where $[\,]^T$ denotes the conjugate transpose of a matrix. We now introduce the Y-parameters by employing the following relationships:

$$[I] = [Y][V]$$
$$[I^*]^T = [V^*]^T[Y^*]^T$$

We can now write equation (5.4.4) as follows:

$$P_{\text{tot}} = \frac{1}{2}[V^*]^T[Y][V] + \frac{1}{2}[V^*]^T[Y^*]^T[V] \tag{5.4.5}$$

which can be factored to give:

$$P_{\text{tot}} = [V^*]^T\left[\frac{1}{2}([Y] + [Y^*]^T)\right][V] \tag{5.4.6}$$

The term $1/2([Y]+[Y^*]^T)$ is called a *Hermitian form* of the admittance matrix (in this case). A Hermitian matrix is a square matrix with complex entries that is equal to its own conjugate transpose. This means that the elements along the principal diagonal are real, and the complex elements on each side of the principal diagonal are complex conjugates.

Passivity requires $P_{tot} \geq 0$ for any $V_i \neq 0$. Which requires that the Hermitian form in equation (5.4.6) must be *positive semidefinite*. This means that all principal minors in the Y-parameter determinant, which is the set of subdeterminants that can be taken symmetrically around the diagonal, are positive or zero.

$$\det\left[\frac{1}{2}([Y]+[Y^*]^T)\right] = \begin{vmatrix} G_{11} & \frac{1}{2}(Y_{12}+Y_{21}^*) \\ \frac{1}{2}(Y_{21}+Y_{12}^*) & G_{22} \end{vmatrix} \tag{5.4.7}$$

where $Y_{ij} = G_{ij} + jB_{ij}$.

Expanding the determinant of equation (5.4.7) gives us three conditions for passivity, as follows:

$$G_{11} \geq 0 \tag{5.4.8}$$

$$G_{22} \geq 0 \tag{5.4.9}$$

$$G_{11}G_{22} - \frac{1}{4}(Y_{21}+Y_{12}^*)(Y_{12}+Y_{21}^*) \geq 0 \tag{5.4.10}$$

Condition equation (5.4.10) can be rewritten as:

$$4(G_{11}G_{22} - G_{12}G_{21}) - |Y_{21} - Y_{12}|^2 \geq 0 \tag{5.4.11}$$

From equation (5.4.11) we can write the passivity condition as:

$$0 \leq U \leq 1 \tag{5.4.12}$$

where U is defined as:

$$U = \frac{|Y_{21} - Y_{12}|^2}{4(G_{11}G_{22} - G_{12}G_{21})} \tag{5.4.13}$$

The quantity U is referred to as *Mason's unilateral power gain* or *Mason's invariant* [2], and, as the name suggests, has a wider significance than just as an indication of passivity.

U usually decreases with increasing frequency. The frequency where $U = 1$ is the frequency where the device ceases to be *active* and becomes *passive* [3]. This frequency is called the maximum frequency of oscillation, f_{max}, which is usually (but not always) higher than the cut-off frequency, f_T, that is traditionally used to specify the upper operating frequency of transistors. We will talk a bit more about f_{max} and f_T when we discuss microwave transistors in Chapter 12.

Equation (5.4.13) can be more succinctly written in matrix form as follows:

$$U = \frac{\left|\det\left[Y - Y^T\right]\right|}{\det\left[Y + Y^*\right]} \tag{5.4.14}$$

or in Z-parameter form as:

$$U = \frac{\left|\det\left[Z - Z^T\right]\right|}{\det\left[Z + Z^*\right]} \tag{5.4.15}$$

which equates to:

$$U = \frac{|Z_{21} - Z_{12}|^2}{4(R_{11}R_{22} - R_{12}R_{21})} \tag{5.4.16}$$

where $Z_{ij} = R_{ij} + jX_{ij}$.

Mason's invariant "U" as defined by equations (5.4.13)–(5.4.16) is the only device characteristic that is invariant under lossless, reciprocal "embeddings," meaning that the active device is embedded within some other lossless, reciprocal network (such as a passive feedback network). This means that U can be used as a figure of merit to compare any three-terminal, active device [4].[1]

U can also be used as a criterion for activity/passivity: if $U > 1$, then the two-port device is active; otherwise, the device is passive.

A special case of the passive network is the *lossless* network, where the total power entering the two-port is zero, that is:

$$P_{\text{tot}} = 0 \tag{5.4.17}$$

It can be shown that the condition for losslessness is that the network contains no resistive elements. The inequalities in conditions (5.4.8)–(5.4.10) are then replaced by an equality, that is:

$$G_{11} = 0 \tag{5.4.18}$$

$$G_{22} = 0 \tag{5.4.19}$$

$$G_{11}G_{22} - \frac{1}{4}(Y_{21} + Y_{12}^*)(Y_{12} + Y_{21}^*) = 0 \tag{5.4.20}$$

Which leads to the condition for losslessness in terms of Mason's invariant "U," as:

$$U = 1 \tag{5.4.21}$$

---

[1] The history and significance of U is well explained in the excellent review paper by Dr M.S. Gupta [4].

5.4.2 UNILATERALITY

At this point it is worth pausing to consider the implications of equation (5.3.4) through equation (5.3.7). If the reverse transfer parameter, Y_{12} or Z_{12}, is zero then the second term in equations (5.3.4)–(5.3.7) disappears. The input and output immittances then simply reduce to:

$$Y_{in} = Y_{11} \qquad (5.4.22)$$
$$Y_{out} = Y_{22} \qquad (5.4.23)$$

and

$$Z_{in} = Z_{11} \qquad (5.4.24)$$
$$Z_{out} = Z_{22} \qquad (5.4.25)$$

In the case of a zero reverse transfer parameter, the input impedance is effectively de-coupled from the load, and the output impedance is effectively de-coupled from the source. In this case, the device is referred to as being *unilateral*.

5.4.3 RECIPROCITY AND SYMMETRY

A two-port is said to be *reciprocal* when the reverse and forward transfer parameters are identical, that is:

A special case of reciprocity is *symmetry*, where the two-port behaves identically when ports 1 and 2 are interchanged. The conditions for reciprocity and symmetry are summarised in Table 5.4 and Table 5.5 respectively.

Table 5.4 Reciprocity Conditions

Z-parameters:	$Z_{12} = Z_{21}$
Y-parameters:	$Y_{12} = Y_{21}$
h-parameters:	$h_{12} = h_{21}$
ABCD-parameters:	$AD - BC = 1$

Table 5.5 Symmetry Conditions

Z-parameters:	$Z_{11} = Z_{22}$
Y-parameters:	$Y_{11} = Y_{22}$
h-parameters:	$\Delta h = h_{11}h_{22} - h_{21}h_{12} = 1$
ABCD-parameters:	$A = D$

The *T*-network used in Example 5.1 is an example of a passive reciprocal network.

5.5 IMMITTANCE PARAMETER REPRESENTATION OF ACTIVE DEVICES

5.5.1 TWO-PORT REPRESENTATION OF TRANSISTORS

The two-port approach is commonly used in transistor circuit analysis and design. The two-port approach only applies under *small-signal* operating conditions. That is to say that the underlying assumption is that the signal voltages and currents applied to the device are small enough so as not to significantly alter the parameters being measured, which are a function of the DC bias voltages and currents applied to the device.

At low to medium frequencies transistors are usually represented using the *h*-parameters. Above about 300 MHz the tendency is to use *Y*-parameters because a precise short circuit is easier to implement than a precise open circuit at higher frequencies. The convention is, when dealing with transistors, to give the parameters the subscript "*ij*," where *i* indicates the terminal or direction of the measurement and *j* denotes the transistor configuration. The direction subscript, *i*, can take one of four possible values, namely, "i" = input, "f" = forward, "r" = reverse, and "o" = output. For example, the *h*-parameters for the three possible transistor configurations are renamed as in Table 5.6.

Consider the same transistor in three possible configurations, as shown in Figure 5.8. At the same frequency and bias point (for simplicity the bias networks are not shown) the three two-ports of Figure 5.8(a-c) will have different relationships between terminal voltages and currents, and hence different sets of immittance parameters.

Table 5.6 *h*-Parameter Transistor Subscripts

h-Parameter	Common Emitter	Common Base	Common Collector
h_{11}	h_{ie}	h_{ib}	h_{ic}
h_{12}	h_{re}	h_{rb}	h_{rc}
h_{21}	h_{fe}	h_{fb}	h_{fc}
h_{22}	h_{oe}	h_{ob}	h_{oc}

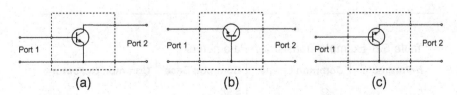

(a)   (b)   (c)

FIGURE 5.8

Transistor two-port configurations. (a) Common emitter. (b) Common base. (c) Common collector.

Each of the configurations shown in Figure 5.8 has specific advantages and disadvantages in terms of circuit performance at different frequency ranges, and it is sometimes advantageous to be able to convert the two-port parameters for one configuration into two-port parameters for the other two. A method of doing this for Y-parameters is presented in Chapter 8. Assuming we started with the common emitter h-parameters for a given transistor, the h-parameters for the same transistor in common base and common collector configurations would be as shown in Table 5.7.

As an indication, the typical h-parameters for a small-signal bipolar junction transistor at low frequencies are as follows:

$$h_{ie} = 1 \text{ k}\Omega$$
$$h_{re} = 3 \times 10^{-4}$$
$$h_{fe} = 250$$
$$h_{oe} = 3 \times 10^{-6} \text{ S}$$

Using the formulae in Table 5.7, the above device will have the parameters shown in Table 5.8.

The one thing we notice from Table 5.8 is that the sign of h_{fe} is positive for "common emitter" and negative for both "common base" and "common collector" configurations. Students familiar with transistor configurations will recognize that this is consistent with the fact that a CE amplifier is "inverting" whereas both CB and CC amplifiers are "noninverting." The polarity of h_{fe} in each case can be explained by considering the definitions of currents shown earlier in Figure 5.5.

Table 5.7 Transistor h-Parameter Conversion Formulae

h-Parameter	Common Emitter	Common Base	Common Collector
h_{11}	h_{ie}	$h_{ib} = \dfrac{h_{ie}}{1 + h_{fe}}$	$h_{ic} = h_{ie}$
h_{12}	h_{re}	$h_{rb} = \dfrac{h_{ie}h_{oe}}{1 + h_{fe}} - h_{re}$	$h_{rc} = 1 - h_{re}$
h_{21}	h_{fe}	$h_{fb} = \dfrac{-h_{fe}}{1 + h_{fe}}$	$h_{fc} = -(1 + h_{fe})$
h_{22}	h_{oe}	$h_{ob} = \dfrac{h_{oe}}{1 + h_{fe}}$	$h_{oc} = h_{oe}$

Table 5.8 Example Transistor h-Parameters

h-Parameter	Common Emitter	Common Base	Common Collector
h_{11}	1 kΩ	3.98 Ω	1 kΩ
h_{12}	3×10^{-4}	-2.88×10^{-4}	1
h_{21}	250	-0.996	-251
h_{22}	3×10^{-6} S	1.20×10^{-8} S	3×10^{-6} S

5.6 IMMITTANCE PARAMETER ANALYSIS OF TWO-PORTS WITH FEEDBACK

5.6.1 THE BENEFITS OF FEEDBACK

The application of feedback to an active two-port effectively adds another degree of freedom to the design process. If the two-port parameters of the transistor in question are not to your liking, you can change them by applying feedback and then proceed with your design using a new set of two-port parameters describing the transistor plus feedback.

In general, feedback is applied to achieve one or more of the following objectives:

1. increase stability;
2. extend the bandwidth by "flattening" the frequency response;
3. alter the input and/or output impedance to improve matching; and
4. to generate negative resistance at one of the ports (in the case of oscillator design).

The above improvements often come at a cost, which is usually in the form of a reduction in power gain.

The study of feedback is a significant branch of electronic engineering in its own right, and is extensively covered elsewhere. We will only provide here a brief description of the two main types of two-port feedback used in RF circuits, namely shunt feedback and series feedback.

5.6.2 SHUNT FEEDBACK

The Y-parameter representation of a two-port can be used to calculate the effect of shunt feedback on an active device. Consider a shunt feedback network represented by the Y-matrix $[Y^{fb}]$, connected to a transistor represented by the Y-matrix, $[Y]$, as shown in Figure 5.9. The overall Y-matrix of two two-ports connected in parallel is simply the sum of the two individual Y-matrices, as follows:

$$[Y] = [Y] + [Y_{fb}] = \begin{bmatrix} Y_{11} + Y_{11}^{fb} & Y_{12} + Y_{12}^{fb} \\ Y_{21} + Y_{21}^{fb} & Y_{22} + Y_{22}^{fb} \end{bmatrix} \tag{5.6.1}$$

5.6.3 SERIES FEEDBACK

The Z-parameter representation of a two-port can be used to calculate the effect of series feedback on an active device. Consider a series feedback network represented by the Z-matrix $[Z^{fb}]$, connected to a transistor represented by the Z-matrix, $[Z]$, as shown in Figure 5.10. The overall Z-matrix of two two-ports connected in series is simply the sum of the two individual Z-matrices, as follows:

$$[Z] = [Z] + [Z_{fb}] = \begin{bmatrix} Z_{11} + Z_{11}^{fb} & Z_{12} + Z_{12}^{fb} \\ Z_{21} + Z_{21}^{fb} & Z_{22} + Z_{22}^{fb} \end{bmatrix} \tag{5.6.2}$$

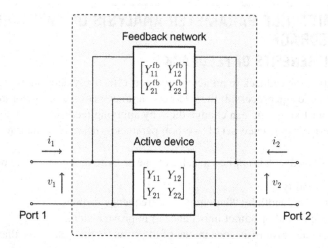

FIGURE 5.9

Two-port network, [Y], with shunt feedback network, [Y^{fb}], added.

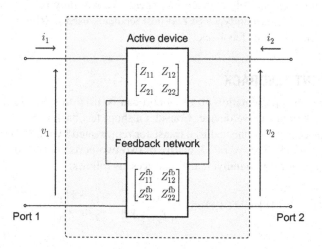

FIGURE 5.10

Two-port network, [Z], with series feedback network, [Z^{fb}], added.

5.6.4 MILLER'S THEOREM

Miller's theorem is a widely used analysis tool that can be used to simplify two-port feedback problems, such as that shown in Figure 5.11. The theorem states that if the voltage gain, A_v between port 1 and port 2 is known, then it is possible to obtain a

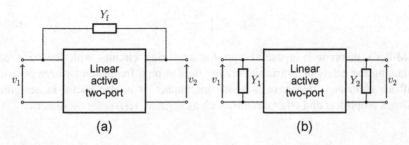

FIGURE 5.11

Miller's theorem. (a) Immittance two-port with shunt feedback. (b) Equivalent circuit, according to Miller's theorem.

circuit like that shown in Figure 5.11(b) which is equivalent to that in Figure 5.11(a) in terms of input impedance and forward voltage gain at a spot frequency.

The values of the equivalent admittances shown in Figure 5.11(b), in terms of Y_f, are given by:

$$Y_1 = Y_f(1 - A_v) \tag{5.6.3}$$

$$Y_2 = Y_f\left(\frac{A_v - 1}{A_v}\right) \tag{5.6.4}$$

where the voltage gain A_v is defined as:

$$A_v = \frac{v_2}{v_1}\bigg|_{i_2=0} \tag{5.6.5}$$

The dual of Miller's theorem applies in the case of series feedback. The values of the equivalent impedances shown in Figure 5.12(b), in terms of Z_f, are given by:

$$Z_1 = Z_f(1 - A_v) \tag{5.6.6}$$

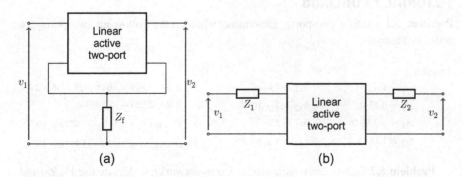

FIGURE 5.12

The dual of Miller's theorem. (a) Immittance two-port with series feedback. (b) Equivalent circuit, according to the dual of Miller's theorem.

$$Z_2 = Z_f \left(\frac{A_v - 1}{A_v} \right)$$ (5.6.7)

Miller's theorem is a useful tool for analyzing circuits with feedback and explaining the behavior of transistor amplifiers at high frequencies where parasitic feedback elements often have a significant impact. It is also useful in designing feedback amplifiers and other circuits, such as negative resistance oscillators.

5.7 TAKEAWAYS

1. Any two-port circuit block can be fully described by a set of *immittance parameters*, that is ratios of signal voltages and currents at the external ports, provided that the two-port in question is linear and contains no independent current or voltage sources.
2. The four most common immittance parameters are the admittance or Y-parameters, the impedance or Z-parameters, the hybrid or h-parameters, and the $ABCD$ or "chain" parameters. Each set of two-port parameters defines two simultaneous linear equations in the port variables.
3. Although active devices, such as transistors, are inherently nonlinear, they can be considered to be linear at a specific bias point provided that the signal amplitudes being used are small (so-called "small-signal" operation). We can therefore use the two-port immittance approach to characterize transistors and other active devices.
4. Miller's theorem provides a way of transforming a shunt or series feedback immittance into equivalent shunt or series admittances, respectively, at the ports of the two-port.

TUTORIAL PROBLEMS

Problem 5.1 (Active two-port). Determine whether the following two-ports are active or passive:

Two-port 1:

$z_{11} = 1.3 \ \Omega$

$z_{12} = 0.25 \ \Omega$

$z_{21} = -0.82 \ \Omega$

$z_{22} = 1.15 \ \Omega$

Two-port 2:

$h_{11} = 1 \ \text{k}\Omega$

$h_{12} = 3 \times 10^{-4}$

$h_{21} = 250$

$h_{22} = 3 \times 10^{-6} \ \text{S}$

Two-port 3:

$A = -8.309 \times 10^{-2} - j5.7038 \times 10^{-2}$

$B = -23.24 - j6.1948$

$C = 6.173 \times 10^{-4} - j2.4748 \times 10^{-3}$

$D = 3.332 \times 10^{-2} - j3.1278 \times 10^{-1}$

Problem 5.2 (Immittance parameters of a π-network). Calculate the Y-, Z-, and $ABCD$-parameters of the π-network in Figure 5.13.

Problem 5.3 (Y-parameters of a bridged T). Derive the Y-parameter representation for the bridged T-network shown in Figure 5.14.

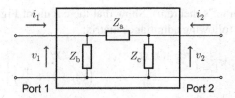

FIGURE 5.13

Pi-network.

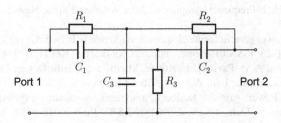

FIGURE 5.14

Bridged T-network.

Problem 5.4 (h-parameters of a cascode). The *cascode* amplifier configuration consists of two transistors cascaded together in CE-CB configuration, as shown in Figure 5.15.

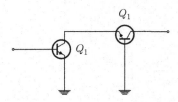

FIGURE 5.15

Bipolar transistor cascode.

If both transistors in Figure 5.15 have the h-parameters given in Table 5.8, calculate the overall h-parameters of the cascode.

Problem 5.5 (Unilateral two-port). A two-port network is said to be unilateral if excitation applied at the output port produces a zero response at the input port. Show that a two-port is unilateral if $AD - BC = 0$.

Problem 5.6 (Miller's theorem). Show that the circuit of Figure 5.11(b) can be characterized by the following admittance matrix:

$$[Y^M] = \begin{bmatrix} Y^a_{11} + Y_f(1 + A_v) & Y^a_{12} \\ Y^a_{21} & \left(Y^a_{22} + Y_f + \frac{Y_f}{A_v}\right) \end{bmatrix} \tag{5.7.1}$$

where A_v is given by equation (5.6.5).

REFERENCES

[1] R. Carson, High Frequency Amplifiers, John Wiley and Sons, New York, 1975, ISBN 0471137057.

[2] S. Mason, Power gain in feedback amplifiers, Trans. IRE Prof. Group Circuit Theory 1 (2) (1954) 20-25, ISSN 0197-6389, http://dx.doi.org/10.1109/TCT.1954.1083579.

[3] G.D. Vendelin, A.M. Pavio, U.L. Rohde, Microwave Circuit Design Using Linear and Nonlinear Techniques, John Wiley and Sons Inc., New York, USA, 1990.

[4] M. Gupta, Power gain in feedback amplifiers: a classic revisited, IEEE Trans. Microw. Theory Tech. 40 (5) (1992) 864-879, ISSN 0018-9480, http://dx.doi.org/10.1109/22.137392.

S-parameters

6

CHAPTER OUTLINE

INTENDED LEARNING OUTCOMES

- *Knowledge*
  - Be aware of the limitations of immittance parameters at higher frequencies.
  - Understand the definition of *S*-parameters as ratios of "power waves" at the ports of a network.
  - Understand the principles of *S*-parameter measurement.
  - Understand the need for calibration to remove error.
  - Understand the various multi-term error models and how they are applied.
- *Skills*
  - Be able to calculate the input reflection coefficient of a two-port network with an arbitrary load termination, and the output reflection coefficient of a two-port network with an arbitrary source termination.
  - Be able to apply *S*-parameters to determine whether a given two-port is "active," "passive," or "lossless."
  - Be able to apply Mason's figure of merit to determine the degree of *Unilaterality* of a network.
  - Be able to calculate the *S*-parameters of two cascaded two-port networks by applying the *T*-parameters.
  - Be able to apply the Signal Flow Graph technique as a means of visualizing and solving *S*-parameters network problems.
  - Be able to, with suitable training, carry out a set of *S*-parameter measurements using a vector network analyzer.

6.1 INTRODUCTION

The use of linear immittance parameters to represent active and passive circuit elements was introduced in Chapter 5. As we increase the frequency of operation, the limitations of immittance parameters, which are ratios of measured voltage and current, become increasingly apparent. The most obvious limitation relates to the fact that immittance parameters are measured with open- and short-circuit terminations, which become increasingly difficult to implement accurately as the frequency range increases. In Chapter 1, we introduced the concept of *reflection coefficient*, being the ratio of reflected voltage wave to incident voltage wave at any discontinuity on a transmission line. It turns out that reflection coefficients are much easier to measure at microwave frequencies than static voltages and currents, so would it not be better if we had a method of characterizing multi-port networks based on some reflection coefficient-like parameters?

In this chapter, we introduce an alternative set of linear two-port parameters which have become the bedrock of microwave active circuit design. *S*-parameters are linear complex coefficients that describe the behavior of a two-port in terms of incident and reflected *power waves* at the network ports. The "S" in *S*-parameters stands for "Scattering" which means that they describe what happens when an incident power wave traveling on a transmission line hits a *discontinuity* [1]. In such cases, a portion of the incident power is transmitted across the discontinuity and the balance of the incident power gets reflected back toward the source, as shown in Figure 2.19. This is true except in two special cases where either (a) all the incident power is absorbed at the discontinuity, which is the case when the terminating impedance is matched to the transmission line system (i.e., there is effectively no discontinuity) or (b) all the incident power is reflected back toward the source, which is the case when the discontinuity is an open circuit or short circuit.

Scattering parameters were first used in work on the theory of transmission lines by Campbell and Foster [2] who defined the scattering matrix in terms of incident and reflected voltage and current waves. In 1960, Penfield introduced the concept of *power waves* for the purposes of discussing noise performance of negative resistance amplifiers [3]. The relationship between power waves and the scattering matrix was then consolidated in 1965 in a landmark paper by Kurokawa [1]. Although the concept of scattering parameters had been around for some time, *S*-parameter design techniques only really started to gain traction in the mid-1960s as improved manufacturing processes extended the operating frequency of bipolar transistors into the hundreds of MHz, and advances in test equipment made it possible to measure *S*-parameters directly at these frequencies. A new circuit analysis and design framework was needed that was compatible with analyzing power flows on transmission lines rather than the static voltages and currents used at lower frequencies. A landmark publication of this period was the February 1967 issue of the Hewlett Packard Journal (Figure 6.1), which contained some of the foundational papers on *S*-parameter design techniques, such as those by Bodway [4] and Froehner [5]. The interested reader will certainly enjoy looking up some of these early articles on the subject to gain a deeper understanding of the historical development of *S*-parameter theory.

6.1.1 LIMITATIONS OF THE IMMITTANCE PARAMETER APPROACH

For the purpose of measurement, *Z*-parameters require open-circuit terminations, *Y*-parameters require short-circuit terminations, and *h*- and *ABCD*-parameters require both types of termination. Since it is easier to fabricate a broadband short-circuit termination than a broad-band open-circuit termination, the *Y*-parameters are preferred at frequencies above about 100 MHz. Even *Y*-parameters, however, become increasingly difficult to measure as the frequency of interest approaches the microwave region (above about 300 MHz). Not only are short- and open-circuit terminations difficult to implement accurately at microwave frequencies, but microwave semiconductor devices can become unstable and may suffer damage if short- or open-circuit terminations are applied to them, as all the RF power emanating from the device will be reflected back into it.

Scattering parameters are based on power measurements made with the active device terminated in the characteristic impedance of the transmission line system

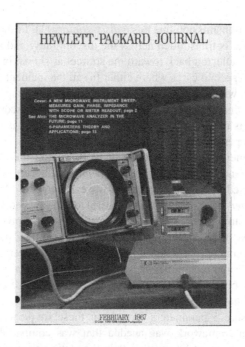

FIGURE 6.1

The front cover of the ground-breaking HP Journal of February 1967 (© Hewlett-Packard Co. 1967. Reproduced with Permission, Courtesy of Keysight Technologies).

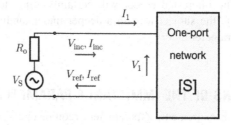

FIGURE 6.2

One-port network with power waves.

and are therefore suited to characterizing devices at frequencies into the millimeter wave region and beyond.

6.1.2 DEFINITION OF THE SCATTERING PARAMETERS

The scattering parameters may be defined in terms of incident and reflected voltages and currents on a transmission line by considering the one-port network shown in Figure 6.2.

In general, the instantaneous port terminal voltage and current are related to the incident and reflected voltage and current components by the following relationships:

$$V_1 = V_{\text{inc}} + V_{\text{ref}} \tag{6.1.1}$$

$$I_1 = I_{\text{inc}} - I_{\text{ref}} \tag{6.1.2}$$

If one considers the case of a transmission line system having a real characteristic impedance, R_0, then the incident and reflected components are related to this characteristic impedance by:

$$\frac{V_{\text{inc}}}{I_{\text{inc}}} = \frac{V_{\text{ref}}}{I_{\text{ref}}} = R_0 \tag{6.1.3}$$

Combining equations (6.1.1)–(6.1.3) yields the following relationships:

$$\left.\begin{aligned} V_{\text{inc}} &= \frac{1}{2}(V_1 + R_0 I_1) \\ V_{\text{ref}} &= \frac{1}{2}(V_1 - R_0 I_1) \end{aligned}\right\} \tag{6.1.4}$$

$$\left.\begin{aligned} I_{\text{inc}} &= \frac{(V_1 + R_0 I_1)}{2R_0} \\ I_{\text{ref}} &= \frac{(V_1 - R_0 I_1)}{2R_0} \end{aligned}\right\} \tag{6.1.5}$$

Either of the pairs of equation (6.1.4) or (6.1.5) are sufficient to fully describe the one-port of Figure 6.2, but these may be replaced by a single pair of equations if we introduce the concept of normalized current and voltage variables [6]. From equations (6.1.4) and (6.1.5), let us define two new normalized variables a and b such that:

$$\frac{V_{\text{inc}}}{\sqrt{R_0}} = I_{\text{inc}}\sqrt{R_0} = a \tag{6.1.6}$$

$$\frac{V_{\text{ref}}}{\sqrt{R_0}} = I_{\text{ref}}\sqrt{R_0} = b \tag{6.1.7}$$

The variables a and b have the dimensions of the square root of power and are known as the *scattering variables*. It is more useful to define these variables in terms of the terminal voltages and currents, V_1 and I_1, so by combining equations (6.1.4)–(6.1.7) we can now write:

$$a = \frac{V_1 + I_1 R_0}{2\sqrt{R_0}} \tag{6.1.8}$$

$$b = \frac{V_1 - I_1 R_0}{2\sqrt{R_0}} \tag{6.1.9}$$

The discussion so far has been based on the case of a real reference impedance, R_0. Penfield [3] extended the definition of the variables a and b by considering the case of a complex reference impedance, Z_0. The variables were thus renamed *power waves* and redefined as follows:

$$a = \frac{V + IZ_0}{2\sqrt{|\text{Re}(Z_0)|}} \tag{6.1.10}$$

$$b = \frac{V - IZ_0}{2\sqrt{|\text{Re}(Z_0)|}} \tag{6.1.11}$$

Let us define a dimensionless ratio "S" for the one-port of Figure 6.2 as:

$$S = \frac{a}{b} \tag{6.1.12}$$

The "S"-parameter in equation (6.1.12) is numerically identical to the reflection coefficient of the one-port in Figure 6.2, making it conceptually easy to understand. Having defined an "S"-parameter for the one-port network, we will now proceed to extend this definition to cover networks with an arbitrary number of ports. Since the subject of this book is active circuit analysis and design, we will focus mainly on the case of one, two, or three port networks, as these are the characterizations most commonly used for active devices, such as diodes and transistors.

The one-port relationship in equation (6.1.12) can be extended to the two-port network shown in Figure 6.3 by replacing a and b by the column vectors $[a]$ and $[b]$. These power wave vectors are related to each other by an S-matrix containing four complex parameters, as follows:

$$\left.\begin{array}{l} b_1 = S_{11}a_1 + S_{12}a_2 \\ b_2 = S_{21}a_1 + S_{22}a_2 \end{array}\right\} \tag{6.1.13}$$

or using matrix notation:

$$\begin{bmatrix} b_1 \\ b_2 \end{bmatrix} = \begin{bmatrix} s_{11} & s_{12} \\ s_{21} & s_{22} \end{bmatrix} \begin{bmatrix} a_1 \\ a_2 \end{bmatrix} \tag{6.1.14}$$

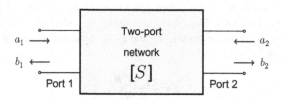

FIGURE 6.3

Two-port network with power waves.

The two-port S-parameters of Figure 6.3 are measured as follows:

$$
\left.
\begin{aligned}
S_{11} &= \left.\frac{b_1}{a_1}\right|_{a_2=0} \\[6pt]
S_{12} &= \left.\frac{b_1}{a_2}\right|_{a_1=0} \\[6pt]
S_{21} &= \left.\frac{b_2}{a_1}\right|_{a_2=0} \\[6pt]
S_{22} &= \left.\frac{b_2}{a_2}\right|_{a_1=0}
\end{aligned}
\right\}
\tag{6.1.15}
$$

In other words, S_{11} is the measured ratio of b_1 to a_1 with a_2 set to zero, and so on.

The S-matrix representation can be extended to describe a network with n pairs of terminals by considering the incident and reflected power waves at each port, as shown in Figure 6.4.

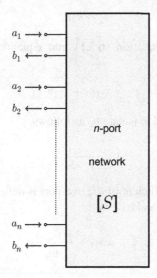

FIGURE 6.4

Generalized n-port network.

In the general case, for any network with n-ports, the power wave variables defined by equations (6.1.10) and (6.1.11) are expressed as nth-order column vectors, $[a]$ and $[b]$, which are related by the $n \times n$ S-matrix as follows:

$$
\begin{bmatrix} b_1 \\ b_2 \\ \vdots \\ b_n \end{bmatrix}
=
\begin{bmatrix}
S_{11} & S_{12} & \cdots & S_{1n} \\
S_{21} & S_{22} & \cdots & S_{2n} \\
\vdots & \vdots & \ddots & \vdots \\
S_{n1} & S_{n2} & \cdots & S_{nn}
\end{bmatrix}
\cdot
\begin{bmatrix} a_1 \\ a_2 \\ \vdots \\ a_n \end{bmatrix}
\tag{6.1.16}
$$

6.2 INPUT AND OUTPUT IMPEDANCE OF A TWO-PORT IN TERMS OF *S*-PARAMETERS

Consider the two-port network driven by a source having reflection coefficient Γ_S, and terminated with an arbitrary load having reflection coefficient Γ_L, as shown in Figure 6.5.

In the general case of a two-port network where $S_{12} \neq 0$ and $S_{21} \neq 0$, the input reflection coefficient Γ_{in} will be dependent on the value of the load and vice versa. Equations for the input and output immittances of a two-port with arbitrary source and load terminations were derived in immittance parameter form in Section 5.3. Similar expressions will now be derived in *S*-parameter form. To derive the input reflection coefficient of the two-port, Γ_{in}, with load Γ_L terminating port 2, we start by stating the following power wave relationship at the load:

$$a_2 = b_2 \Gamma_L \tag{6.2.1}$$

If we now substitute equation (6.2.1) into equation (6.1.13) we obtain the following:

$$b_2 = S_{21} a_1 + S_{22} b_2 \Gamma_L \tag{6.2.2}$$

Which can be rearranged to isolate b_2 as follows:

$$b_2 = \frac{S_{21} a_1}{(1 - S_{22} \Gamma_L)} \tag{6.2.3}$$

The input reflection coefficient of the two-port is defined as $\Gamma_{in} = b_1/a_1$, or, by applying equations (6.1.13) and (6.2.1):

$$\Gamma_{in} = S_{11} + \frac{S_{12} b_2 \Gamma_L}{a_1} \tag{6.2.4}$$

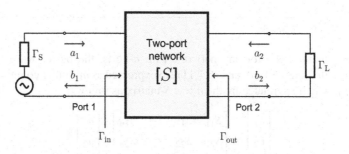

FIGURE 6.5

Two-port network with arbitrary source and load.

Simply combining equations (6.2.3) and (6.2.4) yields the following:

$$\Gamma_{in} = S_{11} + \frac{S_{12}S_{21}\Gamma_L}{1 - S_{22}\Gamma_L} \qquad (6.2.5)$$

Given the definition of $\Delta = S_{11}S_{22} - S_{12}S_{12}$ the reader can easily show that equation (6.2.5) can also be written as:

$$\Gamma_{in} = \frac{S_{11} - \Delta\Gamma_L}{1 - S_{22}\Gamma_L} \qquad (6.2.6)$$

Both forms of this expression (i.e., equations 6.2.5 and 6.2.6) are found in general use.

A similar analysis applied to the output port yields the expression for the output reflection coefficient with an arbitrary source termination:

$$\Gamma_{out} = S_{22} + \frac{S_{12}S_{21}\Gamma_S}{1 - S_{11}\Gamma_S} \qquad (6.2.7)$$

Which can likewise also be written as:

$$\Gamma_{out} = \frac{S_{22} - \Delta\Gamma_S}{1 - S_{11}\Gamma_S} \qquad (6.2.8)$$

These important relationships (equations 6.2.5–6.2.8) will keep popping up again and again throughout this book. We will later show an alternative derivation of them using a signal flow graph technique (Example 6.2).

6.3 CLASSIFICATION OF S-MATRICES

6.3.1 LOSSLESS AND RECIPROCAL NETWORKS

A lossless network is one where the sum of the signal power entering all the network ports is equal to the sum of signal powers leaving all the network ports; that is, no power is being dissipated inside the network itself.

Let P_{in} and P_{out} represent total input and output powers, respectively. These powers are defined, for an n-port network, as:

$$P_{in} = \sum_{i=1}^{n} |a_i|^2 = [a]^{\dagger}[a] \qquad (6.3.1)$$

$$P_{out} = \sum_{i=1}^{n} |b_i|^2 = [b]^{\dagger}[b] \qquad (6.3.2)$$

where $[M]^{\dagger}$ represents the conjugate transpose (Hermitian matrix) of a given matrix M. Given that for a lossless network $P_{in} = P_{out}$ by definition, we combine equations (6.3.1) and (6.3.2) as follows:

$$[a]^\dagger[a] = [b]^\dagger[b] \tag{6.3.3}$$

Noting that $b = [S]a$ we can rewrite equation (6.3.3) as:

$$[a]^\dagger[I][a] = ([S][a])^\dagger([S][a]) \tag{6.3.4}$$

where $[I]$ is the identity matrix. Equation (6.3.4) can be rearranged as:

$$[a]^\dagger[I][a] = [a]^\dagger[S]^\dagger[S][a] \tag{6.3.5}$$

Which implies that:

$$[S]^\dagger[S] = [I] \tag{6.3.6}$$

A consequence of equation (6.3.6) for an $n \times m$ *S*-matrix is that:

$$\sum_{i=1}^{n} |S_{im}|^2 = 1 \tag{6.3.7}$$

for all *m*.

A reciprocal network is a multi-port network in which the power losses are the same between any pair of ports regardless of direction of propagation. In *S*-parameter terms, this means $S_{ij} = S_{ji}$ for any $i \neq j$. Most passive networks such as cables, attenuators, power dividers, and couplers are reciprocal. The only case where a passive device is not reciprocal would be when it contains *anisotropic* materials, which have different electrical properties (such as relative dielectric constant) depending on the direction of signal propagation. One example of an anisotropic material is *ferrite*, which is used in circulators and isolators for the specific purpose of making these nonreciprocal devices.

Transistors and other two-port active devices are typically nonreciprocal, since $S_{12} \neq S_{21}$ in most cases.

An obvious conclusion to the above should be that any reciprocal network has a symmetric *S*-parameter matrix, meaning that all values along the lower-left to upper-right diagonal are equal. It can be shown, in fact, that the *S*-matrix of a reciprocal network must satisfy the condition $[S]^\dagger[S]$. A typical two-port *S*-matrix for a reciprocal network would therefore look like:

$$\begin{bmatrix} S_{11} & S_{12} \\ S_{21} & S_{22} \end{bmatrix} = \begin{bmatrix} 0.8 & 0.5 \\ 0.5 & 0.8 \end{bmatrix}$$

6.3.2 ACTIVITY AND PASSIVITY

The concepts of Activity and Passivity were introduced in terms of immittance parameters in Section 5.4.1. This concept can be extended to the *S*-parameter domain by considering power flows at the ports. The variables *a* and *b* described by equations (6.1.10) and (6.1.11) are known as power waves because they have the dimensions of the square root of power, they are therefore ideally suited to the task of describing power flow between a microwave network and associated generators and loads. Kurokawa [1] has shown by a simple proof that the squared magnitude of the incident power wave, a, is equal to the available power from a linear generator connected to the ith port of the network, that is,

$$P_n = |a_i|^2 \tag{6.3.8}$$

Similarly, the reflected power from the ith port can be written as:

$$P_r = |b_i|^2 \tag{6.3.9}$$

Thus in terms of power wave variables, the power delivered to the one-port network in Figure 6.2 is given by:

$$P = |a_i|^2 - |b_i|^2 \tag{6.3.10}$$

The ratio of reflected to incident power at the device port is known as the power reflection coefficient which is equal to the squared magnitude of the *S*-parameters

$$\frac{|b_i|^2}{|a_i|^2} = |S|^2 \tag{6.3.11}$$

Using the above relationship, equation (6.3.10) for the power flowing into the one-port network can be written in terms of the one-port scattering parameter as follows:

$$P = |a|^2 \left(1 - |S|^2\right) \tag{6.3.12}$$

The relationships discussed so far can be generalized to deal with the case of the n-port network by considering the power flow at each port. The power flowing into the ith port is described by an equation of the same form as equation (6.3.12) for that particular port:

$$P = |a_i|^2 - |b_i|^2 = a_i a_i^* - b_i b_i^* \tag{6.3.13}$$

The total power entering the n-port network from all external sources is therefore given by the sum of the powers entering each port [6].

$$P_{\text{tot}} = \sum_{i=1}^{n} P_i = \sum_{i=1}^{n} (a_i a_i^* - b_i b_i^*) \tag{6.3.14}$$

Equation (6.3.14) can be rewritten in matrix form as follows:

$$P_{\text{tot}} = [a]^\dagger [a] - [b]^\dagger [b] \tag{6.3.15}$$

where, again, $[M]^\dagger$ denotes the conjugate transpose of the matrix M. Employing the fact that $[b] = [S][a]$, equation (6.3.15) can be written as:

$$P_{\text{tot}} = [a]^\dagger [a] - [a]^\dagger [S]^\dagger [S][a]$$

or

$$P_{\text{tot}} = [a]^\dagger \left[[I] - [S]^\dagger [S] \right] [a] \qquad (6.3.16)$$

The matrix $[[I] - [S]^\dagger [S]]$ is known as the *dissipation matrix* and is represented as $[Q]$. The total power flow into the n-port network given by equation (6.3.16) can thus be written in the concise form:

$$P_{\text{tot}} = [a]^\dagger [Q][a] \qquad (6.3.17)$$

For an active network the net flow of signal power *into* the network is negative (i.e., $P_{\text{tot}} < 0$), meaning that there is a net positive flow of signal power out of the network. By contrast, for a passive network the net flow of signal power into the network is positive or zero (i.e., $P_{\text{tot}} \geq 0$).

It is important to distinguish between the everyday use of the word "active" to describe circuits containing "active" devices (transistors, etc.), and the strict definition of activity established using the above criteria. The fact that the network in question may contain transistors or other so-called "active" devices does not define it as being active in this context.

By returning now to a consideration of the dissipation matrix, the activity and passivity criteria can be stated in terms of the S-parameters of the network. For the passive network, equation (6.3.17) becomes:

$$[a]^\dagger [Q][a] \geq 0 \qquad (6.3.18)$$

Equation (6.3.18) means that the dissipation matrix of a passive network must be positive definite (PD) or positive semi-definite (PSD). The necessary and sufficient condition for a Hermitian matrix to be PD is that its determinant and that of its principal minors be non-negative [6].

This can be illustrated by considering the two-port scattering matrix of equation (6.1.13). The dissipation matrix for the two-port network is given by:

$$[Q] = \begin{bmatrix} \left(1 - |S_{11}|^2 - |S_{21}|^2\right) & -\left(S_{11}^* S_{12} + S_{21}^* S_{22}\right) \\ -\left(S_{11} S_{12}^* + S_{21} S_{22}^*\right) & \left(1 - |S_{12}|^2 - |S_{22}|^2\right) \end{bmatrix} \qquad (6.3.19)$$

For the principal minors of the dissipation matrix to be non-negative, the following two conditions must be simultaneously satisfied:

$$1 - |S_{11}|^2 - |S_{21}|^2 \geq 0 \qquad (6.3.20)$$

$$|Q|^2 \geq 0 \qquad (6.3.21)$$

Expanding the determinant of [Q] and simplifying leads to a restatement of equation (6.3.21):

$$\left(1 - |S_{11}|^2 - |S_{12}|^2 - |S_{21}|^2 - |S_{22}|^2\right) + |\Delta|^2 \geq 0 \tag{6.3.22}$$

where

$$\Delta = S_{11}S_{22} - S_{12}S_{21}$$

Replacing equation (6.3.21) by equation (6.3.22), the passivity criteria can be stated thus:

$$|S_{11}|^2 + |S_{21}|^2 \leq 1 \tag{6.3.23}$$

$$\frac{|S_{11}|^2 + |S_{12}|^2 + |S_{21}|^2 + |S_{22}|^2}{1 + |\Delta|^2} \leq 1 \tag{6.3.24}$$

The passivity criteria must be invariant when input and output are interchanged therefore a further condition can be stated based on equation (6.3.23):

$$|S_{22}|^2 + |S_{12}|^2 \leq 1 \tag{6.3.25}$$

If any of the conditions (6.3.23)–(6.3.25) are violated then the device is active. It is possible, however, for condition (6.3.23) or (6.3.25) to be violated even though the magnitudes of the individual scattering parameters are less than unity. Furthermore, condition (6.3.24) can be violated even though the remaining two conditions are satisfied, meaning that this condition is not redundant. This point is best illustrated by a numerical example.

Example 6.1 (Proof of passivity).

Problem. Determine whether the following hypothetical two-port network is active or passive:

$$\begin{bmatrix} S_{11} & S_{12} \\ S_{21} & S_{22} \end{bmatrix} = \begin{bmatrix} 0.5\angle 90 & 0.1\angle -70 \\ 0.8\angle -100 & 0.7\angle 110 \end{bmatrix}$$

Solution. By inspection it can be seen that the magnitudes of all the *S*-parameters are less than unity. At first glance, therefore, we would suspect that this device is passive. Checking conditions (6.3.23), (6.3.25) and (6.3.24), however, we find the following:

$$|S_{11}|^2 + |S_{21}|^2 = 0.89 < 1$$

$$|S_{22}|^2 + |S_{12}|^2 = 0.50 < 1$$

$$\frac{|S_{11}|^2 + |S_{12}|^2 + |S_{21}|^2 + |S_{22}|^2}{1 + |\Delta|^2} = \frac{1.3900}{1.074} = 1.294 > 1$$

Since condition (6.3.24) is violated, therefore, the network would appear to be active.

6.3.3 UNILATERALITY

The reader will by now be familiar with the concept of unilaterality when applied to immittance parameters (Section 5.4.2). This concept can be simply extended to the S-parameter domain. When describing a two-port by its S-matrix, we use the convention that S_{21} is the forward transfer parameter (from port 1 to port 2) and S_{12} is the reverse direction (from port 2 to port 1). With this convention, a unilateral device has the property that $S_{12} = 0$. In other words, a unilateral device will have an infinite reverse isolation.

Perfect unilaterality is not achievable in the real world, but it can sometimes be assumed if the value of S_{21} is sufficiently small. The assumption of unilaterality leads to a great simplification of the design process, particularly for amplifiers, as explained in more detail in Section 13.2.1.

6.4 SIGNAL FLOW GRAPHS

The signal flow graph technique is a useful tool that can help us to visualize and analyze power flow in a microwave network described by S-parameters. A signal flow graph is a pictorial representation of a system of simultaneous equations. The technique was originally developed for use in control theory, but was later applied to S-parameter circuit analysis by the likes of Kuhn [7].

Once a network has been described in terms of a signal flow graph, some basic rules can be applied to analyze and simplify the network. It is often quicker to use the signal flow graph technique than to use algebraic manipulation, although the two approaches yield identical results, as will be demonstrated here.

A signal flow graph is made up of *Nodes* and *Branches*, which are defined as follows:

- *Nodes*:

    1. Each port, i, of a microwave network has two nodes, a_i and b_i.
    2. Node a_i is identified with a wave entering port i, while node b_i is identified with a wave reflected from port i.
    3. The voltage at a node is equal to the sum of all signals entering that node.

- *Branches*:

    1. A branch is a directed path between two nodes, representing signal flow from one node to another.
    2. Every branch has an associated coefficient (i.e., an S-parameter or reflection coefficient).

A signal flow graph representation of power wave flows in a two-port network is shown in Figure 6.6. In this case, a wave of amplitude a_1 incident at port 1 is split and proceeds as follows:

1. A portion of the incident power wave passes through S_{11} and comes out at port 1 as a reflected wave.
2. The remainder of the power wave gets transmitted through S_{21} and emerges from node b_2.
3. If a load with a nonzero reflection coefficient is connected at port 2, the wave emerging from node b_2 will be partly reflected at the load and will re-enter the two-port network at node a_2.
4. Part of the reflected wave entering at node a_2 will be reflected back out of port 2 via S_{22}.
5. The remainder will be transmitted out from port 1 through S_{12}.

It is important not to confuse a signal flow graph with a circuit diagram. The nodes a_1, a_2, b_1, and b_2 in Figure 6.6 are not the same as physical nodes in a circuit diagram, but rather represent voltage waves at the input and output ports of the two-port network. The branches in Figure 6.6 represent the signals flowing between the various nodes and are weighted by complex coefficients (the S-parameters) that represent how the signal is changed in magnitude and phase when moving from one specific node to another.

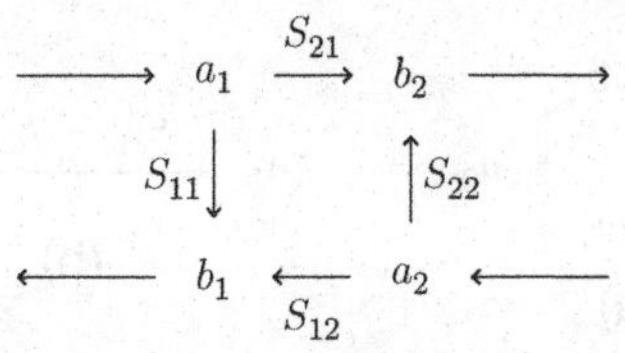

FIGURE 6.6

Signal flow graph representation of a two-port network.

6.4.1 DECOMPOSITION OF SIGNAL FLOW GRAPHS

By applying some simple rules, we can use signal flow graph theory to simplify many S-parameter network problems. The four key rules of signal flow analysis are summarized as follows.

Rule 1 (series rule)

Two branches, whose common node has only one incoming and one outgoing wave (branches in series), may be combined to form a single branch whose coefficient is the product of the coefficients of the original branches. Figure 6.7 shows this rule. The signal flow graph in Figure 6.7(b) is equivalent to Figure 6.7(a).

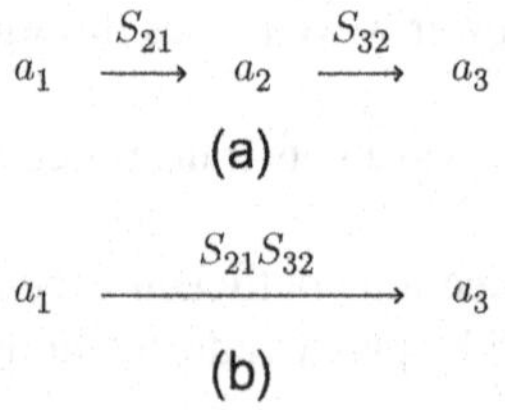

$$a_1 \xrightarrow{\quad S_{21} \quad} a_2 \xrightarrow{\quad S_{32} \quad} a_3$$

(a)

$$a_1 \xrightarrow{\quad S_{21}S_{32} \quad} a_3$$

(b)

FIGURE 6.7

Signal flow graph: Rule 1. (a) Network before application of Rule 1 and (b) network after application of Rule 1.

Rule 2 (parallel rule)

Two branches from one common node to another common node (branches in parallel) may be combined into a single branch whose coefficient is the sum of the coefficients of the original branches. Figure 6.8 shows this rule. The signal flow graph in Figure 6.8(b) is equivalent to Figure 6.8(a).

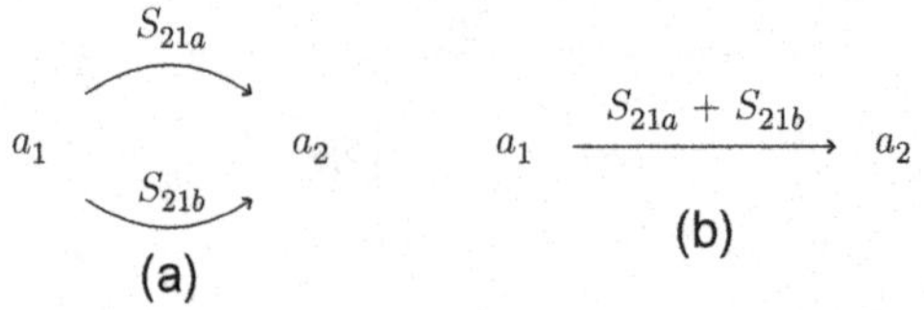

(a)   (b)

FIGURE 6.8

Signal flow graph: Rule 2. (a) Network before application of Rule 2 and (b) network after application of Rule 2.

Rule 3 (self-loop rule)

When a node has a self-loop (a branch that begins and ends at the same node) of coefficient S_{xx}, the self-loop can be eliminated by multiplying coefficients of the branches feeding that node by $1/(1 - S_{xx})$. This can be proven with reference to Figure 6.9 as follows:

$$a_2 = S_{21}a_1 + S_{22}a_2$$

$$a_2(1 - S_{22}) = S_{21}a_1$$

$$\therefore a_2 = \frac{S_{21}}{(1 - S_{22})}a_1$$

$$a_2 = S_{21}a_1 + S_{22}a_2$$

$$a_2(1 - S_{22}) = S_{21}a_1$$

$$\therefore a_2 = \frac{S_{21}}{(1-S_{22})}a_1$$

(a)

(b)

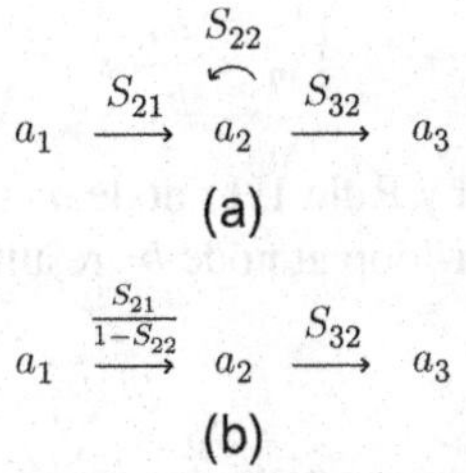

FIGURE 6.9

Signal flow graph: Rule 3. (a) Network before application of Rule 3 and (b) network after application of Rule 3.

Rule 4 (splitting rule)

A node may be split into two separate nodes as long as the resulting flow graph contains, once and only once, each combination of separate (not self-loops) input and output branches that connect to the original node. Figure 6.10 shows this rule. The signal flow graph in Figure 6.10(b) is equivalent to Figure 6.10(a).

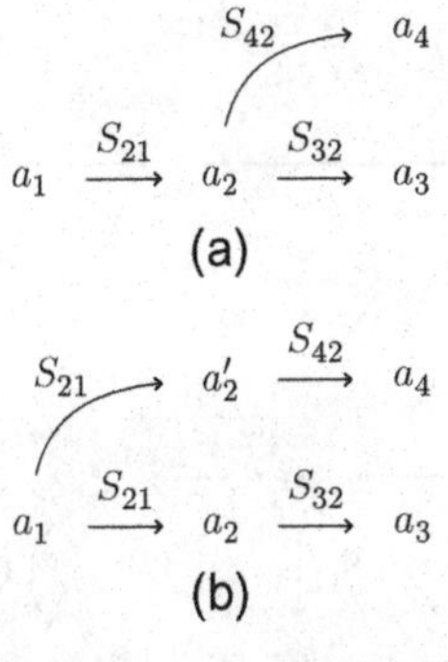

(a)

(b)

FIGURE 6.10

Signal flow graph: Rule 4. (a) Network before application of Rule 4 and (b) network after application of Rule 4.

Example 6.2 (Signal flow analysis of a two-port with load).

Problem. Determine the input reflection coefficient of the two-port with load shown in Figure 6.5, using signal flow graphs.

Solution. We start by drawing the signal flow graph representation of Figure 6.5 as shown in Figure 6.11.

We need to decompose the signal flow graph of Figure 6.11 to determine:

$$\Gamma_{in} = \frac{a_1}{b_1}$$

Applying Rule 4, followed by Rule 1, to node a_2 results in Figure 6.12.

Applying Rule 3 for the self-loop at node b_2 results in Figure 6.13.

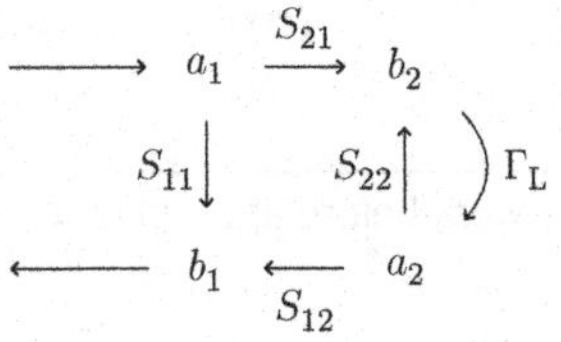

FIGURE 6.11

Signal flow graph representation of a two-port network.

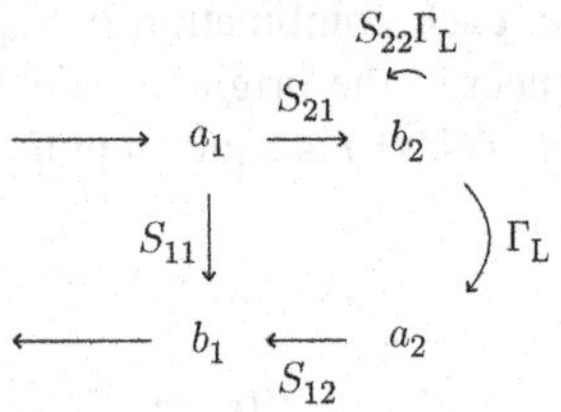

FIGURE 6.12

Rules 4 and 1 applied to Figure 6.11.

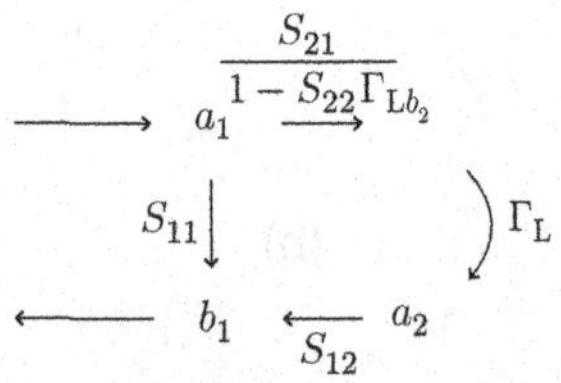

FIGURE 6.13

Rule 3 applied to Figure 6.12.

Now, if we apply Rules 1 and 2 to Figure 6.13 we get the following result:

$$\Gamma_{in} = S_{11} + \frac{S_{12}S_{21}\Gamma_L}{1 - S_{22}\Gamma_L} \qquad (6.2.5)$$

The reader may notice that this is the same equation (6.2.5) we obtained algebraically in Section 6.2.

As an exercise, the reader may wish to apply the signal flow technique in a similar way to prove equation (6.2.7) for the output reflection coefficient.

Mason's rule

Although a bit more complicated than the four rules of signal flow graph analysis so far presented, Mason's rule, or "Mason's nontouching loop rule," to give it its full name, is a useful short-cut that allows us to calculate the transfer function for a signal flow graph by "inspection." Although Mason's rule requires a bit more effort to understand initially, it saves time in network analysis once it is correctly understood and applied.

The rule will be explained with reference to Figure 6.14, which is basically the same two-port with load that was used in Figure 6.11, but with the source reflection coefficient, Γ_S, included.

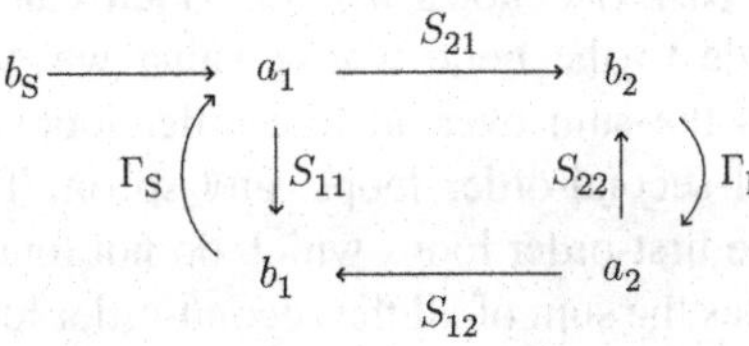

FIGURE 6.14

Two-port network with load (illustration of Mason's rule).

The network of Figure 6.14 has basically only one independent variable, namely b_S, the power wave emanating from the source. We can analyze Figure 6.14 in terms of *paths* and *loops*, which are defined as follows:

- A *path* is a series of directed lines followed in sequence and in the same direction in such a way that no node is touched more than once. The value of the path is the product of all the coefficients encountered in the process of traversing the path. There are three paths in Figure 6.14, namely:
  1. The path from b_S to b_2 having the value S_{21}.
  2. The path from b_S to b_1 having the value S_{11}.
  3. The path from b_S to b_1 having the value $S_{21}\Gamma_L S_{12}$.

- *A first-order loop* is a series of directed lines coming to a closure when followed in sequence and in the same direction with no node passed more than once. The value of the loop is the product of all coefficients encountered in the process of traversing the loop. There are three first-order loops in Figure 6.14, namely:
  1. The loop a_1-b_1 having the value $\Gamma_S S_{11}$.
  2. The loop b_2-a_2 having the value $S_{22}\Gamma_S$.
  3. The loop a_1-b_2-a_2-b_1 having the value $S_{21}\Gamma_L S_{12}\Gamma_S$.
- *A second-order loop* is the product of any two first-order loops which do not touch at any point. There is one second-order loop in Figure 6.14, namely $\Gamma_S S_{11} S_{22}\Gamma_L$.
- *A third-order loop* is the product of any three first-order loops which do not touch. There are no third-order loops in Figure 6.14.
- *An nth-order loop*, in general, is the product of any n first-order loops which do not touch.

The rule can be expressed symbolically as shown in equation (6.4.1).

$$T = \frac{P_1[1 - \sum \mathcal{L}(1)^{(1)} + \sum \mathcal{L}(2)^{(1)} - \cdots] + P_2[1 - \sum \mathcal{L}(1)^{(2)} + \cdots]}{1 - \sum \mathcal{L}(1) + \sum \mathcal{L}(2) - \sum \mathcal{L}(3) + \cdots} \qquad (6.4.1)$$

Each P_i in equation (6.4.1) denotes a path which can be followed from the independent variable node to the node whose value we wish to determine. The notation $\sum \mathcal{L}(1)$ denotes the sum over all first-order loops. The notation $\sum \mathcal{L}(2)$ denotes the sum over all second-order loops, and so on. The notation $\sum \mathcal{L}(1)^{(1)}$ denotes the sum of all the first-order loops which do not touch P_1 at any point. The notation $\sum \mathcal{L}(2)^{(1)}$ denotes the sum of all the second-order loops which do not touch P_1 at any point, and so on. In both these cases, the superscript (1) denotes path 1. By extension the superscript (2) denotes path 2, etc.

To avoid confusion in the application of Mason's rule, it helps to list out all the paths and *n*th-order loops for the circuit in question. We can then apply these to equation (6.4.1) systematically. Let us say, for example, we wanted to determine the transfer function $T = b_2/b_S$ for the circuit of Figure 6.14. We would therefore compile the following list.

- *Paths*: In this example, there is only one path from b_S to b_2, which we shall call P_1. The value of this path is S_{21}.
- *First-order loops*: We can list these as follows:
  1. $\Gamma_S S_{11}$.
  2. $S_{22}\Gamma_S$.
  3. $S_{21}\Gamma_L S_{12}\Gamma_S$.
- *Second-order loops*: These are the product of two nontouching first-order loops. For instance, since loops $S_{11}\Gamma_S$ and $S_{22}\Gamma_L$ do not touch, their product is the one and only second-order loop, that is, $S_{11}\Gamma_S S_{22}\Gamma_L$

We can now write the numerator of equation (6.4.1) as:

$$P_1\left[1 - \sum \mathcal{L}(1)^{(1)} + \sum \mathcal{L}(2)^{(1)} - \cdots\right] + P_2\left[1 - \sum \mathcal{L}(1)^{(2)} + \cdots\right] = S_{21}(1 - 0)$$

(6.4.2)

and the denominator of equation (6.4.1) as:

$$1 - \sum \mathcal{L}(1) + \sum \mathcal{L}(2) = 1 - \Gamma_S S_{11} - S_{22}\Gamma_S - S_{21}\Gamma_L S_{12}\Gamma_S + S_{11}\Gamma_S S_{22}\Gamma_L \quad (6.4.3)$$

Now applying equation (6.4.1), we can write the transfer function $T = b_2/b_S$ as:

$$T = \frac{S_{21}}{1 - \Gamma_S S_{11} - S_{22}\Gamma_L - S_{21}\Gamma_L S_{12}\Gamma_S + S_{11}\Gamma_S S_{22}\Gamma_L} \quad (6.4.4)$$

The denominator of equation (6.4.4) is in the form of $1 - x - y + xy$ which can be factored as $(1 - x)(1 - y)$. We can therefore rewrite equation (6.4.4) as:

$$T = \frac{S_{21}}{(1 - \Gamma_S S_{11})(1 - S_{22}\Gamma_L) - S_{21}\Gamma_L S_{12}\Gamma_S} \quad (6.4.5)$$

6.5 SCATTERING TRANSFER PARAMETERS

It is useful to be able to calculate the S-parameters of two or more two-port networks in cascade. If we rearrange the power wave expression to make the input power waves the independent variables, we get a set of two-port parameters which are referred to as the *transfer scattering parameters* or *T-parameters*.

Referring to the two-port shown in Figure 6.3, if we choose b_1 and a_1 as the independent variables and b_2 and a_2 as the dependent variables we have:

$$\left.\begin{array}{l} a_1 = T_{11}b_2 + T_{12}a_2 \\ b_1 = T_{21}b_2 + T_{22}a_2 \end{array}\right\} \quad (6.5.1)$$

or in matrix notation:

$$\begin{bmatrix} a_1 \\ b_1 \end{bmatrix} = \begin{bmatrix} T_{11} & T_{12} \\ T_{21} & T_{22} \end{bmatrix} \begin{bmatrix} b_2 \\ a_2 \end{bmatrix} \quad (6.5.2)$$

where T_{ij} are known as the *scattering transfer parameters* or T-parameters [8]. Equation (6.5.2) relates the power waves at one-port to the power waves at the other port. The relationship between the S-parameters and the T-parameters may be found by comparing equations (6.5.2) and (6.1.16) which results in [9]:

$$T_{11} = \frac{S_{12}S_{21} - S_{11}S_{22}}{S_{21}} = \frac{-\Delta}{S_{21}}$$

$$T_{12} = \frac{S_{11}}{S_{21}}$$

$$T_{21} = -\frac{S_{22}}{S_{21}}$$

$$T_{22} = \frac{1}{S_{21}}$$

(6.5.3)

Rearrangement of equation (6.5.3) gives the *S*-parameters in terms of the *T*-parameters:

$$S_{11} = \frac{T_{12}}{T_{22}}$$

$$S_{12} = \frac{T_{11}T_{22} - T_{12}T_{21}}{T_{22}} = \frac{\Delta_T}{T_{22}}$$

$$S_{21} = \frac{1}{T_{22}}$$

$$S_{22} = -\frac{T_{21}}{T_{22}}$$

(6.5.4)

where

$$\Delta_T = T_{11}T_{22} - T_{12}T_{21} \tag{6.5.5}$$

Note that an alternative definition of the *T*-parameters is also sometimes used [10,11]. The primary application of *T*-parameters, as just mentioned, is in calculating the overall parameters of two cascaded two-ports, as shown in Figure 6.15. The *T*-parameters can therefore be considered as the power wave equivalent of the *ABCD* immittance parameters introduced in Section 5.1.4.

For two-port "a" in Figure 6.15 we have the following relationship:

$$\begin{bmatrix} a_1^a \\ b_1^a \end{bmatrix} = \begin{bmatrix} T_{11}^a & T_{12}^a \\ T_{21}^a & T_{22}^a \end{bmatrix} \begin{bmatrix} b_2^a \\ a_2^a \end{bmatrix} \tag{6.5.6}$$

Similarly, for two-port "b" we have:

$$\begin{bmatrix} a_1^b \\ b_1^b \end{bmatrix} = \begin{bmatrix} T_{11}^b & T_{12}^b \\ T_{21}^b & T_{22}^b \end{bmatrix} \begin{bmatrix} b_2^b \\ a_2^b \end{bmatrix} \tag{6.5.7}$$

But from Figure 6.15 we can see:

$$\begin{bmatrix} a_1^b \\ b_1^b \end{bmatrix} = \begin{bmatrix} b_2^a \\ a_2^a \end{bmatrix} \tag{6.5.8}$$

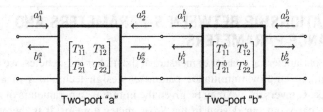

FIGURE 6.15

Cascaded two-ports represented by T-parameters.

Therefore, we can combine equations (6.5.6) and (6.5.7) to obtain:

$$
\begin{bmatrix} a_1^a \\ b_1^a \end{bmatrix} = \begin{bmatrix} T_{11}^a & T_{12}^a \\ T_{21}^a & T_{22}^a \end{bmatrix} \begin{bmatrix} T_{11}^b & T_{12}^b \\ T_{21}^b & T_{22}^b \end{bmatrix} \begin{bmatrix} b_2^b \\ a_2^b \end{bmatrix}
\tag{6.5.9}
$$

or

$$
\begin{bmatrix} a_1^a \\ b_1^a \end{bmatrix} = \begin{bmatrix} T_{11}^{ab} & T_{12}^{ab} \\ T_{21}^{ab} & T_{22}^{ab} \end{bmatrix} \begin{bmatrix} b_2^b \\ a_2^b \end{bmatrix}
\tag{6.5.10}
$$

where

$$
\begin{bmatrix} T_{11}^{ab} & T_{12}^{ab} \\ T_{21}^{ab} & T_{22}^{ab} \end{bmatrix} = \begin{bmatrix} T_{11}^a & T_{12}^a \\ T_{21}^a & T_{22}^a \end{bmatrix} \begin{bmatrix} T_{11}^b & T_{12}^b \\ T_{21}^b & T_{22}^b \end{bmatrix}
\tag{6.5.11}
$$

So, the T-matrix of the cascade can be obtained by multiplying the T-matrices of the individual two-ports in the order in which they are cascaded. For the case of the two cascaded two-ports in Figure 6.15 we then have, from equation (6.5.11):

$$
\left.
\begin{aligned}
T_{11}^{ab} &= T_{11}^a T_{11}^b + T_{12}^a T_{21}^b \\
T_{12}^{ab} &= T_{11}^a T_{12}^b + T_{12}^a T_{22}^b \\
T_{21}^{ab} &= T_{21}^a T_{11}^b + T_{22}^a T_{21}^b \\
T_{22}^{ab} &= T_{21}^a T_{12}^b + T_{22}^a T_{22}^b
\end{aligned}
\right\}
\tag{6.5.12}
$$

Once the overall T-matrix has been obtained by multiplication the S-parameters of the cascade can be calculated by applying equation (6.5.4).

In the special case of a cascade of two two-port networks, characterized by the two-port matrices $[S^a]$ and $[S^b]$, respectively, as shown in Figure 6.15, we can write the overall S-matrix of the cascade as:

$$
\begin{bmatrix} S_{11}^{ab} & S_{12}^{ab} \\ S_{21}^{ab} & S_{22}^{ab} \end{bmatrix} = \frac{1}{1 - S_{22}^a S_{11}^b} \begin{bmatrix} (S_{11}^a - S_{11}^b \Delta^a) & S_{12}^a S_{12}^b \\ S_{21}^b S_{21}^a & (S_{22}^b - S_{22}^a \Delta^b) \end{bmatrix}
\tag{6.5.13}
$$

where $\Delta^a = S_{11}^a S_{22}^a - S_{12}^a S_{21}^a$ and $\Delta^b = S_{11}^b S_{22}^b - S_{12}^b S_{21}^b$.

6.6 RELATIONSHIP BETWEEN *S*-PARAMETERS AND IMMITTANCE PARAMETERS

Although *S*-parameters are easier to measure at the higher frequencies, we sometimes find it useful to employ immittance parameter design techniques in a particular circumstance. Conversely, we may be given the immittance parameters of a transistor and want to carry out our design in the *S*-parameter domain. It is therefore useful to have a means of converting between the two representations. A set of equations relating the *S*-parameters to immittance parameters can derived using straightforward matrix algebra. Taking the Z-parameters as an example, we can write:

$$[V] = [Z][I] \tag{6.6.1}$$

In the case of a two-port network, [Z] can be written as:

$$[Z] = \begin{bmatrix} Z_{11} & Z_{12} \\ Z_{21} & Z_{22} \end{bmatrix} \tag{6.6.2}$$

with the voltage and current vectors being:

$$[V] = \begin{bmatrix} V_1 \\ V_2 \end{bmatrix} \tag{6.6.3}$$

$$[I] = \begin{bmatrix} I_1 \\ I_2 \end{bmatrix} \tag{6.6.4}$$

The instantaneous voltages or currents at the ports is the sum of incident and reflected voltages or currents. We can therefore write equation (6.6.1) in terms of incident and reflected voltage and current matrices as follows:

$$[V^+] + [V^-] = [Z] \left([I^+] - [I^-] \right) \tag{6.6.5}$$

where the respective voltage and current waves are related by:

$$[V^+] = [Z_0][I^+] \tag{6.6.6}$$

and

$$[V^-] = [Z_0][I^-] \tag{6.6.7}$$

Substituting equations (6.6.6) and (6.6.7) into equation (6.6.5) we can write:

$$([Z] + [Z_0]) [I^-] = ([Z] - [Z_0]) [I^+] \tag{6.6.8}$$

The matrix ratios of reflected quantities, $[I^-]$ and $[V^-]$ to the respective incident quantities $[I^+]$ and $[V^+]$ are essentially the *S*-matrix of the two-port. In other words, we can write, in matrix notation:

$$[S] = [V^+]^{-1}[V^-] = [I^+]^{-1}[I^-] \tag{6.6.9}$$

Combining equations (6.6.8) and (6.6.9), we can now define the *S*-matrix purely in terms of [Z] and [Z_o]:

$$[S] = ([Z] + [Z_o])^{-1} ([Z] - [Z_o]) \qquad (6.6.10)$$

In the two-port case, we can expand the matrix relationship equation (6.6.10) into individual expressions for each of the two-port *S*-parameters, as follows:

$$S_{11} = \frac{(Z_{11} - Z_o)(Z_{22} + Z_o) - Z_{12}Z_{21}}{(Z_{11} + Z_o)(Z_{22} + Z_o) - Z_{12}Z_{21}}$$

$$S_{12} = \frac{2Z_{12}Z_o}{(Z_{11} + Z_o)(Z_{22} + Z_o) - Z_{12}Z_{21}}$$

$$S_{21} = \frac{2Z_{21}Z_o}{(Z_{11} + Z_o)(Z_{22} + Z_o) - Z_{12}Z_{21}}$$

$$S_{22} = \frac{(Z_{11} + Z_o)(Z_{22} - Z_o) - Z_{12}Z_{21}}{(Z_{11} + Z_o)(Z_{22} + Z_o) - Z_{12}Z_{21}}$$

A similar matrix analysis to the above could be carried out in terms of the *Y*-parameters, *h*-parameters, and *ABCD*-parameters. The resulting conversion formulae are given in Appendix A.

6.7 MEASUREMENT OF *S*-PARAMETERS

Measurement of device and circuit parameters at microwave frequencies is a specialist area. The reader will be familiar with common test equipment, such as oscilloscopes and spectrum analyzers that are used to measure low-frequency voltages and currents and signal parameters. We will now introduce an instrument that is specific to device and circuit parameter measurements at RF and microwave frequencies, namely the *Network Analyzer*. At our frequencies of interest, the Network Analyzer is specifically and more-or-less exclusively used to measure *S*-parameters of devices and subcircuits.

A microwave network analyzer is a highly complex instrument that combines the most advanced RF electronic circuitry, precision mechanical engineering and sophisticated software. Consequently, although there are a large number of general test equipment manufacturers, there are only a handful of companies manufacturing microwave network analyzers. The largest of these are Keysight Technologies (formerly Hewlett Packard, and later Agilent Technologies Inc.), Rohde and Schwarz and Anritsu (formerly Wiltron). Typical current models of network analyzer from these three manufacturers are shown in Figure 6.16.

If we consider the power wave relationships for a two-port network, as given by equation (6.1.16) we can see that S_{11} and S_{22} are simply the reflection coefficients of the input and output ports with the opposite port terminated with the characteristic impedance. These parameters may therefore be measured by any device that can measure a one-port reflection coefficient, such as the slotted line. On the other

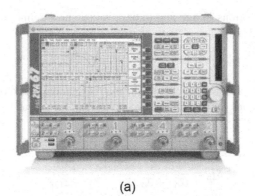

(a)

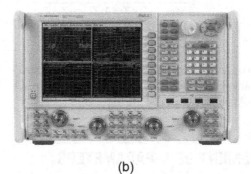

(b)

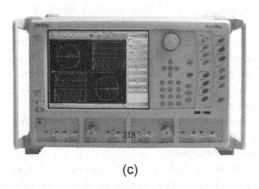

(c)

FIGURE 6.16

Contemporary vector network analyzers (VNAs). (a) Rohde and Schwarz ZVA series VNA
(reproduced by permission of Rohde and Schwarz). (b) Agilent PNA-X, model N5247A
(© Keysight Technologies, Inc. 2014. Reproduced with Permission, Courtesy of Keysight
Technologies). (c) Anritsu VectorStar Vector Network Analyzer (reproduced by permission
of Anritsu EMEA Ltd).

hand, S_{12} and S_{21} represent the insertion loss (or insertion gain) of the network and may be measured using any device that can measure the magnitude and phase of gain/attenuation. All four *S*-parameters can be measured at any spot frequency by a device called a *vector voltmeter*, which is a two-channel RF voltmeter that can measure the phase relationship as well as magnitude ratio between two input signals.

Such traditional manual measurement techniques have long given way to sophisticated, highly automated test equipment that can measure *S*-parameters in both magnitude and phase over a wide frequency range, and can automatically remove the numerous errors implicit in high-frequency measurements.

6.7.1 THE NETWORK ANALYZER

There are basically two categories of network analyzer, namely [12,13]:

- *Scalar network analyzer (SNA)*: measures amplitude ratios only.
- *Vector network analyzer (VNA)*: measures both amplitude and phase ratios.

In general, a VNA is simply referred to as a "network analyzer" by default, as these are the most common type of network analyzer in use today. The term *automatic network analyzer* (ANA) is also used to describe a VNA which is somehow programmable by means of a built in computer, that is to say all of them these days. Finally, Keysight Technologies (formerly Hewlett Packard, and later Agilent Technologies Inc.) now classify their ANA product range into "high-Performance Network Analyzers" (PNA) and "Economy Network Analyzers" (ENA) categories.

An *SNA* can measure only scalar (i.e., magnitude) quantities. It cannot provide any phase information. Although not as capable as a VNA, the SNA is still useful where phase information is not important, such as measuring the gain of an amplifier or the passband of a filter.

Where phase information is required, such as in measuring the *S*-parameters of a transistor, a *VNA* is required. The other main difference between an SNA and VNA is, of course, in terms of cost.

A simplified block diagram of a VNA is shown in Figure 6.17. Because we need to make both magnitude and phase measurements we need to downconvert both the reference and test signals to a lower intermediate frequency where it is possible to make such a phase comparison. With today's generation of VNAs, both the phase and magnitude comparison will be carried out in the digital domain using a powerful digital signal processor. The results can then be displayed on a built-in display in any number of different formats, such as rectangular or polar plots, and stored digitally for later analysis.

In order to allow maximum flexibility in the choice of frequency range, the VNA is often constructed as two separate units which may actually be physically separate boxes. These are the "*S*-parameter Test Set" and the "Display Section" as shown in Figure 6.17. It should be pointed out that Figure 6.17 is an idealized block diagram

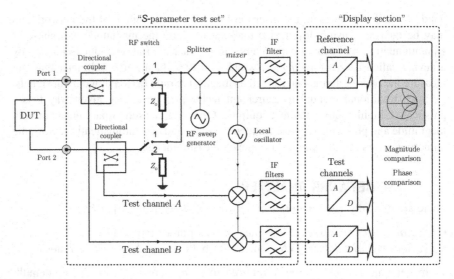

"*S*-parameter test set"   "Display section"

FIGURE 6.17

Vector network analyzer: conceptual block diagram.

which illustrates the basic functional blocks of a conceptual VNA. Real VNAs will differ widely in the details of how these functions are implemented, depending on manufacturer and specification.

With reference to Figure 6.17, we can describe the various functional elements of a conceptual VNA as follows.

The RF sweep generator provides the main stimulus signal that is swept over the RF range of interest. The sweep generator is usually implemented as wide band, high performance, frequency synthesizer under computer control.

The signal provided by the sweep generator is split into two channels by means of an RF power splitter. One channel is fed to one of the external ports via an RF switch and used to drive the device under test (DUT). The other channel is used as a reference channel within the VNA.

The two directional couplers in Figure 6.17 are used to sample the incident and reflected power waves entering and emerging from the DUT. These sample signals constitute two test channels "A" and "B." If the RF switch is in position 1, the DUT will be driven from port 1 of the analyzer. The test signal will pass through the DUT and will appear at port 2 of the VNA. In this case, test channel "A" will represent the transmitted power wave and test channel "B" will represent the reflected power wave. By calculating the ratio of these two test channels with the reference channel, the analyzer can determine the complex values of S_{21} and S_{11}. If the RF switch is now moved to position 2, the DUT will be driven from port 2 of the analyzer. The test signal will pass through the DUT and will appear at port 1. Calculating the same ratios will now yield the complex values of S_{12} and S_{22}.

In order to facilitate the above processing, the reference channel and the two test channels must be downconverted to a suitable IF where they can be digitized. This is achieved by mixing with a local oscillator, with the same local oscillator being used to downconvert all three channels. The IF channels are digitized and a digital signal processor measures the amplitude ratio and phase difference between the reference and test IF channels at each frequency data point. The resulting data can then be displayed in a range of different graphical formats on the built in display.

6.7.2 VNA MEASUREMENT PROCEDURE

In order to measure S_{11} and S_{21}, the RF switch in Figure 6.17 is set to position 1 so that the DUT is driven from port 1. The reflected power wave b_1 is sampled by the port 1 directional coupler and the transmitted power wave b_2 is sampled by the port 2 directional coupler. The reference channel signal represents the incident power wave, a_1. The digital signal processor in the display section calculates the following ratios:

$$S_{11} = \frac{|b_1|}{|a_1|} \angle \Phi_{11} \qquad (6.7.1)$$

$$S_{21} = \frac{|b_2|}{|a_1|} \angle \Phi_{21} \qquad (6.7.2)$$

In order to measure S_{12} and S_{22}, the RF switch is set to position 2. The DUT is then driven from port 2. The reflected power wave b_2 is sampled by the port 2 directional coupler and the transmitted power wave b_1 is sampled by the port 1 directional coupler. This time, the reference channel signal represents the incident power wave, a_2. The digital signal processor in the display section now calculates the following ratios:

$$S_{22} = \frac{|b_2|}{|a_2|} \angle \Phi_{22} \qquad (6.7.3)$$

$$S_{12} = \frac{|b_1|}{|a_2|} \angle \Phi_{12} \qquad (6.7.4)$$

The above switching of the RF switch is under control of the main computer in the display section. This allows the VNA to automatically measure all four parameters without the user having to physically reverse the port connections to the DUT.

6.7.3 NETWORK ANALYZER ERROR CORRECTION

The main sources of error in *S*-parameter measurements using a network analyzer are [14–16]:

(i) Errors due to finite directivity of the couplers (directivity error).
(ii) Errors due to the source not being a perfect Z_o (source mismatch error).

(iii) Errors due to the reference and test channels not having identical amplitude and phase response with frequency (frequency tracking error).

(iv) Crosstalk between reference and test channels.

The error effects (iii) and (iv) are generally grouped together under the term "frequency response error."

In order to take these error effects into account, it is necessary to have a mathematical model which relates the true device characteristics to those measured.

6.7.4 ONE-PORT ERROR MODEL (THE "THREE-TERM" MODEL)

Consider the case where we wish to measure the reflection coefficient of a single impedance or device port. This could, for example, mean the measurement of S_{11} or S_{22} of a two-port network. We can consider all the three sources of error associated with such a one-port measurement to be combined together in the form of an "error two-port" inserted between a perfect (error less) test set and the true value of the load we wish to measure. Consider that we wish to measure the actual value of S_{11} of the DUT, which we shall denote S_{11}^a, whereas the actual measured value we will denote S_{11}^m. What stands between the actual and measured values of S_{11} is the *error two-port*, represented by means of a signal flow graph [7,17] in Figure 6.18.

The relationship between the error terms E_{ij} of Figure 6.18 and the sources of error listed at the beginning of this section are as follows:

- E_{11} = Combined effect of coupler directivity error and connector and cable loss.
- $E_{12} \cdot E_{21}$ = Combined effect of frequency response and cable losses.
- E_{22} = Source mismatch error.

Relationships between input and output reflection coefficients of a two-port with arbitrary terminations have been derived in Section 6.2, as follows:

$$\Gamma_{in} = S_{11} + \frac{S_{21}S_{12}\Gamma_L}{1 - S_{22}\Gamma_L} \tag{6.2.5}$$

$$\Gamma_{out} = S_{22} + \frac{S_{21}S_{12}\Gamma_S}{1 - S_{11}\Gamma_S} \tag{6.2.7}$$

where

Γ_S = source reflection coefficient;
Γ_L = load reflection coefficient;

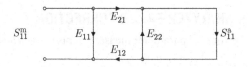

FIGURE 6.18

One-port error model.

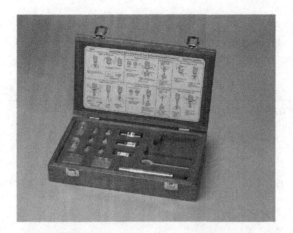

FIGURE 6.19

The Agilent 85052C Precision Calibration Kit (© Keysight Technologies, Inc. 2014.
Reproduced with Permission, Courtesy of Keysight Technologies).

Γ_{in} = Input reflection coefficient of two-port; and
Γ_{out} = Output reflection coefficient of two-port.

Applying equation (6.2.5) to Figure 6.18 we can relate the true value of the load
we are trying to measure (S_{11}^a) to the measured value (S_{11}^m) as follows:

$$S_{11}^m = E_{11} + \frac{S_{11}^a (E_{21} E_{12})}{1 - S_{11}^a E_{22}} \tag{6.7.5}$$

Rearranging equation (6.7.5) we obtain S_{11}^a in terms of S_{11}^m and the error terms as
follows:

$$S_{11}^a = \frac{S_{11}^m - E_{11}}{E_{22}(S_{11}^m - E_{11}) + E_{12}E_{21}} \tag{6.7.6}$$

So, in order to find S_{11}^a we need to somehow determine the values of the error
terms, E_{ij}, in equation (6.7.6). This can be done by measuring S_{11}^m for three loads
of known reflection coefficient. This process is called *calibration* even though it is
carried out every time a measurement is to be made and not once in a while as the
name calibration might suggest. A typical "Cal kit," containing high precision short-
circuit, open-circuit, and matched (Z_o) terminations is shown in Figure 6.19.

The calibration and error correction processes involved in accurate network analyzer measurements used to be a time-consuming manual procedure, but the powerful
computing capabilities built into modern network analyzers mean that the process is
now largely automated. With a modern network analyzer the error terms are automatically calculated and stored inside the instrument and then used to carry out the necessary correction calculations "behind the scenes." Since the error terms E_{12} and E_{21}
always appear as the product $E_{12}E_{21}$, only the product term needs to be determined.

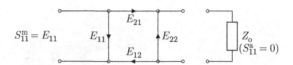

FIGURE 6.20

One-port error model: matched load.

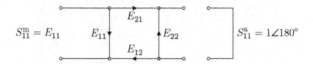

FIGURE 6.21

One-port error model: short-circuit termination.

The calibration procedure is as follows:

1. Measure S_{11}^m for a "perfect" matched load, as shown in Figure 6.20. In this case, $S_{11}^m = 0$ and equation (6.7.5) becomes:

$$S_{11}^m = E_{11} \tag{6.7.7}$$

The computer therefore stores this measured value of S_{11}^m as the error parameter E_{11}.

2. Measure S_{11}^m for a short-circuit termination, as shown in Figure 6.21. In this case, $S_{11a} = 1\angle 180°$ and equation (6.7.5) becomes:

$$S_{11}^m = E_{11} - \frac{(E_{21}E_{12})}{1 + E_{22}} \tag{6.7.8}$$

3. Measure S_{11}^m for an open-circuit termination, as shown in Figure 6.22. In this case, $S_{11a} = 1\angle 0°$ and equation (6.7.5) becomes:

$$S_{11}^m = E_{11} + \frac{(E_{21}E_{12})}{1 - E_{22}} \tag{6.7.9}$$

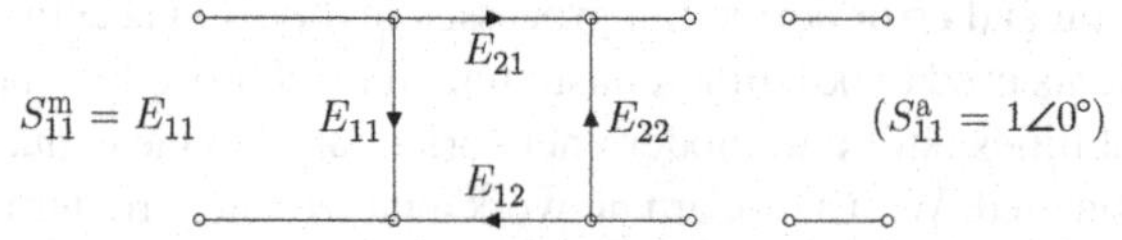

FIGURE 6.22

One-port error model: open-circuit termination.

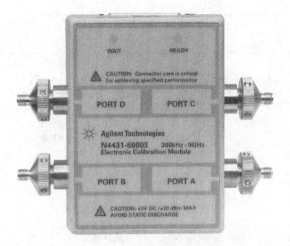

FIGURE 6.23

The Agilent N4431B RF electronic calibration (ECal) module (© Keysight Technologies, Inc. 2014. Reproduced with Permission, Courtesy of Keysight Technologies).

The computer simultaneously solves equations (6.7.8) and (6.7.9) to obtain E_{22} and the product $E_{12} \cdot E_{21}$. These three complex error terms (E_{11}, E_{22}, and $E_{21} \cdot E_{12}$) are stored for each frequency and used to correct each S_{11}^{m} measurement made.

The procedure described above can be carried out manually, with the user physically connecting the various terminations in response to prompts from the network analyzer's calibration program. Alternatively, the calibration process can be carried out using an electronic calibration or "ECAL" module like the one shown in Figure 6.23.

The ECal device contains the three standard terminations, plus a thru line, that can be connected to alternate ports of the VNA under computer control, with minimal operator interaction. This results in faster and more repeatable calibrations.

6.7.5 TWO-PORT ERROR MODEL (THE "12-TERM" MODEL)

When all the *S*-parameters of a two-port network are being measured, a similar error model is used to correct the measured value of S_{22} by determining the error terms associated with port 2. Transmission tracking, leakage, and load match error terms are also incorporated for each direction of signal flow resulting in a 12-term error model as shown in Figure 6.24 (forward direction) and Figure 6.25 (reverse direction).

The directivity error is caused primarily by the coupler leakage or "coupler directivity." This error is also increased by cable and connector match errors between the measurement coupler and the DUT. The reflection and transmission tracking is caused by reflectometer and mixer tracking, as well as cable length imbalance between the measured ports. The leakage error is through the LO path of the mixers.

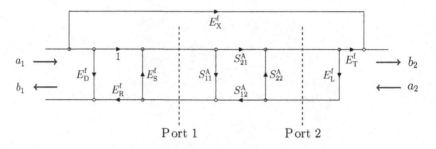

FIGURE 6.24

Two-port error model: forward.

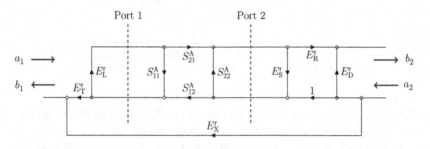

FIGURE 6.25

Two-port error model: reverse.

It is not the leakage of the switch and this model assumes the switch leakage is negligible.

$$E_D^f = \text{forward directivity error;}$$

$$E_S^f = \text{forward source mismatch error;}$$

$$E_R^f = \text{forward reflection tracking error;}$$

$$E_L^f = \text{forward load match error;}$$

$$E_T^f = \text{forward transmission tracking error; and}$$

$$E_X^f = \text{forward isolation error.}$$

$$E_D^r = \text{reverse directivity error;}$$

$$E_S^r = \text{reverse source mismatch error;}$$

$$E_R^r = \text{reverse reflection tracking error;}$$

$$E_L^r = \text{reverse load match error;}$$

E_T^r = reverse transmission tracking error; and

E_X^r = reverse isolation error.

Each actual S-parameter is a function of all four measured S-parameters and may be calculated from the following relationships [18]:

$$S_{11}^a = \frac{\left(\dfrac{S_{11}^m - E_D^f}{E_R^f}\right)\left(1 + \dfrac{S_{22}^m - E_D^r}{E_R^r}E_S^r\right) - E_L^f\left(\dfrac{S_{21}^m - E_X^f}{E_T^f}\right)\left(\dfrac{S_{12}^m - E_X^r}{E_T^r}\right)}{\left(1 + \dfrac{S_{11}^m - E_D^f}{E_R^f}E_S^f\right)\left(1 + \dfrac{S_{22}^m - E_D^r}{E_R^r}E_S^r\right) - E_L^f E_L^r\left(\dfrac{S_{21}^m - E_X^f}{E_T^f}\right)\left(\dfrac{S_{12}^m - E_X^r}{E_T^r}\right)}$$

(6.7.10)

$$S_{21}^a = \frac{\left(\dfrac{S_{21}^m - E_X^f}{E_T^f}\right)\left[1 + \dfrac{S_{22}^m - E_D^r}{E_R^r}\left(E_S^r - E_L^f\right)\right]}{\left(1 + \dfrac{S_{11}^m - E_D^f}{E_R^f}E_S^f\right)\left(1 + \dfrac{S_{22}^m - E_D^r}{E_R^r}E_S^r\right) - E_L^f E_L^r\left(\dfrac{S_{21}^m - E_X^f}{E_T^f}\right)\left(\dfrac{S_{12}^m - E_X^r}{E_T^r}\right)}$$

(6.7.11)

$$S_{12}^a = \frac{\left(\dfrac{S_{12}^m - E_X^r}{E_T^r}\right)\left[1 + \dfrac{S_{11}^m - E_D^f}{E_R^f}\left(E_S^f - E_L^r\right)\right]}{\left(1 + \dfrac{S_{11}^m - E_D^f}{E_R^f}E_S^f\right)\left(1 + \dfrac{S_{22}^m - E_D^r}{E_R^r}E_S^r\right) - E_L^f E_L^r\left(\dfrac{S_{21}^m - E_X^f}{E_T^f}\right)\left(\dfrac{S_{12}^m - E_X^r}{E_T^r}\right)}$$

(6.7.12)

$$S_{22}^a = \frac{\left(\dfrac{S_{22}^m - E_D^r}{E_R^r}\right)\left(1 + \dfrac{S_{11}^m - E_D^f}{E_R^f}E_S^f\right) - E_L^r\left(\dfrac{S_{21}^m - E_X^f}{E_T^f}\right)\left(\dfrac{S_{12}^m - E_X^r}{E_T^r}\right)}{\left(1 + \dfrac{S_{11}^m - E_D^f}{E_R^f}E_S^f\right)\left(1 + \dfrac{S_{22}^m - E_D^r}{E_R^r}E_S^r\right) - E_L^f E_L^r\left(\dfrac{S_{21}^m - E_X^f}{E_T^f}\right)\left(\dfrac{S_{12}^m - E_X^r}{E_T^r}\right)}$$

(6.7.13)

These formidable equations are all computed automatically at each measurement frequency by the modern VNA, based on the forward and reverse error parameters E_N^f and E_N^r obtained in the calibration process. This is an illustration of how modern computing power makes it possible to make swept frequency measurements with an accuracy that would have been previously impractical.

6.8 TAKEAWAYS

1. As frequency increases (above about 500 MHz) immittance parameters become increasingly difficult to measure, as they require short- and open-circuit terminations. S-parameters are measured with the device terminated in the system characteristic impedance, making them easier to measure at microwave and millimeter wave frequencies.

2. *S*-parameters are based on the concept of *power flows* in to and out of the device being measured. *S*-parameters are actually the ratios of something called *power waves* which are quantities having the units of the square root of power.
3. The signal flow graph technique is a useful means of visualizing and solving *S*-parameters network problems.
4. The transfer scattering parameters, or "*T*-parameters," can be used to calculate the overall *S*-parameters of any number of cascaded two-ports.
5. With a defined characteristic impedance, well-known formulae exist to convert *S*-parameters to immittance parameters and vice versa.
6. *S*-parameters are measured using a device called a Vector Network Analyser (VNA). Modern VNAs are powerful enough to carry out all the necessary error compensation automatically, allowing very accurate *S*-parameter measurement over a wide frequency range.

TUTORIAL PROBLEMS

Problem 6.1 (*S*-parameters of shunt/series passive elements). Show that the two-port *S*-matrices of the series and shunt immittances shown in Figure 6.26 are as follows:

1. Shunt admittance (Figure 6.26(a)):

$$[S] = \begin{bmatrix} \Gamma_1 & (1 - \Gamma_1) \\ (1 - \Gamma_1) & \Gamma_1 \end{bmatrix} \tag{6.8.1}$$

where $\Gamma_1 = \frac{-Y_1}{Y_1 + 2Y_o}$.

2. Series impedance (Figure 6.26(b)):

$$[S] = \begin{bmatrix} \Gamma_2 & (1 - \Gamma_2) \\ (1 - \Gamma_2) & \Gamma_2 \end{bmatrix} \tag{6.8.2}$$

where $\Gamma_2 = \frac{Z_2}{Z_2 + 2Z_o}$.

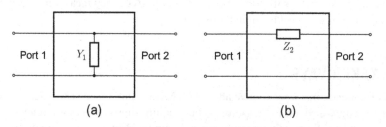

(a)   (b)

FIGURE 6.26

S-parameters of passive networks. (a) Shunt impedance. (b) Series impedance.

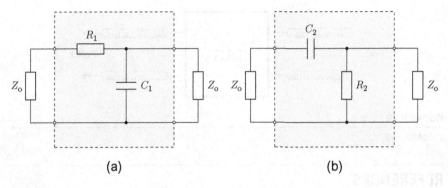

FIGURE 6.27

Low-pass and high-pass two-ports. (a) *L*-section: Type 1. (b) *L*-section: Type 2.

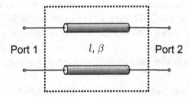

FIGURE 6.28

Problem 6.4.

Problem 6.2 (*S*-parameters of low-pass and high-pass filters). Determine the *S*-parameters of the low-pass and high-pass filters two-port circuits shown in Figure 6.27, in terms of R_1, C_1, R_2, and C_2.

Problem 6.3 (*S*-parameters of a transmission line segment). Determine the two-port *S*-matrix of the lossless transmission line segment shown in Figure 6.28, in terms of the length of the line, l, and its phase constant, β.

Problem 6.4 (De-embedding a two-port). When measuring any two-port DUT, we are often faced with the problem of "de-embedding" the true *S*-parameters of the DUT from the measured *S*-parameters of the device plus test fixture. This is shown in Figure 6.29.

Using the result obtained in Problem 6.3, derive an expression that relates the measured *S*-parameters of the combined network, $S^\dagger$ to the actual *S*-parameters of the DUT.

Problem 6.5 (Lossless, reciprocal two-port). Show that for a lossless, reciprocal two-port network, $|S_{21}|^2 = 1 - |S_{11}|^2$.

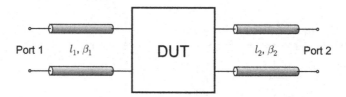

FIGURE 6.29

Problem 6.4.

REFERENCES

[1] K. Kurokawa, Power waves and the scattering matrix, IEEE Trans. Microw. Theory Tech. 13 (6) (1965) 194-202.

[2] G. Campbell, R. Foster, Maximum output networks for telephone substation and repeater circuits, Trans. AIEE 39 (1920) 231-280.

[3] P. Penfield, Noise in negative-resistance amplifiers, IRE Trans. Circuit Theory 7 (2) (1960) 166-170, ISSN 0096-2007, http://dx.doi.org/10.1109/TCT.1960.1086655.

[4] G. Bodway, Two-port power flow analysis using generalised scattering parameters, Microw. J. 10 (1967) 61-69.

[5] W. Froehner, Quick amplifier design with scattering parameters, Electronics 40 (1967) 100.

[6] H. Carlin, A. Giordano, Network Theory: An Introduction to Reciprocal and Non-Reciprocal Circuits, Prentice-Hall Inc., Englewood Cliffs, NJ, 1964.

[7] N. Kuhn, Simplified signal flow graph analysis, Microw. J. 6 (11) (1963) 59-66.

[8] Agilent Technologies Inc, De-Embedding and Embedding *S*-Parameter Networks Using a Vector Network Analyzer, Application Note 1364-1, 2004.

[9] Hewlett Packard, S-parameters Design, Application Note 154, 1972.

[10] R. Ludwig, G. Bogdanov, RF Circuit Design, second ed., Pearson Education Inc., Upper Saddle River, NJ, 2009.

[11] G. Gonzalez, Microwave Transistor Amplifiers, Analysis and Design, second ed., Prentice Hall Inc., Englewood Cliffs, NJ, 1997.

[12] G. Bryant, Institution of Electrical Engineers, Principles of Microwave Measurements, IEE Electrical Measurement Series, P. Peregrinus Limited, 1993, ISBN 9780863412967.

[13] R. Collier, A. Skinner, Microwave Measurements, third ed., IET Electrical Measurement Series, Institution of Engineering and Technology, 2007, ISBN 9780863417351.

[14] S. Adam, Microwave Theory and Applications, Prentice-Hall Inc., Englewood Cliffs, NJ, 1969.

[15] B. Oliver, J. Cage, Electronic Measurements and Instrumentation, Inter-University Electronics Series, McGraw-Hill Inc., New York, 1971.

[16] J. Fitzpatrick, Error models for systems measurement, Microw. J. 22 (5) (1978) 63-66.

[17] G. Bodway, Circuit design and characterisation of transistors by means of three-port scattering parameters, Microw. J. 11 (5) (1968) 4-27.

[18] Agilent Technologies Inc., Applying Error Correction to Network Analyzer Measurements, Application Note 1287-3, 2002.

Gain and stability of active networks

INTENDED LEARNING OUTCOMES

- *Knowledge*
  - Understand the definitions of conditional and unconditional stability and be able to determine the stability of a microwave active device using defined stability criteria.
  - Understand that an active two-port network has the possibility of becoming unstable, and therefore requires different treatment from a passive two-port.
  - Understand the various definitions of power at the input and output ports, and the corresponding definitions of "power gain" using either immittance parameters or *S*-parameters.

- *Skills*
  - Be able to calculate the transducer gain, available gain, operating power gain, and maximum available power gain of any two-port network with given source and load terminations based on immittance parameters or *S*-parameters.
  - Be able to calculate the optimum terminations to achieve simultaneous conjugate matching, and therefore maximum available gain, from an unconditionally stable transistor.
  - Be able to determine whether any two-port network is unconditionally stable or potentially unstable given a set of two-port immittance parameters or *S*-parameters.

7.1 INTRODUCTION

In Chapters 5 and 6, we distinguished between passive and active two-port networks described by immittance parameters and *S*-parameters, respectively. Active networks require more sophisticated analysis because the presence of positive *power gain* raises the issue of *stability*, a concept which does not apply to passive networks.

In this chapter, we will show that certain values of input and output terminations can result in instability, which restricts the range of terminating impedances that can be used. We will derive *stability circles* as a means of mapping areas of stable operation onto a Smith Chart representing the load or source plane. As a shortcut to determining stability without plotting these circles, the concept of *stability criteria* will also be introduced.

The fact that there is more than one definition of "power" at the input and output ports of an active two-port network, means that there are multiple definitions of power gain. In this chapter, we cover the various definitions of power gain, their application and how to calculate them using both immittance parameters and *S*-parameters.

At higher frequencies, we prefer to work with *Y*-parameters, so in the interest of brevity we will restrict our discussion to *Y*-parameter derivations in this chapter. Needless to say, all the following expressions have *Z*- or *h*-parameter equivalents, which can be found elsewhere in the literature.

7.2 POWER GAIN IN TERMS OF IMMITTANCE PARAMETERS
7.2.1 VOLTAGE GAIN OF AN ACTIVE TWO-PORT

As a prelude to discussing power gain, we need to derive an expression for the voltage gain obtainable from an active two-port, which generally depends upon the source and load admittance, as well as the intrinsic parameters of the device itself. Assume that the two-port in Figure 7.1 represents a transistor correctly biased and configured so as to make the two-port *active*, as defined in Chapter 5.

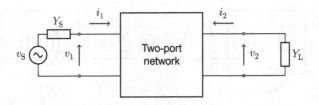

FIGURE 7.1

Two-port power gain definitions.

The voltage gain of the two-port in Figure 7.1 is given by:

$$A_v = \frac{v_2}{v_1} \tag{7.2.1}$$

From Figure 7.1 we can write:

$$i_2 = -v_2 Y_L \tag{7.2.2}$$

But i_2 is also related to v_1 and v_2 via the Y-parameters of the two-port, so we can also write:

$$i_2 = Y_{21} v_1 + v_2 Y_{22} \tag{7.2.3}$$

Solving equations (7.2.2) and (7.2.3) as simultaneous equations and applying equation (7.2.1) gives the small signal voltage gain for the two-port as follows:

$$A_v = \frac{v_2}{v_1} = \frac{-Y_{21}}{Y_L + Y_{22}} \tag{7.2.4}$$

The reader may wonder why the voltage gain given in equation (7.2.4) is dependent only on the load admittance and is independent of the source admittance. This is because we have chosen to define voltage gain in terms of v_1, rather than v_S.

7.2.2 POWER GAIN OF AN ACTIVE TWO-PORT

Since the definition of terminal voltages v_1 and v_2 is unambiguous, there is only one definition of voltage gain, namely equation (7.2.4). In the case of power gain, however, there is more than one possible definition of power at the device terminals and therefore more than one definition of power gain. In particular, we need to distinguish between the *power available* from a "source," and power actually *delivered* to the "load." This applies at both the input and output ports, where the "source" and "load" are appropriately defined.

Depending on whether our focus is on matching at the load side, matching the source side, or attempting to do both simultaneously we end up with a different

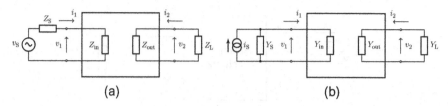

FIGURE 7.2

Two-port power gain definitions. (a) Z-parameters. (b) Y-parameters.

definition of "power gain." In order to clarify this, consider Figure 7.2(a), which shows the input and output impedances of the active two-port in question, as well as source and load impedances. The Norton equivalent of Figure 7.2(a), for Y-parameter analysis, is shown in Figure 7.2(b).

The various available and delivered powers are defined, with reference to Figure 7.2, as follows:

1. *Power absorbed by the amplifier input (P_{in}):*
   This is the power actually delivered to the input port of the amplifier, irrespective of whether the source is conjugately matched to the input port. This is defined, in terms of Z- or Y-parameters as:

Z-Parameters	Y-Parameters
$P_{in} = \|i_1\|^2 \mathrm{Re}(Z_{in})$ (7.2.5)	$P_{in} = \|v_1\|^2 \mathrm{Re}(Y_{in})$ (7.2.6)
$P_{in} = \|v_S\|^2 \dfrac{\mathrm{Re}(Z_{in})}{\|Z_S + Z_{in}\|^2}$ (7.2.7)	$P_{in} = \|i_S\|^2 \cdot \dfrac{\mathrm{Re}(Y_{in})}{\|Y_S + Y_{in}\|^2}$ (7.2.8)

2. *Power available from the generator (P_{AVS}):*
   This is the maximum power available from the source when it is conjugately matched to the input impedance of the amplifier. By setting $Z_S = Z_{in}^*$ in equation (7.2.7) or $Y_S = Y_{in}^*$ in equation (7.2.8) we get:

Z-Parameters	Y-Parameters
$P_{AVS} = \dfrac{\|v_S\|^2}{4\mathrm{Re}(Z_S)}$ (7.2.9)	$P_{AVS} = \dfrac{\|i_S\|^2}{4\mathrm{Re}(Y_S)}$ (7.2.10)

3. *Power absorbed by the load (P_L):*
   This is the power actually delivered to the load, irrespective of whether the load is conjugately matched to the output of the two-port. This is defined as:

Z-Parameters		**Y-Parameters**	
$P_L = \|i_2\|^2 \cdot \mathrm{Re}(Z_L)$	(7.2.11)	$P_L = \|v_2\|^2 \cdot \mathrm{Re}(Y_L)$	(7.2.12)
$P_L = \|v_2\|^2 \cdot \dfrac{\mathrm{Re}(Z_L)}{\|Z_{\mathrm{out}} + Z_L\|^2}$	(7.2.13)	$P_L = \|i_2\|^2 \cdot \dfrac{\mathrm{Re}(Y_L)}{\|Y_{\mathrm{out}} + Y_L\|^2}$	(7.2.14)

4. *Power available from the amplifier output (P_{AVN}):*
   This is the maximum power available from the amplifier output when it is conjugately matched to the load. By setting $Z_S = Z_{\mathrm{in}}^*$ in equation (7.2.9) or $Y_S = Y_{\mathrm{in}}^*$ in equation (7.2.10) we get:

Z-Parameters		**Y-Parameters**	
$P_{AVN} = \dfrac{\|v_2\|^2}{4\mathrm{Re}(Z_{\mathrm{out}})}$	(7.2.15)	$P_{AVN} = \dfrac{\|i_2\|^2}{4\mathrm{Re}(Y_{\mathrm{out}})}$	(7.2.16)

The reader may notice that the available and delivered powers defined above are related as follows:

$$P_{\mathrm{in}} = P_{AVS} \cdot \frac{4\mathrm{Re}(Z_S)\mathrm{Re}(Z_{\mathrm{in}})}{\|Z_S + Z_{\mathrm{in}}\|^2} = P_{AVS} \cdot \frac{4\mathrm{Re}(Y_S)\mathrm{Re}(Y_{\mathrm{in}})}{\|Y_S + Y_{\mathrm{in}}\|^2} \qquad (7.2.17)$$

and

$$P_L = P_{AVN} \cdot \frac{4\mathrm{Re}(Z_{\mathrm{out}})\mathrm{Re}(Z_L)}{\|Z_{\mathrm{out}} + Z_L\|^2} = P_{AVN} \cdot \frac{4\mathrm{Re}(Y_{\mathrm{out}})\mathrm{Re}(Y_L)}{\|Y_{\mathrm{out}} + Y_L\|^2} \qquad (7.2.18)$$

Using the power definitions (equations 7.2.5–7.2.16), we can state the four most common definitions of two-port power gain as follows [1]:

1. *Transducer power gain*:

$$G_T = \frac{P_L}{P_{AVS}} \qquad (7.2.19)$$

2. *Available power gain*:

$$G_A = \frac{P_{AVN}}{P_{AVS}} \qquad (7.2.20)$$

3. *Operating power gain*:

$$G_0 = \frac{P_L}{P_{\mathrm{in}}} \qquad (7.2.21)$$

For the sake of brevity, we will focus exclusively on Y-parameter representation for the subsequent analysis, although all of the expressions derived have their equivalent in Z-parameter (and indeed other immittance parameter) representation.

The transducer gain, G_T, refers to the general case of any arbitrary source and load termination. The expression for transducer gain must therefore contain both source admittance, Y_S, and load admittance, Y_L. Applying equations (7.2.10) and (7.2.12) to equation (7.2.19) we can write the transducer gain as follows:

$$G_T = \frac{P_L}{P_{AVS}} = 4 \cdot \frac{|v_2|^2}{|i_s|^2} \cdot \text{Re}(Y_S)\text{Re}(Y_L) \tag{7.2.22}$$

We now need to replace the term $|v_2|^2/|i_s|^2$ in equation (7.2.22) by a term involving only the Y-parameters. With reference to Figure 7.2(b) we can write the following:

$$v_1 = \frac{i_s}{Y_S + Y_{in}} \tag{7.2.23}$$

By replacing Y_{in} in equation (7.2.23) with equation (5.3.4) we can write:

$$v_1 = \frac{i_s}{Y_S + \left[Y_{11} - \dfrac{Y_{12}Y_{21}}{Y_L + Y_{22}} \right]} \tag{7.2.24}$$

$$= \frac{i_s(Y_L + Y_{22})}{(Y_S + Y_{11})(Y_L + Y_{22}) - Y_{12}Y_{21}} \tag{7.2.25}$$

We note that v_2 is related to v_1 and i_2 by the definition of the Y-parameters in Section 5.1.1, that is:

$$i_2 = Y_{22}v_2 + Y_{21}v_1 \tag{7.2.26}$$

Rearranging equation (7.2.26) gives:

$$v_2 = \frac{i_2 - Y_{21}v_1}{Y_{22}} \tag{7.2.27}$$

From Figure 7.2(b) we can see that i_2 is also equal to $-v_2Y_L$, so we can write:

$$v_2 = \frac{-v_2Y_L - Y_{21}v_1}{Y_{22}} \tag{7.2.28}$$

$$= \frac{-Y_{21}v_1}{Y_{22} + Y_L} \tag{7.2.29}$$

Substituting equation (7.2.24) into equation (7.2.28) gives:

$$\frac{v_2}{i_s} = \frac{-Y_{21}}{(Y_S + Y_{11})(Y_L + Y_{22}) - Y_{12}Y_{21}} \tag{7.2.30}$$

Finally, we substitute equation (7.2.30) back into equation (7.2.22) to get the required expression for transducer gain in terms of Y-parameters as follows:

$$G_T = \frac{4|Y_{21}|^2 \text{Re}(Y_L)\text{Re}(Y_S)}{|Y_{11} + Y_S|^2 |Y_{22} + Y_L|^2} \qquad (7.2.31)$$

We can gain a little more insight into G_T by rearranging equation (7.2.31) as follows:

$$G_T = \frac{2\text{Re}(Y_S)}{|Y_{11} + Y_S|^2} \cdot |Y_{21}|^2 \cdot \frac{2\text{Re}(Y_L)}{|Y_{22} + Y_L|^2} \qquad (7.2.32)$$

From which it is apparent that the transducer gain is comprised of three factors:

(i) An intrinsic gain component: $|Y_{21}|^2$.

(ii) A source mismatch factor: $\dfrac{2\text{Re}(Y_S)}{|Y_{11} + Y_S|^2}$.

(iii) A load mismatch factor: $\dfrac{2\text{Re}(Y_L)}{|Y_{22} + Y_L|^2}$.

The first is the intrinsic gain of the two-port when terminated with the reference impedance that is used to measure the Y-parameters (i.e., a short circuit).

The second factor accounts for the degree of mismatch at the input port. This can be demonstrated by considering what happens as $Y_S \to \infty$, in which case this factor tends to unity. A similar argument applies to the load mismatch factor.

We often wish to optimize the gain obtainable from a given device, which implies conjugate matching at either the input or output ports or both. We should recall from Section 5.3 that, for a nonunilateral device ($Y_{12} \neq 0$), the input admittance of the two-port is a function of the load admittance and the output admittance is a function of the source admittance. This interdependence means that simultaneously conjugately matching both the input and output ports in order to obtain the maximum possible gain is a nontrivial task, but we will come to this in due course.

First, we will proceed to derive the other power gains (equations 7.2.20 and 7.2.21), in terms of Y-parameters starting with the available power gain, G_A, for which we apply equations (7.2.10) and (7.2.16) to equation (7.2.20) to obtain:

$$G_A = \frac{P_{AVN}}{P_{AVS}} = \frac{|i_2|^2}{|i_S|^2} \cdot \frac{\text{Re}(Y_S)}{\text{Re}(Y_{out})} \qquad (7.2.33)$$

We can see from Figure 7.2(b) that $i_S = -v_1(Y_S + Y_{11})$ (note that we use Y_{11}, not Y_{in} because i_2 is defined here as the short-circuit output current, i.e., $Y_L = \infty$ and so $Y_{in} = Y_{22}$). We can therefore write:

$$\frac{i_2}{i_S} = \frac{i_2}{-v_1(Y_S + Y_{11})} = \frac{-Y_{21}}{Y_S + Y_{11}} \qquad (7.2.34)$$

If we now replace i_2/i_S in equation (7.2.33) with equation (7.2.34) we have:

$$G_A = \frac{|Y_{21}|^2}{|Y_S + Y_{11}|^2} \cdot \frac{\text{Re}(Y_S)}{\text{Re}(Y_{out})} \qquad (7.2.35)$$

We now apply a similar reasoning to obtain the operating power gain, G_o, in terms of Y-parameters. By applying equations (7.2.6) and (7.2.12) to equation (7.2.21) we have:

$$G_o = \frac{P_L}{P_{in}} = \frac{|v_2|^2}{|v_1|^2} \cdot \frac{\text{Re}(Y_L)}{\text{Re}(Y_{in})} \qquad (7.2.36)$$

From Figure 7.2(b) we can see that $i_2 = -v_2(Y_L + Y_{22})$. Again, i_1 is defined as the short-circuit input current so $Y_S = \infty$ and $Y_{out} = Y_{11}$. We can therefore write:

$$\frac{v_2}{v_1} = \frac{i_2}{-v_1(Y_L + Y_{22})} = \frac{-Y_{21}}{Y_L + Y_{22}} \qquad (7.2.37)$$

We now replace v_2/v_1 in equation (7.2.36) with equation (7.2.37) to give:

$$G_o = \frac{|Y_{21}|^2}{|Y_L + Y_{22}|^2} \cdot \frac{\text{Re}(Y_L)}{\text{Re}(Y_{in})} \qquad (7.2.38)$$

From the above three gain equations (7.2.31), (7.2.35), and (7.2.38), we can conclude that, for a given two-port, the available power gain, G_A, depends only on the source admittance, Y_S, the operating power gain, G_o, depends only on the load admittance, Y_L, and the transducer power gain, G_T, is a function of both Y_L and Y_S. One way of interpreting this is to say that the available power gain, G_A, is calculated for a chosen value of Y_S on the assumption that the resulting value of Y_{out} will then be conjugately matched by setting $Y_L = Y_{out}^*$. Conversely, the operating power gain, G_o, is calculated for chosen value of Y_L on the assumption that the resulting value of Y_{in} will then be conjugately matched by setting $Y_S = Y_{in}^*$.

Since the input power P_{in} will always be less than or equal to the available source power P_{AVS}, and the available output power P_{AVN} is always greater than or equal to the power actually delivered to the load P_L, we can state the following inequalities relating the three gains:

$$G_T \leq G_o \qquad (7.2.39)$$

and

$$G_T \leq G_A \qquad (7.2.40)$$

It follows from equations (7.2.39) and (7.2.40) that when we maximize G_T, by suitable choice of Y_S and Y_L, then both G_o and G_A will also be maximized. In other words,

$$GT_{\max} = G_{\text{Omax}} = G_{\text{Amax}} = G_{\max} \qquad (7.2.41)$$

where $G_{\max}$ is the unique value of *maximum available gain* (MAG) that is a single figure of merit for a given two-port under conditions of simultaneous conjugate matching at both ports [2,3]. We therefore postulate an admittance pair, Y_{mL} and Y_{mS}, that simultaneously maximizes the transducer gain in equation (7.2.31). Under conditions of simultaneous conjugate matching of a nonunilateral two-port, Y_{mL} and Y_{mS} are related by equations (5.3.4) and (5.3.5) as follows:

$$Y_{\text{mS}}^* = Y_{11} - \frac{Y_{12}Y_{21}}{Y_{\text{mL}} + Y_{22}} \qquad (7.2.42)$$

$$Y_{\text{mL}}^* = Y_{22} - \frac{Y_{12}Y_{21}}{Y_{\text{mS}} + Y_{11}} \qquad (7.2.43)$$

The optimal generator and load admittances could be found by solving the two equations above simultaneously. This approach has been adopted by Carson [4]. Alternatively, one of the gain expressions (equation 7.2.35 or 7.2.38) can be optimized directly by, in the case of equation (7.2.38), finding the value $Y_L = Y_{\text{mL}}$ where $\partial G_{\text{o}}/\partial Y_L$ becomes zero. Y_{mS} is then determined as the complex conjugate of the corresponding two-port input admittance given by equation (7.2.42). This latter approach results in the following equation for $G_{\max}$ [5]:

$$G_{\max} = \frac{|Y_{21}|^2}{2G_{11}G_{22}(1+M) - \text{Re}(Y_{21}Y_{12})} \qquad (7.2.44)$$

where M is given by:

$$M = \sqrt{1 - \frac{\text{Re}(Y_{21}Y_{12})}{G_{11}G_{22}} - \left[\frac{\text{Im}(Y_{21}Y_{12})}{2G_{11}G_{22}}\right]^2} \qquad (7.2.45)$$

where G_{ij} denotes the real part of the Y-parameter Y_{ij}, that is, $Y_{ij} = G_{ij} + jB_{ij}$.

Through a rather lengthy derivation [5] the optimum source and load admittance terminations, Y_{mS} and Y_{mL}, can be shown to be:

$$Y_{\text{mS}} = G_{11}M + j\left[\frac{\text{Im}(Y_{21}Y_{12})}{2G_{22}} - B_{11}\right] \qquad (7.2.46)$$

and

$$Y_{\text{mL}} = G_{22}M + j\left[\frac{\text{Im}(Y_{21}Y_{12})}{2G_{11}} - B_{22}\right] \qquad (7.2.47)$$

For M to be real, the Y-parameters of the device must satisfy the condition:

$$1 - \frac{\text{Re}(Y_{21}Y_{12})}{G_{11}G_{22}} - \left[\frac{\text{Im}(Y_{21}Y_{12})}{2G_{11}G_{22}}\right]^2 > 0 \qquad (7.2.48)$$

Otherwise, the two-port has no finite optimal power gain and it cannot be simultaneously conjugately matched at both ports. In the next section, we will show that a device that satisfies condition (7.2.48) and can therefore be simultaneously conjugately matched at both ports is referred to as being *unconditionally stable*.

Example 7.1 (Simultaneous conjugate matching using Y-parameters).

Problem. Determine the MAG and optimum terminations for a transistor having the following Y-parameters:

$$Y_{11} = (10 + j15) \text{ mS}$$
$$Y_{12} = (-0.015 - j0.145) \text{ mS}$$
$$Y_{21} = (42 - j65) \text{ mS}$$
$$Y_{22} = (0.3 - j1.5) \text{ mS}$$

Solution. We first calculate the quantity M using equation (7.2.45), as follows:

$$M = \sqrt{1 - \frac{\text{Re}(Y_{21}Y_{12})}{G_{11}G_{22}} - \left[\frac{\text{Im}(Y_{21}Y_{12})}{2G_{11}G_{22}}\right]^2}$$

$$= \sqrt{1 - \frac{-10.055}{10 \times 0.3} - \left[\frac{-5.115}{2 \times 10 \times 0.3}\right]^2}$$

$$= 1.8995$$

The MAG can now be calculated using equation (7.2.44), as follows:

$$G_{\text{max}} = \frac{|Y_{21}|^2}{2G_{11}G_{22}(1 + M) - \text{Re}(Y_{21}Y_{12})}$$

$$= \frac{5,989}{2 \times 10 \times 0.3 \times (1 + 1.8995) + 10.055}$$

$$= 218.16 = 23.4 \text{ dB}$$

Finally we calculate the optimum terminations using equations (7.2.46) and (7.2.47), as follows:

$$Y_{\text{mS}} = G_{11}M + j\left[\frac{\text{Im}(Y_{21}Y_{12})}{2G_{22}} - B_{11}\right]$$

$$= 10^{-2} \times 1.8995 + j\left[\frac{-5.115 \times 10^{-6}}{2 \times 0.3 \times 10^{-3}} - 15 \times 10^{-3}\right]$$

$$= (18.995 - j23.525) \text{ mS}$$

and

$$Y_{\text{mL}} = G_{22}M + j\left[\frac{\text{Im}(Y_{21}Y_{12})}{2G_{11}} - B_{22}\right]$$

$$= 0.3 \times 10^{-3} \times 1.8995 + j\left[\frac{-5.115 \times 10^{-6}}{2 \times 10^{-2}} + 1.5 \times 10^{-3}\right]$$

$$= (0.570 - j1.244) \text{ mS}$$

7.3 STABILITY IN TERMS OF IMMITTANCE PARAMETERS

The maximum available gain, G_{max}, defined in the previous section is not always obtainable, as certain combinations of input and output terminations may cause the active device to oscillate. As a rule, we should only seek to apply simultaneous conjugate matching once we have established that the device is *unconditionally stable* which, by definition, means that there are no values of passive load or source termination that will result in a negative conductance at the other port.

In 1952, Llewellyn proposed a set of conditions for unconditional stability of a transistor which formed the basis for much of the later work on two-port stability factors [6]. Llewellyn's stability criterion is that the following three conditions must be simultaneously satisfied:

$$\text{Re}(Y_{11}) \geq 0 \tag{7.3.1}$$

$$\text{Re}(Y_{22}) \geq 0 \tag{7.3.2}$$

$$\text{Re}(Y_{11})\text{Re}(Y_{22}) - M \geq L \tag{7.3.3}$$

where $(Y_{12}Y_{21}) = M + jN$ and $L = |Y_{12}Y_{21}|$. (note that the M defined here is not the same M as defined in equation (7.2.45))

Conditions (7.3.1) and (7.3.2) should be obvious, in that if we see negative resistance/conductance at either port 1 or 2 then the device is potentially unstable. It would be more useful to have a single parameter that can be used to determine unconditional stability. We can derive this by taking equation (5.3.4) and considering the conditions which would make the real part of Y_{in} go negative. So, starting with equation (5.3.4) and applying the definition of M, N, and L above, and adopting the general convention $Y_{nm} = G_{nm} + jB_{nm}$ we have:

$$Y_{in} = G_{11} + jB_{11} - \frac{M + jN}{G_{22} + jB_{22} + G_L + jB_L} \tag{7.3.4}$$

Realizing the denominator of equation (7.3.4) gives:

$$Y_{in} = G_{11} + jB_{11} - \frac{(M + jN)[G_{22} + G_L - j(B_{22} + B_L)]}{(G_{22} + G_L)^2 + (B_{22} + B_L)^2} \tag{7.3.5}$$

We are only interested in the real part of Y_{in} (i.e., G_{in}) so, from equation (7.3.5) we have:

$$G_{in} = G_{11} - \frac{M(G_{22} + G_L) + N(B_{22} + B_L)}{(G_{22} + G_L)^2 + (B_{22} + B_L)^2} \tag{7.3.6}$$

$$G_{in} = \frac{(G_{22} + G_L)^2 + (B_{22} + B_L)^2 - M(G_{22} + G_L)/G_{11} + N(B_{22} + B_L)/G_{11}}{[(G_{22} + G_L)^2 + (B_{22} + B_L)^2]/G_{11}} \tag{7.3.7}$$

The denominator of equation (7.3.7) is always positive if G_{11} is positive (if G_{11} is negative then the device is already potentially unstable), so we need only consider the numerator of equation (7.3.7), which we shall label as P. This can be rearranged as follows:

$$P = \left[G_L + \left(G_{22} - \frac{M}{2G_{11}}\right)\right]^2 + \left[B_L + \left(B_{22} - \frac{N}{2G_{11}}\right)\right]^2 - \frac{M^2 + N^2}{4G_{11}} \qquad (7.3.8)$$

From equation (7.3.8) we can deduce that P can go negative only if the first two terms are smaller than the last term. To minimize the first two terms, we set $G_L = 0$ and $B_L = -\left(B_{22} - \frac{N}{2G_{11}}\right)$. This results in

$$P_{\min} = \left[G_L + \left(G_{22} - \frac{M}{2G_{11}}\right)\right]^2 - \frac{M^2 + N^2}{4G_{11}} \qquad (7.3.9)$$

In order for the input conductance of the two-port to remain positive, $P_{\min}$ must remain positive. The stability criterion, from equation (7.3.9), therefore becomes:

$$\left[G_L + \left(G_{22} - \frac{M}{2G_{11}}\right)\right]^2 - \frac{M^2 + N^2}{4G_{11}} > 0 \qquad (7.3.10)$$

Recalling that $M^2 + N^2 = L^2$ and rearranging equation (7.3.10) we have the following stability criterion:

$$G_{11}G_{22} > \frac{M + L}{2} \qquad (7.3.11)$$

Equation (7.3.11) leads to the so-called *Linvill stability factor*, which was originally specified in terms of h-parameters [7], but can be derived directly from equation (7.3.11) as:

$$C = \frac{L}{2G_{11}G_{22} - M} \qquad (7.3.12)$$

Substituting for M and L, and switching back to the notation originally used by Linvill we have:

$$C = \frac{|Y_{12}Y_{21}|}{2\text{Re}(Y_{11})\text{Re}(Y_{22}) - \text{Re}(Y_{12}Y_{21})} \qquad (7.3.13)$$

The device is unconditionally stable for $0 < C < 1$. The inverse of the Linvill stability factor is called the *Rollett stability factor*, K, which tends to be more commonly used:

$$K = \frac{2\text{Re}(Y_{11})\text{Re}(Y_{22}) - \text{Re}(Y_{12}Y_{21})}{|Y_{12}Y_{21}|} \qquad (7.3.14)$$

It is the Rollett stability factor that the reader is more likely to encounter in the literature, either in immittance parameter form, as in equation (7.3.14), or in S-parameter form, which will be defined in the next section.

A necessary but not sufficient condition for unconditional stability requires that $K > 1$. If $K < 1$, the device is referred to as *potentially unstable* and may oscillate with some combinations of passive source and load immittances. This does not mean that the device cannot be used as an amplifier, it simply means that the source and load terminations must be carefully chosen to lie outside regions of instability. This topic is covered in more detail in the following sections.

7.4 STABILITY IN TERMS OF *S*-PARAMETERS
7.4.1 INTRODUCTION

We introduced the concept of active two-port stability in terms of immittance parameters in the preceding sections. We now proceed to address the problem of stability from the *S*-parameter perspective.

In the preceding section, we defined two categories of two-port stability, namely:

- *Unconditional stability*: A two-port is said to be inherently stable or unconditionally stable at a given frequency if no combination of passive terminations exists which will induce it to oscillate at that frequency.
- *Conditional stability (potential instability)*: A two-port is said to be *potentially unstable* or *conditionally stable* at a given frequency if a combination of passive terminations can be found that will induce sustained oscillation at that frequency.

7.4.2 STABILITY CIRCLES

For potentially unstable devices there will be some combination of source and load terminations that will make the device unstable, meaning that there is a passive value of Γ_L that will result in $|\Gamma_{in}| > 1$ and vice versa. In this section, we will show that the boundaries between stable and unstable regions on the source and load planes are in the form of a circle, known as a *stability circle*. The centers and radii of the stability circles are functions of the device *S*-parameters [8].

Considering Figure 7.3, and looking first at the load reflection coefficient plane, we define values of load termination which cause instability as those which cause the magnitude of the input reflection coefficient of the device to be greater than or equal to unity [9].

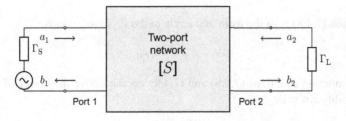

FIGURE 7.3

Two-port network with arbitrary source and load.

The input reflection coefficient of any two-port with the output port terminated with a load Γ_L is given by:

$$\Gamma_{in} = S_{11} + \frac{S_{12}S_{21}\Gamma_L}{1 - S_{22}\Gamma_L} = \frac{S_{11} - \Delta\Gamma_L}{1 - S_{22}\Gamma_L} \qquad (7.4.1)$$

If we set $|\Gamma_{in}| = 1$ in equation (7.4.1) we establish a boundary in the Γ_L plane that separates stable from unstable regions. From equation (7.4.1) we can see that all values of Γ_L that lie on his boundary satisfy the following:

$$|S_{11} - \Delta\Gamma_L| = |1 - S_{22}\Gamma_L| \qquad (7.4.2)$$

In order to simplify the subsequent algebra we define the following variables:

$$C_2 = S_{22} - \Delta S_{11}^* \qquad (7.4.3)$$

$$D_2 = |S_{22}|^2 - |\Delta|^2 \qquad (7.4.4)$$

We now proceed by squaring both sides of equation (7.4.2) and collecting terms, resulting in:

$$|S_{11}|^2 + |\Gamma_L|^2(|\Delta|^2 - |S_{22}|^2) + 2\text{Re}(\Gamma_L)\text{Re}(C_2^*) + 2\text{Im}(\Gamma_L)\text{Im}(C_2^*) = 1 \qquad (7.4.5)$$

Which simplifies to:

$$|S_{11}| + |\Gamma_L|^2(|\Delta|^2 - |S_{22}|^2) + \Gamma_L C_2^* + \Gamma_L^* C_2 = 1 \qquad (7.4.6)$$

Therefore,

$$|\Gamma_L|^2 - \Gamma_L\frac{C_2^*}{D_2^2} - \Gamma_L^*\frac{C_2}{D_2^2} = \frac{|S_{11}|^2 - 1}{D_2^2} \qquad (7.4.7)$$

Adding $|C_2|^2/D_2^2$ to both sides of equation (7.4.7) results in:

$$|\Gamma_L|^2 + \frac{|C_2|^2}{D_2^2} - \Gamma_L\frac{C_2^*}{D_2^2} - \Gamma_L^*\frac{C_2}{D_2^2} = \frac{(|S_{11}|^2 - 1)D_2^2 + |C_2|^2}{D_2^2} \qquad (7.4.8)$$

Equation (7.4.8) is of the form of a circle in the Γ_L plane, that is:

$$|\Gamma_L - C_{SL}|^2 = |\gamma_{SL}|^2 \qquad (7.4.9)$$

By comparing equations (7.4.8) and (7.4.9) we can derive the center of the load plane stability circle as:

$$\boxed{C_{SL} = \frac{C_2^*}{|S_{22}|^2 - |\Delta|^2}} \qquad (7.4.10)$$

and the radius as:

$$\gamma_{SL} = \frac{\sqrt{(|S_{11}|^2 - 1)(|S_{22}|^2 - |\Delta|^2) + |C_2|^2}}{|S_{22}|^2 - |\Delta|^2} \qquad (7.4.11)$$

Equation (7.4.11) may be simplified to yield:

$$\gamma_{SL} = \frac{|S_{12}S_{21}|}{|S_{22}|^2 - |\Delta|^2} \qquad (7.4.12)$$

We can also define source plane stability circles by considering the boundary of load plane stability defined by $|\Gamma_{out}| = 1$ where Γ_{out} is defined by:

$$\Gamma_{out} = S_{22} + \frac{S_{12}S_{21}\Gamma_S}{1 - S_{11}\Gamma_S} = \frac{S_{22} - \Delta\Gamma_S}{1 - S_{11}\Gamma_S} \qquad (7.4.13)$$

Following a similar analysis as above, starting with equation (7.4.13), we can derive the center and radius of the source plane stability circle as follows:

$$C_{SS} = \frac{C_1^*}{|S_{11}|^2 - |\Delta|^2} \qquad (7.4.14)$$

and

$$\gamma_{SS} = \frac{|S_{12}S_{21}|}{|S_{11}|^2 - |\Delta|^2} \qquad (7.4.15)$$

where

$$C_1 = S_{11} - \Delta S_{22}^* \qquad (7.4.16)$$

The stability circles represent the boundaries of permitted terminations in their respective reflection coefficient planes if stable operation is to be assured.

An important question arises as to whether the stable region is represented by the interior or exterior of the stability circle. This may be answered by considering what happens when we set $\Gamma_S = 0$, which, from equation (7.4.13), will result in $|\Gamma_{out}| = |S_{22}|$. We can therefore say that, if $|S_{22}| < 1$ for that particular device, then the origin of the source plane Smith Chart must lie in the stable region. Putting this another way, if $|S_{22}| < 1$ and the source plane stability circle encompasses the origin, then the inside of the stability circle will represent the stable region, as shown in Figure 7.4(a). If $|S_{22}| < 1$ and the source plane stability circle does not encompass the origin, then the inside of the stability circle will represent the unstable region, as shown in Figure 7.4(b). In general, therefore, the necessary condition for the source plane stability circle to encompass the origin is:

$$\gamma_{SS} > |C_{SS}| \qquad (7.4.17)$$

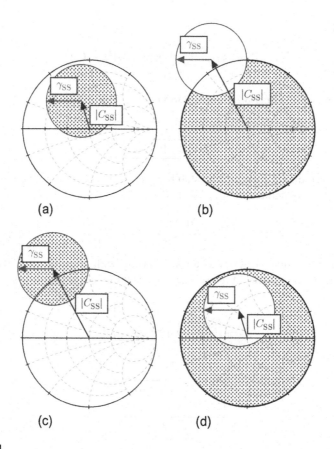

FIGURE 7.4

Source plane stability circles (stable region shaded). (a) $|S_{22}| < 1$, $\gamma_{ss} > |C_{ss}|$. (b) $|S_{22}| < 1$, $\gamma_{ss} < |C_{ss}|$. (c) $|S_{22}| > 1$, $\gamma_{ss} < |C_{ss}|$. (d) $|S_{22}| > 1$, $\gamma_{ss} > |C_{ss}|$.

If $|S_{22}| > 1$, however, the situation just described will be reversed. If $|S_{22}| > 1$ and the source plane stability circle does not encompass the origin then the stable region is represented by the inside of the stability circle. This situation is shown in Figure 7.4(c). Finally, if $|S_{22}| > 1$ and the source plane stability circle does encompass the origin then the stable region will be represented by the source plane Smith Chart area outside of the stability circle, as shown in Figure 7.4(d).

Similar logic applies to the load plane, with similar results, so we will not repeat it here.

The reader may conclude, from the above discussion, that $|S_{11}| > 1$ or $|S_{22}| > 1$ does not necessarily imply that the device is "unstable," only that the device is *conditionally stable* and, with the right choice of terminations, can still be used as an amplifier. This is interesting, as $|S_{11}| > 1$ implies that, with a matched load at the output, the device is actually presenting a negative resistance at its input port (and vice versa for $|S_{22}| > 1$).

Interested readers are referred to an extensive and detailed treatment of the properties of stability circles given by Grivet [10] and a useful description of their application given by Froehner [11].

Example 7.2 (Stability circles).

Problem. Consider that you are required to design a single stage, 1 GHz amplifier using an MRF902 transistor.[1] At 1 GHz with bias conditions of $V_{CE} = 5$ V, $I_c = 5$ mA the device has the following S-parameters:

$$\begin{bmatrix} S_{11} & S_{12} \\ S_{21} & S_{22} \end{bmatrix} = \begin{bmatrix} 0.64\angle -158° & 0.087\angle 28° \\ 4.13\angle 88° & 0.39\angle -68° \end{bmatrix}$$

We need to determine the range of acceptable load and source terminations that can be applied without inducing instability in the device. We can do this by drawing source and load plane stability circles.

Solution. We start by calculating the parameters Δ, C_1, and C_2 as follows:

$$\begin{aligned} \Delta &= S_{11}S_{22} - S_{12}S_{21} \\ &= 0.64 \times 0.39\angle(-158° - 68°) - 0.087 \times 4.13\angle(28° + 88°) \\ &= 0.250\angle -266° - 0.359\angle 116° \\ &= \boxed{0.144\angle -96°} \end{aligned}$$

$$\begin{aligned} C_1 &= S_{11} - S_{22}^*\Delta \\ &= 0.64\angle -158° - (0.39\angle 68° \times 0.144\angle -96°) \\ &= \boxed{0.677\angle -162°} \end{aligned}$$

$$\begin{aligned} C_2 &= S_{22} - S_{11}^*\Delta \\ &= 0.39\angle -68° - (0.64\angle 158° \times 0.144\angle -96°) \\ &= \boxed{0.455\angle -77°} \end{aligned}$$

We can now calculate the source plane stability circle centers and radii using equations (7.4.14) and (7.4.15) as follows:

$$\begin{aligned} C_{SS} &= \frac{C_1^*}{|S_{11}|^2 - |\Delta|^2} \\ &= \frac{0.677\angle 162°}{0.410 - 0.021} \\ &= \boxed{1.740\angle 162°} \end{aligned}$$

---

[1] This example is based on an example originally used in Peter Yip's book [12, p. 108], with kind permission of Dr Yip.

$$\gamma_{SS} = \frac{|S_{12}S_{21}|}{|S_{11}|^2 - |\Delta|^2}$$

$$= \frac{0.359}{0.410 - 0.021}$$

$$= \boxed{0.924}$$

Finally, we calculate the load plane stability circle centers and radii using equations (7.4.10) and (7.4.11) as follows:

$$C_{SL} = \frac{C_2^*}{|S_{22}|^2 - |\Delta|^2}$$

$$= \frac{0.455\angle 77°}{0.152 - 0.021}$$

$$= \boxed{3.473\angle 77°}$$

$$\gamma_{SL} = \frac{|S_{12}S_{21}|}{|S_{22}|^2 - |\Delta|^2}$$

$$= \frac{0.359}{0.152 - 0.021}$$

$$= \boxed{2.740}$$

These stability circles are shown in Figure 7.5.

Since neither of the stability circles in Figure 7.5 encompasses the origin, we conclude that the unstable regions are represented by the interior of the respective circles (i.e., the shaded areas in Figure 7.5). This gives us a wide scope for choosing arbitrary values of Γ_S and Γ_L. We do, however, need to take account of the fact that Γ_{in} is a function of Γ_L and Γ_{out} is a function of Γ_S, according to equations (6.2.5) and (6.2.7), so we need to check that any load termination we choose does not result in a corresponding source termination lying in the source plane unstable area (and vice versa).

One way to proceed with a design would be to select a suitable value of Γ_L to achieve, say, a specified value of operating power gain (the calculation of which is detailed in the next section) and then conjugately match the source to the resulting value of Γ_{in}. We need to determine the value of $\Gamma_S = \Gamma_{in}^*$ corresponding to our chosen load and check that it lies outside the unstable region on the Γ_{in} plane.

Let us say we choose a load value of $\Gamma_{L1} = 0.7\angle 60°$, which is in the stable portion of the load plane. The corresponding value of Γ_{S1} is found by applying equation (6.2.5), that is:

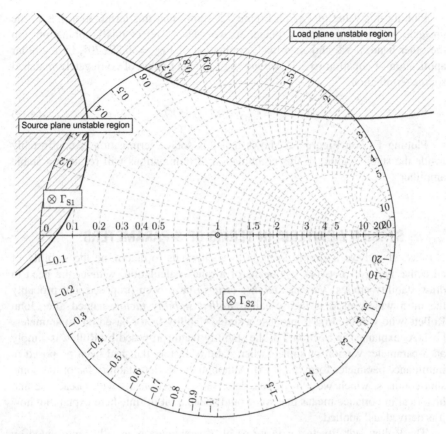

FIGURE 7.5

Source and load plane stability circles for Example 7.2.

$$\Gamma_{in1} = S_{11} + \frac{S_{12}S_{21}\Gamma_{L1}}{1 - S_{22}\Gamma_{L1}}$$

$$= 0.64\angle - 158° + \frac{0.087\angle 28° \times 4.13\angle 88° \times 0.7\angle 60°}{1 - 0.39\angle - 68° \times 0.7\angle 60°}$$

$$= 0.64\angle - 158° + \frac{0.252\angle 176°}{0.731\angle 3.01°}$$

$$= 0.956\angle - 168°$$

Then we have:

$$\Gamma_{S1} = \Gamma_{in1}^* = 0.956\angle 168° \qquad (7.4.18)$$

Plotting this value of Γ_{S1} on Figure 7.5 shows that this termination lies in the unstable region of the source plane, so we need to reject this combination. On the other hand, if we choose a load termination of $\Gamma_{L2} = 0.20\angle70°$, by a similar application of equation (6.2.5), we get the following value of conjugately matched source termination:

$$\Gamma_{S2} = \Gamma_{in2}^* = 0.715\angle159.7 \qquad (7.4.19)$$

Plotting Γ_{S2} on Figure 7.5 shows that this source termination is comfortably inside the stable region, so this combination of terminations will result in a stable amplifier.

7.4.3 **STABILITY CRITERIA IN TERMS OF *S*-PARAMETERS**

It is useful to have a simple formula that allows us to determine the stability of an active two-port network by means of a quick calculation, without the need to draw stability circles. Various *stability criteria* have been proposed, but probably the most widely used is the so-called *Rollett stability factor*, named after John Rollett, who was the first to propose a single stability factor based on *S*-parameters [13]. As explained in Section 7.3, the stability factor proposed by Rollett is simply an *S*-parameter version of some earlier stability factors that had been proposed in immittance parameter form [6,14,15]. Although the Rollett stability factor has some shortcomings, which we will discuss later in this section, its widespread use and historical importance means that we should spend a bit of time here explaining how it is derived and applied.

The Rollett stability factor in terms of *S*-parameters is usually represented by the upper case "*K*," to distinguish it from the immittance parameter version which is usually represented by the lower case "*k*." It is left as an exercise for the reader to demonstrate that these two factors are numerically equal for a given device (see Problem 7.3).

K can be derived in terms of *S*-parameters by considering the stability circles for an unconditionally stable device. Figure 7.6 shows an example of a source stability circle lying entirely outside the unit circle in the Γ_S plane, meaning that there are no terminations in the unstable (shaded) region that can be realized with passive circuit elements.

From Figure 7.6 we can define the source plane stability criterion as:

$$\gamma_{SS} - |C_{SS}| > 1 \qquad (7.4.20)$$

Substituting equations (7.4.10) and (7.4.11) into equation (7.4.20) results in:

$$\frac{|S_{22}^2 - \Delta S_{11}^*| - |S_{21}S_{12}|}{|S_{22}|^2 - |\Delta|^2} > 1 \qquad (7.4.21)$$

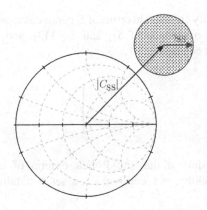

FIGURE 7.6

Stability circle for an unconditional stable device.

Rearranging equation (7.4.21) and squaring both sides:

$$||S_{22}^2 - \Delta S_{11}^*| - |S_{12}S_{21}||^2 > ||S_{22}|^2 - |\Delta|^2|^2 \tag{7.4.22}$$

Expanding equation (7.4.22) results in:

$$2|S_{12}S_{21}||S_{22}^2 - \Delta S_{11}^*| < |S_{22}^2 - \Delta S_{11}^*|^2 + |S_{12}S_{21}|^2 - ||S_{22}|^2 - |\Delta|^2|^2 \tag{7.4.23}$$

We note that the term $|S_{11} - S_{22}\Delta|^2$ can be expressed as:

$$|S_{22}^2 - \Delta S_{11}^*|^2 = |S_{12}S_{21}|^2 + (1 - |S_{11}|^2)(|S_{22}|^2 - |\Delta|^2) \tag{7.4.24}$$

Substituting equation (7.4.24) into equation (7.4.23) yields:

$$(|S_{22}|^2 - |\Delta|^2)^2 \left[\left[(1 - |S_{11}|^2) - (|S_{22}|^2 - |\Delta|^2) \right] - 4|S_{12}S_{21}|^2 \right] > 0 \tag{7.4.25}$$

Which can be simplified to yield the following:

$$|S_{21}S_{12}| < 1 - |S_{11}|^2 - |S_{22}|^2 + |\Delta|^2 \tag{7.4.26}$$

This leads to the definition of the Rollett stability factor as:

$$\boxed{K = \frac{1 - |S_{11}|^2 - |S_{22}|^2 + |\Delta|^2}{2|S_{21}S_{12}|}} \tag{7.4.27}$$

and the Rollett stability criterion, which is:

$$K > 1 \tag{7.4.28}$$

A similar analysis for the load plane yields a result identical to equation (7.4.26).

The complete stability criterion in terms of S-parameters, as published by Rollett, takes into account the magnitude of S_{11} and S_{22} [13], and can be stated as the following simultaneous conditions:

$$\left.\begin{aligned} K &> 1 \\ |S_{11}| &< 1 \\ |S_{22}| &< 1 \end{aligned}\right\} \tag{7.4.29}$$

Woods [16] later showed that the Rollett criteria of equation (7.4.29) are insufficient to prove stability, and proposed a new set of stability criteria, namely:

$$\left.\begin{aligned} K &> 1 \\ &\text{and} \\ |\Delta| &< 1 \end{aligned}\right\} \tag{7.4.30}$$

where $\Delta = S_{11}S_{22} - S_{12}S_{21}$.

Aside from Woods, a number of other researchers have pointed out deficiencies in the basic Rollett Stability Criteria of equation (7.4.29) and have come up with various other combinations of parameters that need to be considered in addition to K. Bodway [17] presented an alternative stability criterion as the following simultaneous conditions:

$$\left.\begin{aligned} K &> 1 \\ B_1 &> 0 \\ B_2 &> 0 \end{aligned}\right\} \tag{7.4.31}$$

where

$$B_1 = 1 + |S_{11}|^2 - |S_{22}|^2 - |\Delta|^2 \tag{7.4.32}$$

and

$$B_2 = 1 - |S_{11}|^2 + |S_{22}|^2 - |\Delta|^2 \tag{7.4.33}$$

Recently, the Rollett stability criteria have come under renewed scrutiny, and a new stability criterion which seems to be growing in popularity is the "Edwards-Sinsky" criterion, after the authors who first published it in a 1992 paper [18].

Edwards and Sinsky employ a geometrical method to arrive at two stability parameters, referred to as μ_1 and μ_2, which are defined as follows:

$$\mu_1 = \frac{1 - |S_{11}|^2}{|S_{22} - \Delta S_{11}^*| + |S_{12}S_{21}|} \tag{7.4.34}$$

$$\mu_2 = \frac{1 - |S_{22}|^2}{|S_{11} - \Delta S_{22}^*| + |S_{12}S_{21}|} \tag{7.4.35}$$

μ_1 measures the radius from the center of the Smith Chart (Z_o) to the nearest unstable point in the output plane. μ_2 does the same for the input plane. In order for the unstable region to be outside the unit circle, the Edwards-Sinsky stability criteria are, therefore,

$$\mu_1 > 1 \tag{7.4.36}$$

and simultaneously

$$\mu_2 > 1 \tag{7.4.37}$$

Edwards and Sinsky showed that these two criteria are equivalent, that is, either $\mu_1 > 1$ or $\mu_2 > 1$ are sufficient to prove that an active two-port network is unconditionally stable.

One advantage of the Edwards-Sinsky stability criteria, equation (7.4.36) or (7.4.37), is that only a single parameter needs to be evaluated. There are no auxiliary conditions, as in the case of the Rollett criteria. In addition, the magnitudes of μ_1 and μ_2 give an indication of the *stability margin*.

Interestingly, for a theoretical topic that would appear to be very mature and settled, transistor stability has produced a number of recent papers that attempt to take a fresh look at the search for a single "sufficient" stability criterion [19–21].

Example 7.3 (Transistor stability).

Problem. Given a transistor with the following S-parameters at a specific frequency and bias point:

$$\begin{bmatrix} S_{11} & S_{12} \\ S_{21} & S_{22} \end{bmatrix} = \begin{bmatrix} 0.535\angle 108 & 0.102\angle 71 \\ 2.761\angle 55 & 0.683\angle -188 \end{bmatrix}$$

Determine whether the device is unconditionally stable or potentially unstable, using the various criteria introduced in this chapter.

Solution. Most of the stability criteria make use of the parameter Δ, so we calculate this parameter first:

$$\begin{aligned}
\Delta &= S_{11}S_{22} - S_{12}S_{21} \\
&= 0.535 \times 0.683\angle(108° - 188°) - 0.102 \times 2.761\angle(71° + 55°) \\
&= 0.365\angle -80° - 0.282\angle 126° \\
&= \boxed{0.631\angle -69°}
\end{aligned}$$

We next calculate the Rollett stability factor, K, as follows:

$$K = \frac{1 - |S_{11}|^2 - |S_{22}|^2 + |\Delta|^2}{2|S_{21}S_{12}|}$$

$$= \frac{1 - 0.535^2 - 0.683^2 + 0.631^2}{2 \times 0.282}$$

$$= \frac{0.247 + 0.398}{0.564}$$

$$= \boxed{1.144}$$

Next we calculate B_1 and B_2 for the Bodway criteria:

$$B_1 = 1 + |S_{11}|^2 - |S_{22}|^2 - |\Delta|^2 \qquad (7.4.38)$$

$$= 1 + 0.286 - 0.466 - 0.398 \qquad (7.4.39)$$

$$= \boxed{0.422} \qquad (7.4.40)$$

and

$$B_2 = 1 - |S_{11}|^2 + |S_{22}|^2 - |\Delta|^2 \qquad (7.4.41)$$

$$= 1 - 0.286 + 0.466 - 0.398 \qquad (7.4.42)$$

$$= \boxed{0.782} \qquad (7.4.43)$$

Finally, we calculate the Edwards-Sinsky parameters as follows:

$$\mu_1 = \frac{1 - |S_{11}|^2}{|S_{22} - \Delta S_{11}^*| + |S_{12}S_{21}|}$$

$$= \frac{0.714}{0.358 + 0.282} = \frac{0.714}{0.640}$$

$$= \boxed{1.116}$$

$$\mu_2 = \frac{1 - |S_{22}|^2}{|S_{11} - \Delta S_{22}^*| + |S_{12}S_{21}|}$$

$$= \frac{0.534}{0.141 + 0.282} = \frac{0.534}{0.423}$$

$$= \boxed{1.262}$$

	Rollett	Woods	Bodway	Edwards-Sinsky
Criterion	$K = 1.144$	$K = 1.144$ $\|\Delta\| = 0.631$	$K = 1.144$ $B_1 = 0.422$ $B_2 = 0.782$	$\mu_1 = 1.116$ $\mu_2 = 1.262$
Conclusion	Unconditionally stable	Unconditionally stable	Unconditionally stable	Unconditionally stable

The above table illustrates the fact that the various stability criteria do not always concur. The current tendency is to take the Edwards-Sinsky criteria as being the most rigorous. The above table would suggest, however, that the Bodway criteria are more conservative. The reader may wish to extend this exercise by attempting to determine whether there are in fact any passive load terminations that will result in $|\Gamma_{in}| > 1$ or a passive source termination that will induce $|\Gamma_{out}| > 1$.

7.5 **POWER GAIN IN TERMS OF *S*-PARAMETERS**

We will now revisit the discussion of power gain we covered in Section 7.2.2, only this time we will approach the topic from the *S*-parameter perspective.

Consider the active two-port of Figure 7.3, terminated with an arbitrary source and load, described by their reflection coefficients, Γ_S and Γ_L.

In Section 7.2.2, we showed, using *Y*-parameters, that the maximum available gain will be found at a unique combination of source and load termination, Y_{mS} and Y_{mL}, via a process known as *simultaneous conjugate matching*. We also showed in Sections 7.3 and 7.4 that simultaneous conjugate matching is only possible if the device is *unconditionally stable*. For conditionally stable devices we have to settle for a gain which is lower than the theoretical maximum in order to have terminating impedances that are in the stable regions on the Smith Chart (as defined by the stability circles). Again, we note that, for a nonunilateral device, matching at the input and output ports are interrelated via equations (6.2.5) and (6.2.7). Conjugately matching the input and output ports of a nonunilateral two-port simultaneously is therefore a nontrivial task.

The various powers available and delivered at the input and output ports of a two-port device were defined in equations (7.2.5) to (7.2.18) in terms of Z- and Y-parameters. The equivalent definitions in terms of the power waves shown in Figure 7.3 are as follows:

1. *Power absorbed by the amplifier input (P_{in}):*
   This is the power actually delivered to the input port of the amplifier, irrespective of whether the source is conjugately matched to the input port. This is defined as:

$$P_{in} = \frac{1}{2}|a_1|^2 - \frac{1}{2}|b_1|^2 = \frac{1}{2}|a_1|^2(1 - |\Gamma_{in}|^2) \tag{7.5.1}$$

2. *Power available from the generator (P_{AVS}):*

This is the maximum power delivered from the source when it is conjugately matched to the input port, that is:

$$P_{AVS} = \frac{1}{2}|b_S|^2 - \frac{1}{2}|a_S|^2 = \frac{1}{2} \cdot \frac{|b_S|^2}{(1 - |\Gamma_S|^2)} \tag{7.5.2}$$

3. *Power absorbed by the load (P_L):*
   This is the power actually delivered to the load, irrespective of whether the load is conjugately matched to the output of the two-port. This is defined as the difference between incident and reflected power at the load, that is:

$$P_L = \frac{1}{2}|b_2|^2 - \frac{1}{2}|a_2|^2 = \frac{1}{2}|b_2|^2(1 - |\Gamma_L|^2) \tag{7.5.3}$$

4. *Power available from the amplifier output (P_{AVN}):*
   This is the maximum power available from the amplifier output when it is conjugately matched to the load:

$$P_{AVN} = \frac{|v_2|^2}{4\mathrm{Re}(Z_{out})} \tag{7.5.4}$$

The transducer power gain, G_T, is the most general case which applies when source and load terminations are arbitrarily chosen (i.e., not necessarily matched to the respective ports). From equation (7.2.19) and employing the power definitions (equations 7.5.2 and 7.5.3) we can write:

$$G_T = \frac{P_L}{P_{AVS}} = \frac{|b_2|^2}{|b_S|^2} \cdot (1 - |\Gamma_L|^2)(1 - |\Gamma_S|^2) \tag{7.5.5}$$

At this point, we employ the signal flow analysis technique that was introduced in Section 6.4 to determine the ratio $|b_2|^2/|b_S|^2$ in equation (7.5.5). Figure 7.7 shows the signal flow within the two-port with load.

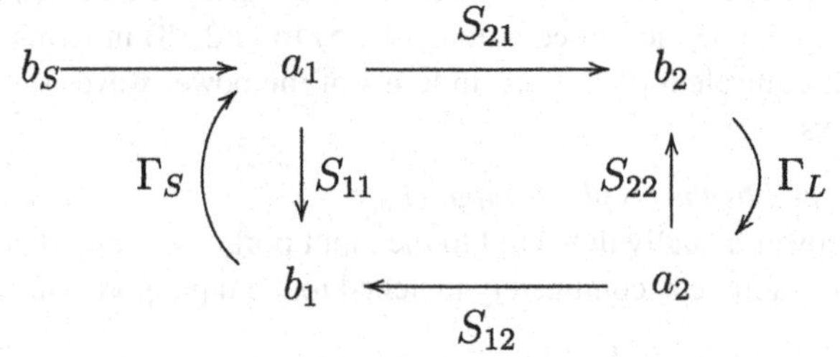

FIGURE 7.7

Mason's rule applied to a two-port with load.

Applying Mason's rule to Figure 7.7 we get [22]:

$$\frac{b_2}{b_S} = \frac{S_{21}}{1 - (S_{11}\Gamma_S + S_{22}\Gamma_L + S_{21}\Gamma_L S_{12}\Gamma_S) + S_{11}\Gamma_S S_{22}\Gamma_L} \tag{7.5.6}$$

The denominator of equation (7.5.6) can be simplified to give:

$$\frac{b_2}{b_S} = \frac{S_{21}}{(1 - S_{11}\Gamma_S)(1 - S_{22}\Gamma_L) - S_{21}S_{12}\Gamma_S\Gamma_L} \tag{7.5.7}$$

Replacing $|b_2|^2/|b_S|^2$ in equation (7.5.5) by equation (7.5.7) gives:

$$G_T = \frac{|S_{21}|^2(1 - |\Gamma_L|^2)(1 - |\Gamma_S|^2)}{|(1 - S_{11}\Gamma_S)(1 - S_{22}\Gamma_L) - S_{21}S_{12}\Gamma_S\Gamma_L|^2} \tag{7.5.8}$$

Equation (7.5.8) expresses the transducer gain in terms of arbitrary source and load terminations. Alternatively we can express G_T in terms of Γ_{in} and Γ_{out} by further rearranging the denominator of equation (7.5.8) and employing equations (6.2.5) and (6.2.7). This results in two equivalent equations for transducer power gain, that is:

$$G_T = \frac{(1 - |\Gamma_S|^2)}{|1 - \Gamma_{in}\Gamma_S|^2}|S_{21}|^2\frac{(1 - |\Gamma_L|^2)}{|1 - S_{22}\Gamma_L|^2} \tag{7.5.9}$$

and

$$G_T = \frac{(1 - |\Gamma_S|^2)}{|1 - S_{11}\Gamma_S|^2}|S_{21}|^2\frac{(1 - |\Gamma_L|^2)}{|1 - \Gamma_{out}\Gamma_L|^2} \tag{7.5.10}$$

Equations (7.5.9) and (7.5.10) are *S*-parameter equivalents of equation (7.2.32), having a similar structure consisting of the product of three factors, namely:

(i) An intrinsic gain component: $|S_{21}|^2$.

(ii) A source mismatch factor: $\dfrac{(1 - |\Gamma_S|^2)}{|1 - S_{11}\Gamma_S|^2}$.

(iii) A load mismatch factor: $\dfrac{(1 - |\Gamma_L|^2)}{|1 - S_{22}\Gamma_L|^2}$.

The available power gain, G_A, is a special case of transducer gain with the output conjugately matched and the source termination as the variable. Setting $\Gamma_L = \Gamma_{out}^*$ in equation (7.5.10) gives:

$$G_A = \frac{|S_{21}|^2(1 - |\Gamma_S|^2)}{|1 - |\Gamma_{out}|^2||1 - S_{11}\Gamma_S|^2} \tag{7.5.11}$$

We can rewrite equation (7.5.11) with Γ_S as the only independent variable by replacing Γ_{out} in equation (7.5.11) by equation (6.2.8) as follows:

$$G_A = \frac{|S_{21}|^2(1 - |\Gamma_S|^2)}{\left|1 - \left|\frac{S_{22} - \Delta\Gamma_S}{1 - S_{11}\Gamma_S}\right|^2\right| |1 - S_{11}\Gamma_S|^2} \tag{7.5.12}$$

Which simplifies to:

$$G_A = \frac{|S_{21}|^2(1 - |\Gamma_S|^2)}{|1 - S_{11}\Gamma_S|^2 - |S_{22} - \Delta\Gamma_S|^2} \tag{7.5.13}$$

The operating power gain is a special case of transducer gain with the input conjugately matched and the load termination as the variable. Setting $\Gamma_S = \Gamma_{in}^*$ in equation (7.5.9) gives:

$$G_0 = \frac{|S_{21}|^2(1 - |\Gamma_L|^2)}{|1 - |\Gamma_{in}|^2||1 - S_{22}\Gamma_L|^2} \tag{7.5.14}$$

We can rewrite equation (7.5.14) with Γ_L as the only independent variable by replacing Γ_{in} in equation (7.5.14) by equation (6.2.6) as follows:

$$G_0 = \frac{|S_{21}|^2(1 - |\Gamma_L|^2)}{\left|1 - \left|\frac{S_{11} - \Delta\Gamma_L}{1 - S_{22}\Gamma_L}\right|^2\right| |1 - S_{22}\Gamma_L|^2} \tag{7.5.15}$$

Which simplifies to:

$$G_0 = \frac{|S_{21}|^2(1 - |\Gamma_L|^2)}{|1 - S_{22}\Gamma_L|^2 - |S_{11} - \Delta\Gamma_L|^2} \tag{7.5.16}$$

In the special case where we terminate the two-port in the system characteristic impedance, Z_0, used to measure the original S-parameter, we have $\Gamma_S = \Gamma_L = 0$, and the gains defined in equations (7.5.8), (7.5.13), and (7.5.16) are reduced to [23]:

$$\text{Transducer gain: } G_T = |S_{21}|^2 \tag{7.5.17}$$

$$\text{Available gain: } G_A = \frac{|S_{21}|^2}{1 - |S_{22}|^2} \tag{7.5.18}$$

$$\text{Operating gain: } G_0 = \frac{|S_{21}|^2}{1 - |S_{11}|^2} \tag{7.5.19}$$

Example 7.4 (Two-port power gains).

Problem. A transistor in common emitter configuration has the following measured S-parameters at a specific frequency and bias point, and with a 50 Ω characteristic impedance:

$$\begin{bmatrix} S_{11} & S_{12} \\ S_{21} & S_{22} \end{bmatrix} = \begin{bmatrix} 0.452\angle -88° & 0.2\angle -18° \\ 3.015\angle 49° & 0.787\angle -45° \end{bmatrix}$$

The transistor is driven by a source impedance $Z_S = 40 + j30 \; \Omega$ and is driving a load resistance of 75 Ω.

Calculate the following gains:

(i) Transducer power gain (G_T).
(ii) Available gain (G_A).
(iii) Operating power gain (G_o).

Solution. We start by calculating the source and load reflection coefficients as follows:

$$\Gamma_S = \frac{Z_S - Z_o}{Z_S + Z_o}$$
$$= \frac{40 + j30 - 50}{40 + j30 + 50} = \frac{-10 + j30}{90 + j30}$$
$$= \frac{-10 + j30}{90 + j30} = 0 + j0.334$$
$$= 0.334\angle 90°$$

$$\Gamma_L = \frac{Z_L - Z_o}{Z_L + Z_o}$$
$$= \frac{75 - 50}{75 + 50} = \frac{25}{125} = 0.200 + j0$$
$$= 0.2\angle 0°$$

Next we calculate Γ_{in} and Γ_{out} for the given source and load terminations, using equations (6.2.5) and (6.2.7):

$$\Gamma_{in} = S_{11} + \frac{S_{12}S_{21}\Gamma_L}{1 - S_{22}\Gamma_L}$$
$$= 0.452\angle -88 + \frac{(0.2\angle -18) \times (3.015\angle 49) \times (0.2\angle 0)}{1 - (0.787\angle -45) \times (0.2\angle 0)}$$
$$= 0.452\angle -88 + \frac{0.121\angle 31}{0.896\angle 7.1} = 0.452\angle -88 + 0.135\angle 23.9$$
$$= \boxed{0.421\angle -70.7}$$

$$\Gamma_{out} = S_{22} + \frac{S_{12}S_{21}\Gamma_S}{1 - S_{11}\Gamma_S}$$
$$= 0.787\angle -45 + \frac{(0.2\angle -18) \times (3.015\angle 49) \times (0.334\angle 90°)}{1 - (0.452\angle -88) \times (0.334\angle 90)}$$

$$= 0.787\angle - 45 + \frac{0.201\angle 121}{0.849\angle - 0.3} = 0.787\angle - 45 + 0.237\angle 120.7$$

$$= \boxed{0.559\angle - 39.3}$$

Next we calculate some common parameters we will need in the gain calculations, as follows:

$$(1 - |\Gamma_S|^2) = 0.889$$

$$(1 - |\Gamma_L|^2) = 0.960$$

$$(1 - S_{11}\Gamma_S) = 1 - 0.452\angle - 88° \times 0.334\angle 90° = 0.849\angle - 0.36$$

$$(1 - S_{22}\Gamma_L) = 1 - 0.787\angle - 45° \times 0.2\angle 0° = 0.896\angle 7.2$$

Now we are in a position to calculate the transducer power gain using equation (7.5.8) as follows:

$$
\begin{aligned}
G_T &= \frac{|S_{21}|^2(1 - |\Gamma_L|^2)(1 - |\Gamma_S|^2)}{|(1 - S_{11}\Gamma_S)(1 - S_{22}\Gamma_L) - S_{21}S_{12}\Gamma_S\Gamma_L|^2} \\
&= \frac{9.090 \times 0.960 \times 0.889}{|0.849 \times 0.896\angle(7.2 - 0.36)° - 3.015 \times 0.2 \times 0.334 \times 0.2\angle(49 - 18 + 90)°|^2} \\
&= \frac{7.758}{|0.761\angle 6.84° - 0.040\angle 121°|^2} = \frac{7.758}{0.605} \\
&= 12.810 \ (11.1 \ \text{dB})
\end{aligned}
$$

Next we calculate the available power gain using equation (7.5.11).

$$
\begin{aligned}
G_A &= \frac{|S_{21}|^2(1 - |\Gamma_S|^2)}{|1 - |\Gamma_{out}|^2||1 - S_{11}\Gamma_S|^2} \\
&= \frac{9.090 \times 0.889}{0.687 \times 0.722} = \frac{8.080}{0.495} \\
&= 16.31 \ (12.1 \ \text{dB})
\end{aligned}
$$

Finally we calculate the operating power gain using equation (7.5.14).

$$
\begin{aligned}
G_o &= \frac{|S_{21}|^2(1 - |\Gamma_L|^2)}{|1 - |\Gamma_{in}|^2||1 - S_{22}\Gamma_L|^2} \\
&= \frac{9.090 \times 0.960}{0.823 \times 0.802} = \frac{8.727}{0.660} \\
&= 13.22 \ (11.2 \ \text{dB})
\end{aligned}
$$

The reader may wish to confirm that G_A and G_o could alternatively be calculated using equations (7.5.13) and (7.5.16), respectively, with identical results.

We note that the transducer gain is the lowest of the three gains in this example, as suggested by equations (7.2.39) and (7.2.40).

7.5.1 MAXIMUM AVAILABLE GAIN AND CONJUGATE TERMINATIONS

A common design goal is to extract the maximum possible power gain from a given active device, known as the maximum available gain (or MAG). In Section 7.2.2, we showed that the maximum power gain is obtainable at a unique combination of source and load terminations [4,17]. Under these conditions, the three gain definitions embodied in equations (7.2.19)–(7.2.21), all attain the same maximum value of gain which is denoted by G_{max}. The terminations required to realize G_{max} can be found by ensuring that input and output ports of the device are simultaneously conjugately matched.

From equation (7.5.8) we can prove that G_T (and, by implication, G_A and G_o) is maximized when:

$$\Gamma_{in} = \Gamma_S^* \qquad (7.5.20)$$

and, simultaneously:

$$\Gamma_{out} = \Gamma_L^* \qquad (7.5.21)$$

By applying equations (7.5.20) and (7.5.21) to equations (6.2.5) and (6.2.7) we can define the optimum terminations by the following simultaneous equations:

$$\Gamma_S^* = S_{11} + \frac{S_{12}S_{21}\Gamma_L}{1 - S_{22}\Gamma_L} \qquad (7.5.22)$$

$$\Gamma_L^* = S_{22} + \frac{S_{12}S_{21}\Gamma_S}{1 - S_{11}\Gamma_S} \qquad (7.5.23)$$

By solving equations (7.5.22) and (7.5.23) simultaneously we obtain the expressions the values of source and load terminations for simultaneous conjugate matching. These values are referred to as Γ_{ms} and Γ_{ml} and are defined as [17,22]:

$$\Gamma_{ms} = C_1^* \left[\frac{B_1 \pm \sqrt{B_1^2 - 4|C_1|^2}}{2|C_1|^2} \right] \qquad (7.5.24)$$

$$\Gamma_{ml} = C_2^* \left[\frac{B_2 \pm \sqrt{B_2^2 - 4|C_2|^2}}{2|C_2|^2} \right] \qquad (7.5.25)$$

where the variables B_1, B_2, C_1, and C_2 are defined by equations (7.4.32), (7.4.33), (7.4.16), and (7.4.3), respectively.

The negative sign in equation (7.5.24) applies if $B_1 > 0$ and the negative sign in equation (7.5.25) if $B_2 > 0$.

By substituting equations (7.5.24) and (7.5.25) into any one of the equations and by making use of equation (7.4.27) the value of the MAG can be determined as [17,22]:

$$\text{MAG} = \frac{|S_{21}|}{|S_{12}|}\left[K \pm \sqrt{K^2 - 1}\right] \qquad (7.5.26)$$

Here the negative sign in equation (7.5.26) applies if $B_1 > 0$ and $B_2 > 0$. As an interesting aside, under these conditions the reverse isolation of the device is also at a maximum and is given by:

$$\text{RI} = \frac{|S_{12}|}{|S_{21}|}\left[K \pm \sqrt{K^2 - 1}\right] \qquad (7.5.27)$$

With the same conditions applying to the polarity of the square root term in equation (7.5.27) as in equation (7.5.26). The MAG is only realizable if the active device is unconditionally stable as defined in Section 7.4. If this is not the case, then we have to settle for the *maximum stable gain* (MSG) which is defined by [4]

$$\text{MSG} = \frac{|S_{21}|}{|S_{12}|} \qquad (7.5.28)$$

Example 7.5 (Simultaneous conjugate matching using S-parameters).

Problem. We will illustrate the calculation of Γ_{ms} and Γ_{ml} and MAG for an unconditionally stable transistor by reproducing the example given in the classic paper by William H. Froehner [11]. The paper refers to the design of an amplifier operating at 750 MHz using a 2N3570 transistor. At bias conditions $V_{\text{CE}} = 10$ V and $I_c = 4$ mA, the measured S-parameters of the transistor are given as:

$$\begin{bmatrix} S_{11} & S_{12} \\ S_{21} & S_{22} \end{bmatrix} = \begin{bmatrix} 0.277\angle -59° & 0.078\angle 93° \\ 1.920\angle 64° & 0.848\angle -31° \end{bmatrix}$$

Solution. We start by calculating the various parameters Δ, B_1, B_2, C_1, and C_2 as follows:

$$\Delta = S_{11}S_{22} - S_{12}S_{21}$$
$$= 0.277 \times 0.848\angle(-59° - 31°) - 0.078 \times 1.920\angle(93° + 64°)$$
$$= 0.235\angle -90° - 0.150\angle 157°$$
$$= \boxed{0.324\angle -64°}$$

$$B_1 = 1 + |S_{11}|^2 - |S_{22}|^2 - |\Delta|^2$$
$$= 1 + 0.077 - 0.719 - 0.105$$
$$= \boxed{0.253}$$

$$C_1 = S_{11} - S_{22}^*\Delta$$
$$= 0.277\angle -59° - (0.848\angle 31° \times 0.324\angle -64°)$$
$$= \boxed{0.120\angle -135.4°}$$

$$B_2 = 1 - |S_{11}|^2 + |S_{22}|^2 - |\Delta|^2$$
$$= 1 - 0.077 + 0.719 - 0.105$$
$$= \boxed{1.537}$$
$$C_2 = S_{22} - S_{11}^* \Delta$$
$$= 0.848\angle - 31° - (0.277\angle 59° \times 0.324\angle - 64°)$$
$$= \boxed{0.768\angle - 33.8°}$$

The original Froehner paper used the Rollett stability criteria, so we will also, as follows:

$$K = \frac{1 - |S_{11}|^2 - |S_{22}|^2 + |\Delta|^2}{2|S_{21}S_{12}|}$$
$$= \frac{1 - 0.277^2 - 0.848^2 + 0.324^2}{2 \times 0.150}$$
$$= \frac{0.310}{0.300}$$
$$= \boxed{1.033}$$

Since $K > 1$ and $|\Delta| < 1$ we conclude that the device is unconditionally stable (the reader may wish to confirm this using the Edwards-Sinsky criteria). We can therefore proceed to calculate Γ_{ms} and Γ_{ml} for simultaneous conjugate matching. Since both $B_1 > 0$ and $B_2 > 0$, we use the negative sign in equations (7.5.24), (7.5.25), and (7.5.26), as follows:

$$\Gamma_{ms} = C_1^* \left[\frac{B_1 - \sqrt{B_1^2 - 4|C_1|^2}}{2|C_1|^2} \right]$$
$$= 0.120\angle 135.4° \times \left[\frac{0.253 - \sqrt{0.064 - 0.0576}}{0.0288} \right]$$
$$= \boxed{0.730\angle 135.4°}$$

$$\Gamma_{ml} = C_2^* \left[\frac{B_2 - \sqrt{B_2^2 - 4|C_2|^2}}{2|C_2|^2} \right]$$
$$= 0.768\angle 33.8° \times \left[\frac{1.537 - \sqrt{2.362 - 2.359}}{1.180} \right]$$
$$= \boxed{0.951\angle 33.8°}$$

The MAG is calculated by:

$$MAG = \frac{|S_{21}|}{|S_{12}|}\left[K - \sqrt{K^2 - 1}\right]$$

$$= \frac{1.920}{0.078}\left[1.033 - \sqrt{1.067 - 1}\right]$$

$$= 19.052 \ (12.8 \ dB)$$

In Froehner's paper, he goes on to complete the amplifier design by using single stub matching networks to realize Γ_{ms} and Γ_{ml}. Interested readers may find this paper worth consulting.

7.5.2 CONSTANT POWER GAIN CIRCLES

We occasionally need to design an amplifier with a specific value of gain, other than the MAG or MSG. This is especially true when there are constraints on the value of load or source termination, such as when the input port needs to be deliberately mismatched to minimize the noise figure, as will be discussed in Chapter 16.

For a nonunilateral amplifier, where the load termination has an influence on the input reflection coefficient of the amplifier and vice versa, we generally vary one of the terminating impedances, load or source, to achieve the required gain on the assumption that the other port will then be conjugately matched. To recap on what was discussed in the preceding section, therefore, we have the following possibilities:

1. *Operating power gain*: We vary the load termination, Γ_L, to achieve the specified gain, then conjugately match the input port (i.e., set $\Gamma_S = \Gamma_{in}^*$).
2. *Available power gain*: We vary the source termination, Γ_S, to achieve the specified gain, then conjugately match the output port (i.e., set $\Gamma_L = \Gamma_{out}^*$).

In such cases, it is useful to have a graphical representation of the relationship between gain and source/load terminating impedance in order to graphically evaluate the trade-off between gain and other performance metrics. Hence, in this section, we will derive circles of *constant gain* in the source and load reflection coefficient planes.

Circles of constant operating power gain

The operating power gain, G_o, is a function of the load termination, under the assumption that the input port match will always be conjugately matched (i.e., $\Gamma_S = \Gamma_{in}^*$). Recalling equation (7.5.16) for the operating power gain:

$$G_o = \frac{|S_{21}|^2(1 - |\Gamma_L|^2)}{|1 - S_{22}\Gamma_L|^2 - |S_{11} - \Delta\Gamma_L|^2}$$

We can rewrite this in the following form, by expanding the denominator:

$$G_o = \frac{|S_{21}|^2(1 - |\Gamma_L|^2)}{1 - |S_{11}|^2 + |\Gamma_L|^2 D_2 - 2Re(C_2\Gamma_L)} \tag{7.5.29}$$

where C_2 and D_2 are defined by:

$$C_2 = S_{22} - \Delta S_{11}^*$$
$$D_2 = (|S_{22}|^2 - |\Delta|^2)$$

We will now define a normalized operating power gain parameter, g_0 as:

$$g_0 = \frac{G_0}{|S_{21}|^2} \tag{7.5.30}$$

Then, using equation (7.5.29) we can write:

$$g_0 = \frac{1 - |\Gamma_L|^2}{1 - |S_{11}|^2 + |\Gamma_L|^2 D_2 - 2\mathrm{Re}(C_2\Gamma_L)} \tag{7.5.31}$$

At this point, we recall that the equation of a circle on the Γ_L plane is of the form:

$$|\Gamma_L - C_{gL}|^2 = |\gamma_{gL}|^2 \tag{7.5.32}$$

where C_{gL} is the center and γ_{gL} is the radius of the constant operating gain circle. We can thus rearrange equation (7.5.31) to be in the form of equation (7.5.32), as follows:

$$\left| \Gamma_L - \frac{g_0 C_2^*}{1 + g_0 D_2} \right|^2 = \left| \frac{\sqrt{(1 - 2K|S_{12}S_{21}|g_0 + |S_{12}S_{21}|^2 g_0^2)}}{1 + g_0 D_2} \right|^2 \tag{7.5.33}$$

By comparing equation (7.5.33) with equation (7.5.32) we can discern the centers and radii of the constant operating power gain circles on the load reflection coefficient plane as follows:

$$\boxed{C_{gL} = \frac{g_0 C_2^*}{1 + g_0 D_2}} \tag{7.5.34}$$

$$\boxed{\gamma_{gL} = \frac{\sqrt{(1 - 2K|S_{12}S_{21}|g_0 + |S_{12}S_{21}|^2 g_0^2)}}{1 + g_0 D_2}} \tag{7.5.35}$$

Circles of constant available power gain

Alternatively, we may need to determine the effect of varying the source reflection coefficient with the output port conjugately matched (i.e., $\Gamma_L = \Gamma_{out}^*$). In which case, we need to use the available power gain, G_A.

The derivation of circles of constant available power gain follows the same logic as the previous derivation of constant operating power gain circles. Recalling equation (7.5.13) for the available power gain:

$$G_A = \frac{|S_{21}|^2(1 - |\Gamma_S|^2)}{|1 - S_{11}\Gamma_S|^2 - |S_{22} - \Delta\Gamma_S|^2}$$

We can expand the denominator and rewrite equation (7.5.13) in the form:

$$G_A = \frac{|S_{21}|^2(1 - |\Gamma_S|^2)}{1 - |S_{22}|^2 + |\Gamma_S|^2 D_1 - 2\mathrm{Re}(C_1\Gamma_S)} \tag{7.5.36}$$

where C_1 and D_1 are defined by:

$$C_1 = S_{11} - \Delta S_{22}^*$$

$$D_1 = (|S_{11}|^2 - |\Delta|^2)$$

We will define the normalized available power gain parameter, g_a as:

$$g_a = \frac{G_A}{|S_{21}|^2} \tag{7.5.37}$$

Then, using equation (7.5.36) we can write:

$$g_a = \frac{1 - |\Gamma_S|^2}{1 - |S_{22}|^2 + |\Gamma_S|^2 D_1 - 2\mathrm{Re}(C_1\Gamma_S)} \tag{7.5.38}$$

Which we can rearrange into the form of the equation of a circle on the Γ_S, similar to equation (7.5.32), that is:

$$|\Gamma_S - C_{gS}|^2 = |\gamma_{gS}|^2 \tag{7.5.39}$$

Rearranging equation (7.5.38) to be in the form of equation (7.5.39) we get:

$$\left|\Gamma_S - \frac{g_p C_1^*}{1 + g_a D_1}\right|^2 = \left|\frac{\sqrt{(1 - 2K|S_{12}S_{21}|g_a + |S_{12}S_{21}|^2 g_a^2)}}{1 + g_a D_1}\right|^2 \tag{7.5.40}$$

Once again, by comparing equation (7.5.40) with equation (7.5.39) we can determine the centers and radii of the constant available power gain circles on the source reflection coefficient plane as follows:

$$\boxed{C_{gS} = \frac{g_a C_1^*}{1 + g_a D_1}} \tag{7.5.41}$$

$$\boxed{\gamma_{gS} = \frac{\sqrt{(1 - 2K|S_{12}S_{21}|g_a + |S_{12}S_{21}|^2 g_a^2)}}{1 + g_a D_1}} \tag{7.5.42}$$

At this point it is worth noting a few interesting properties of the gain circle equations (7.5.34), (7.5.35), (7.5.41), and (7.5.42), as follows:

1. First, we note that the centers of load plane constant gain circles always lie on a line drawn between the point C_2^* and the origin of the load plane. Similarly, the centers of source plane constant gain circles always lie on a line drawn between the point C_1^* and the origin of the source plane.

2. The radius of constant gain circles decreases with increasing gain. In case of an unconditionally stable device, the maximum gain is therefore achieved when $\gamma_{gL} = 0$ or $\gamma_{gS} = 0$. Applying these conditions to either equation (7.5.35) or (7.5.42) results in:

$$g_{max} = \frac{1}{|S_{12}S_{21}|}\left(K - \sqrt{K^2 - 1}\right) \tag{7.5.43}$$

Which, applying equation (7.5.37), gives:

$$G_{max} = \frac{|S_{21}|}{|S_{12}|}\left(K - \sqrt{K^2 - 1}\right) \tag{7.5.44}$$

Readers will note that equation (7.5.43) corresponds to MAG derived for an unconditional stable amplifier in equation (7.5.26).

3. In case of a potentially unstable device, that is when $G_o = \infty$, and $G_A = \infty$, the constant gain circles described by equations (7.5.34), (7.5.35), (7.5.41), and (7.5.42) become equal to the source and load plane stability circles, described by equations (7.4.10), (7.4.12), (7.4.14), and (7.4.15).

7.6 TAKEAWAYS

1. The key difference between an active and passive two-port network is that the former has the possibility of becoming *unstable*, whereas the latter does not.

2. The fact that there is more than one definition of "power" at the input and output ports of an active two-port network, means that there are multiple definitions of power gain. The three most important definitions of power gain are the transducer power gain, G_T; available power gain, G_A; and operating power gain, G_o.

3. Active two-ports can be classified as either *unconditionally stable* or *potentially unstable*, depending upon whether there is any combination of passive source and load terminations that can induce instability.

4. For potentially unstable devices there will be some combination of source and load terminations that will make the device unstable. The boundaries between stable and unstable regions on the source and load planes are in the form of a circle, known as a *stability circle*. The centers and radii of the stability circles are functions of the device S-parameters.

5. A number of *stability criteria* have been presented, based on either immittance parameters or S-parameters, which allow us to determine whether a given two-port is unconditionally stable or potentially unstable.

6. For unconditionally stable devices, it is possible to simultaneously conjugately match both the input and output ports so as to achieve that maximum available power gain from the device. Equations are available that allow the calculation of the necessary the source and load terminations in terms of either Y-parameters or S-parameters.

TUTORIAL PROBLEMS

Problem 7.1 (Gain and stability using Y-parameters). A transistor has the following Y-parameters in the common emitter configuration and with a particular bias point at 800 MHz:

$$Y_{ie} = (6 - j7) \text{ mS}$$
$$Y_{fe} = (0.1 - j0.5) \text{ mS}$$
$$Y_{re} = (50 - j45) \text{ mS}$$
$$Y_{oe} = (2 - j1) \text{ mS}$$

1. Determine whether the device is inherently stable.
2. Calculate the maximum available power gain in dB.
3. Calculate the optimum source and load terminations.

Problem 7.2 (Passive two-port). Show that a passive, reciprocal two-port (i.e., one containing only capacitors, inductors, and resistors) is always absolutely stable, and therefore can always be simultaneously matched at both ports.

Problem 7.3 (Rollett stability factor). Show that the Rollett stability factor in terms of S-parameters (K) and the immittance parameter version (k) are numerically equal for the following device:

$$\begin{bmatrix} S_{11} & S_{12} \\ S_{21} & S_{22} \end{bmatrix} = \begin{bmatrix} 0.60\angle 30° & 0.15\angle -90° \\ 1.80\angle -60° & 0.45\angle 45° \end{bmatrix}$$

Problem 7.4 (Transistor stability). Determine the approximate frequency at which the following device becomes potentially unstable.

f (GHz)	S_{11}	S_{12}	S_{21}	S_{22}
2.0	$0.71\angle -75°$	$0.05\angle 70°$	$3.6\angle 75°$	$0.56\angle -35°$
4.0	$0.75\angle -83°$	$0.07\angle 65°$	$2.9\angle 74°$	$0.66\angle -40°$
6.0	$0.79\angle -90°$	$0.11\angle 60°$	$2.6\angle 73°$	$0.76\angle -45°$

Problem 7.5 (Stability circles). By means of stability circles, determine a suitable combination of source and load terminations that would allow stable operation

for the device in Problem 7.4 at 6 GHz. Calculate the transducer power gain of the device with these terminations.

Problem 7.6 (Simultaneous conjugate matching using S-parameters). Determine the MAG and simultaneous conjugate matching terminations for the device in Problem 7.4 at 2 GHz. What would be the maximum gain available from the same device at 4 and 6 GHz?

Problem 7.7 (Gain of an active two-port using S-parameters). The BFU690 in common emitter configuration has the following S-parameters at 4 GHz with bias conditions of 2.5 V and 10 mA:

$$\begin{bmatrix} S_{11} & S_{12} \\ S_{21} & S_{22} \end{bmatrix} = \begin{bmatrix} 0.829\angle 154.8° & 0.070\angle 19° \\ 1.909\angle 40° & 0.332\angle -172° \end{bmatrix}$$

The transistor is driven by a source impedance $Z_S = 50 + j10\ \Omega$ and is driving a load resistance of $Z_L = 25 - j50\ \Omega$.

Calculate the following gains:

(i) Transducer power gain (G_T).
(ii) Available gain (G_A).
(iii) Operating power gain (G_o).

Problem 7.8 (Programming exercise). Write a MATLAB program to determine the stability of a two-port given a set of two-port S-parameters. Your program should calculate MAG and conjugate matching terminations or MSG and stability circles as appropriate.

Problem 7.9 (Simultaneous conjugate matching using Y-parameters). Repeat Example 7.5 using Y-parameters for the same device. Compare your results with those of Example 7.5.

REFERENCES

[1] G. Vendelin, Design of Amplifiers and Oscillators by the S-Parameter Method, John Wiley and Sons, New York, 1982.
[2] J. Linvill, J. Gibbons, Transistors and Active Circuits, McGraw-Hill, New York, 1961.
[3] R. Spence, Linear Active Networks, John Wiley and Sons, New York, 1970.
[4] R. Carson, High Frequency Amplifiers, John Wiley and Sons, New York, 1975, ISBN 0471137057.
[5] P. Meincke, J. Vidkjaer, Introduction to Wireless RF System Design, John Wiley and Sons, London, 2010, ISBN 9780470696811.
[6] F. Llewellyn, Some fundamental properties of transmission systems, Proc. IRE 40 (3) (1952) 271-283, ISSN 0096-8390, http://dx.doi.org/10.1109/JRPROC.1952.273783.
[7] J. Linvill, L. Schimf, The design of tetrode transistor amplifiers, Bell Syst. Tech. J. 35 (1956) 813Û-840.
[8] K.L. Kotzebue, Microwave amplifier design with potentially unstable FETs, IEEE Trans. Microw. Theory Tech. 27 (1) (1979) 1-3.

[9] R. Pengelly, Microwave Field Effect Transistors: Theory, Design and Applications, John Wiley and Sons, Chichester, 1982.

[10] P. Grivet, Microwave Circuits and Amplifiers, Academic Press Inc., London, 1976.

[11] W.H. Froehner, Quick amplifier design with scattering parameters, Electronics 40 (1967) 100.

[12] P. Yip, High-Frequency Circuit Design and Measurements, Chapman and Hall, London, 1990.

[13] J. Rollett, Stability and power gain invariants of linear two-ports, IRE Trans. Circuit Theory 9 (1962) 29-32.

[14] A. Stern, Stability and power gain of tuned transistor amplifiers, Proc. IRE 45 (3) (1957) 335-343, ISSN 0096-8390, http://dx.doi.org/10.1109/JRPROC.1957.278369.

[15] E. Bolinder, Survey of some properties of linear networks, IRE Trans. Circuit Theory 4 (3) (1957) 70-78, ISSN 0096-2007, http://dx.doi.org/10.1109/TCT.1957.1086385.

[16] D. Woods, Reappraisal of the unconditional stability criteria for active 2-port networks in terms of S parameters, IEEE Trans. Circuits Syst. 23 (2) (1976) 73-81, ISSN 0098-4094, http://dx.doi.org/10.1109/TCS.1976.1084179.

[17] G. Bodway, Two-port power flow analysis using generalised scattering parameters, Microw. J. 10 (1967) 61-69.

[18] M. Edwards, J. Sinsky, A new criterion for linear 2-port stability using a single geometrically derived parameter, IEEE Trans. Microw. Theory Tech. 40 (12) (1992) 2303-2311, ISSN 0018-9480, http://dx.doi.org/10.1109/22.179894.

[19] P. Bianco, G. Ghione, M. Pirola, New simple proofs of the two-port stability criterium in terms of the single stability parameter μ_1 (μ_2), IEEE Trans. Microw. Theory Tech. 49 (6) (2001) 1073-1076, ISSN 0018-9480, http://dx.doi.org/10.1109/22.925493.

[20] S. Alechno, Reflections on transistor stability, RF Design (2003).

[21] G. Lombardi, B. Neri, Criteria for the evaluation of unconditional stability of microwave linear two-ports: a critical review and new proof, IEEE Trans. Microw. Theory Tech. 47 (6) (1999) 746-751, ISSN 0018-9480, http://dx.doi.org/10.1109/22.769346.

[22] G. Gonzalez, Microwave Transistor Amplifiers, Analysis and Design, second ed., Prentice Hall Inc., Englewood Cliffs, NJ, 1997.

[23] H. Carlin, A. Giordano, Network Theory: An Introduction to Reciprocal and Non-Reciprocal Circuits, Prentice-Hall Inc., Englewood Cliffs, NJ, 1964.

Three-port analysis techniques

8

CHAPTER OUTLINE

INTENDED LEARNING OUTCOMES

- *Knowledge*
  - Understand the use of three-port representation for microwave transistors in the design of feedback circuits.
  - Understand the power of feedback mappings in circuit design, the different classifications and how to apply them.
  - Understand and be able to interpret reverse feedback mappings.

- *Skills*
  - Be able to calculate the shunt and series feedback three-port S-parameters from measured two-port S-parameters.
  - Be able to calculate two-port S-parameters for common base/gate and common collector/drain configurations, given the common emitter/source two-port S-parameters for a device (configuration conversion).
  - Be able to calculate the two-port S-parameters of a transistor with arbitrary shunt and series feedback.
  - Be able to construct and interpret feedback mappings and reverse feedback mappings.
  - Be able to determine the optimum reactive feedback termination required to generate negative resistance in a given transistor.

8.1 INTRODUCTION

We established in Chapters 5 and 6 that we can use S-parameters to represent any linear network with an arbitrary number of ports, n. It may seem odd, therefore, to devote an entire chapter to the specific case of three-port networks. The reason for this chapter is that the three-port representation has a particular significance in microwave active circuit design since many designs involve applying feedback to transistors, which are three terminal devices. A body of specialist design techniques has been developed, based on the three-port S-parameter representation of a transistor, that, due to their elegance and power, are worth covering in some detail in a chapter of their own.

Since transistors are inherently three terminal devices, it would be logical to measure and characterize them by means of a three-port S-matrix, and some work has been done on three-port S-parameter measurement techniques [1–4]. It is still the convention, however, to characterize transistors as two-ports with one of the device terminals (usually the emitter in the case of a bipolar transistor or source in the case of an FET) connected to a common ground. Most test jigs used for measuring S-parameters are designed to accommodate and bias the device in common emitter/source (ces) configuration, and most published manufacturers' data have been measured in this configuration. Fortunately, standard formulae exist for calculating three-port S-parameters given a set of CE two-port S-parameters [5], and these will be demonstrated in this chapter.

Although the bulk of microwave circuit design covered in this book is based on the assumption of a two-port active device, treating the transistor as a three-port device opens up the possibility of "tailoring" the S-parameters of the resulting two-port (transistor plus feedback), so as to achieve design objectives that would not be possible with the basic device. For example, a transistor will have better high-frequency performance when it has simultaneously greater $|S_{21}|$ and smaller $|S_{12}|$ [1]. Using three-port design techniques we can engineer both these parameters to be more amenable to what we are trying to achieve. Another example is adding feedback to achieve the condition $|S_{11}| > 1$, which is required in negative resistance oscillator design, but is not normally exhibited with a "bare" transistor in common emitter (CE) configuration [6].

In this chapter, we will spend a brief amount of time on three-port immittance parameters (specifically Y-parameters) but will focus mostly on three-port S-parameters, as these have a wider range of applications in contemporary microwave circuit design.

8.2 THREE-PORT IMMITTANCE PARAMETERS

We will start with an immittance parameter approach, by considering a transistor as being a floating three-port with all three terminals independently referenced to ground, as shown in Figure 8.1. We can characterize the device in terms of a three-port Y-matrix, referred to as the *indefinite admittance matrix* or "IAM" [7] and defined as:

$$
\begin{bmatrix} i_1 \\ i_2 \\ i_3 \end{bmatrix} = \begin{bmatrix} Y_{11} & Y_{12} & Y_{13} \\ Y_{21} & Y_{22} & Y_{23} \\ Y_{31} & Y_{32} & Y_{33} \end{bmatrix} \begin{bmatrix} v_1 \\ v_2 \\ v_3 \end{bmatrix}
\tag{8.2.1}
$$

It can be shown, by a simple application of Kirchoff's current law [7], that the sum of any row or column of the IAM is zero. Therefore, if four of the nine Y-parameters are known, provided that three are not in the same row or column, then the remainder of the parameters can be deduced.

It follows from the above that if the Y-parameters for a transistor in any one of the three possible configurations (CE, common base [CB], and common collector [CC] in the case of a BJT) are known, then the Y-parameters for the transistor in the other

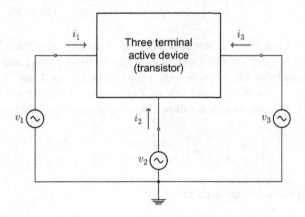

FIGURE 8.1

Indefinite admittance matrix definition.

two configurations can be determined by completing the IAM and then choosing the appropriate port definitions. This is best illustrated by means of an example.

Example 8.1 (Calculation CB and CC transistor Y-parameters).

Problem. Calculate the CB and CC two-port Y-parameters for a bipolar transistor with the following CE Y-parameters:

$$\begin{bmatrix} Y_{ie} & Y_{re} \\ Y_{fe} & Y_e \end{bmatrix} = \begin{bmatrix} (3+j3) & (-3-j25) \\ (25-j2) & (20+j50) \end{bmatrix}$$

Note:

1. We are using the same port nomenclature as for the h-parameters in Table 5.6 on page 159.
2. All conductance values are in millisiemens (mS).

Solution. Let us define this CE two-port as the IAM shown in Figure 8.1 with port 2 connected to ground. In other words, port 1 is the base port and port 3 is the collector port. We now insert the above Y-parameters into the appropriate row and column locations in the IAM, with unknown values indicated by "?" as follows:

$$\begin{bmatrix} Y_{11} & Y_{12} & Y_{13} \\ Y_{21} & Y_{22} & Y_{23} \\ Y_{31} & Y_{32} & Y_{33} \end{bmatrix} = \begin{bmatrix} (3+j3) & ? & (-3-j25) \\ ? & ? & ? \\ (25-j2) & ? & (20+j50) \end{bmatrix}$$

The unknown values can be calculated on the basis that the sum of any row or column of the IAM is zero. Thus we can write:

$$\begin{bmatrix} Y_{11} & Y_{12} & Y_{13} \\ Y_{21} & Y_{22} & Y_{23} \\ Y_{31} & Y_{32} & Y_{33} \end{bmatrix} = \begin{bmatrix} (3+j3) & (0+j22) & (-3-j25) \\ (-28-j1) & (45+j26) & (-17-j25) \\ (25-j2) & (-45-j48) & (20+j50) \end{bmatrix}$$

We have defined the CE terminal ports as 1 and 3 with port 2 grounded. It therefore follows that the CB configuration has terminal ports 2 and 3 with port 1 grounded. Similarly, CC is configured with terminal ports 1 and 2 with port 3 grounded. Picking out the relevant sets of Y-parameters from the IAM above, we determine the CB Y-parameters as follows (note the use of transistor subscripts):

$$\begin{bmatrix} Y_{ib} & Y_{rb} \\ Y_{fb} & Y_{ob} \end{bmatrix} = \begin{bmatrix} (45+j26) & (-17-j25) \\ (-45-j48) & (20+j50) \end{bmatrix}$$

and the CC Y-parameters are determined as:

$$\begin{bmatrix} Y_{ic} & Y_{rc} \\ Y_{fc} & Y_{oc} \end{bmatrix} = \begin{bmatrix} (3+j3) & (0-j22) \\ (-28-j1) & (45+j26) \end{bmatrix}$$

8.3 THREE-PORT *S*-PARAMETERS

At microwave frequencies, *S*-parameters are more widely used than immittance parameters, for reasons that were explained in Chapter 6. The three-port *S*-parameter representation of microwave transistors was first introduced by Bodway [1,8] who obtained three-port *S*-parameters by direct measurement using a special test fixture.

For the three-port network of Figure 8.2 we can write:

$$\left. \begin{aligned} b_1 &= s_{11}a_1 + s_{12}a_2 + s_{13}a_3 \\ b_2 &= s_{21}a_1 + s_{22}a_2 + s_{23}a_3 \\ b_3 &= s_{31}a_1 + s_{32}a_2 + s_{33}a_3 \end{aligned} \right\} \tag{8.3.1}$$

where a_i and b_i are the scattering power wave variables defined in Chapter 6.

Note a convention we will use throughout this chapter, and the rest of the book, namely that upper case S_{ij} will be used exclusively to represent the *S*-parameters of a two-port, whereas lower case s_{ij} will be used to represent the *S*-parameters of networks with three or more ports.

The most important application of three-port design techniques in microwave circuit design is to determine the behavior of transistors with externally applied feedback. In this context, Figure 8.3 shows the two essential ways in which a transistor can be embedded in an external feedback network, namely via a *series* connection (Figure 8.3(a)) or via a *shunt* connection (Figure 8.3(b)). These have also been described as "Z-embedding" or "Y-embedding," respectively [9,10]. Any practical shunt and series feedback circuit can be considered as a special case of one of these two classifications.

Neither topology in Figure 8.3 is shown with a ground connection, but any point on the peripheral circle in either figure could be grounded. In the case of Figure 8.3(a), all points on the peripheral circle are electrically identical, so there is, in effect, only one possible grounded topology. In the case of Figure 8.3(b), there are three points where a ground connection could be made, corresponding to one

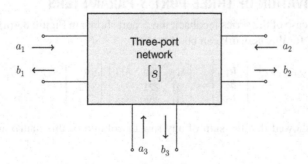

FIGURE 8.2

Three-port network with power waves.

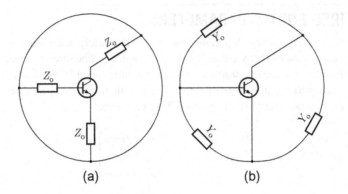

FIGURE 8.3

Generic feedback topologies. (a) Series feedback topology. (b) Shunt feedback topology.

terminal of the transistor. The location of the ground connection will determine the "configuration" of the resultant feedback circuit (CE, CB, or CC).

Figure 8.4 shows the basic design flow in a three-port design process, beginning with the measured CE/common source (CS) two-port transistor in Figure 8.4(c) and progressing via either series or shunt three-port representation (Figure 8.4(a) and (d), respectively), and finally to the corresponding *reduced two-port* with feedback (Figure 8.4(b) and (e), respectively).

In the "series" feedback case (Figure 8.4(a)), the transistor has all three terminals floating with respect to ground, whereas in the shunt feedback case (Figure 8.4(d)), the emitter remains grounded throughout. If a configuration other than CE is required, then the necessary configuration conversion must be made by rearranging the ports in Figure 8.4(a), irrespective of whether series or shunt feedback is going to be applied to the device.

8.3.1 DERIVATION OF THREE-PORT S-PARAMETERS

Consider the case of the series feedback three-port shown in Figure 8.4(a). The three-port S-matrix for this network can be written as:

$$
\begin{bmatrix} b_1 \\ b_2 \\ b_3 \end{bmatrix} = \begin{bmatrix} s_{11} & s_{12} & s_{13} \\ s_{21} & s_{22} & s_{23} \\ s_{31} & s_{32} & s_{33} \end{bmatrix} \begin{bmatrix} a_1 \\ a_2 \\ a_3 \end{bmatrix}
\tag{8.3.2}
$$

Bodway showed that the sum of any row or column of this matrix is unity, that is [8]:

$$
\sum_{j=1}^{3} s_{ij} = \sum_{i=1}^{3} s_{ij} = 1
\tag{8.3.3}
$$

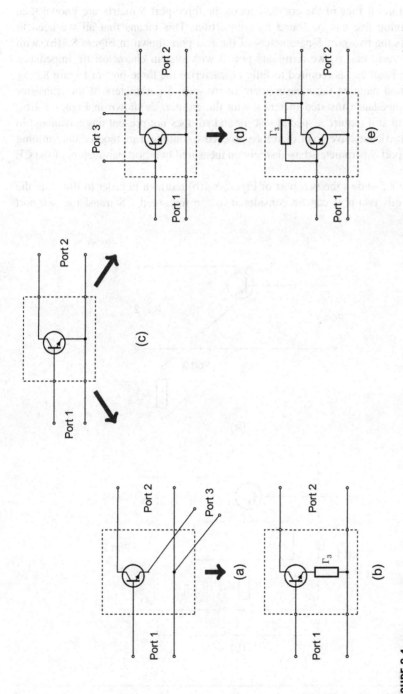

FIGURE 8.4

Design flow for series and shunt feedback starting with two-port CE *S*-parameters.

Therefore, if four of the coefficients of the three-port S-matrix are known then the remaining five can be found by subtraction. This means that all we need to measure is the two-port S-parameters of the two-port shown in Figure 8.4(b), with Γ_3 set to zero; that is, we terminate port 3 with system characteristic impedance (Z_0). This is all that is required to fully characterize the three-port of Figure 8.4(a). This method requires the measurement of two-port S-parameters of the transistor with an impedance inserted in series with the emitter, as shown in Figure 8.4(b). This would still require a special test jig and so does not present any advantage in cost or practicality. We shall therefore proceed to outline a method of determining the three-port S-parameters based solely on measured two-port parameters of the CE transistor.

Figure 8.5 shows the two-port of Figure 8.4(b) redrawn in order to illustrate the fact that this two-port can be considered as the measured CS transistor two-port

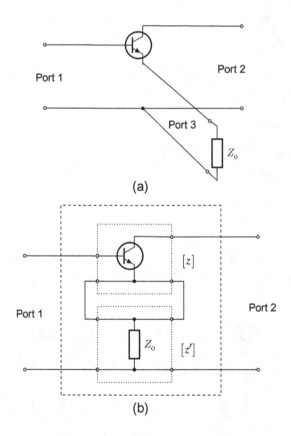

FIGURE 8.5

Series feedback circuit analysis.

connected in series with a passive two-port, comprising a shunt matched termination Z_o (i.e., $\Gamma_3 = 0$).

The transistor in Figure 8.5(b) can be represented by its normalized two-port impedance matrix. If we are starting with measured two-port *S*-parameters then we need to convert them to *z*-parameters using the transformations given in Appendix A:

$$[z] = \begin{bmatrix} z_{11} & z_{12} \\ z_{21} & z_{22} \end{bmatrix} \tag{8.3.4}$$

The normalized impedance matrix of the lower two-port in Figure 8.5(b), consisting of a shunt matched termination, Z_o, is given by:

$$[z'] = \begin{bmatrix} 1 & 1 \\ 1 & 1 \end{bmatrix} \tag{8.3.5}$$

Since these two two-port networks are in series, the overall normalized impedance matrix is the sum of the two *z*-matrices in equations (8.3.4) and (8.3.5), that is,

$$[z_T] = \begin{bmatrix} (z_{11} + 1) & (z_{12} + 1) \\ (z_{21} + 1) & (z_{22} + 1) \end{bmatrix} \tag{8.3.6}$$

The *S*-matrix of the complete two-port of Figure 8.5 can then be found by transformation of equation (8.3.6) back into the *S*-domain, and the remaining five three-port *S*-parameters can be found by application of Bodway's relationship equation (8.3.3):

$$\left.\begin{aligned} s_{13} &= 1 - s_{11} - s_{12} \\ s_{31} &= 1 - s_{11} - s_{21} \\ s_{23} &= 1 - s_{21} - s_{22} \\ s_{32} &= 1 - s_{12} - s_{22} \\ s_{33} &= 1 - s_{31} - s_{32} \end{aligned}\right\} \tag{8.3.7}$$

The above method describes how to calculate the series feedback three-port *S*-parameters for a given transistor for which the two-port *S*-parameters are available. Alternatively, we can make use of a definition of the series three-port *S*-parameter, s_{33}, derived by Khanna [11]:

$$s_{33} = \frac{\xi}{4 - \xi} \tag{8.3.8}$$

where ξ is the sum of the two-port *S*-parameters [12], that is,

$$\xi = \sum_{I,j=1}^{2} S_{Ij} = S_{11} + S_{12} + S_{21} + S_{22} \tag{8.3.9}$$

Using the relationship in equation (8.3.8) we can now write the complete series three-port *S*-matrix, explicitly in terms of the original CE two-port *S*-matrix as follows:

$$
\begin{bmatrix} s_{11} & s_{12} & s_{13} \\ s_{21} & s_{22} & s_{23} \\ s_{31} & s_{32} & s_{33} \end{bmatrix} = \begin{bmatrix} \left(S_{11} + \dfrac{\Delta_{11}\Delta_{12}}{4-\xi}\right) & \left(S_{12} + \dfrac{\Delta_{11}\Delta_{21}}{4-\xi}\right) & \dfrac{2\Delta_{11}}{4-\xi} \\[2mm] \left(S_{21} + \dfrac{\Delta_{22}\Delta_{12}}{4-\xi}\right) & \left(S_{22} + \dfrac{\Delta_{22}\Delta_{21}}{4-\xi}\right) & \dfrac{2\Delta_{22}}{4-\xi} \\[2mm] \dfrac{2\Delta_{12}}{4-\xi} & \dfrac{2\Delta_{21}}{4-\xi} & \dfrac{\xi}{4-\xi} \end{bmatrix} \tag{8.3.10}
$$

where

$$
\begin{aligned}
\Delta_{11} &= 1 - S_{11} - S_{12} \\
\Delta_{12} &= 1 - S_{11} - S_{21} \\
\Delta_{21} &= 1 - S_{12} - S_{22} \\
\Delta_{22} &= 1 - S_{21} - S_{22}
\end{aligned}
$$

This alternative method has the advantage that the calculations can be performed entirely in the S-domain, with no conversion back and forth between S and immittance domains being required.

The shunt feedback three-port shown in Figure 8.6(a) does not have the same three-port S-matrix as in the series feedback case of Figure 8.5(a), so Bodway's relationship equation (8.3.3) does not apply in this case. The sum of each row or column of the shunt three-port S-matrix no longer equals unity, but Bodway has provided the following relationships between the three-port S-parameters for the shunt feedback case:

$$
\left.\begin{aligned}
s_{13} &= 1 + s_{11} - s_{12} \\
s_{31} &= 1 - s_{21} + s_{11} \\
s_{23} &= s_{21} - s_{22} - 1 \\
s_{32} &= s_{12} - s_{22} - 1 \\
s_{33} &= s_{31} - s_{32} - 1
\end{aligned}\right\} \tag{8.3.11}
$$

A rigorous proof of the relationships in (8.3.11) can be found in Medley's book [13].

As in the series feedback case, if any four of the three-port S-parameters are known then the remaining five can be determined from the relationships equation (8.3.11) above. The measurement of the two-port in Figure 8.6(b) would therefore be sufficient to fully characterize the three-port.

The two-port of Figure 8.6(b) can be considered as a single two-port, comprising a series matched admittance, $Y_0 = 1/Z_0$, in parallel with the CE transistor, measured as a two-port. The transistor can be represented by its normalized CS admittance matrix by employing the transformations in Appendix A, namely:

$$
[y] = \begin{bmatrix} y_{11} & y_{12} \\ y_{21} & y_{22} \end{bmatrix} \tag{8.3.12}
$$

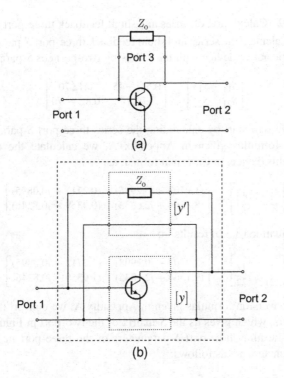

FIGURE 8.6

Shunt feedback circuit analysis.

The normalized admittance matrix of the upper two-port in Figure 8.6(b), consisting of a shunt matched termination is given by:

$$[y'] = \begin{bmatrix} 1 & -1 \\ -1 & 1 \end{bmatrix}$$ (8.3.13)

Since these two two-port networks are in parallel, the overall normalized admittance matrix is given by the sum of the two *y*-matrices in equations (8.3.12) and (8.3.13), as follows:

$$[y_T] = \begin{bmatrix} (y_{11} + 1) & (y_{12} - 1) \\ (y_{21} - 1) & (y_{22} + 1) \end{bmatrix}$$ (8.3.14)

The *S*-matrix for the two-port of Figure 8.6(b) can then be found by converting equation (8.3.14) back into the *S*-domain using transformations given in Appendix A. The remaining five three-port *S*-parameters can be found by application of the relationships in (8.3.11).

Example 8.2 (Calculation of series and shunt feedback three-port parameters).

Problem. Calculate the series and shunt feedback three-port S-parameters for the following generic active device with the following two-port ces S-parameters:

$$\begin{bmatrix} S_{11} & S_{12} \\ S_{21} & S_{22} \end{bmatrix} = \begin{bmatrix} 0.5\angle -45 & 0.1\angle 70 \\ 3.0\angle 70 & 0.8\angle -70 \end{bmatrix}$$

Solution. We will start by calculating the series three-port S-parameters. Using the conversion formulas given in Appendix A, we calculate the normalized z-parameters for this device, in Cartesian format, as:

$$\begin{bmatrix} z_{11} & z_{12} \\ z_{21} & z_{22} \end{bmatrix} = \begin{bmatrix} (1.9703 - j0.2998) & (0.2712 + j0.0895) \\ (8.1364 + j2.6835) & (0.9359 - j0.8246) \end{bmatrix}$$

Applying equation (8.3.6) results in:

$$\begin{bmatrix} z'_{11} & z'_{12} \\ z'_{21} & z'_{22} \end{bmatrix} = \begin{bmatrix} (2.9703 - j0.2998) & (1.2712 + j0.0895) \\ (9.1364 + j2.6835) & (1.9359 - j0.8246) \end{bmatrix}$$

Using the conversion formulas given in Appendix A, we convert the above back into S-parameters, which gives us the S-matrix of the two-port in Figure 8.5(b), that is, four out of the nine three-port S-parameters of the three-port in Figure 8.5(a). These four parameters are as follows:

$$\begin{bmatrix} s_{11} & s_{12} \\ s_{21} & s_{22} \end{bmatrix} = \begin{bmatrix} (1.063\angle -41.2°) & (0.304\angle 93.8°) \\ (2.272\angle 106.1°) & (1.323\angle -45.7°) \end{bmatrix}$$

We can now apply equation (8.3.3) to the above to calculate the remaining three-port S-parameters of the three-port in Figure 8.5 as follows:

$$s_{13} = 1 - (1.063\angle -41.2°) - (0.304\angle 93.8°) = 0.453\angle 61.0°$$
$$s_{31} = 1 - (1.063\angle -41.2°) - (2.272\angle 106.1°) = 1.700\angle -60.7°$$
$$s_{23} = 1 - (2.272\angle 106.1°) - (1.323\angle -45.7°) = 1.423\angle -60.2°$$
$$s_{32} = 1 - (0.304\angle 93.8°) - (1.323\angle -45.7°) = 0.559\angle -165.9°$$
$$s_{33} = 1 - (1.700\angle -60.7°) - (0.559\angle -165.9°) = 0.843\angle 85.0°$$

We can now assemble the full series feedback three-port S-matrix as follows:

$$\begin{bmatrix} s_{11} & s_{12} & s_{13} \\ s_{21} & s_{22} & s_{23} \\ s_{31} & s_{32} & s_{33} \end{bmatrix} = \begin{bmatrix} (1.063\angle -41.2°) & (0.304\angle 93.8°) & (0.453\angle 61.0°) \\ (2.272\angle 106.1°) & (1.323\angle -45.7°) & (1.423\angle -60.2°) \\ (1.700\angle -60.7°) & (0.651\angle 81.6°) & (0.843\angle 85.0°) \end{bmatrix}$$

$$(8.3.15)$$

Now turning to the shunt feedback case, we use the conversion formulas given in Appendix A, to calculate the normalized y-parameters for this device, in Cartesian format, as follows:

$$\begin{bmatrix} y_{11} & y_{12} \\ y_{21} & y_{22} \end{bmatrix} = \begin{bmatrix} (0.2121 + j0.3018) & (0.0351 - j0.0768) \\ (1.0521 - j2.3051) & (0.0244 + j0.5889) \end{bmatrix}$$

Applying equation (8.3.14) results in:

$$\begin{bmatrix} y'_{11} & y'_{12} \\ y'_{21} & y'_{22} \end{bmatrix} = \begin{bmatrix} (1.2121 + j0.3018) & (-0.9649 - j0.0768) \\ (0.0521 - j2.3051) & (1.0244 + j0.5889) \end{bmatrix}$$

Using the conversion formulas given in Appendix A, we convert the above back into *S*-parameters, which gives us the *S*-matrix of the two-port in Figure 8.6(b), that is four out of the nine three-port *S*-parameters of the three-port in Figure 8.6(a). These four parameters are as follows:

$$\begin{bmatrix} s_{11} & s_{12} \\ s_{21} & s_{22} \end{bmatrix} = \begin{bmatrix} (0.344\angle111.8°) & (0.427\angle8.4°) \\ (1.016\angle95.2°) & (0.202\angle100.4°) \end{bmatrix} \qquad (8.3.16)$$

We can now apply equation (8.3.11) to the above to calculate the remaining three-port *S*-parameters of the three-port in Figure 8.6(b) as follows:

$$s_{13} = 1 + (0.344\angle111.8°) - (0.427\angle8.4°) = 0.519\angle29.7°$$

$$s_{31} = 1 - (1.016\angle95.2°) + (0.344\angle111.8°) = 1.187\angle-35.7°$$

$$s_{23} = (1.016\angle95.2°) - (0.202\angle100.4°) - 1 = 1.332\angle142.4°$$

$$s_{32} = (0.427\angle8.4°) - (0.202\angle100.4°) - 1 = 0.559\angle-165.9°$$

$$s_{33} = (0.519\angle29.7°) - (0.559\angle-165.9°) - 1 = 0.752\angle-47.7°$$

We can now assemble the full shunt feedback three-port *S*-matrix as follows:

$$\begin{bmatrix} s_{11} & s_{12} & s_{13} \\ s_{21} & s_{22} & s_{23} \\ s_{31} & s_{32} & s_{33} \end{bmatrix} = \begin{bmatrix} (0.344\angle111.8°) & (0.427\angle8.4°) & (0.519\angle29.7°) \\ (1.016\angle95.2°) & (0.202\angle100.4°) & (1.332\angle142.4°) \\ (1.187\angle-35.7°) & (0.559\angle-165.9°) & (0.752\angle-47.7°) \end{bmatrix}$$

$$(8.3.17)$$

This example illustrates the fact that the three-port *S*-parameters for the series and shunt feedback three-ports are different, as illustrated by comparing equation (8.3.15) with equation (8.3.17).

8.3.2 CALCULATION OF REDUCED TWO-PORT *S*-PARAMETERS

The two- to three-port conversion presented in the previous section allows the calculation of two-port parameters for a transistor with arbitrary shunt or series feedback terminations. Such a transistor plus feedback combination is referred to as a *reduced two-port* [8]. Once the parameters of the reduced two-port have been found, standard two-port design techniques can be used to complete the design. The feedback termination can be thought of as providing an extra degree of freedom to the standard design procedure since feedback allows a wide range of two-port S-parameters to be realized.

The two-port S-parameters of the reduced two-port can be expressed in terms of the three-port S-parameters, s_{ij}, and a third-port termination, Γ_3, as follows [8]:

$$S'_{ij} = s_{ij} + \frac{s_{i3}s_{3j}\Gamma_3}{1 - s_{33}\Gamma_3} \tag{8.3.18}$$

or in matrix form as:

$$\begin{bmatrix} S'_{11} & S'_{12} \\ S'_{21} & S'_{22} \end{bmatrix} = \begin{bmatrix} s_{11} & s_{12} \\ s_{21} & s_{22} \end{bmatrix} + \frac{\Gamma_3}{1 - s_{33}\Gamma_3} \begin{bmatrix} s_{13}s_{31} & s_{13}s_{32} \\ s_{23}s_{31} & s_{23}s_{32} \end{bmatrix} \tag{8.3.19}$$

In the special case where port 3 of the series feedback three-port in Figure 8.5(a) is terminated with a short circuit (i.e., $\Gamma_3 = 1\angle 180°$), this reduces to:

$$\begin{bmatrix} S_{11} & S_{12} \\ S_{21} & S_{22} \end{bmatrix} = \begin{bmatrix} \left(s_{11} - \frac{s_{13}s_{31}}{1 + s_{33}}\right) & \left(s_{12} - \frac{s_{13}s_{32}}{1 + s_{33}}\right) \\ \left(s_{21} - \frac{s_{23}s_{31}}{1 + s_{33}}\right) & \left(s_{22} - \frac{s_{23}s_{32}}{1 + s_{33}}\right) \end{bmatrix} \tag{8.3.20}$$

Similarly, when port 3 of the shunt feedback three-port in Figure 8.6(a) is terminated with an open circuit (i.e., $\Gamma_3 = 1\angle 0°$), equation (8.3.19) reduces to:

$$\begin{bmatrix} S_{11} & S_{12} \\ S_{21} & S_{22} \end{bmatrix} = \begin{bmatrix} \left(s_{11} + \frac{s_{13}s_{31}}{1 - s_{33}}\right) & \left(s_{12} + \frac{s_{13}s_{32}}{1 - s_{33}}\right) \\ \left(s_{21} + \frac{s_{23}s_{31}}{1 - s_{33}}\right) & \left(s_{22} + \frac{s_{23}s_{32}}{1 - s_{33}}\right) \end{bmatrix} \tag{8.3.21}$$

Equations (8.3.20) and (8.3.21) simply express the original two-port S-matrix (which is CE/CS by default) in terms of the series and shunt feedback three-port S-parameters, respectively. It is left as an exercise for the reader to verify equations (8.3.20) and (8.3.21) numerically using the three-port S-parameter data in equations (8.3.15) and (8.3.17).

(*Note*: Once again, we need to remember that the three-port S-parameters for the series and shunt feedback three-ports are different).

Example 8.3 (Calculation of reduced two-port S-parameters).

Problem. Calculate the reduced two-port S-parameters of the device in Example 8.2 with a series feedback inductance of 1 nH at 1.8 GHz:

Solution.

$$X_{fb} = \omega L$$

$$= 2\pi \times 1.8 \times 10^9 \times 10^{-9}$$

$$= 11.31 \ \Omega$$

Therefore,

$$\Gamma_3 = \frac{Z_{fb} - Z_o}{Z_{fb} + Z_o}$$

$$= \frac{j11.31 - 50}{j11.31 + 50}$$

$$= 1\angle 154.5°$$

Applying the series feedback three-port S-parameters given in equation (8.3.15) to equation (8.3.19) gives us the reduced two-port S-matrix of the device plus feedback, as follows:

$$\begin{bmatrix} S_{11} & S_{12} \\ S_{21} & S_{22} \end{bmatrix} = \begin{bmatrix} 0.598\angle -32.3 & 0.121\angle 99.4 \\ 2.511\angle 69.7 & 0.778\angle -56.5 \end{bmatrix}$$

8.4 CONFIGURATION CONVERSION

The two-port S-parameters of any transistor configuration may be determined by starting with the series feedback three-port of Figure 8.4(a) then simply rearranging the three-port S-matrix and adding a short circuit ($\Gamma_3 = 1\angle 180°$) to the appropriate port. Depending on the type of device used (BJT or FET), the remaining two-ports represent either ces, common base/gate (cbg), or common collector/drain (ccd) configuration depending on which port has been shorted. This provides a simple means of determining the two-port S-parameters for the three possible transistor configurations from one set of measurements, without having to measure a different set of S-parameters for each configuration.

Consider the BJT shown in Figure 8.7, which is characterized as a series feedback three-port network with the following S-matrix:

$$\begin{bmatrix} b_1 \\ b_2 \\ b_3 \end{bmatrix} = \begin{bmatrix} s_{11} & s_{12} & s_{13} \\ s_{21} & s_{22} & s_{23} \\ s_{31} & s_{32} & s_{33} \end{bmatrix} \begin{bmatrix} a_1 \\ a_2 \\ a_3 \end{bmatrix} \qquad (8.4.1)$$

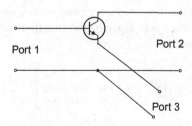

FIGURE 8.7

Series CE three-port definitions for a transistor.

We can "reduce" the three-port of Figure 8.7 to one of the three possible *reduced two-port* configurations by connecting one of the ports to ground, as follows:

1. Grounding port 1 gives the cbg reduced two-port.
2. Grounding port 2 gives the ccd reduced two-port.
3. Grounding port 3 gives the ces reduced two-port.

In signal terms, "grounding" means applying a short-circuit termination ($\Gamma = 1\angle 180°$) to the port in question.

By convention, two-port S-parameters are usually measured in the ces configuration, and manufacturers only supply S-parameter data in this configuration. In the following sections, we will derive some useful conversion formulae that allow the reduced two-port S-parameters for the other two configurations to be calculated given the three-port S-parameters of the circuit in Figure 8.7.

8.4.1 COMMON BASE/GATE CONFIGURATION

To convert from ces configuration to cbg we swap ports 1 and 3 of Figure 8.4(a). In terms of the original ces three-port S-parameters, the three-port S-matrix for the cbg configuration then becomes:

$$\left[s^{\mathrm{cbg}} \right] = \begin{bmatrix} s_{33} & s_{32} & s_{31} \\ s_{23} & s_{22} & s_{21} \\ s_{13} & s_{12} & s_{11} \end{bmatrix} \tag{8.4.2}$$

Applying a short circuit at the new port 3 (the base/gate terminal) results in the reduced two-port S-matrix of cbg configuration. In terms of the ces three-port S-parameters, this can be written as follows:

$$\begin{bmatrix} S_{11}^{\mathrm{cbg}} & S_{12}^{\mathrm{cbg}} \\ S_{21}^{\mathrm{cbg}} & S_{22}^{\mathrm{cbg}} \end{bmatrix} = \begin{bmatrix} \left(s_{33} - \dfrac{s_{31}s_{13}}{1 + s_{11}} \right) & \left(s_{32} - \dfrac{s_{31}s_{12}}{1 + s_{11}} \right) \\ \left(s_{23} - \dfrac{s_{21}s_{13}}{1 + s_{11}} \right) & \left(s_{22} - \dfrac{s_{21}s_{12}}{1 + s_{11}} \right) \end{bmatrix} \tag{8.4.3}$$

8.4.2 COMMON COLLECTOR/DRAIN CONFIGURATION

To convert from ces configuration to ccd we swap ports 2 and 3 of Figure 8.4(a). In terms of the original ces three-port S-parameters, the three-port S-matrix for the ccd configuration then becomes:

$$\left[s^{\mathrm{ccd}} \right] = \begin{bmatrix} s_{11} & s_{13} & s_{12} \\ s_{31} & s_{33} & s_{32} \\ s_{21} & s_{23} & s_{22} \end{bmatrix} \tag{8.4.4}$$

Applying a short circuit at the new port 3 (the collector/drain terminal) results in the reduced two-port S-matrix of ccd configuration. In terms of the ces three-port S-parameters, this can be written as follows:

$$
\begin{bmatrix} S_{11}^{\text{ccd}} & S_{12}^{\text{ccd}} \\ S_{21}^{\text{ccd}} & S_{22}^{\text{ccd}} \end{bmatrix} = \begin{bmatrix} \left(s_{11} - \dfrac{s_{12}s_{21}}{1+s_{22}} \right) & \left(s_{13} - \dfrac{s_{12}s_{23}}{1+s_{22}} \right) \\ \left(s_{31} - \dfrac{s_{32}s_{21}}{1+s_{22}} \right) & \left(s_{33} - \dfrac{s_{32}s_{23}}{1+s_{22}} \right) \end{bmatrix}
\tag{8.4.5}
$$

Equations (8.4.3) and (8.4.5) allow the direct calculation of the two-port S-parameters for a transistor in cbg configuration or ccd configuration, respectively, given the CE three-port S-parameters of the device.

8.5 FEEDBACK MAPPINGS

8.5.1 INTRODUCTION

Equation (8.3.18) is a *bilinear transformation* [14] that allows us to calculate the reduced two-port S-parameters for a three-terminal device with any arbitrary third-port termination, Γ_3. A cursory investigation of equation (8.3.18) will reveal that changing the value of Γ_3 can alter the reduced two-port S-parameters dramatically. Just by way of example, if we set $\Gamma_3 = s_{33}$, equation (8.3.18) yields an infinite value of S_{ij}'.

In this section, we will introduce a graphical technique that provides a powerful visualization of the effect of feedback on the reduced two-port S-parameters, and allows us to determine the required feedback termination to engineer a specific set of reduced two-port S-parameters to achieve the desired objective. For the purposes of circuit design we often need to find values of Γ_3 which maximize or minimize a given reduced two-port parameter, S_{ij}, and for this purpose a graphical technique is most appropriate.

Since the function equation (8.3.18) is *bilinear*, it will map circles and straight lines on the Γ_3 plane into circles and straight lines on the S_{ij} plane. Thus the Smith Chart representing all passive values of third-port termination can be mapped onto the S_{ij} plane. Due to the translation, magnification, and rotation effects of the bilinear transformation, the Γ_3 Smith Chart may be severely distorted in shape by the mapping process. In other words, circles may be mapped to straight lines and vice versa.

The feedback mapping technique was first proposed by Bodway [8] and has been studied by others [13,15,16]. In the early days, feedback mapping was carried out by direct computation of equation (8.3.18) and plotting values of S_{ij}' in a point-by-point manner [15]. The centers and radii of feedback mapping circles can, however, be derived explicitly in closed form, as will be shown in this section.

The third-port reflection coefficient, Γ_3, is related to the normalized third-port terminating impedance, z_3, by another well-known bilinear transformation, namely:

$$
\Gamma_3 = \frac{z_3 - 1}{z_3 + 1}
\tag{8.5.1}
$$

We can substitute equation (8.5.1) into equation (8.3.18) to obtain S'_{ij} in terms of z_3 as follows:

$$S'_{ij} = s_{ij} + \frac{s_{i3}s_{3j}\left\{\dfrac{z_3 - 1}{z_3 + 1}\right\}}{1 - s_{33}\left\{\dfrac{z_3 - 1}{z_3 + 1}\right\}} \tag{8.5.2}$$

Rearranging equation (8.5.2) we get:

$$S'_{ij} = s_{ij} + \frac{s_{i3}s_{3j}(z_3 - 1)}{(z_3 + 1) - s_{33}(z_3 - 1)} \tag{8.5.3}$$

From which z_3 can be expressed in terms of S'_{ij} as follows:

$$z_3 = \frac{s_{i3}s_{3j} + (S'_{ij} - s_{ij})(s_{33} + 1)}{s_{i3}s_{3j} + (S'_{ij} - s_{ij})(s_{33} - 1)} \tag{8.5.4}$$

We can rearrange equation (8.5.4) into the form of yet another bilinear transformation as follows:

$$z_3 = \frac{A_{ij}S'_{ij} + C_{ij}}{B_{ij}S'_{ij} + D_{ij}} \tag{8.5.5}$$

where A_{ij}, B_{ij}, C_{ij}, and D_{ij} are feedback mapping coefficients defined as:

$$\left.\begin{aligned} A_{ij} &= (s_{33} + 1) \\ B_{ij} &= (s_{33} - 1) \\ C_{ij} &= s_{i3}s_{3j} - s_{ij}\,(s_{33} + 1) \\ D_{ij} &= s_{i3}s_{3j} - s_{ij}\,(s_{33} - 1) \end{aligned}\right\} \tag{8.5.6}$$

By separating equation (8.5.5) into real and imaginary parts we get expressions for the centers and radii of constant normalized third-port resistance and reactance circles in the S'_{ij} plane. The centers are as follows:

$$\Gamma_{rij} = \left[\frac{B^*_{ij}(C_{ij} - 2rD_{ij}) + D_{ij}A^*_{ij}}{2\left(|B_{ij}|^2 r - \mathrm{Re}(A_{ij}B^*_{ij})\right)}\right] \tag{8.5.7}$$

$$\Gamma_{xij} = -\left[\frac{2(B^*_{ij}D_{ij})x - j(D_{ij}A^*_{ij} - C_{ij}B^*_{ij})}{2\left(|B_{ij}|^2 x - \mathrm{Im}(A_{ij}B^*_{ij})\right)}\right] \tag{8.5.8}$$

The radii are given by:

$$\gamma_{rij} = \sqrt{|\Gamma_{rij}|^2 - \frac{|D_{ij}|^2 r - \text{Re}(D_{ij}^* C_{ij})}{|B_{ij}|^2 r - \text{Re}(A_{ij} B_{ij}^*)}} \tag{8.5.9}$$

$$\gamma_{xij} = \sqrt{|\Gamma_{xij}|^2 - \frac{|D_{ij}|^2 x - \text{Im}(D_{ij}^* C_{ij})}{|B_{ij}|^2 x - \text{Im}(A_{ij} B_{ij}^*)}} \tag{8.5.10}$$

where $i = 1, 2$ and $j = 1, 2$.

This is best illustrated by an example.

Example 8.4 Calculation of series feedback mapping parameters.

Problem. Calculate the constant resistance and constant reactance series feedback mapping circles on the reduced two-port S_{11}' plane for the device used in Example 8.2 for r and $x = 0, 0.5, 1$, and 2.

Solution. The series feedback three-port S-parameters have already been calculated in Example 8.2 and are given in equation (8.3.15). We use these three-port parameters to calculate the mapping coefficients for the S_{11}' plane using equation (8.5.6) as follows:

$$A_{11} = (s_{33} + 1) = (0.843\angle 85.0° + 1) = 1.3629\angle 38.0°$$

$$B_{11} = (s_{33} - 1) = (0.843\angle 85.0° - 1) = 1.2505\angle 137.8°$$

$$C_{11} = s_{13}s_{31} - s_{11}(s_{33} + 1)$$
$$= 0.453\angle 61.0° \times 1.700\angle -60.7° - 1.063\angle -41.2 \times 1.3629\angle 38.0°$$
$$= 0.7701\angle 0.3° - 1.4487\angle -3.2°$$
$$= 0.6817\angle 172.8°$$

$$D_{11} = s_{13}s_{31} - s_{11}(s_{33} - 1)$$
$$= 0.453\angle 61.0° \times 1.700\angle -60.7° - 1.063\angle -41.2 \times 1.2505\angle 137.8°$$
$$= 0.7701\angle 0.3° - 1.3292\angle 96.6°$$
$$= 1.6107\angle -55.0°$$

Applying equations (8.5.7) and (8.5.9) we calculate the center and radius of the mapped $r = 0$ in the S_{11}' plane as:

$$\Gamma_{0ij} = \left[\frac{B_{ij}^* C_{ij} + D_{ij} A_{ij}^*}{2 \left(|B_{ij}|^2 r - \text{Re}(A_{ij} B_{ij}^*) \right)} \right]$$

$$= \left[\frac{1.2505\angle - 137.8° \times 0.6817\angle 172.8° + 1.6107\angle - 55.0° \times 1.3629\angle - 38.0°}{-2\text{Re}(1.3629\angle 38.0° \times 1.2505\angle - 137.8°)} \right]$$

$$= \frac{1.8004\angle - 71.1°}{-0.5802}$$

$$= 3.103\angle 108.9° = -1.0051 + j2.9357$$

$$\gamma_{0ij} = \sqrt{|\Gamma_{rij}|^2 - \frac{\text{Re}(D_{ij}^* C_{ij})}{\text{Re}(A_{ij} B_{ij}^*)}}$$

$$= \sqrt{3.103^2 - \frac{\text{Re}(1.6107\angle 55.0° \times 0.6817\angle 172.8°)}{\text{Re}(1.3629\angle 38.0° \times 1.2505\angle - 137.8°)}}$$

$$= \sqrt{9.6286 - \frac{-0.7376}{-0.2901}}$$

$$= 2.662$$

Applying equations (8.5.8) and (8.5.10) we calculate the center and radius of the mapped $x = 0$ in the S'_{11} plane as:

$$\Gamma_{0ij} = -\left[\frac{-j(D_{ij} A_{ij}^* - C_{ij} B_{ij}^*)}{-2\text{Im}(A_{ij} B_{ij}^*)} \right]$$

$$= -j \left[\frac{(1.6107\angle - 55.0° \times 1.3629\angle - 38.0° - 0.6817\angle 172.8° \times 1.2505\angle - 137.8°)}{-2\text{Im}(1.3629\angle 38.0° \times 1.2505\angle - 137.8°)} \right]$$

$$= \frac{2.8018\angle 163.1}{3.3589}$$

$$= 0.8341\angle 163.1° = -0.7981 + j0.2425$$

$$\gamma_{0ij} = \sqrt{|\Gamma_{xij}|^2 - \frac{\text{Im}(D_{ij}^* C_{ij})}{\text{Im}(A_{ij} B_{ij}^*)}}$$

$$= \sqrt{0.8341^2 - \frac{\text{Im}(1.6107\angle 55.0° \times 0.6817\angle 172.8°)}{\text{Im}(1.3629\angle 38.0° \times 1.2505\angle - 137.8°)}}$$

$$= \sqrt{0.6957 - \frac{-0.8199}{-1.6794}}$$

$$= 0.4556$$

By following the same method of calculation, we determine the feedback mapping constant resistance and constant reactance circle parameters for $r, x = 0.5, 1.0, 2.0$ as set out in Tables 8.1 and 8.2.

Table 8.1 Feedback Mapping Constant Resistance Circle Parameters

r	Center	Radius
0.0	$3.103\angle 108.9°$	2.6537
0.5	$1.5522\angle -40.1°$	0.7188
1.0	$1.4020\angle -29.9°$	0.4157
2.0	$1.3346\angle -22.4°$	0.2255

Table 8.2 Feedback Mapping Constant Reactance Circle Parameters

x	Center	Radius
0.0	$0.8339\angle 163.2°$	0.4588
0.5	$0.9771\angle 164.9°$	0.3131
1.0	$1.0516\angle 165.6°$	0.2376
2.0	$1.1280\angle 166.2°$	0.1603

Equations (8.5.7) to (8.5.10) enable the complete Smith Chart representing the passive Γ_3 to be mapped onto the S_{ij} plane. Figure 8.8 shows series feedback mapping onto all four S_{ij} reduced two-port planes for a typical microwave transistor. Note that S_{11} and S_{22} mappings are overlaid on a Smith Chart whereas S_{12} and S_{21} mappings are overlaid on polar plots.

Figure 8.8 shows the power of feedback mapping to help us visualize the effects of feedback on an active device. We can see that the values of all four reduced two-port S-parameters can be manipulated to achieve our specific design goal by choosing the appropriate value of feedback termination, Γ_3. For example, the feedback mappings in Figure 8.8(a) and (d) extend beyond the boundary of the Smith Chart in the S_{11} and S_{22} planes, indicating that there exists a range of values of passive Γ_3 that can induce the transistor to exhibit negative resistance at the input port (i.e., $|S_{11}| > 1$) or the output port (i.e., $|S_{22}| > 1$). In the case of amplifier design, we may wish to obtain the minimum values of $|S_{11}|$ and $|S_{22}|$ as this gives the best input and output matching, respectively. The mappings of Figure 8.8(a) and (d) reveal the optimum values of Γ_3 in this case. The effect of the chosen feedback termination on the "gain" of the device (approximated by $|S_{21}|$) can be seen in Figure 8.8(c). Another example is the application of feedback to minimize $|S_{12}|$ so as to create an *active isolator*. Figure 8.8(b) shows that, with a specific value of Γ_3, we can achieve a reduced

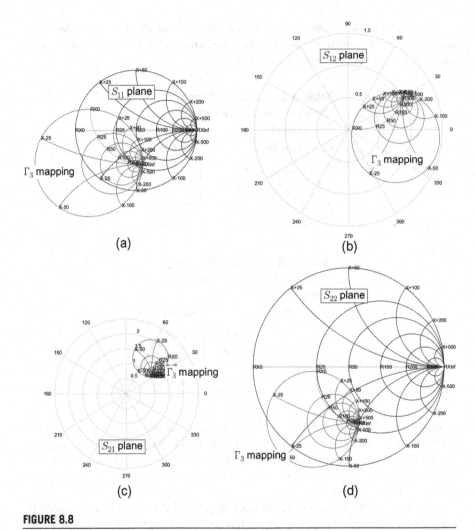

FIGURE 8.8

Typical series feedback mappings in the S_{ij} planes. (a) S_{11} plane mapping. (b) S_{12} plane mapping. (c) S_{21} plane mapping. (d) S_{22} plane mapping.

two-port $|S_{12}|$ very close to zero. We will elaborate on some of these examples in the following sections.

8.5.2 CLASSIFICATION OF FEEDBACK MAPPINGS

Feedback mappings can be classified into three classes depending on the radius of the $|\Gamma_3| = 1$ circle when mapped onto the respective S-plane [17]. The three classes shall be referred to as "bounded," "unbounded," and "inverted," for reasons which will become apparent. The type of mapping is significant because it determines the maximum magnitude of S_{ij} obtainable with a passive Γ_3. This is of particular significance in negative resistance oscillator design, where we normally apply feedback with the express intention of maximizing $|S_{11}|$ [18].

In this section, it will be shown that by studying the properties of the feedback mapping equations derived in the previous section, a scheme of feedback mapping classification can be arrived at which requires only a knowledge of the magnitude of the three-port S-parameter, s_{33}. Figures 8.9 to 8.11 show the three possible shapes of the feedback mapping. For clarity, only the $r_3 = 0$, $r_3 = 1$, and $x_3 = 0$ circles of the Γ_3 Smith Chart have been drawn. It is the nature of the $r_3 = 0$ ($|\Gamma_3| = 1$) circle which determines whether the mapping will be bounded, unbounded, or inverted. For a bounded mapping the radius of the $|\Gamma_3| = 1$ circle in the S'_{ij} plane is positive, as shown in Figure 8.9. Note that the $|\Gamma_3| = 1$ circle is also the $r_3 = 0$ circle, where r_3 is the normalized resistive part of the feedback termination.

For an unbounded mapping the $|\Gamma_3| = 1$ circle maps to a straight line in the S'_{ij} plane, therefore the radius of the mapped $|\Gamma_3| = 1$ circle is infinite, as shown in Figure 8.10.

For an inverted mapping the radius of the $|\Gamma_3| = 1$ circle in the S'_{ij} plane is negative, meaning that the mapped Γ_3 Smith Chart is turned "inside out" as shown in Figure 8.11.

Considering equation (8.5.7) we can see that the center of the $r_3 = 0$ circle are given by:

$$\Gamma_{0ij} = \frac{B^*_{ij}C_{ij} + D_{ij}A^*_{ij}}{-2\mathrm{Re}(A_{ij}B^*_{ij})} \tag{8.5.11}$$

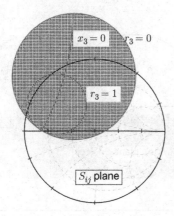

FIGURE 8.9

Bounded feedback mapping ($|\Gamma_3| = 1$ circle radius is positive).

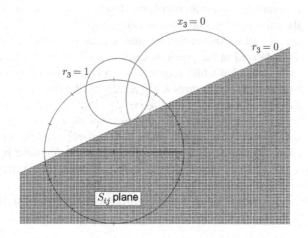

FIGURE 8.10

Unbounded feedback mapping ($|\Gamma_3| = 1$ circle radius is infinite).

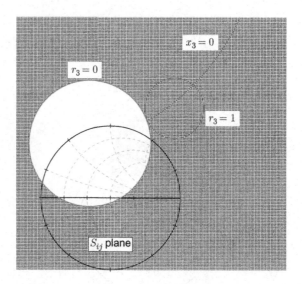

FIGURE 8.11

Inverted feedback mapping ($|\Gamma_3| = 1$ circle radius is negative).

By replacing the mapping constants A_{ij}, B_{ij}, C_{ij}, and D_{ij} in equation (8.5.11) by their three-port parameter equivalents given in equation (8.5.6), we can define center of the $r_3 = 0$ circle as:

$$\Gamma_{0ij} = \frac{2s_{ij}\text{Im}(s_{33}) - s_{i3}s_{3j}}{1 - |s_{33}|^2} \tag{8.5.12}$$

From equation (8.5.9) we can see that the radius of the $r_3 = 0$ circle is given by:

$$\gamma_{0ij} = \sqrt{\frac{|B_{ij}^* + D_{ij}A_{ij}^*|^2 - 2\text{Re}(A_{ij}B_{ij}^*)\text{Re}(D_{ij}^*C_{ij})}{2\text{Re}(A_{ij}B_{ij}^*)^2}} \tag{8.5.13}$$

In the case of the unbounded Smith Chart of Figure 8.10, the $r_3 = 0$ circle maps to a straight line on the S_{ij} plane. In other words, the radius of the $r_3 = 0$ circle is infinite. This occurs when the denominator of equation (8.5.13) is zero, that is, when:

$$\text{Re}(A_{ij}B_{ij}^*) = 0 \tag{8.5.14}$$

Employing the definitions of the mapping constants in terms of the three-port parameters given in equation (8.5.6) we can re-write equation (8.5.14) as:

$$\text{Re}[(s_{33} + 1)(s_{33}^* - 1)] = 0 \tag{8.5.15}$$

which is equivalent to:

$$\text{Re}[|s_{33}|^2 - 2j\text{Im}(s_{33}) - 1] = 0 \tag{8.5.16}$$

which reduces to:

$$|s_{33}|^2 - 1 = 0 \tag{8.5.17}$$

and therefore

$$|s_{33}| = 1 \tag{8.5.18}$$

The magnitude of the three-port parameter, s_{33}, is therefore critical in determining the type of mapping. The inverted mapping of Figure 8.11 corresponds to the $r_3 = 0$ circle having a negative center while all the remaining r_3 circles have positive centers on the S_{ij} plane. By contrast, all the r_3 circles of the bounded mapping (Figure 8.9) have positive centers. We can determine the shape of the mapped Smith Chart on the S_{ij} plane solely by the magnitude of s_{33}, as follows [17]:

$$|s_{33}| < 1: \text{Bounded mapping}$$
$$|s_{33}| = 1: \text{Unbounded mapping} \tag{8.5.19}$$
$$|s_{33}| > 1: \text{Inverted mapping}$$

The bounded Smith Chart of Figure 8.9 indicates that all passive values of Γ_3 correspond to a finite value of S_{ij} which may or may not represent a negative resistance depending on whether the mapping lies inside or outside the S_{ij} circle. On the other hand, the unbounded and inverted mappings of Figures 8.10 and 8.11 suggest that a passive value of Γ_3 exists that will give rise to an infinite value of S_{ij}.

The importance of feedback mappings lies in the fact that any required value of $|S_{ij}|$ obtainable with shunt or series feedback can be determined easily. In the case of a bounded mapping, the maximum value of $|S_{ij}|$ can be determined from Figure 8.9 as:

$$|S'_{ij}|\text{max} = |\Gamma_{oij}| + \gamma_{oij} \tag{8.5.20}$$

If $|S'_{ij}|\text{max}$ is less than unity for a given device in a given configuration then that device/feedback combination is unsuitable for oscillator design.

In all cases, it is preferable to use a lossless feedback termination (i.e., a pure reactance, $|\Gamma_3| = 1$) since microwave resistors introduce unwanted electrical noise, In the case of Figures 8.9 and 8.10, the maximum value of $|S_{ij}|$ obtainable with a passive third-port termination is obtained with a pure reactance, whereas in the case of Figure 8.11 an infinite value of $|S_{ij}|$ can be obtained with a lossy third-port termination.

The value of feedback termination, Γ_3, which results in an infinite value of S_{ij} is known as a "pole" of S_{ij} in the Γ_3 plane [17].

The poles of a given shunt or series feedback configuration can be determined by utilizing equation (8.3.18). If we set $1/S_{ij} = 0$ and solve for Γ_3 we get the following:

$$\frac{1}{S_{ij}} = \frac{1 - s_{33}\Gamma_3}{s_{ij} + \Gamma_3(s_{i3}s_{3j} - s_{ij}s_{33})} = 0 \tag{8.5.21}$$

which implies:

$$1 - s_{33}\Gamma_3 = 0$$

leading to:

$$\Gamma_3 = \frac{1}{s_{33}} = \Pi_3 \tag{8.5.22}$$

Equation (8.5.22) shows that there is one unique pole, Π_3, in the Γ_3 plane for all the reduced two-port S-parameters.

The significance of the pole positions can be summarized as follows:

1. If the pole lies inside the Γ_3 Smith Chart (i.e., if $|s_{33}| > 1$) then an infinite value of S_{ij} can be produced with a passive feedback termination and the feedback mapping in the S_{ij} plane will be inverted.
2. If the pole lies on the boundary of the Γ_3 Smith Chart (i.e., if $|s_{33}| = 1$) then an infinite value of S_{ij} can be produced with a lossless feedback termination and the feedback mapping in the S_{ij} plane will be unbounded.

3. If the pole lies outside the Γ_3 Smith Chart (i.e., if $|s_{33}| = 1$) then only finite values of S_{ij} can be produced with a passive feedback termination and the feedback mapping in the S_{ij} plane will be bounded.

Table 8.3 summarizes the relationship between s_{33} and optimum terminations, based on feedback mapping type.

Table 8.3 Classification of Feedback Mappings

$|s_{33}| < 1$	*Bounded mapping:* Pole lies outside the $|\Gamma_3| = 1$ circle Theoretical optimum termination is *active*
$|s_{33}| = 1$	*Unbounded mapping:* Pole lies on the $|\Gamma_3| = 1$ circle Theoretical optimum termination is *purely reactive*
$|s_{33}| > 1$	*Inverted mapping:* Pole lies inside the $|\Gamma_3| = 1$ circle Theoretical optimum termination is *passive*

8.6 APPLICATION OF THREE-PORT DESIGN TECHNIQUES

We will now give some examples of the feedback mapping technique in active circuit design.

8.6.1 GENERATING NEGATIVE RESISTANCE IN TRANSISTORS

One approach to negative resistance transistor oscillator design is to treat the transistor as a two-port network, induce negative resistance at one port and couple the output power from the other port. This design methodology is discussed in more detail in Chapter 15, but let us just assume, for the purposes of this section, that we need to design a single transistor two-port subcircuit having $|S_{ii}| > 1$. This condition usually requires external feedback, although it can sometimes be achieved by simply changing the device configuration. When feedback is used, a single series feedback element is usually applied. A very common approach with bipolar transistors is to use inductive series feedback with the transistor in the CB configuration [18,19], but negative resistance has also been demonstrated with series feedback applied in the CE configuration [6,20].

Figure 8.12 shows series feedback two-port subcircuits created by adding a passive series feedback termination, Γ_3, to the common port of a transistor in the three possible configurations, with port 2 of the subcircuit terminated in the system

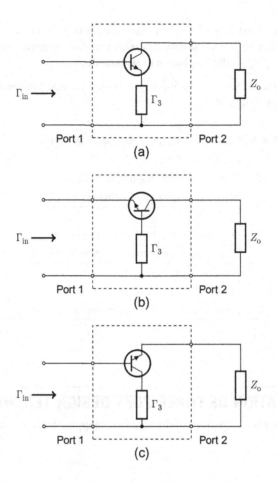

FIGURE 8.12

Series feedback two-port subcircuits. (a) CE with series feedback. (b) CB with series feedback. (c) CC with series feedback.

characteristic impedance, Z_0. Bias circuitry has been omitted for simplicity. To implement an oscillator using the subcircuits in Figure 8.12, we need to couple a passive resonator to port 1.

The input reflection coefficient of the CE two-port subcircuit in Figure 8.12(a) is calculated given the three-port S-parameters, s_{ij}, and an arbitrary feedback termination, Γ_3, using equation (8.3.18) as follows:

$$\Gamma_{in} = s_{11} + \frac{s_{13}s_{31}\Gamma_3}{1 - s_{33}\Gamma_3} \tag{8.6.1}$$

Equation (8.6.1) applies equally to the other two configurations in Figure 8.12 with appropriate port re-assignment, as per Section 8.4.

A series feedback termination which is as close as practicable to the "Pole," as defined by equation (8.5.22), will generate the maximum magnitude of Γ_{in}, and will therefore be optimum from the point of view of generating negative resistance at port 1. It is important to note that the term "optimum" is here simply taken to mean the feedback termination that results in a maximum magnitude of Γ_{in}. This may not necessarily result in negative resistance at the ports of a given transistor as the maximum may still be less than unity for that configuration, under the specific bias conditions. Furthermore, whereas a theoretically infinite value of $|\Gamma_{in}|$ can be achieved if the port 3 termination is equal to the Pole value, this is neither achievable nor desirable in practice. We do not usually want the impedance looking into port 1 of the active two-port subcircuit to be exactly $-Z_o$. Having said this, selecting a series feedback termination close to the Pole in the Γ_3 plane is still the best approach, as this will maximize the likelihood of achieving the $|\Gamma_{in}| > 1$ over the widest frequency range. We can therefore still refer to such a feedback termination, Γ_{3opt}, as "optimum" from the point of view of practical negative resistance oscillator design.

In practice, we avoid using resistive elements in the feedback circuit, so as to minimize thermal noise and losses. We will therefore restrict our analysis to purely reactive feedback terminations, that is $|\Gamma_{3opt}| = 1$. We can now refine the question of determining Γ_{3opt} to that of finding the phase angle of the Pole, Φ_{3opt}. Our optimum termination, Γ_{3opt}, now becomes:

$$\Gamma_{3opt} = 1\angle\Phi_{3opt} \tag{8.6.2}$$

where Φ_{3opt} is given by:

$$\Phi_{3opt} = \arctan\left[\frac{\text{Im}(\Pi_3)}{\text{Re}(\Pi_3)}\right] \tag{8.6.3}$$

where $\text{Re}(\Pi_3)$ and $\text{Im}(\Pi_3)$ represent, respectively, the real and imaginary parts of the Pole, Π_3.

From Table 8.3 we conclude that, if the mapping is bounded then only finite values of Γ_{in} can be achieved with a passive feedback termination. In this case, an "active" termination (i.e., $|\Gamma_3| > 1$) would be required to achieve an infinite value of Γ_{in}. If the mapping is unbounded, however, a theoretically infinite value of Γ_{in} can be achieved using a purely reactive third-port termination having the exact value of Π_3. If the mapping is inverted an infinite value of Γ_{in} could, in theory, be obtained with a passive, lossy, feedback termination having the exact value of Π_3.

It turns out that equation (8.6.2) is a good working approximation to the optimum third-port termination for all three classes of mapping listed above, even though, in the case of an inverted mapping, a greater value of $|\Gamma_{in}|$ could be obtained by using a lossy third-port termination.

The three-port S-parameter, s_{33}, may be calculated directly from the original device two-port S-parameters using equation (8.3.8). If we now combine equations (8.5.22), (8.6.2), (8.6.3), (8.3.8), and (8.3.9) we arrive at the following equation for the optimum lossless series feedback termination for negative resistance generation, solely in terms of the original device two-port S-parameters:

$$\Gamma_{3opt} = 1\angle \arctan\left(\frac{-4\text{Im}(\xi)}{4\text{Re}(\xi) - |\xi|^2}\right) \tag{8.6.4}$$

If we envisage the argument of the arctan in equation (8.6.4) as representing a vector in two-dimensional space with the vertical axis represented by $(-4\text{Im}(\xi))$ and the horizontal axis represented by $(4\text{Re}(\xi) - |\xi|^2)$, we can then see that a negative value of $\text{Im}(\xi)$ puts Φ_{3opt} in the upper half plane (between $0°$ and $180°$) and a positive value of $\text{Im}(\xi)$ puts Φ_{3opt} in the lower half plane (between $0°$ and $-180°$). The sign of $(4\text{Re}(\xi) - |\xi|^2)$ will determine whether Φ_{3opt} lies in the left- or right-hand half planes, which is of little significance for our purposes. Since Φ_{3opt}, the phase angle of the feedback reflection coefficient, determines whether the feedback reactance is capacitive or inductive, the implications of equation (8.6.4) can be summarized in Table 8.4.

Table 8.4 Optimum Reactive Terminations, Based on $\text{Im}(\xi)$

$\text{Im}(\xi) > 0$	Optimum feedback termination is *capacitive*
$\text{Im}(\xi) = 0$	Optimum feedback termination is *resistive*
$\text{Im}(\xi) < 0$	Optimum feedback termination is *inductive*

Thus, we have established a useful "rule of thumb" which can be stated as follows: "the type of series feedback reactance required to generate negative resistance in a given transistor configuration depends only on the sign of the imaginary part of the sum of the device two-port S-parameters for that configuration" [6].

Example 8.5 (Generating negative resistance in a transistor using feedback).

Problem. Carry out a computational exercise to test the validity of equation (8.6.4), using the S-parameters of the general purpose SiGe RF transistor, type BFP620 at the bias conditions of $V_{cc} = 2.0$ V, $I_c = 8$ mA (Table 8.5).

Compare the behavior of CE, CB, and CC configurations.

Table 8.5 S-Parameters for BFP620 at $V_{cc} = 2.0$ V, $I_c = 8$ mA

F (GHz)	S_{11}	S_{12}	S_{21}	S_{22}
1	$0.567\angle - 93.5°$	$0.065\angle 50.5°$	$13.24\angle 113.6°$	$0.583\angle - 60.9°$
2	$0.423\angle - 150°$	$0.092\angle 40.1°$	$7.67\angle 84.1°$	$0.302\angle - 90.9°$
3	$0.413\angle 172.7°$	$0.115\angle 34.7°$	$5.24\angle 64.8°$	$0.166\angle - 120.6°$

Solution. First, we need to calculate the CB and CC two-port S-parameters using the formulas given in Section 8.4, which means that we need to determine the CE series three-port S-parameters, using equation (8.3.10).

Table 8.6 shows Φ_{3opt} calculated using equation (8.6.4) and Im(ξ) calculated using equation (8.3.9) for this device in the three possible configurations (CE, CB, and CC) at the stated bias conditions.

Table 8.6 Computed Parameters for BFP620 at $V_{cc} = 2.0$ V, $I_c = 8$ mA

F	CE		CB		CC	
(GHz)	Im(ξ)	Φ_{3opt}	Im(ξ)	Φ_{3opt}	Im(ξ)	Φ_{3opt}
1	11.109	−165.2°	−0.254	14.5°	−0.220	12.6°
2	7.175	−150.0°	−0.457	26.5°	−0.416	23.8°
3	4.719	−134.1°	−0.466	30.6°	−0.490	29.8°

Computing Γ_{in} using equation (8.6.1) with each value of $\Gamma_{3opt} = 1\angle\Phi_{3opt}$ given in Table 8.6, we get the results shown in Table 8.7.

Table 8.7 Computed $|\Gamma_{in}|$ at Port 1 for BFP620 at $V_{cc} = 2.0$ V, $I_c = 8$ mA

F (GHz)	CE	CB	CC
1	2.34	14.17	38.00
2	6.53	4.51	5.74
3	38.15	1.76	2.04

From Table 8.7 we can see that all of the optimum feedback terminations defined by the Φ_{3opt} values in Table 8.6 result in negative resistance (i.e., $|\Gamma_{in}| > 1$) at port 1 of the BFP620 at $V_{cc} = 2.0$ V, $I_c = 8$ mA when it is placed in any of the two-port subcircuits in Figure 8.12. Table 8.6 also supports the prevailing understanding that inductive series feedback should be used to generate negative resistance in the CB configuration, but that capacitive series feedback should be used to generate negative resistance in the CE and CC configurations.

8.6.2 THE ACTIVE ISOLATOR

The *isolator* is a commonly used component in microwave subsystems to shield the output of one subsystem from the effects of variation in load conditions. Isolators are applied, for example, to prevent a microwave source being detuned by a mismatched load.

An ideal isolator is a nonreciprocal device, with the following nonsymmetric two-port S-matrix:

$$\begin{bmatrix} S_{11} & S_{12} \\ S_{21} & S_{22} \end{bmatrix} = \begin{bmatrix} 0 & 0 \\ 1 & 0 \end{bmatrix} \tag{8.6.5}$$

In other words, an ideal isolator would transmit all the power entering port 1 to port 2, while absorbing all the power entering port 2. Practical isolators are usually fabricated using fixed magnets and special ferrite materials, which makes the device quite bulky and unsuitable for fabrication in MMIC form.

We can apply feedback mappings to help us design an *active isolator* based on a single transistor, the goal being to apply feedback so as to minimize S_{12}. The feedback can be either shunt or series type.

The value of feedback termination, Γ_3, required to achieve unilateralization can be determined by setting $S_{12} = 0$ in equation (8.3.18) and solving for Γ_3:

$$\Gamma_3 = \frac{s_{12}}{s_{12}s_{33} - s_{13}s_{32}} \tag{8.6.6}$$

It should be noted that equation (8.6.6) applies equally to either shunt or series feedback, depending on whether shunt or series three-port S-parameters are being used. In practice, shunt feedback is more often used to achieve unilateralization [21, 22]. We can illustrate the design of such circuits by means of Example 8.6.

Example 8.6 (Unilateralization of a transistor).

Problem. The BFR93 transistor has the following CE S-parameters at 1.8 GHz and with bias conditions $V_{cc} = 5$ V and $I_c = 10$ mA:

$$\begin{bmatrix} S_{11} & S_{12} \\ S_{21} & S_{22} \end{bmatrix} = \begin{bmatrix} 0.382\angle161.7° & 0.202\angle67° \\ 2.053\angle59.3° & 0.275\angle-53.5° \end{bmatrix}$$

Unilateralize this transistor at 1.8 GHz, using shunt feedback.

Solution. We start by calculating the shunt feedback three-port S-parameters of the transistor, using the formulae given in Section 8.3.1. The normalized y-parameters of the transistor are calculated using the conversion formulas given in Appendix A as follows:

$$\begin{bmatrix} y_{11} & y_{12} \\ y_{21} & y_{22} \end{bmatrix} = \begin{bmatrix} (1.1993 + j0.2919) & (-0.0309 - j0.3771) \\ (-0.8247 - j3.7562) & (0.0627 + j0.5885) \end{bmatrix}$$

By following a similar procedure as used in Example 8.2, which we will not repeat in detail here, we arrive at the shunt feedback three-port S-parameters of the device as follows:

$$\begin{bmatrix} s_{11} & s_{12} & s_{13} \\ s_{21} & s_{22} & s_{23} \\ s_{31} & s_{32} & s_{33} \end{bmatrix} = \begin{bmatrix} (0.592\angle138.7°) & (0.348\angle39.3°) & (0.334\angle30.8°) \\ (1.323\angle83.2°) & (0.427\angle99.9°) & (1.181\angle130.8°) \\ (1.006\angle-66.5°) & (0.687\angle-163.1°) & (0.725\angle-85.4°) \end{bmatrix}$$

We can now apply equation (8.6.6), using the above three-port parameter values, to calculate the required value of shunt feedback termination, thus:

$$\Gamma_3 = \frac{s_{12}}{s_{12}s_{33} - s_{13}s_{32}}$$

$$= \frac{0.348\angle 39.3°}{0.348\angle 39.3° \times 0.725\angle -85.4° - 0.334\angle 30.8° \times 0.687\angle -163.1°}$$

$$= \frac{0.348\angle 39.3°}{0.3294\angle -2.1°}$$

$$= 1.056\angle 41.4°$$

Since $|\Gamma_3|$ is greater than unity in this case, the required feedback termination is unrealizable with passive components. We will therefore choose the closest possible passive feedback termination, that is, we set $|\Gamma_3| = 1$, resulting in a feedback reflection coefficient of $\Gamma_3 = 1\angle 41.4°$, which represents an inductance of around 11.7 nH at 1.8 GHz. The complete active isolator circuit is shown in Figure 8.13.

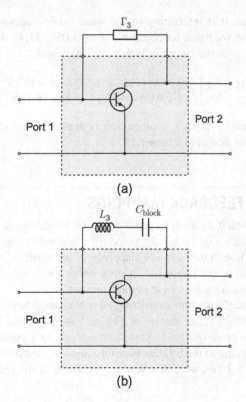

(a)

(b)

FIGURE 8.13

Unilateralization of a transistor using shunt feedback.

Note the need for a DC blocking capacitor, C_{block}, that must be inserted to separate the collector and base DC bias voltages. This capacitor must be large enough so that its reactance will be negligible at 1.8 GHz, which normally means that its reactance should be less than 13 Ω (one-tenth of the inductor reactance). This would put the value of C_{block} at something greater than 7 pF.

With a feedback termination of $\Gamma_3 = 1\angle 41.4°$, the reduced two-port S-parameters of the circuit in Figure 8.13 at 1.8 GHz can be calculated from equation (8.3.18) to be:

$$\begin{bmatrix} S_{11} & S_{12} \\ S_{21} & S_{22} \end{bmatrix} = \begin{bmatrix} 0.108\angle 136.2° & 0.026\angle -8.0° \\ 2.969\angle 70.0° & 0.904\angle -18.6° \end{bmatrix}$$

So, we have been able to reduce $|S_{12}|$ by almost a factor of 10 by applying inductive shunt feedback. In dB terms, this is the difference between 13 dB of isolation for the basic transistor and 32 dB for the transistor with feedback. Interestingly, we also notice that adding inductive feedback has also improved S_{11} (lower magnitude) and S_{21} (higher magnitude) at the expense of S_{22}, which has increased.

For completeness, it is interesting to see what would happen if it was in fact possible to apply the feedback termination $\Gamma_3 = 1.056\angle 41.4°$. If so, the reduced two-port S-parameters of the circuit in Figure 8.13(a) would be:

$$\begin{bmatrix} S_{11} & S_{12} \\ S_{21} & S_{22} \end{bmatrix} = \begin{bmatrix} 0.0864\angle 155.5° & 0.0004\angle -127.6° \\ 3.0394\angle 67.5° & 0.9424\angle -23.9° \end{bmatrix}$$

This transistor plus feedback combination is applied as an (almost) unilateral device in an amplifier design in Chapter 13.

8.7 REVERSE FEEDBACK MAPPINGS

An alternative approach to analyzing the effect of feedback on a transistor is to plot circles of constant $|S_{ij}|$ on the Γ_3 plane. In other words, we refer to these as *reverse mapping* although they are not mappings of the entire S_{ij}, and only provide magnitude information, not phase information about the S-parameters obtainable. They are, nonetheless, a useful tool in certain circumstances.

Consider a reduced two-port consisting of a transistor with a shunt or series feedback termination, Γ_3, as shown in Figure 8.4(b) or (e). We have already established that any point on the S_{ij} plane corresponds to a point on the Γ_3 plane and vice versa according to the bilinear transformation of equation (8.3.18).

From equation (8.3.18), we can state the magnitude of the reduced two-port S-parameter as follows:

$$|S'_{ij}| = \left| \frac{s_{ij} - (s_{ij}s_{33} - s_{i3}s_{3j})\Gamma_3}{1 - s_{33}\Gamma_3} \right|$$

Which can be rewritten as:

$$|S'_{ij}| = \left| \frac{s_{ij} - \Delta_{ij}\Gamma_3}{1 - s_{33}\Gamma_3} \right| \tag{8.7.1}$$

where $\Delta_{ij} = s_{ij}s_{33} - s_{i3}s_{3j}$.

Expanding equation (8.7.1) results in the following:

$$|\Gamma_3| \left[|S'_{ij}|^2|s_{33}|^2 - |\Delta_{ij}|^2 \right] - 2\mathrm{Re}(\Gamma_3) \left[|S'_{ij}|^2\mathrm{Re}\left(s_{33}^*\right) - \mathrm{Re}\left(s_{ij}^*\Delta_{ij}\right) \right] -$$
$$2\mathrm{Im}\,(\Gamma_3) \left[|S'_{ij}|^2\mathrm{Im}\left(s_{33}^*\right) - \mathrm{Im}(s_{ij}^*\Delta_{ij}) \right] +$$
$$\left(|S'_{ij}|^2 - |s_{ij}|^2 \right) = 0 \tag{8.7.2}$$

Dividing equation (8.7.2) throughout by the term $[|S'_{ij}|^2|s_{33}|^2 - |\Delta_{ij}|^2]$ results in:

$$|\Gamma_3| - 2\mathrm{Re}(\Gamma_3) \left[\frac{|S'_{ij}|^2\mathrm{Re}(s_{33}^*) - \mathrm{Re}(s_{ij}^*\Delta_{ij})}{|S'_{ij}|^2|s_{33}|^2 - |\Delta_{ij}|^2} \right] -$$
$$2\mathrm{Im}(\Gamma_3) \left[\frac{|S'_{ij}|^2\mathrm{Im}(s_{33}^*) - \mathrm{Im}(s_{ij}^*\Delta_{ij})}{|S'_{ij}|^2|s_{33}|^2 - |\Delta_{ij}|^2} \right] +$$
$$\frac{|S'_{ij}|^2 - |s_{ij}|^2}{|S'_{ij}|^2|s_{33}|^2 - |\Delta_{ij}|^2} = 0 \tag{8.7.3}$$

Equation (8.7.3) describes a circle in the reduced two-port S_{ij} plane with a center at:

$$\Gamma_{\mathrm{T}ij} = \frac{|S'_{ij}|^2 s_{33}^* - s_{ij}\Delta_{ij}^*}{|S'_{ij}|^2|s_{33}|^2 - |\Delta_{ij}|^2} \tag{8.7.4}$$

and a radius given by:

$$\gamma_{\mathrm{T}ij} = \sqrt{ \left[|\Gamma_{\mathrm{T}ij}|^2 - \frac{|S'_{ij}|^2 - |s_{ij}|^2}{|S'_{ij}|^2\,|s_{33}|^2 - |\Delta_{ij}|^2} \right] } \tag{8.7.5}$$

Further manipulation of equation (8.7.5) results in the following, simpler expression:

$$\gamma_{\mathrm{T}ij} = \frac{|S'_{ij}||\Delta_{ij}^* - s_{33}^*s_{ij}|}{|S'_{ij}|^2|s_{33}|^2 - |\Delta_{ij}|^2} \tag{8.7.6}$$

If the required value of $|S_{ij}|$ is known then the constant $|S_{ij}|$ circle allows the corresponding values of Γ_3 to be determined. It is important to note that, since every

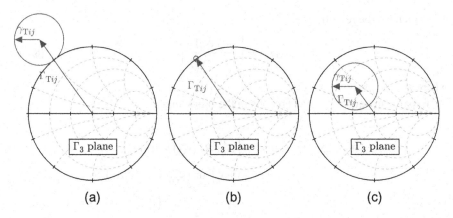

FIGURE 8.14

Reverse feedback mappings. (a) Bounded ($|s_{33}| < 1$). (b) Unbounded ($|s_{33}| = 1$). (c) Inverted ($|s_{33}| > 1$).

point on a constant $|S_{ij}|$ circle on the Γ_3 plane represents a complex value of Γ_3, there are an infinite number of different values of Γ_3 for any given value of $|S_{ij}|$.

Let us now consider the specific case of the $|S_{ij}| = 1$ circle, which, in the case of S_{11} and S_{22}, defines the boundary between positive and negative resistance at the input and output ports of the device, respectively. In the case of a bounded mapping the center of the $|S_{ij}| = 1$ reverse mapping circle lies outside the Γ_3 Smith Chart as shown in Figure 8.14(a).

In this case, the center and radius of the Γ_3 for $|S_{ij}|_{max}$ circle satisfy the following condition:

$$|\Gamma_{Tij}| - \gamma_{Tij} = 1 \qquad (8.7.7)$$

For an unbounded mapping, the pole lies on the boundary of the Γ_3 Smith Chart as shown in Figure 8.14(b). The radius of the Γ_3 for $|S_{ij}|_{max}$ circle is therefore zero and the following condition is satisfied with regard to the center:

$$|\Gamma_{Tij}| = 1 \qquad (8.7.8)$$

For an inverted mapping, the center of the $|S_{ij}| = 1$ reverse mapping circle lies inside the Γ_3 Smith Chart as shown in Figure 8.14(c). In this case, the center and radius of the Γ_3 for $|S_{ij}|_{max}$ circle satisfy the following condition:

$$|\Gamma_{Tij}| + \gamma_{Tij} \leq 1 \qquad (8.7.9)$$

Reverse feedback mappings are useful when we know exactly what magnitude of S_{ij} we require, and we are not concerned about its phase. This is the case, for example, when we wish to determine the maximum gain available from a device, when we wish to achieve a specific input or output VSWR, or when we are looking to generate negative resistance (in oscillator design).

8.8 TAKEAWAYS

1. Because transistors are inherently three-terminal devices, it is advantageous, when designing transistor circuits involving external feedback, to describe the transistor in terms of three-port parameters.
2. The three-port Y-matrix, otherwise known as the "indefinite admittance matrix," or "IAM," is easily calculated from the CE or CS two-port Y-matrix of the transistor. The three-port S-matrix can similarly be calculated from the CE or CS two-port S-matrix of the transistor.
3. In the case of three-port S-parameters, we need to distinguished between series feedback and shunt feedback three-port S-matrices, which have a different definition and application.
4. Converting two-port S-parameters for one transistor configuration into another, such as converting CE to CB of CC, can be easily achieved using three-port S-parameters.
5. Feedback mapping is a powerful graphical technique that allows the designer to see the effect of feedback on the two-port S-parameters of a transistor, and choose optimum feedback terminations in order to achieve a design objective, such as maximizing or minimizing the magnitude of a particular S-parameter.
6. For a given transistor, there will be one particular value of feedback termination that will result in theoretically infinite values of all four two-port S-parameters. This is referred to as the "Pole" of that particular S-matrix in the Γ_3 plane, and its value is equal to $1/s_{33}$.
7. Reverse feedback mappings are a quick way of determining values of feedback termination when we know exactly what magnitude of S_{ij} we require.

TUTORIAL PROBLEMS

Problem 8.1 (Y-parameter configuration conversion). The BFU690 heterojunction bipolar transistor has the following CE S-parameters at 4 GHz with bias conditions of 2.5 V and 10 mA:

$$\begin{bmatrix} S_{11} & S_{12} \\ S_{21} & S_{22} \end{bmatrix} = \begin{bmatrix} 0.829\angle 154.8° & 0.070\angle 19° \\ 1.909\angle 40° & 0.332\angle -172° \end{bmatrix}$$

Calculate the CB and CC two-port Y-parameters for the above device.

Problem 8.2 (Using feedback to minimize input return loss). You are required to use the device of Problem 8.1 to design a circuit having the lowest possible return loss at the input port. In other words, you need to design an active two-port network having $|S_{11}|$ to be as small as possible. Design such a circuit by applying a series and/or shunt feedback mapping technique.

Problem 8.3 (Unilateralization using series feedback). Repeat the exercise in Example 8.6 (unilateralization of a transistor) this time employing series feedback.

Problem 8.4 (Feedback mapping). Draw series feedback mapping circles on the S_{21} plane for the BFP620 SiGe BJT at 15 GHz with bias conditions of 2.5 V and 20 mA:

$$\begin{bmatrix} S_{11} & S_{12} \\ S_{21} & S_{22} \end{bmatrix} = \begin{bmatrix} 0.7929\angle 31° & 0.2634\angle -69.7° \\ 1.503\angle -82° & 0.3733\angle 26.6° \end{bmatrix}$$

Using the mapping you have produced, determine by how much the gain of the device can be increased by using a reactive series feedback termination, while still maintaining stable operation.

Problem 8.5 (Reverse mapping). Draw the reverse mapping circles on the S_{21} plane for the device in Problem 8.4 for $|S_{21}| = 2.0$ and thereby determine the two possible values of reactive series feedback termination that can be used to achieve this value of gain.

REFERENCES

[1] Y. Satoda, G. Bodway, Three-port scattering parameters for microwave transistor measurement, IEEE J. Solid State Circuits 3 (3) (1968) 250-255, ISSN 0018-9200, http://dx.doi.org/10.1109/JSSC.1968.1049894.

[2] U. Mahalingam, S. Rustagi, G. Samudra, Three-port RF characterization of MOS transistors, in: 65th ARFTG Conference Digest, 2005, Spring, 2005, pp. 56-61, http://dx.doi.org/10.1109/ARFTGS.2005.1500569.

[3] D. Woods, General renormalisation transforms for 3-port S parameters, Electron. Lett. 6 (25) (1970) 823-825, ISSN 0013-5194, http://dx.doi.org/10.1049/el:19700568.

[4] D. Woods, Transistor S parameter measurement: 2-port to 3-port transforms, Electron. Lett. 7 (16) (1971) 451-452, ISSN 0013-5194, http://dx.doi.org/10.1049/el:19710306.

[5] R. Collin, Foundations for Microwave Engineering, second ed., John Wiley and Sons, New York, NY, 2005, ISBN 9788126515288.

[6] C. Poole, I. Darwazeh, A simplified approach to predicting negative resistance in a microwave transistor with series feedback, in: IEEE MTT-S International Microwave Symposium, IEEE MTT-S International Microwave Symposium, Seattle, USA, 2013.

[7] R. Carson, High Frequency Amplifiers, John Wiley and Sons, New York, NY, 1975, ISBN 0471137057.

[8] G. Bodway, Circuit design and characterisation of transistors by means of three-port scattering parameters, Microw. J. 11 (5) (1968) 4-27.

[9] J. Cote, Matrix analysis of oscillators and transistor applications, IRE Trans. Circuit Theory 5 (1958) 181-188.

[10] K. Johnson, Large signal GaAs MESFET oscillator design, IEEE Trans. Microw. Theory Tech. 27 (3) (1979) 217-227.

[11] A. Khanna, Three-port S-parameters ease GaAs MESFET designing, Microwav. RF (1985) 81-84.

[12] D. Eungdamrong, D. Misra, Working with transistor S-parameters, RF Des. (2002) 38-42.

[13] M. Medley, Microwave and RF Circuits : Analysis, Synthesis and Design, Artech House, Boston, 1993.

[14] E. Kreyszig, Advanced Engineering Mathematics, John Wiley & Sons, 2010, ISBN 9780470458365.

[15] R. Dougherty, Feedback analysis and design techniques, Microw. J. (1985) 133-148.

[16] R. Pengelly, Microwave Field Effect Transistors: Theory, Design and Applications, John Wiley and Sons, Chichester, 1982.

[17] C. Poole, MESFET oscillator design based on feedback mapping classification, in: 33rd Midwest Symposium on Circuits and Systems, vol. 1, Calgary, Canada, 1990, pp. 609-612.

[18] G. Basawapatna, R. Stancliff, A unified approach to the design of wideband microwave solid state oscillators, IEEE Trans. Microw. Theory Tech. 27 (5) (1979) 379-385.

[19] G. Gonzalez, Foundations of Oscillator Circuit Design, Artech House, Boston, MA, 2007.

[20] P. Yip, High-Frequency Circuit Design and Measurements, Chapman and Hall, London, 1990.

[21] M. Gupta, Power gain in feedback amplifiers: a classic revisited, IEEE Trans. Microw. Theory Tech. 40 (5) (1992) 864-879, ISSN 0018-9480, http://dx.doi.org/10.1109/22.137392.

[22] S. Mason, Power gain in feedback amplifiers, Trans. IRE Prof. Group Circuit Theory 1 (2) (1954) 20-25, ISSN 0197-6389, http://dx.doi.org/10.1109/TCT.1954.1083579.

Lumped element matching networks

CHAPTER OUTLINE

INTENDED LEARNING OUTCOMES

- *Knowledge*
  - Understand the theory of *L*-section matching networks both analytically and graphically.
  - Understand forbidden regions on the Smith Chart for various types of *L*-section.
  - Understand the theory of three-element matching networks (π-section and *T*-section).
  - Understand bandwidth performance of lumped element matching networks.
- *Skills*
  - Be able to select the appropriate *L*-section matching network (type 1 or type 2) depending on the values of impedances to be matched.

Microwave Active Circuit Analysis and Design. http://dx.doi.org/10.1016/B978-0-12-407823-9.00009-3

- Be able to design an *L*-section matching network to match two arbitrary impedances using either the analytical or Smith Chart (graphical) approach.
- Be able to design a *T* or *π* matching network to match two arbitrary impedances.

9.1 INTRODUCTION
9.1.1 THE NEED FOR IMPEDANCE MATCHING

A basic understanding of the importance of impedance matching and how to achieve it is an essential skill for any RF circuit designer.

Impedance matching refers to the practice of arranging the input impedance of an electrical network so as to maximize the power transferred from a given source. Maximum power transfer was covered in Section 1.4.1 with the assumption that we have control over the value of the "load." In many cases, however, we have no control over either the source impedance or the input impedance of the electrical network in question, which means that we have to insert a *matching network* between the two in order to achieve the required match. It is the design of such matching networks that is the subject of this chapter and Chapter 10.

Although, in most cases, we require maximum power to be transferred from a given source to a given load, this is not always the case. In some circumstances, we may want to deliberately "mismatch" the source or load terminations of an active device, such as the input matching of a low-noise amplifier in order to achieve minimum noise figure (covered further in Chapter 14). Our definition of impedance matching therefore needs to be broadened to encompass the matching of any arbitrary source and/or load impedance. We use the terms *source* and *load* here in their broadest sense, where power is originating from the source and is being transferred to, the load. In reality, we are free to define any arbitrary points in the circuit as source or load depending on the context.

Power dissipation in the load implies that the load contains resistive elements. Similarly, "maximum power transfer" implies that we require zero power dissipation in the matching network itself. As a result, we will restrict our discussion to matching network design employing only reactive components: capacitors, inductors, and lossless transmission lines.

We shall distinguish between "lumped" and "distributed" matching networks. The former refers to circuits that employ discrete components (capacitors and inductors). The latter refers to circuits employing transmission lines. This distinction arises from the fact that the design methodology differs quite significantly. This chapter deals only with lumped element matching network design. Chapter 10 covers distributed element matching network design. This chapter, and Chapter 10 focus on matching at a single frequency. When designing wideband matching networks, different approaches need to be used, and we will not go into great detail on this subject.

There are a number of textbooks that the reader may consult which give a good coverage of broadband matching network design [1–4].

9.2 *L*-SECTION MATCHING NETWORKS

9.2.1 *L*-SECTION MATCHING OF A RESISTIVE SOURCE TO A RESISTIVE LOAD

Let us now proceed to design an actual matching network. Assume that we need to connect a resistive load R_L to a resistive source R_S, as shown in Figure 9.1(a), where, in the general case, $R_S \neq R_L$.

We need to place a matching network between R_S and R_L, as in Figure 9.1(b), to ensure that the source is terminated by an impedance equal to itself, thereby ensuring maximum power transfer.

Since we restrict ourselves to using only reactive elements in the matching network, the matching of the two resistive elements proceeds as follows:

1. First, we place a reactive element, represented by the susceptance jB, in parallel with R_L, such that the resistive part of the resulting combination is equal to R_S.
2. We then cancel the reactive part of the combination ($jB \parallel R_L$) by adding the equal and opposite series reactive element jX.

The specific type of matching network shown in Figure 9.1(b) is called an "*L*-section," as the two matching network elements form the shape of the letter "L." The *L*-section is one of the simplest and most widely used lumped element matching networks and is also a building block for more sophisticated matching networks. It is therefore well worth spending a bit of time to gain an understanding of the theory behind it.

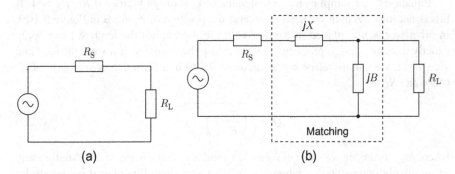

FIGURE 9.1

Matching arbitrary R_L and R_S.

We can analyze the circuit of Figure 9.1(b) as follows: since we know that, to satisfy the matching condition, the total impedance of the parallel combination ($jB \parallel R_L$) should be the complex conjugate of $R_S + jX$, we can write:

$$R_S + jX = \frac{1}{(1/R_L) + jB} = \frac{R_L - jBR_L^2}{1 + B^2 R_L^2} \tag{9.2.1}$$

From equation (9.2.1) we get R_S and X in terms of R_L and B as follows:

$$R_S = \frac{R_L}{1 + B^2 R_L^2} \tag{9.2.2}$$

and

$$X = \left(\frac{-BR_L^2}{1 + B^2 R_L^2} \right) \tag{9.2.3}$$

Recalling equation (1.6.5) on page 37, we note that the unloaded Q of the parallel combination of R_L and B is given by $Q = BR_L$. We can therefore express equations (9.2.2) and (9.2.3) in terms of the Q of the parallel combination, as follows:

$$R_S = \frac{R_L}{1 + Q^2} \tag{9.2.4}$$

and (ignoring the sign)

$$X = \frac{QR_L}{1 + Q^2} \tag{9.2.5}$$

From equation (9.2.4) we note that:

$$Q = \sqrt{\left(\frac{R_L}{R_S} \right) - 1} \tag{9.2.6}$$

Equation (9.2.6) implies that a real value of Q is only obtained if $R_L/R_S > 1$. If this is not the case then we need to reverse the position of X and B in Figure 9.1(b), in other words B is placed in parallel with the source, not the load. We can apply exactly the same design procedure, only treating the source as if it were the load and vice versa. We can therefore write equation (9.2.6) in a form that covers any values of R_S and R_L as:

$$Q = \sqrt{\left(\frac{R_{high}}{R_{low}} \right) - 1} \tag{9.2.7}$$

where R_{high} is the higher value of R_S and R_L, and R_{low} is the lower value. Another way of intuitively understanding where the parallel arm should be placed is to consider that, if $R_L > R_S$ then R_L needs to be reduced by adding a parallel resistance. On the other hand, if $R_L < R_S$ then it needs to be increased by adding a series resistance.

We can now set out the basic design procedure for an *L*-section to match resistive loads as follows:

1. Calculate the Q for a given R_S and R_L using equation (9.2.6) (note the orientation of the parallel arm based on whether (R_L/R_S) is greater or less than unity).
2. Calculate B from:

$$B = \pm \frac{Q}{R_L} \qquad (9.2.8)$$

3. Calculate X from equation (9.2.5).

Note that the sign of B in step 2 above may be chosen arbitrarily, since the load is purely resistive and we are free to choose B to be either capacitive or inductive. The difference being that the type of reactance chosen for B will determine whether the *L*-section has a high- or low-pass frequency characteristic away from the center frequency. If a positive value of B is chosen (i.e., parallel capacitance) then the *L*-section will have a low-pass characteristic. If a negative value of B is chosen (i.e., parallel inductance) then the *L*-section will have a high-pass characteristic.

We will now define two types of *L*-matching network, which we will refer to as "type 1" and "type 2," depending on the location of the parallel element with respect to the load, as shown in Figure 9.2. Based on the preceding discussion it will be apparent that the "type 1" and "type 2" classification is itself arbitrary, as it depends on our definition of "source" and "load," which can be interchanged at will. We will stick with this terminology, however, for the sake of clarity.

In general, the load and source will be complex. We can generalize the above technique to cover a complex Z_L and Z_S as shown in Figure 9.2 by considering only the resistive parts of Z_L and Z_S first and then absorbing the reactive parts into the resulting matching components X and B.

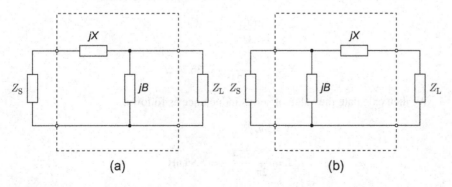

(a)   (b)

FIGURE 9.2

L-section matching networks. (a) *L*-section: type 1. (b) *L*-section: type 2.

For simplicity, we will restrict our analysis to the most common situation, namely where we need to match the complex load Z_L to the system characteristic impedance, Z_0. As before, the choice of whether to use a type 1 or type 2 matching network will depend on the relationship between the resistive part of the load, R_L, and Z_0. As was shown for the case of purely resistive loads, the parallel element, jB, should be placed in parallel with whichever is larger of R_L or Z_0, in other words:

If $R_L > Z_0$: use type 1 L-section (shunt element is next to the load).
If $R_L < Z_0$: use type 2 L-section (shunt element is next to the source).

Example 9.1 (*L*-section matching: resistive source and load).

Problem. Design an L-section matching network to match a resistive load of 30 Ω to a 75 Ω source at 1 GHz.

Solution. Since $R_L < R_S$ we need to use a type 2 L-section; that is, the shunt element is placed across the source and we calculate the Q using equation (9.2.6) with R_S and R_L interchanged:

$$Q = \sqrt{\left(\frac{75}{30}\right) - 1} = 1.225$$

We then calculate the shunt susceptance, B, from equation (9.2.8):

$$B = \frac{1.225}{75} = 0.0163 \text{ S}$$

On this occasion we have chosen B to be capacitive, therefore:

$$B = \omega C$$

$$C = \frac{0.0163}{2\pi \times 10^9} = 2.6 \text{ pF}$$

We now apply equation (9.2.5) to determine the series reactance X:

$$X = \frac{QR_L}{1 + Q^2}$$

$$X = \frac{1.225 \times 75}{1 + 1.225^2} = 36.74 \text{ }\Omega$$

We then calculate the value of series inductance as follows:

$$X = \omega L$$

$$L = \frac{36.74}{2\pi \times 10^9} = 5.85 \text{ nH}$$

The final matching circuit, together with source and load, is shown in Figure 9.3.

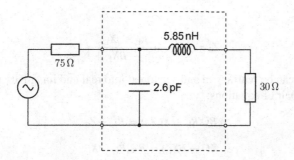

FIGURE 9.3

L-section matching: Example 9.1.

9.2.2 GENERALIZED ANALYTICAL DESIGN OF TYPE 1 *L*-SECTION

We will now proceed to analyze a generalized type 1 *L*-section to match a complex load $Z_L = R_L + jX_L$ to the system characteristic impedance, Z_o, as shown in Figure 9.4.

From Figure 9.4 we can write down three equations with two unknowns:

$$Z_L = R_L + jX_L \qquad \text{(known)} \qquad (9.2.9)$$

$$Z_{in} = jX + \frac{1}{jB + (R_L + jX_L)^{-1}} \qquad (B, X: \text{unknown}) \qquad (9.2.10)$$

$$Z_S = Z_o \qquad \text{(known)} \qquad (9.2.11)$$

Matching is defined as $Z_{in} = Z_S^*$. This implies $Z_{in} = Z_o$. Therefore, from equation (9.2.10) above:

$$Z_o = jX + \frac{1}{jB + (R_L + jX_L)^{-1}} \qquad (9.2.12)$$

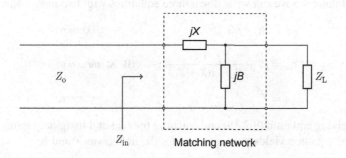

FIGURE 9.4

L-section matching network: type 1.

or

$$Z_0 = jX + \frac{R_L + jX_L}{jBR_L - BX_L + 1} \tag{9.2.13}$$

The above can be rearranged and separated into real and imaginary parts to yield the following pair of equations:

$$B(XR_L - X_L Z_0) = R_L - Z_0 \tag{9.2.14}$$

$$X(1 - BX_L) = BZ_0 R_L - X_L \tag{9.2.15}$$

If we solve equation (9.2.14) for X, and insert the result into equation (9.2.15), we find a quadratic equation for B with the following solution:

$$B = \frac{X_L \pm \sqrt{\frac{R_L}{Z_0} \left(R_L^2 + X_L^2 - Z_0 R_L \right)}}{R_L^2 + X_L^2} \tag{9.2.16}$$

The requirement that $R_L > Z_0$ ensures that the term under the square root in the numerator of equation (9.2.16) is real. Note that two solutions for B are possible and both solutions are physically realizable given that B can be positive or negative. Once B is determined, X can be found using equation (9.2.15):

$$X = \frac{BZ_0 R_L - X_L}{1 - BX_L} \tag{9.2.17}$$

The existence of two pairs of solutions for both B and X implies that there are two different matching network solutions.

9.2.3 GENERALIZED ANALYTICAL DESIGN OF TYPE 2 L-SECTION

Consider a generalized type 2 L-section of Figure 9.5, consisting of a parallel susceptance, B and a series reactance, X:

From Figure 9.5 we can write down three equations with two unknowns:

$$Z_L = R_L + jX_L \qquad \text{(known)} \tag{9.2.18}$$

$$Y_{in} = jB + \frac{1}{R_L + j(X + X_L)} \qquad \text{(B, X: unknown)} \tag{9.2.19}$$

$$Z_S = Z_0 \qquad \text{(known)} \tag{9.2.20}$$

Multiplying equation (9.2.19) and equating the real and imaginary terms on both sides of the equation yields two equations for the unknowns X and B.

$$BZ_0(X + X_L) = Z_0 - R_L \tag{9.2.21}$$

$$(X + X_L) = BZ_0 R_L \tag{9.2.22}$$

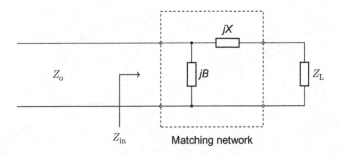

FIGURE 9.5

L-section matching network: type 2.

Solving the above for *B* and *X* yields:

$$B = \pm \frac{1}{Z_0} \sqrt{\frac{(Z_0 - R_L)}{R_L}} \qquad (9.2.23)$$

$$X = \pm \sqrt{R_L(Z_0 - R_L)} - X_L \qquad (9.2.24)$$

Once again, the requirement that $R_L < Z_0$ ensures that the terms under the square roots in the expressions for *B* and *X* and are real. Again, with two solution pairs for *B* and *X*, there are two different matching network solutions.

9.2.4 *L*-SECTION DESIGN USING THE SMITH CHART

We will now proceed to describe a procedure for designing *L*-section matching networks using the Smith Chart. Visualizing the design process in this way can provide a greater insight than the purely analytical methods outlined so far. Visualizing the design procedure on the Smith Chart will also reveal regions of the Smith Chart which cannot be accessed by a particular *L*-section topology. These are called *Forbidden Regions*. Whichever approach is adopted: analytical or graphical, the end result will, of course, be the same.

First, we should identify two particular circles on the Smith Chart that are significant when designing *L*-sections:

- *Unit resistance circle*: the locus of all impedances of the form $z = 1 \pm jx$.
- *Unit conductance circle*: the locus of all admittances of the form $y = 1 \pm jb$.

These two circles are shown in Figure 9.6. They are the only two circles on the Smith Chart that pass through the origin, which means that any matching network we design to match an arbitrary complex load to Z_0 will have to intersect one or the other of these circles in order to get to Z_0.

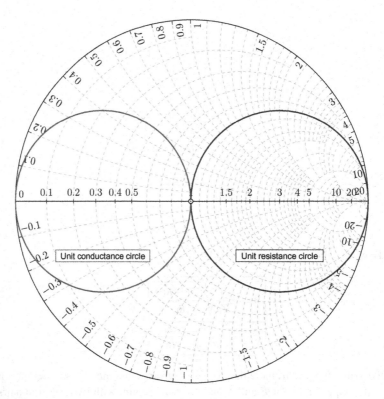

FIGURE 9.6

Unit conductance and resistance circles.

We first locate the load impedance on the Smith Chart at point "A" in Figure 9.7. We can move away from this starting point in one of four directions, depending on whether the first matching element is series or shunt, *L* or *C*. In any case, since we are only adding one purely reactive element, we can only move along the constant resistance circle or constant conductance circle passing through point A.

We can summarize Figure 9.7 as follows:

- *Adding a **series** component*:
  - Series *L*: move *clockwise* along the *constant resistance* circle through A.
  - Series *C*: move *counter-clockwise* along the *constant resistance* circle through A.
- *Adding a **shunt** component*:
  - Shunt *L*: move *counter-clockwise* along the *constant conductance* circle through A.
  - Shunt *C*: move *clockwise* along the *constant conductance* circle through A.

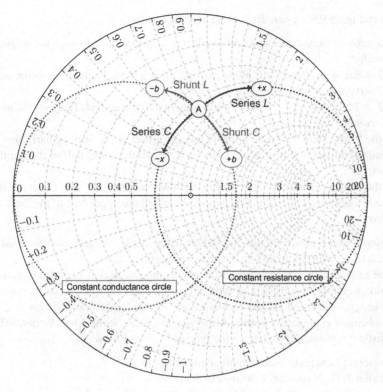

FIGURE 9.7

The effect of adding shunt and series reactive elements to an arbitrary load.

The ultimate goal is to arrive at the origin ($Z_{in} = Z_0$). It should be apparent from Figure 9.7 that, except in the very special case where point "A" already lies on the unit resistance circle or unit conductance circle, we cannot reach the origin simply by adding a single reactive element in series or parallel with the load. Hence the need for a two-element *L*-section to provide the necessary degrees of freedom to match an arbitrary load to Z_0.

If we consider the case of a type 1 *L*-section, the first element, the one next to the load, is a shunt susceptance. In this case, we rotate along the constant conductance circle until we intersect the unit resistance circle. The value of the shunt susceptance required, *jB*, is represented by the change in susceptance of moving from point A to the intersection with the unit resistance circle on the Smith Chart. The impedance of the parallel combination of the load and *jB* is then added to the reactance of the matching element *jX* by rotating along the unit resistance circle until we reach the origin of the Smith Chart (matched condition).

From Figure 9.6 we can draw some general conclusions:

1. Only the unit resistance and unit conductance circles pass through the origin, therefore:
    (a) Adding a series L or C alone can only match those loads lying on the *unit resistance circle*.
    (b) Adding a shunt L or C alone can only match those loads lying on the *unit conductance circle*.
2. If the load is inductive (i.e., it lies in the upper half of the Smith Chart) then we need to add series C or shunt C to move it toward the lower half of the Smith Chart.
3. If the load is capacitive (i.e., it lies in the lower half of the Smith Chart) then we need to add series L or shunt L to move it toward the upper half of the Smith Chart.

Points 2 and 3 above should be obvious to the reader: in order to cancel out the load reactance we need to add an equal and opposite reactance.

The broader conclusion from all the preceding discussion is that the choice of L-section topology (i.e., type 1 or type 2, L or C) depends fundamentally on where the load, z_L, is located on the Smith Chart. Richard Li's book [5] gives a very comprehensive explanation of this. According to Li, the Smith Chart be divided into four distinct regions, depending on the type of load, as follows:

1. Region 1: Low resistance or high conductance loads.
2. Region 2: High resistance or low conductance loads.
3. Region 3: Low resistance and low conductance loads.
4. Region 4: Low resistance and low conductance loads.

These regions are shown in Figure 9.8.

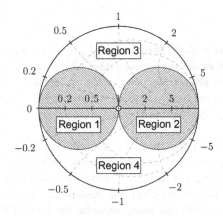

FIGURE 9.8

Smith Chart regions.

In terms of the normalized load $z_L = r + jx, y_L = g + jb$ we can define the four regions in Figure 9.8 as shown in Table 9.1 [5].

Table 9.1 Smith Chart Region Classification

Region 1	Region 2	Region 3	Region 4
Low Resistance or High Conductance	High Resistance or Low Conductance	Low Resistance and Low Conductance	Low Resistance and Low Conductance
$r < 1$	$r > 1$	$r < 1$	$r < 1$
$x < |0.5|$	$-\infty < x < +\infty$	$x > 0$	$x < 0$
$g > 1$	$g < 1$	$g < 1$	$g < 1$
$-\infty < b < +\infty$	$b < |0.5|$	$b < 0$	$b > 0$

9.2.5 FORBIDDEN REGIONS ON THE SMITH CHART

From the preceding discussion we can conclude that not every reactive L-section topology can perform the required matching of an arbitrary complex load. Depending on the particular configuration of reactive elements, the Smith Chart will be divided into a *Matchable region* and a *Forbidden region* [6] which, respectively, contain all the loads that can and cannot be matched by that particular L-section topology.

For example, consider the type 2 L-section shown in Figure 9.9(a). Let us start at the source, which we will assume to be the system characteristic impedance, Z_0. So we start at the origin of the Smith Chart. Addition of a shunt capacitance across the source will result in us moving *clockwise* away from the origin along the constant

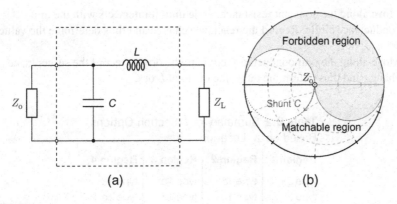

(a)   (b)

FIGURE 9.9

L-section matching: forbidden region (shaded). (a) L-section example. (b) Matchable and forbidden regions on the Smith Chart.

conductance circle, as shown in Figure 9.9(b). Any value of shunt capacitance in this configuration can only take us to a point on the constant conductance circle in Figure 9.9(b).

If we now consider adding the series element, which we have chosen to be an inductor in this example, we can only start at some point on the constant conductance circle in Figure 9.9(b), depending on the value of shunt C we have chosen. We can then only move counter clockwise along a constant resistance circle from that point, which means that we can only access load impedances within the unshaded region of Figure 9.9(b). We refer to the shaded region in Figure 9.9(b) as the *forbidden region* since we cannot match loads in this region of the Smith Chart with this particular *L*-section.

This analysis can be extended to cover the other seven types of lossless *L*-section, as set out in Appendix C.

9.2.6 FOUR-STEP DESIGN PROCEDURE FOR GENERALIZED *L*-SECTIONS

1. Normalize ($z_L = Z_L/50$) and locate z_L on the Smith Chart.
2. Select the appropriate *L*-section topology, based on where Z_L lies in relation to the various forbidden regions, as summarized in Table 9.2.

For a type 1 L-section:

3. Move along the constant conductance circle until it intersects with the unit resistance circle. Record the susceptance change and thus determine the value of shunt *L* or *C*.
4. Move along the unit resistance circle to the origin, record the reactance change and thus determine the value of series *L* or *C*.

For a type 2 L-section:

3. Move along the constant resistance circle until it intersects with the unit conductance circle. Record the reactance change and thus determine the value of series *L* or *C*.
4. Move along the unit conductance circle to the origin, record the susceptance change and thus determine the value of shunt *L* or *C*.

Table 9.2 Summary of *L*-Section Options, Based on z_L Location

Region 1	Region 2	Region 3	Region 4
type 2a	type 1a	type 1b	type 1a
type 2b	type 1b	type 1d	type 1c
		type 2b	type 2a
		type 2d	type 2c

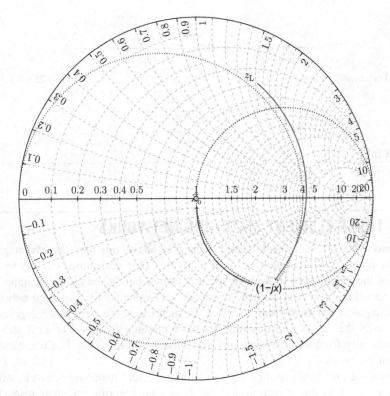

FIGURE 9.10

L-section: type 1b example: Smith Chart design procedure.

Example 9.2 (*L*-section matching using the Smith Chart).

Problem. Use a Smith Chart to design an *L*-section matching network to match a load $Z_L = (25 + j75)$ Ω to a 50 Ω transmission line at 10 GHz.

Solution. First, we normalize $z_L = (25 + j75)/50 = 0.5 + j1.5$ and locate this point on the Smith Chart.

We note that z_L is located in region 3, so we select a type 1b *LC L*-section.

We move clockwise along the constant conductance circle in Figure 9.10 until it intersects the unit resistance circle at $1 - jx$. Susceptance change $1/ - 100 - 1/75 = -0.023$ S. Therefore, $C = 0.023/(2\pi f) = 0.371$ pF.

We now move clockwise along the unit resistance circle to the origin. Reactance change $= 100$ Ω. Therefore, $L = 100/2\pi f = 1.59$ nH.

The final matching network is shown in Figure 9.11.

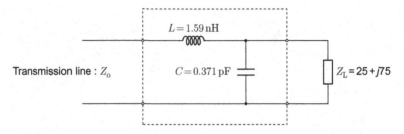

FIGURE 9.11

L-section: type 1b example.

9.3 THREE ELEMENT MATCHING NETWORKS

The two-element *L*-sections discussed in the previous section have the advantage of simplicity, but their limitation lies in the fact that *Q* is determined entirely by the ratio of the impedances to be matched and cannot be chosen by the designer. In some situations, however, we need to specify a *Q*-value for the matching network, for example in a narrow-band amplifier. In such cases, we need more degrees of freedom in the design and this requires more circuit elements. The next step up in complexity from the two-element *L*-sections just described is the three-element matching network.

There are two basic configurations of three-element matching network, which are referred to as the π-section and the *T*-section, according to their respective topologies, as shown in Figure 9.12. The choice of whether to use a π-section matching network or a *T*-section matching network will depend primarily on whether the load/source presents as a series or parallel combination, as well as such considerations as the shape of the frequency response required and whether DC continuity is required through the matching network. For interested readers, Richard Li's book [5] contains a comprehensive discussion of the criteria for choosing between π and *T*-networks.

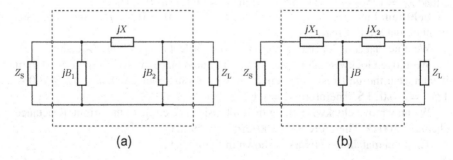

(a)   (b)

FIGURE 9.12

Three element matching networks. (a) π-matching network. (b) *T*-matching network.

9.3.1 THE π-SECTION MATCHING NETWORK

A π-section matching network consists of three elements arranged as shown in Figure 9.12(a). The π-section can be considered as consisting of a type 2 L-section followed by a type 1 L-section in cascade as shown in Figure 9.13, where the central element, X, has been split into two reactive elements of the same type, that is, $X = X_1 + X_2$. We also postulate an "invisible" load resistance, R_x, interposed between the two L subnetworks in Figure 9.13. The purpose of L-section 1, therefore, is to match the source to R_x. Similarly, the purpose of L-section 2 is to match R_x to the load. The individual L-sections can be designed according to the principles set out in the previous section, provided we know the value of R_x, that is. The value of R_x can be chosen arbitrarily, but it should be smaller than both R_S and R_L, since it is connected to the series arms of the two L-sections [7]. If we start with a required value of Q; however, this will determine the choice of R_x.

Consider the deconstructed π-section matching network shown in Figure 9.13. Applying equation (9.2.7) we get the loaded Q of L-network 1, which matches R_S to R_x. We have already stipulated that R_x must be smaller than R_S so we have:

$$Q_{L1} = \sqrt{\left(\frac{R_S}{R_x} - 1\right)} \tag{9.3.1}$$

Applying the same logic, the loaded Q of L-network 2, which matches R_x to R_L (which is larger than R_x), is given by:

$$Q_{L2} = \sqrt{\left(\frac{R_L}{R_x} - 1\right)} \tag{9.3.2}$$

Since the loaded Q of the overall circuit is determined by the branch of the circuit having the highest loaded Q-value, we can write the overall Q of the circuit in Figure 9.13, Q_π, as [7]:

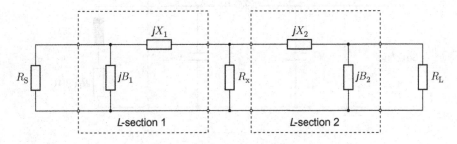

FIGURE 9.13

π-section matching network as a cascade of two L-networks.

$$Q_\pi = \sqrt{\left(\frac{R_{high}}{R_x} - 1\right)} \qquad (9.3.3)$$

where R_{high} is the larger of R_S and R_L. By inspection of equations (9.3.1)–(9.3.3) we can see that the overall Q of the π-section will be equal to the highest Q of the two constituent L-sections.

Given a required value of overall Q_π, we determine the value of the virtual resistor, R_x, from a simple rearrangement of equation (9.3.3):

$$R_x = \frac{R_{high}}{Q_\pi^2 + 1} \qquad (9.3.4)$$

With known values of Q_{L1}, Q_{L2}, and R_x we proceed with the π-section design by designing each constituent L-section, using the techniques set out in Section 9.2.1, and then combining the central elements.

The reader will note that we have so far only considered the resistive parts of the source and load, R_S and R_L. This is because any reactive parts of the source and load can be absorbed into the parallel branches of the π-section; B_1 and B_2.

The best way to illustrate the design technique for a π-section is with an example.

Example 9.3 (π-section matching network).

Problem. Design a π-section matching network having a Q of 4, to match an antenna having an impedance of $20 + j25$ Ω at 2 GHz to a 50 Ω transmission line. Your matching network should be able to pass DC current.

Solution. Since the matching network should be able to pass DC current, we need to use the topology shown in Figure 9.14.

Since we are using a π-section, it is more convenient for us to work with an equivalent load admittance, as we need to absorb the reactive part of the load into the π-section element jB_2. We therefore convert the load to an equivalent parallel

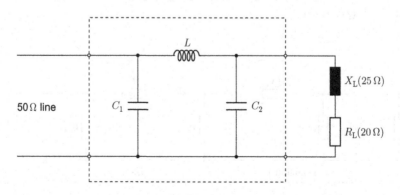

FIGURE 9.14

Π-section matching network example.

configuration, so that we can simply add the susceptances. We therefore calculate the load admittance as:

$$Y_L = \frac{1}{20 + j25} = 0.0195 - j0.0244 \text{ S}$$

We can now redraw Figure 9.14 broken down into two L-sections in Figure 9.15, which also shows the virtual resistance, R_x.

Assuming that we absorb the equivalent load susceptance, B_L, into C_2 of the π-section, the equivalent purely resistive load then becomes:

$$R_L = \frac{1}{G_L} = \frac{1}{0.0195} = 51.28 \text{ }\Omega$$

We can now calculate the value of R_x using equation (9.3.4):

$$R_x = \frac{R_{high}}{Q_\pi^2 + 1} = \frac{51.28}{4^2 + 1} = 3 \text{ }\Omega$$

We now employ the procedures learned in Section 9.2.1 to design the two constituent L-sections in Figure 9.15. Since $R_L > R_S$ we know that the L-section comprising L_2 and C_2, has the highest Q of the two, that is,

$$Q_{L2} = Q_\pi = 4$$

We now calculate the shunt susceptance, B_2, from equation (9.2.8), remembering to compensate for the negative load susceptance, B_L, by adding an equivalent positive susceptance to B_2:

$$B_2 = \frac{4}{51.28} + 0.0244 = 0.078 + 0.0244 = 0.1024 \text{ S}$$

We can now calculate the capacitance C_2 as follows:

$$C_2 = \frac{B_2}{\omega} = \frac{0.1024}{2\pi \times 2 \times 10^9} = 8.15 \text{ pF}$$

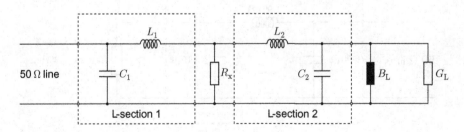

FIGURE 9.15

π-section matching network example, decomposed into L-sections.

We now apply equation (9.2.5) to determine the reactance of L_2:

$$X_2 = \frac{QR_L}{1 + Q_\pi^2} = \frac{4 \times 51.28}{1 + 4^2} = 12 \ \Omega$$

So, the inductance, L_2, is calculated as:

$$L_2 = \frac{X_2}{\omega} = \frac{12}{2\pi \times 2 \times 10^9} = 0.96 \ \text{nH}$$

We now repeat the above design procedure for the L-section comprising L_1 and C_1. We firstly need to calculate Q_{L1} using equation (9.3.1):

$$Q_{L1} = \sqrt{\left(\frac{50}{3} - 1\right)} = 3.96$$

$$B_1 = \frac{Q_{L1}}{R_S} = \frac{3.95}{50} = 0.079 \ S$$

$$C_1 = \frac{B_1}{\omega} = \frac{0.079}{2\pi \times 2 \times 10^9} = 6.30 \ \text{pF}$$

$$X_1 = \frac{Q_{L1}R_S}{1 + Q_{L1}^2} = \frac{3.96 \times 50}{1 + 3.96^2} = 11.87 \ \Omega$$

So, the inductance, L_1, is calculated as:

$$L_1 = \frac{X_1}{\omega} = \frac{11.87}{2\pi \times 2 \times 10^9} = 0.945 \ \text{nH}$$

By combining L_1 and L_2 to give $L = 1.90$ nH, and removing R_x, we arrive at the final configuration shown in Figure 9.16.

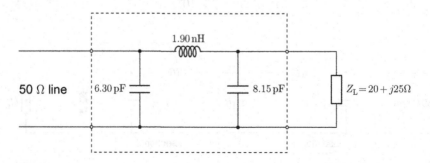

FIGURE 9.16

π-section matching network example: final configuration.

9.3.2 THE *T*-SECTION MATCHING NETWORK

The *T*-section consists of three elements arranged as shown in Figure 9.12(b), and can be considered as the dual of the π-section.

The *T*-section can be decomposed into a type 1 *L*-section followed by a type 2 *L*-section, as shown in Figure 9.17, where the central element, B, has been split into two elements, that is, $B = B_1 + B_2$.

As in the case of the π-network, we introduce a virtual load R_x between the two *L* subnetworks in Figure 9.17 and to which they are both simultaneously matched. This time R_x has to be larger than R_S and R_L since it is connected across the parallel arms of the two *L*-sections [7]. As with the π-network, the value of R_x is determined by the required Q of the circuit.

The loaded Q of the *T*-section is determined by the *L*-section having the highest Q, which is the *L*-section with the smaller of R_S and R_L. The loaded Q of the *T*-section is therefore determined by equation (9.2.7) to be:

$$Q_T = \sqrt{\left(\frac{R_x}{R_{\text{low}}}\right) - 1} \qquad (9.3.5)$$

where R_x is the virtual resistor in Figure 9.17 and R_{low} is the smaller of R_S and R_L. We can rearrange equation (9.3.5) to get an expression for R_x in terms of the required Q of the *T*-section as follows:

$$R_x = R_{\text{low}}\left(Q_T^2 + 1\right) \qquad (9.3.6)$$

We will now illustrate the design of a *T*-network by an example. For the sake of comparison we will use the same circumstances as in Example 9.3.

Example 9.4 (*T*-section matching network).

Problem. Design a *T*-section matching network having a Q of 4, to match an antenna having an impedance of $20 + j25\ \Omega$ at 2 GHz to a 50 Ω transmission line. Your matching network should be able to pass DC current.

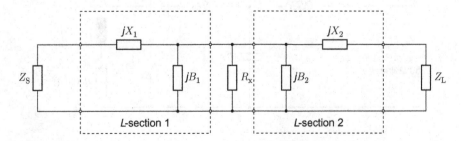

FIGURE 9.17

T-section matching network as a cascade of type 1 *L*-networks.

Solution. Since the matching network should be able to pass DC current, we need to use the topology shown in Figure 9.18.

For clarity, we can redraw Figure 9.18 broken down into two L-sections in Figure 9.19, which shows the virtual resistance, R_x, the value of which can be calculated using equation (9.3.6):

$$R_x = 20\left(4^2 + 1\right) = 340 \ \Omega$$

We now proceed to design the type 2 L-section comprising L_2 and C_2, using the procedures learned in Section 9.2.1.

Since $R_L < Z_o$ in this case, the Q of the L-section C_2, L_2 is equal to the overall Q of the T-section, that is,

$$Q_{L2} = Q_T = 4$$

We can now calculate the shunt susceptance, B_2, from equation (9.2.8), noting that the "load," in this case, is R_x:

$$B_2 = \frac{Q_{L2}}{R_x} = \frac{4}{340} = 0.0118 \ \text{S}$$

The value of C_2 can now be calculated as:

$$C_2 = \frac{B_2}{\omega} = \frac{0.0118}{2\pi \times 2 \times 10^9} = 0.94 \ \text{pF}$$

We now apply equation (9.2.5) to determine the series reactance X_2, noting that we need to subtract the inductive reactance of the load, X_L:

$$X_2 = \frac{Q_{L2}R_x}{1 + Q_{L2}^2} - X_L = \frac{4 \times 340}{1 + 4^2} - 25 = 55 \ \Omega$$

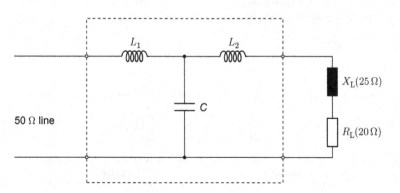

FIGURE 9.18

T-section matching network example.

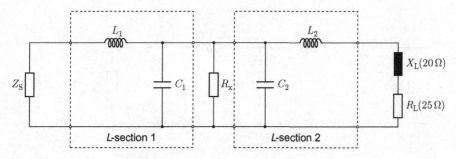

FIGURE 9.19

T-section matching network example, decomposed into L-sections.

We can now calculate L_2 as:

$$L_2 = \frac{X_2}{\omega} = \frac{55}{2\pi \times 2 \times 10^9} = 4.38 \text{ nH}$$

We now repeat the above design procedure for the type 1 *L*-section comprising L_1 and C_1 to give:

$$Q_{L1} = \sqrt{\left(\frac{340}{50}\right) - 1} = 2.41$$

$$B_1 = \frac{Q_{L1}}{R_x} = \frac{2.41}{340} = 0.0071 \text{ S}$$

$$C_1 = \frac{B_1}{\omega} = \frac{0.0071}{2\pi \times 2 \times 10^9} = 0.56 \text{ pF}$$

$$X_1 = \frac{Q_{L1}R_x}{1 + Q^2} = \frac{2.41 \times 340}{1 + 2.41^2} = 120.4 \text{ }\Omega$$

We can now calculate L_1 as:

$$L_1 = \frac{X_1}{\omega} = \frac{120.4}{2\pi \times 2 \times 10^9} = 9.58 \text{ nH}$$

We now complete the design by removing R_x and combining C_1 and C_2 to give $C = 1.5$ pF. The final design is shown in Figure 9.20.

9.3.3 π TO *T* TRANSFORMATION

Finally, it is worth mentioning a useful set of equations that allows us to convert any given π-network of generalized impedances into an equivalent *T*-network and vice versa. This is the well-known *Star-Delta transformation* that the reader may have

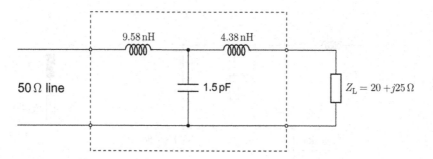

FIGURE 9.20

T-section matching network example: final configuration.

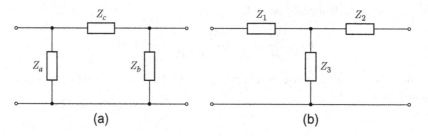

FIGURE 9.21

π to *T* transformation. (a) π-network. (b) *T*-network.

encountered in elementary circuit theory, which we will restate here with reference to the π- and *T*-networks shown in Figure 9.21.

The impedances of the π-matching network in Figure 9.21(a) (Z_a, Z_b, and Z_c) are related to the impedances of the *T*-network in Figure 9.21(b) (Z_1, Z_2, and Z_3) by the following relationships:

T to π transformation:

$$Z_a = \frac{(Z_1 Z_2 + Z_1 Z_3 + Z_2 Z_3)}{Z_2} \tag{9.3.7}$$

$$Z_b = \frac{(Z_1 Z_2 + Z_1 Z_3 + Z_2 Z_3)}{Z_1} \tag{9.3.8}$$

$$Z_c = \frac{(Z_1 Z_2 + Z_1 Z_3 + Z_2 Z_3)}{Z_3} \tag{9.3.9}$$

π to T transformation:

$$Z_1 = \frac{Z_a Z_c}{Z_a + Z_b + Z_c} \tag{9.3.10}$$

$$Z_2 = \frac{Z_b Z_c}{Z_a + Z_b + Z_c} \tag{9.3.11}$$

$$Z_3 = \frac{Z_a Z_b}{Z_a + Z_b + Z_c} \tag{9.3.12}$$

9.4 BANDWIDTH OF LUMPED ELEMENT MATCHING NETWORKS

The bandwidth of any lumped element matching network can be determined from the loaded Q of the circuit, Q_L, by applying equation (1.6.15), that is,

$$\Delta f = \frac{f_0}{Q_L} \tag{9.4.1}$$

In the case of L-sections, the Q and therefore the bandwidth, are solely a function of the load and source resistances. The bandwidth of an L-section can therefore be determined by applying equation (9.2.7) as follows:

$$\Delta f = \frac{f_0}{\sqrt{\left(\dfrac{R_{high}}{R_{low}}\right) - 1}} \tag{9.4.2}$$

The advantage of the three-element networks, (T and π), by contrast, is that Q can be chosen, to some extent, as an independent design parameter. This means that we have some degree of choice of bandwidth, independent of the load and source resistances, provided that the chosen Q is larger than that which is available with an L-network. This means that T- or π-networks are only really suitable for narrowband applications. If wider bandwidth are required, a matching network based on cascaded L-sections may be used, as discussed in, for example, Bowick's book [7].

9.5 TAKEAWAYS

1. In the case of RF circuits, maximum power transfer from source to load requires that $Z_{source} = Z_{load}^*$ which, in general, will not be the case. A *matching network* will therefore need to be placed between the source and load.
2. The *L-section* is one of the simplest and most widely used lumped element matching networks.
3. Not all possible values of load impedance can be matched by any given L-section configuration. Such unmatchable loads lie within a *forbidden region* on the Smith Chart that is specific to that particular L-section configuration.
4. The Q of an L-section matching network is exactly determined by the value of load and source resistances, and cannot be independently designed.

5. The simplest lumped element matching network that allows Q to be independently determined consists of three elements. There are two types of three-element matching network: the "π-network" and the "T-network." These two networks can be considered as duals of each other.

6. A simple set of equations exist to convert any given π-section matching network of generalized impedances into an equivalent T-network and vice versa.

TUTORIAL PROBLEMS

Problem 9.1 (Choice of L-section (forbidden regions)). Determine the type of L-section matching network required to match the following loads to a 50 Ω source:

1. $Z_1 = 70 + j65\ \Omega$
2. $Z_2 = 12\ \Omega$ resistor in series with an inductor $L_2 = 6$ nH at 500 MHz
3. $Y_3 = 28 + j26$ mS
4. $Z_4 = 68\ \Omega$ resistor in parallel with a capacitor $C_4 = 2.2$ pF at 500 MHz

Problem 9.2 (L-section matching network design). Design an L-section to match a 12 GHz radio transmitter having an output impedance of $Z_{\text{out}} = 100 + j75\ \Omega$, to an antenna having an input impedance of $Z_{\text{in}} = 60 + j20\ \Omega$.

Problem 9.3 (T-section matching network design) Design a T-section matching network to match a load $Z_L = 150\ \Omega$ to a source $Z_S = 50 - j35\ \Omega$ at a frequency of 750 MHz, with a quality factor $Q = 3.5$.

Problem 9.4 (π-section matching network design). Repeat Problem 9.3 using a π-section.

Problem 9.5 (π- and T-network equivalence). Verify, by means of equations (9.3.7)–(9.3.12), that the π-section matching network designed in Example 9.3 and the T-section matching network designed in Example 9.4 are exactly equivalent at 2 GHz.

REFERENCES

[1] G. Matthaei, E. Jones, L. Young, Microwave Filters Impedance Matching Networks and Coupling Structures, McGraw-Hill, New York, NY, 1961.
[2] D. Pozar, Microwave Engineering, second ed., John Wiley and Sons Inc., New York, NY, 1998.
[3] R. Collin, Foundations for Microwave Engineering, second ed., John Wiley and Sons Inc., New York, NY, USA, 2005, ISBN 9788126515288.
[4] R. Thomas, A Practical Introduction to Impedance Matching, Arthech House Inc., Boston, MA, 1976.
[5] R. Li, RF Circuit Design, Wiley Publishing, 2009, ISBN 0470167580, 9780470167588.
[6] P. Smith, Electronic Applications of the Smith Chart, Noble Publishing Corporation, 2000, ISBN 978-1884932397.
[7] C. Bowick, RF Circuit Design, Newnes Elsevier, Burlington, MA, USA, 2008.

Distributed element matching networks

10

CHAPTER OUTLINE

INTENDED LEARNING OUTCOMES

- *Knowledge*
  - Understand the advantages and disadvantages of distributed element matching, when compared with lumped element matching networks.
  - Be aware that distributed element matching networks can be designed either analytically (using a computer) or graphically using the Smith Chart.
  - Understand the principles behind stub matching networks and be aware that there are always at least two stub matching solutions to a given matching problem, which differ in terms of their bandwidth.

- Understand the theory behind the quarter-wave ($\lambda/4$) transformer, its properties and applications.
- Understand bandwidth performance of distributed element matching networks.
- *Skills*
  - Be able to design a single stub matching network to match an arbitrary load.
  - Be able to design a double stub matching network to match an arbitrary load.
  - Be able to design a quarter-wave transformer matching network to match an arbitrary load.
  - Be able to calculate the bandwidth of a single stub, double stub, or quarter-wave transformer matching network.

10.1 INTRODUCTION

In the previous chapter, we presented impedance matching based on capacitors and inductors, components that are familiar to any student of low-frequency electronics. As we increase the operating frequency (i.e., reduce the wavelength) we find that the physical dimensions of the components become comparable with the wavelength of our signal, and we enter a realm where lumped components behave in unexpected ways. At higher frequencies, therefore, we need to start thinking in terms of transmission lines, so we need a body of theory that tells us how to design matching networks using "distributed" (i.e., transmission line) elements. That is what this chapter aims to achieve.

It is worth re-stating that the key consideration, when choosing between a lumped or distributed approach, is the relationship between the signal wavelength and the physical dimensions of the lumped components. In the case of Monolithic Microwave Integrated Circuit (MMIC) technology, lumped components can be made extremely small. We therefore find that lumped elements tend to be used inside MMICs operating even at very high microwave frequencies (well into the hundreds of GHz) whereas, at these frequencies, the external circuitry connecting the MMIC will necessarily have to be implemented using distributed elements.

In this chapter, we focus on matching two impedances at a single frequency using distributed elements. When designing wideband matching networks different approaches need to be used, and these are well covered in other texts [1–4].

10.2 IMPEDANCE TRANSFORMATION WITH LINE SECTIONS

The simplest possible distributed matching network is just a single length of transmission line, of characteristic impedance Z_0, connected between load and source, as shown in Figure 10.1.

As we saw in Chapter 2, a length of transmission line has the effect of transforming any load impedance into some other impedance, determined by the length of the

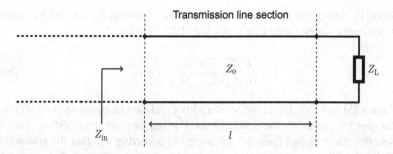

Transmission line section

FIGURE 10.1

Single line matching.

line section and the line characteristic impedance. This technique only works for certain relationships between the load and source immittances, and so it is more a matter of luck as to whether it can be used in a particular instance. It is worth discussing here, however, for the sake of providing insight into distributed element matching in general. This technique is sometimes referred to as "single line matching" [5].

From Chapter 2, we can say that the input impedance of the lossless transmission line section in Figure 10.1, having arbitrary length, l, characteristic impedance Z_o, and terminated with the load Z_L, is given by equation (2.7.15):

$$Z_{in} = R_{in} + jX_{in} = Z_o \left(\frac{Z_L + jZ_o \tan(\beta l)}{Z_o + jZ_L \tan(\beta l)} \right) \qquad (2.7.15)$$

This can be expressed in admittance form as:

$$Y_{in} = G_{in} + jB_{in} = Y_o \left(\frac{Y_L + jY_o \tan(\beta l)}{Y_o + jY_L \tan(\beta l)} \right) \qquad (10.2.1)$$

In order to match two impedances by using a length of transmission line connected between them we need to transform Y_L into the desired input admittance, Y_{in} by varying the quantities Y_o and l. The required relationship between Y_L and Y_{in} referred to above can be found by equating the real and imaginary parts of equation (10.2.1) and solving for l and Y_o. We equate the real parts of both sides of equation (10.2.1) and rearrange to obtain:

$$\tan(\beta l) = Y_o \left(\frac{G_{in} - G_L}{G_{in}B_L + G_L B_{in}} \right) \qquad (10.2.2)$$

Equating imaginary parts of both sides of equation (10.2.1), substituting equation (10.2.2), and solving for Y_o give:

$$Y_o = \sqrt{G_L G_{in} \left[1 + \frac{B_{in}^2 G_L - B_L^2 G_{in}}{G_L G_{in}(G_{in} - G_L)} \right]} \qquad (10.2.3)$$

Since Y_0 must be real for a practical line, the condition for single line matching to be realizable can be taken from equation (10.2.3) to be:

$$\frac{B_{in}^2 G_L - B_L^2 G_{in}}{G_L G_{in}(G_{in} - G_L)} > -1 \qquad (10.2.4)$$

Carson [5] has shown that the realizability condition of equation (10.2.4) may be represented by circles on the Smith Chart. For the case when condition (10.2.4) is not satisfied these circles have the advantage of revealing whether the addition of a parallel stub will move the load admittance into an area where single line matching is possible. Due to this limitation the single line matching technique is rarely used in practice. Interested readers are, however, referred to Carson's book [5] for a more detailed description of this topic.

10.3 SINGLE STUB MATCHING

The "single stub" matching technique is an important basic technique of narrow-band matching with distributed elements [4,6]. This technique makes use of the fact that a length of transmission line terminated in either an open or short circuit will behave as a pure susceptance whose value depends on the nature of the termination (open or short) and the length of the line section. Such a section of line is called a *stub*. Stubs are widely used in planar microwave circuits, such as microstrip and MMIC implementations as they are extremely easy to fabricate. In these media, the stub is connected in parallel with the main line. Because a parallel connection is used, the design procedure is best carried out in terms of admittances. The single stub matching network actually consists of two parts: the stub itself, of length l, and a length of transmission line, length d, connected between the load and the point at which the stub is attached. The purpose of the line section, d, is to transform the complex load admittance into a normalized admittance which has a real part equal to unity. The stub is then designed to cancel the residual susceptance, resulting in a perfect match.

A typical single stub matching network implemented in microstrip is shown schematically in Figure 10.2. Assuming we fix the line characteristic impedance, Z_0, there are basically two variables to play with: the distance between the load and the stub, d, and the length of the stub itself, l. In this section, we will present a systematic approach to creating a perfect match to any complex load by the correct choice of these two variables.

For the purposes of analysis, we will refer to a schematic representation of a generalized transmission line system, which is independent of physical construction, connected to an arbitrary complex load admittance, Y_L. The open-circuit stub configuration is shown in Figure 10.3(a) and the short-circuit stub configuration is shown in Figure 10.3(b). The lines all have characteristic impedance Z_0 (characteristic admittance Y_0).

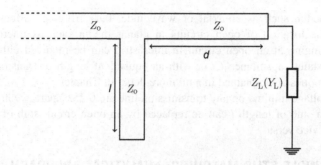

FIGURE 10.2

Open-circuit stub matching implemented in microstrip.

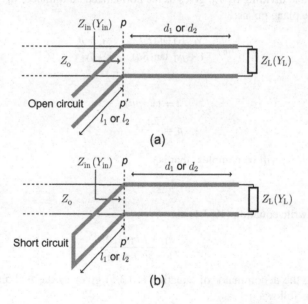

FIGURE 10.3

Generic single stub matching networks. (a) Open-circuit stub. (b) Short-circuit stub.

The choice of short- or open-circuit stub will often depend on the medium of construction. Open-circuit stubs are easier to implement than short-circuit stubs in planar media such as microstrip or MMIC substrate, as there is no need to penetrate the substrate to connect to the ground plane. On the other hand, short-circuit stubs are generally preferred if the medium is co-axial line, parallel wire line, or waveguide, where the physical location of the short circuit can be more precisely defined.

Radiation effects at an open circuit can make it difficult to precisely define the physical location of an open-circuit termination. This is a greater problem with "closed" media, such as co-axial or waveguide, but "fringing" effects can also influence the location of open circuits in planar media such as microstrip. The effect of fringing at an open circuit in microstrip can be modeled either with an equivalent shunt capacitance C, or with an equivalent length of transmission line. These techniques are explained in a bit more detail in Chapter 3.

Since voltage minima on any transmission line are 0.25λ apart, we note that any short-circuit stub of length l can be replaced by an open-circuit stub of length $l \pm (n\lambda)/4$ and vice versa.

10.3.1 SINGLE STUB MATCHING: ANALYTICAL APPROACH

We start by defining the input admittance of the line at the plane pp' in Figure 10.3(a) or (b), at a distance d from the load as Y_{in}. Applying equation (10.2.1), and normalizing and dividing by Y_o, gives us the normalized admittance, y_{in}, looking into the line at the plane pp' as:

$$y_{\text{in}} = \frac{y_L + j\tan(\beta d)}{1 + jy_L \tan(\beta d)} = \frac{y_L + jt}{1 + jy_L t} \tag{10.3.1}$$

where

$$t = \tan(\beta d) \tag{10.3.2}$$

$$\beta = \frac{2\pi}{\lambda} \tag{10.3.3}$$

In general, y_L will be complex, that is,

$$y_L = g_L + jb_L \tag{10.3.4}$$

so we can rewrite equation (10.3.1) as:

$$y_{\text{in}} = \frac{g_L + j(b_L + t)}{(1 - b_L t) + jg_L t} \tag{10.3.5}$$

Realizing the denominator of equation (10.3.5) gives us the real and imaginary parts of y_{in} as follows:

$$g_{\text{in}} = \frac{g_L(1 + t^2)}{(1 - b_L t)^2 + g_L^2 t^2} \tag{10.3.6}$$

$$b_{\text{in}} = \frac{b_L(1 - t^2) + t(1 - b_L^2 - g_L^2)}{(1 - b_L t)^2 + g_L^2 t^2} \tag{10.3.7}$$

The essential idea behind single stub matching is that at some point on the line normalized susceptance looking into the line, g_{in}, will be unity. At this point, a parallel stub can be added to exactly cancel the line susceptance, b_{in}, resulting in a perfect match, that is, $y_{\text{in}} = 1 + j0$. The distance between the load and this point, pp' in Figure 10.3(a) or (b), where the stub is to be attached, can be calculated by setting g_{in} equal to unity in equation (10.3.6), which results in the following quadratic in t:

$$t^2(g_L^2 + b_L^2 - g_L) - 2b_L t + (1 - g_L) = 0 \qquad (10.3.8)$$

This can be solved for t as follows:

$$t = \frac{b_L \pm \sqrt{g_L((1 - g_L)^2 + b_L^2)}}{g_L^2 + b_L^2 - g_L} \qquad (10.3.9)$$

From which we can determine the value of d, as a fraction of a wavelength, as follows:

$$\frac{d}{\lambda} = \begin{cases} \dfrac{1}{2\pi} \tan^{-1}(t) & \text{if } t \geq 0 \\[2mm] \dfrac{1}{2\pi}(\tan^{-1}(t) + \pi) & \text{if } t < 0 \end{cases} \qquad (10.3.10)$$

It can easily be shown that in the special case of purely resistive loads (i.e., $b_L = 0$), there is a single solution to equation (10.3.8), namely:

$$t = \sqrt{\frac{1}{g_L}} \qquad (10.3.11)$$

In the more general case of complex loads, we can see from equations (10.3.9) and (10.3.10) that there are two values of line length, d, that satisfy the condition $g_{in} = 1$. Let us refer to these as d_1 and d_2.

For each value of d determined from equations (10.3.9) and (10.3.10), the susceptance of the stub required to cancel the line susceptance can be found by entering the value, d_1 or d_2 (i.e., the values corresponding t_1 or t_2), into equation (10.3.7). The stub susceptance is simply the same magnitude but opposite polarity of the b_{in} value thus calculated.

The length of the stub needed to realize the susceptance $b_{stub} = -b_{in}$ is given by the familiar equations for normalized susceptance of a lossless line terminated in open circuit and short circuit, respectively (which can be easily derived by setting $y_L = 0$ or $y_L = \infty$ in equation (10.3.5)):

$$b_{open} = \tan \beta l_{open} \qquad (10.3.12)$$
$$b_{short} = -\cot \beta l_{short} \qquad (10.3.13)$$

where l_{open} and l_{short} refer to the lengths of the stubs in Figure 10.3(a) and (b), respectively. Note that since there are two values of d that satisfy equation (10.3.9), there will be a different value of either open-circuit or short-circuit stub length corresponding to each value of d. In other words, there will be four possible matching network solutions, as follows:

1. d_1 with an open-circuit stub.
2. d_1 with a short-circuit stub.
3. d_2 with an open-circuit stub.
4. d_2 with a short-circuit stub.

The choice of line length and stub type will normally be made on the basis of practical considerations. If any of the equations above result in negative electrical lengths, or an electrical length that is impractically short, we can simply add $\pm(n\lambda)/2$ to get an equivalent positive electrical length. The reader may wish to verify that this is the case with reference to equation (10.3.12) or (10.3.13).

Example 10.1 (Single stub matching: direct calculation).

Problem. Determine the single stub matching parameters (d and l) required to match the load $25 - j50$ Ω to a 50 Ω system.

Solution. First, we normalize the load:

$$z_L = \frac{(25 - j50)}{50} = 0.5 - j$$

We then calculate the normalized load admittance:

$$y_L = \frac{1}{(0.5 - j)} = 0.4 + j0.8$$

We now employ equation (10.3.9) to calculate the value of d_1 as follows:

$$t_1 = \frac{0.8 + \sqrt{0.4 \times [(1 - 0.4)^2 + 0.8^2]}}{0.4^2 + 0.8^2 - 0.4}$$

$$t_1 = \frac{0.8 + 0.633}{0.4}$$

$$t_1 = 3.583$$

$$\beta d_1 = \tan^{-1}(3.583) = 1.299$$

$$d_1 = \frac{1.299\lambda}{2\pi} = \boxed{0.207\lambda}$$

We now apply t_1 to equation (10.3.7) to determine b_{in1} at the point of stub attachment, d_1:

$$b_{in1} = \frac{0.8 \times (1 - 3.583^2) + 3.583 \times (1 - 0.8^2 - 0.4^2)}{(1 - 0.8 \times 3.583)^2 + 0.4^2 \times 3.583^2}$$

$$b_{in1} = \frac{-8.755}{5.538} = -1.581$$

We then set the susceptance of an open- or short-circuit stub equal to $-b_{in1}$ in equations (10.3.12) and (10.3.13) to determine the stub lengths as follows:

$$1.581 = \tan(\beta l_{open1})$$

$$l_{open1} = \frac{\tan^{-1}(1.581)}{2\pi}\lambda$$

$$l_{open1} = \boxed{0.160\lambda}$$

$$1.581 = -\cot(\beta l_{short1})$$

$$l_{short1} = \frac{1}{2\pi} \tan^{-1}\left(\frac{-1}{1.581}\right)\lambda$$

$$l_{short1} = -0.090 + 0.5 = \boxed{0.410\lambda}$$

We apply the same methodology to determine the alternative solution, comprising d_2 and l_2:

$$t_2 = \frac{0.8 - \sqrt{0.4 \times [(1-0.4)^2 + 0.8^2]}}{0.4^2 + 0.8^2 - 0.4}$$

$$t_2 = \frac{0.8 - 0.633}{0.4}$$

$$t_2 = 0.418$$

$$\beta d_2 = \tan^{-1}(0.418) = 0.396$$

$$d_2 = \frac{0.396\lambda}{2\pi} = \boxed{0.063\lambda}$$

We now apply t_2 to equation (10.3.7) to determine b_{in2} at the point of stub attachment, d_2:

$$b_{in2} = \frac{0.8 \times (1 - 0.418^2) + 0.418 \times (1 - 0.8^2 - 0.4^2)}{(1 - 0.8 \times 0.418)^2 + 0.4^2 \times 0.418^2}$$

$$b_{in2} = \frac{0.743}{0.470} = 1.581$$

We then set the susceptance of an open- or short-circuit stub equal to $-b_{in2}$ in equations (10.3.12) and (10.3.13) to determine the stub lengths as follows:

$$-1.581 = \tan(\beta l_{open2})$$

$$l_{open2} = \frac{\tan^{-1}(-1.581)}{2\pi}\lambda$$

$$l_{open2} = -0.160 + 0.5 = \boxed{0.340\lambda}$$

$$-1.581 = -\cot(\beta l_{short2})$$

$$l_{short2} = \frac{1}{2\pi} \tan^{-1}\left(\frac{1}{1.581}\right)\lambda$$

$$l_{short2} = \boxed{0.090\lambda}$$

10.3.2 SINGLE STUB MATCHING: GRAPHICAL APPROACH

The rather laborious calculations illustrated in the previous section can be avoided by using the Smith Chart. Since we are adding a parallel stub, we use the Smith Chart in admittance mode; that is, the constant resistance circles are interpreted as circles of constant conductance and the constant reactance circles are interpreted as circles of constant susceptance.

We could choose to use a combined impedance/admittance chart, like the one in Figure 4.9, if we wanted to see both impedance and admittance coordinates simultaneously. This is not usually necessary, however, and in the following examples we will use the standard Smith Chart in admittance mode for clarity and convenience.

The procedure consists of five steps and is best illustrated by a numerical example. To restate our objective: to design a single stub matching network that will match a complex load, Z_L, to a 50 Ω line system.

We will proceed to match the same load ($Z_L = 25 - j50$ Ω) as was used in Example 10.1 so that the reader can compare the results of both analytical and graphical approaches. Note that the line characteristic impedance is 50 Ω throughout. We proceed as follows:

Step 1: Normalize the load: $z_L = (25 - j50)/50 = 0.5 - j$ and plot this point on the Smith Chart (indicated as "z_L" in Figure 10.4).
Step 2: Draw the constant voltage standing wave ratio (VSWR) circle through this point and thereby determine the normalized load admittance as $y_L = 0.4 + j0.8$, which is 180° around the VSWR circle (indicated as "y_L" in Figure 10.4). Note that from this point on, all points on the Smith Chart are to be interpreted as *normalized admittances*.
Step 3: Starting at y_L, we move around the constant VSWR circle clockwise toward the generator until we cross the unit conductance circle at point "P" in Figure 10.4. The normalized admittance looking into the line at point "P" is now $1 + jb$. Note that if we keep going clockwise around the constant VSWR circle we will cross the unit conductance circle again at point "Q." The normalized admittance looking into the line at point "Q" is $1 - jb$. The lengths of line, from the load to the point of stub attachment, in each of these cases is d_1 and d_2, respectively. In our example, we can determine the line lengths, in terms of guided wavelength, by reading off the "wavelengths toward generator" scale on the outer boundary of the Smith Chart. This gives us the following:

$$d_1 = (0.178 - 0.115)\lambda = 0.063\lambda$$
$$d_2 = (0.321 - 0.115)\lambda = 0.206\lambda$$

Step 4: We now need to determine the susceptance of the shunt stub to be attached at either point "P" or point "Q." This should be equal in magnitude and opposite in sign to $\pm b$, as appropriate. The addition of the stub will cancel the transformed load susceptance at the attachment point, resulting in a perfect match at that point, that is $y_{in} = 1 + j0$. We determine the input susceptance of the line at the point of stub attachment by finding the constant susceptance circle which intersects the constant conductance circle at points "P" and "Q" in Figure 10.4. The values in our example are $b = 1.6$ and $b = -1.6$, respectively.

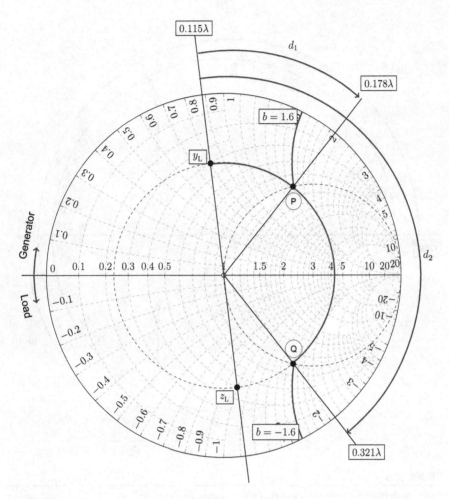

FIGURE 10.4

Single stub matching: graphical method.

Step 5: Now that we know the input susceptance of the line where the stub is to be attached (points "P" or "Q"), we can use the Smith Chart to determine the length of the short- or open-circuit stub required. This procedure is illustrated for point "P" in Figure 10.5 and for point "Q" in Figure 10.6. Note that the stub susceptance is of opposite sign to the line input susceptance. The length of the stub is found by moving clockwise around the chart starting at the open-circuit ($y = 0$) or short-circuit ($y = \infty$) points (labeled "O/C" and "S/C," respectively, in Figures 10.5 and 10.6) until the desired location on the periphery of the Smith Chart is reached. This location is found by drawing a radial through the point where the

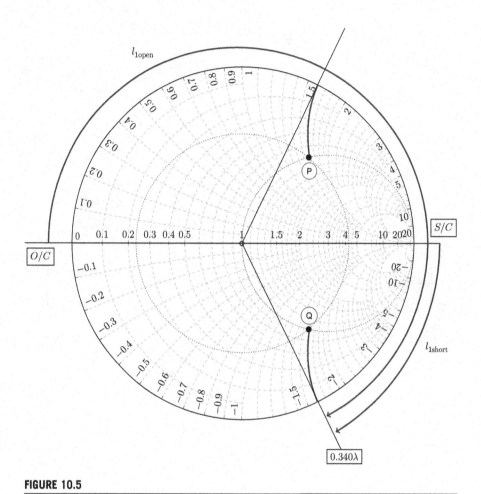

FIGURE 10.5

Single stub matching: determination of stub lengths for point "P."

constant reactance circle intersects the boundary of the Smith Chart. We can then read off the value on the "wavelengths toward generator" scale. The respective stub lengths are l_{1open} and l_{1short} in the case of point "P" (Figure 10.5) and l_{2open} and l_{2short} in the case of point "Q" (Figure 10.6). In our example, these values have been determined with reference to Figures 10.5 and 10.6 as follows:

$$l_{1open} = 0.340\lambda$$
$$l_{1short} = (0.340 - 0.25)\lambda = 0.09\lambda$$
$$l_{2open} = 0.160\lambda$$
$$l_{2short} = (0.160 + 0.25)\lambda = 0.410\lambda$$

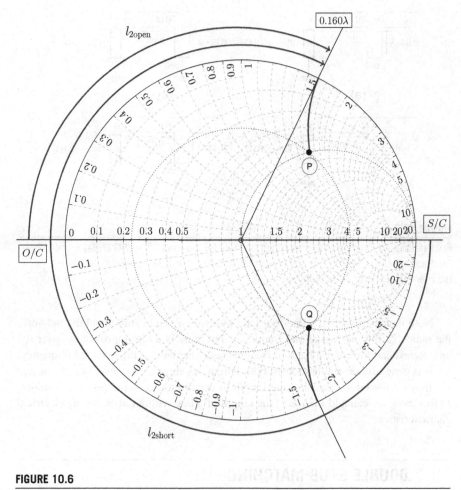

FIGURE 10.6

Single stub matching: determination of stub lengths for point "Q".

Notwithstanding the arbitrary assignment of "Solution 1" and "Solution 2" categories, the reader can see that these results are the same as those obtained by numerical calculation in Example 10.1. The reader may also note that the short-circuit stub lengths are equal to the open-circuit stub lengths ±0.25λ.

Although two possible solutions exist corresponding to points "P" and "Q," the one which represents the shortest line length, subject to implementation constraints, is usually preferred as this is less sensitive to frequency variation and thus has the greater bandwidth. This idea is elaborated on in Section 10.7. To summarize the outcome of this single stub matching exercise, we can draw the four possible matching networks, represented in 50 Ω microstrip form, in Figure 10.7.

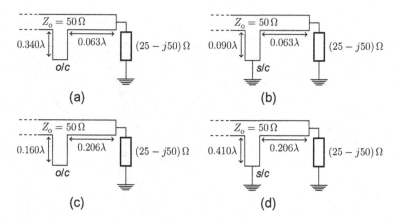

FIGURE 10.7

Four possible microstrip stub matching networks for the load $25 - j50$ Ω. (a) Point "P": open-circuit stub. (b) Point "P": short-circuit stub. (c) Point "Q": open-circuit stub. (d) Point "Q": short-circuit stub.

Note that in matching a line to a load by using a single stub matching network, the match is only perfect at one frequency, but there is a big improvement over the unmatched condition for a band of frequencies either side of the matched frequency.

It is generally desirable to keep both lengths as short as possible, but in practice we may be forced to extend a line length to accommodate some layout constraint. In this case, we can add $(n\lambda)/2$ to the length of a line without affecting its electrical characteristics.

10.4 DOUBLE STUB MATCHING

One limiting characteristic of single stub matching is that the location of the matching stub is a function of the load impedance. This may not be an issue in most cases, but in some practical circumstances there may be constraints on the physical positioning of the stub. In such cases, we can use an alternative matching technique called *double stub matching*, which employs two stubs, spaced a fixed distance apart. The advantage of this double stub matching approach is that, although the distance between the two stubs is generally kept constant, the first stub may be placed at any distance from the load. For this reason, the double stub technique is often used to implement variable tuners, where the match can be adjusted by varying only the stub lengths, with the stub locations on the main line remaining fixed.

The best way to understand double stub matching is to start at the end of the process and work our way toward the beginning. Consider what is happening at the plane bb′ in Figure 10.8 where Stub 2 is to be attached. Since stubs are purely reactive, Stub 2 can only contribute susceptance at point bb′. Therefore, in order

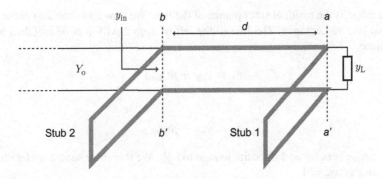

FIGURE 10.8

Double stub matching, first stub at load.

for the attachment of Stub 2 at bb′ to result in a perfect match, the normalized conductance looking into the matching network at point bb′ on the line, before we attach Stub 2, must be unity. Another way of saying this is that the normalized admittance, $y_{bb'}$, looking into the line at point bb′ before Stub 2 is attached should be:

$$y_{bb'} = 1 \pm jb_{in} \qquad (10.4.1)$$

The purpose of adding Stub 2, therefore, is solely to cancel the residual line susceptance, $\pm b_{in}$. This is the same principle behind single stub matching. The difference here is that the condition of unit conductance at the point bb′ is achieved, not by varying the distance to the load, d, as was the case with single stub matching, but by varying the susceptance of another stub, Stub 1, located a fixed distance (d) away, toward the load, from Stub 2. Initially, we have located Stub 1 at the load in Figure 10.8, to simplify the analysis. Later in this section we will show that, depending on the separation between the stubs, there is a maximum value of the load admittance, which can be matched to the line by using a double stub matching network, when the first stub is connected at the load. It can also be shown that if the load admittance exceeds the above-mentioned maximum value, the load can be matched to the line by inserting a length of transmission line between the first stub and the load, or by altering the distance between the stubs.

Let us define $y_{bb'}$ as the normalized input admittance looking into the line at the plane bb′ in the absence of Stub 2. If the susceptance of Stub 1 has been correctly chosen (using a method we will explain shortly) then $y_{bb'}$ can be written as:

$$y_{bb'} = l + jb_{bb'}$$

or

$$y_{bb'} = l - jb_{bb'}$$

where $\pm b_{bb'}$ is the residual susceptance of the line. We now add Stub 2 to cancel this residual line susceptance. The line to the left of Stub 2 will then be matched to the load since,

$$y_{in} = y_{bb'} + jb_{Stub2}$$

or

$$y_{in} = y_{bb'} - jb_{Stub2}$$

where b_{Stub2} is equal and opposite in sign to $b_{bb'}$. We therefore have a perfect match at bb', that is, $y_{in} = 1$.

As with single stub matching, we will outline both analytical and graphical design approaches.

10.4.1 DOUBLE STUB MATCHING: ANALYTICAL APPROACH

The normalized admittance at the load in Figure 10.8, with Stub 1 attached is:

$$y_{aa'} = g_L + j(b_L + b_{Stub1})$$

where b_L is the load susceptance and b_{Stub1} is the susceptance of Stub 1. After this impedance has been transformed through the length of line, d, and prior to the attachment of Stub 2, the normalized admittance at point bb' in Figure 10.8 is:

$$y_{bb'} = \frac{g_L + j(b_L + b_{Stub1} + t)}{1 + jt(g_L + j(b_L + b_{Stub1}))} \tag{10.4.2}$$

where again

$$t = \tan(\beta d) \tag{10.4.3}$$

and

$$\beta = \frac{2\pi}{\lambda} \tag{10.4.4}$$

But we know that the normalized conductance at point bb' must be unity, since this is where we will attach Stub 2 (which can only add susceptance) to achieve a perfect match. We therefore set the real part of equation (10.4.2) equal to 1, which gives us the following relationship between g_L, b_L, and t:

$$g_L^2 - g_L \frac{1 + t^2}{t^2} + \frac{(1 - t(b_L + b_{Stub1}))^2}{t^2} = 0 \tag{10.4.5}$$

Rearranging equation (10.4.5) yields the requisite value of b_{Stub1} as:

$$b_{Stub1} = -b_L + \frac{1 \pm \sqrt{g_L(1 + t^2) - g_L^2 t^2}}{t} \tag{10.4.6}$$

The residual line susceptance at the point bb′, with Stub 1 attached, is given by the imaginary part of equation (10.4.2):

$$b_{bb'} = \frac{(1 - t(b_L + b_{Stub1}))(b_L + b_{Stub1} + t) - g_L^2 t}{(1 - t(b_L + b_{Stub1}))^2 + t^2 g_L^2} \qquad (10.4.7)$$

Now, substituting the value of b_{Stub1} given by equation (10.4.6) into equation (10.4.7) gives us the residual line susceptance at bb′ solely in terms of g_L and t:

$$b_{bb'} = \frac{\mp\sqrt{g_L(1 + t^2) - g_L^2 t^2} - g_L}{g_L t} \qquad (10.4.8)$$

The susceptance of Stub 2 is chosen to cancel the residual line susceptance at bb′, therefore,

$$b_{Stub2} = -b_{bb'}$$

We are now in a position to define b_{Stub2} solely in terms of g_L and t as follows:

$$b_{Stub2} = -\left(\frac{\mp\sqrt{g_L(1 + t^2) - g_L^2 t^2} - g_L}{g_L t}\right) \qquad (10.4.9)$$

where the $\mp$ in equation (10.4.9) is paired with the $\pm$ in equation (10.4.6).

Once we have determined the required stub susceptances from equations (10.4.6) and (10.4.9), the electrical length of the stubs can be calculated for either short- or open-circuited stub by applying the familiar relationships (equations 10.3.12 and 10.3.13).

Example 10.2 (Double stub matching: direct calculation).

Problem. A load consists of a 4 nH inductor which has a series internal resistance of 19.2 Ω. This load needs to be matched to a 50 Ω lossless co-axial transmission line by means of a double stub matching network, consisting of two short-circuit stubs, spaced 0.375λ apart. The stub nearest to the load is 0.1λ away from it. Determine the possible combinations of stub lengths which are required to match the load to the line at the operating frequency of 1835 MHz. You may assume all line sections and stubs are 50 Ω.

Solution. The load impedance is first calculated:

$$Z_L = 19.2 + j(2\pi \times 1.835 \times 10^9 \times 4 \times 10^{-9})$$
$$Z_L = 19.2 + j46.17 \; \Omega$$

The normalized load impedance and admittance are, therefore,

$$z_L = \frac{(19.2 + j46.17)}{50} = 0.384 + j0.923$$

$$y_L = \frac{1}{(0.384 + j0.923)}$$

$$= \frac{0.384 - j0.923}{0.9994}$$

$$= 0.384 - j0.923$$

Stub 1 is attached at a distance $l_x = 0.1\lambda$ from the load, so we need to work out the admittance $y_{aa'}$ at this point of attachment. We can employ equations (10.4.2) and (10.4.3) which give us the conductance and susceptance of a load, y_L, translated a distance, l_x, toward the generator. In this case, $t = \tan(0.2\pi) = 0.727$, therefore,

$$g_{aa'} = \frac{0.384(1 + 0.727^2)}{(1 + 0.923 \times 0.727)^2 + 0.384^2 \times 0.727^2}$$

$$g_{aa'} = \frac{0.587}{2.870}$$

$$g_{aa'} = \boxed{0.204}$$

and

$$b_{aa'} = \frac{-0.923(1 - 0.727^2) + 0.727(1 - (-0.923)^2 - 0.384^2)}{(1 + 0.923 \times 0.727)^2 + 0.384^2 \times 0.727^2}$$

$$b_{aa'} = \frac{-0.435}{2.870}$$

$$b_{aa'} = \boxed{-0.154}$$

Using this value of $y_{aa'} = 0.204 - j0.154$ as the translated load, we now calculate the required value of Stub 1 using equation (10.4.6), this time using $t = \tan(0.375 \times 2\pi) = -1$:

$$b_{Stub1} = 0.154 + \frac{1 \pm \sqrt{0.204(1 + (-1)^2) - 0.204^2 \times (-1)^2}}{-1}$$

$$b_{Stub1} = -0.846 \pm (-0.606)$$

We therefore have two solutions for Stub 1, namely:

$$b_{Stub1} = \boxed{-1.452}$$

and

$$b_{Stub1} = \boxed{-0.240}$$

We calculate the admittance of Stub 2 by applying equation (10.4.9):

$$b_{Stub2} = - \left(\frac{\mp\sqrt{2 \times 0.204 - 0.204^2} - 0.204}{-0.204} \right)$$

$$b_{Stub2} = - \left(\frac{\mp 0.605 - 0.204}{-0.204} \right)$$

We therefore have two solutions for Stub 2, namely:

$$b_{Stub2} = \boxed{1.967}$$

and

$$b_{Stub2} = \boxed{-3.967}$$

We can now apply equations (10.3.12) and (10.3.13) to determine the length of short-circuit and open-circuit Stubs 1 and 2 for both solutions as summarized in Table 10.1.

Table 10.1 Double Stub Matching: Analytical Solutions

	Normalized Susceptance	Length (s/c Stub)	Length (o/c Stub)
Solution 1	Stub 1 = −1.452	Stub 1 = 0.096λ	Stub 1 = 0.346λ
	Stub 2 = −3.967	Stub 2 = 0.039λ	Stub 2 = 0.289λ
Solution 2	Stub 1 = −0.240	Stub 1 = 0.212λ	Stub 1 = 0.462λ
	Stub 2 = 1.967	Stub 2 = 0.425λ	Stub 2 = 0.175λ

10.4.2 DOUBLE STUB MATCHING: GRAPHICAL APPROACH

We will now proceed to explain how a double stub matching network can be designed graphically using the Smith Chart. Once again, since we are adding elements in parallel, the reader should assume that the Smith Chart is being used in admittance mode, unless otherwise stated.

Consider the arrangement shown in Figure 10.8. The total admittance at the point aa' is

$$y_{aa'} = y_L + jb_{Stub1}$$

where y_L is the admittance of the load and b_{Stub1} is the susceptance of Stub 1.

Since Stub 2 can only contribute susceptance, $y_{bb'}$ must be some point on the unit conductance circle on the Smith Chart (circle 1 in Figure 10.10). We therefore deduce that $y_{aa'}$ must lie on a circle of equal radius but having its center rotated d wavelength

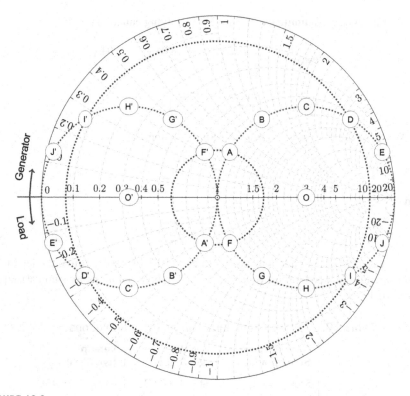

FIGURE 10.9

The principle behind double stub matching.

toward the load. This principle can be illustrated by considering Figure 10.9 where we have chosen d to be $\lambda/4$. Picking an arbitrary point, A, which lies on the unit conductance circle, we see that the effect of adding the line d is to rotate this point around the constant VSWR circle through A, 0.25λ toward the load to a point A'. Another point B on the unit conductance circle is similarly rotated 0.25λ to point B'. The same procedure can be carried out for all the other points C to J in Figure 10.9.

If the locus is then drawn through all the primed points in Figure 10.9, it is found to be a circle of the same radius as the unit conductance circle whose center O' has been rotated 0.25λ toward the load from O.

A similar transformation exists for any other value of d. In each case, the locus of points on the unit conductance circle maps into a circle of the same radius, whose center lies d wavelengths counter-clockwise toward the load from the center of the unit conductance circle. Having established this principle, the method of determining the stub lengths using the Smith Chart may be illustrated with reference to Figure 10.10.

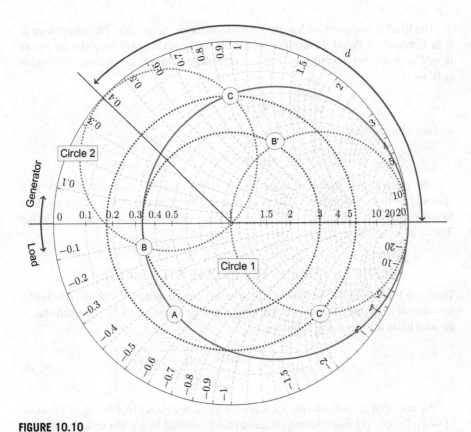

FIGURE 10.10

Double stub matching: graphical method.

In Figure 10.10, circle 2 represents the unit conductance circle transformed through some distance d wavelengths toward the load by means of the fixed line section, d, in Figure 10.8. Let us pick a load admittance represented by the point A in Figure 10.10. The constant conductance circle (solid circle) through point A cuts circle 2 at two locations, B and C. Hence, to achieve a match, we need to transform A to point B or C by choosing a value of Stub 1 which adds the correct susceptance to move point A in the right direction along the constant conductance circle to either point B or point C.

The first stub, therefore, must present a normalized susceptance equal to the difference between the susceptance at point A and the susceptance at point B or C. We determine these values by looking at which constant susceptance circles these points lie on and subtracting one susceptance value from another. With the addition of Stub 1 with a value thus calculated, the input admittance at point B or C becomes $y_{aa'}$.

The fixed line section, of length d, now transforms $y_{aa'}$ to $y_{bb'}$. The admittance at B or C when translated d wavelengths toward the generator will be given by points B′ or C′, respectively, on the unit conductance circle. This means that the admittance at B′ is:

$$y_{bb'} = 1 + jb_{B'}$$

and the admittance at C′ is:

$$y_{bb'} = 1 + jb_{C'}$$

The second stub therefore must present a normalized susceptance equal and opposite to the susceptance value at B′ and C′, these values being determined by looking at the susceptance scale on the Smith Chart. This procedure is further illustrated in Example 10.3.

10.4.3 FORBIDDEN REGIONS FOR DOUBLE STUB MATCHING

There are limitations on the load admittance that can be matched with the configuration shown in Figure 10.8, as we will now show. equation (10.4.5) is a quadratic in g_L which has the following solution:

$$g_L = \frac{1+t^2}{t^2}\left[1 \pm \sqrt{1 - \frac{4t^2(1 - t(b_L + b_{Stub1}))^2}{(1+t^2)^2}}\right] \qquad (10.4.10)$$

We note that the term inside the square root in equation (10.4.10) is of the form $(1 - x)$. In order to meet the requirement that g_L should be a real number, the value of x must lie between 0 and 1. In the case of equation (10.4.10), this leads to the following boundaries on the value of g_L:

$$0 \le g_L \le \frac{1+t^2}{t^2} \qquad (10.4.11)$$

With reference to standard trigonometric identities, equation (10.4.11) can be restated as:

$$0 \le g_L \le \frac{1}{\sin^2 \beta d} \qquad (10.4.12)$$

The limiting conditions set by equation (10.4.12) define a *forbidden region* on the Smith Chart within which loads are unmatchable with a double stub tuner. The forbidden region is bounded by a constant conductance circle whose value depends solely on the electrical stub separation, d/λ. Let us say, for example, we set $d = \lambda/8$. Then $\beta d = \pi/4$ and $\sin^2 \beta d = 0.5$. This means that the forbidden region will be bounded by the $g = 2$ circle, as shown in Figure 10.11. We can understand why any load lying within the forbidden region in Figure 10.11 (i.e., having $g_L > 2$) cannot be matched with this tuner by remembering that altering the value of Stub

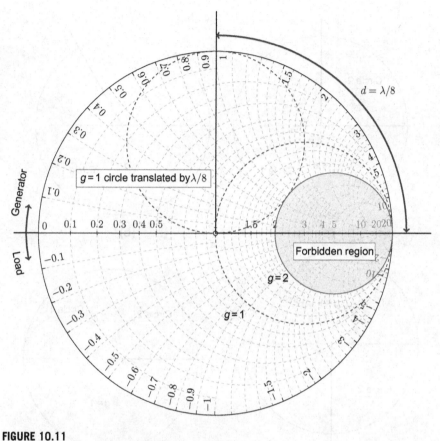

FIGURE 10.11

Double stub matching: forbidden region for $d = \lambda/8$.

1, located at the load, has the effect of moving the load point along the constant conductance circle until an intersection with the translated unit conductance circle occurs (refer back to Figure 10.10). In other words, only constant conductance circles lying outside the forbidden region in Figure 10.11, can intersect with the translated $g = 1$ circle.

The relationship between d and the size of the forbidden region is further shown in Figure 10.12 where circle 1 is the unit conductance circle and circle 2 is this circle transferred d wavelengths toward the load. The forbidden region is shown shaded. If we reduce the stub separation, d, the forbidden region will be reduced in size, but will only become zero when the stub separation is zero (in which case we are no longer talking about a double stub tuner). This is shown for the example of $d = \lambda/12$ in Figure 10.12(b). Conversely, if we increase stub separation the forbidden region will

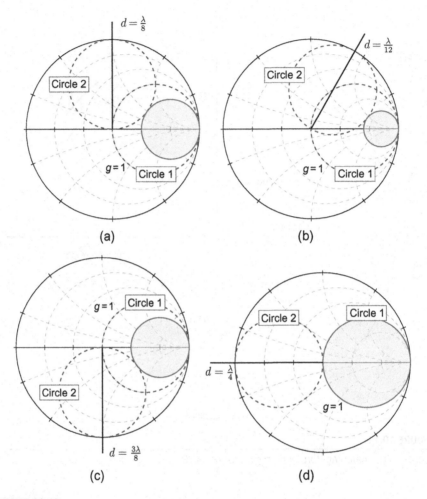

FIGURE 10.12

Double stub matching: forbidden region (shaded) as a function of stub separation.
(a) $d = \frac{\lambda}{8}$. (b) $d = \frac{\lambda}{12}$. (c) $d = \frac{3\lambda}{8}$. (d) $d = \frac{\lambda}{4}$.

increase in size until it reaches a maximum, bounded by the $g = 1$ circle, when the stub separation, d, is $\lambda/4$, as shown in Figure 10.12(d).

If g_L lies within the forbidden region (i.e., it is greater than the value set by equation (10.4.12)) then we can still use the double stub technique with the first stub located a minimum distance, l_x, away from the load toward the generator, as shown in Figure 10.13.

The added line length has the effect of transforming the load admittance so that the transformed value of g_L is reduced to a value below the limit set by

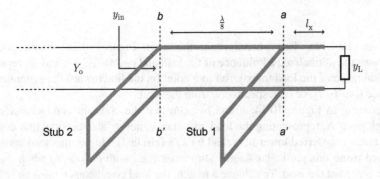

FIGURE 10.13

Double stub matching, first stub located l_x from the load.

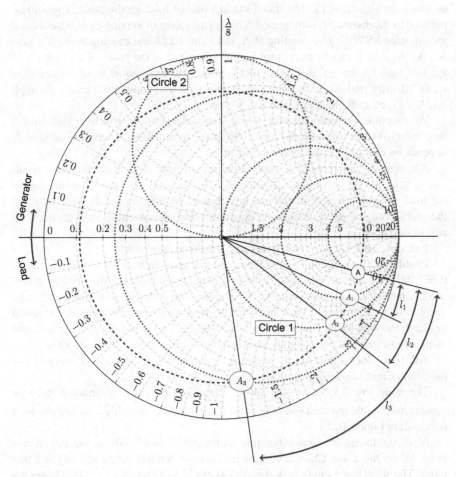

FIGURE 10.14

Double stub matching: location of Stub 1 with respect to the load.

equation (10.4.12). This procedure is shown in Figure 10.14, where point A represents the normalized admittance of the load and points A_1, A_2, and A_3 represent the admittance of the load transferred to a point on the line toward the generator at a distance $l_x = l_1$, l_2, or l_3, respectively, from the load.

Referring to Figure 10.14, it can be seen that the constant conductance circle through point A, representing the load admittance, never intersects the unit conductance circle transferred toward the load by $\lambda/8$ (circle 2). Hence, this load cannot be matched using this particular double stub tuner (i.e., with $d = \lambda/8$) when the first stub is placed at the load. To achieve a match, the load conductance must be reduced to such a value that the constant conductance circle through the new load point on the Smith Chart intersects circle 2. This can be achieved by shifting the point of application of the first stub toward the generator, in other words by adding the length of line l_x in Figure 10.13. The transformed value of load conductance is gradually reduced as the distance l_x is increased. This is equivalent to moving clockwise around the constant VSWR circle, starting at A, in Figure 10.14 and moving through points A_1, A_2, A_3, etc. It can be seen from Figure 10.14 that if the point of application of the first stub (points aa' in Figure 10.13) is at a distance equal to or greater than l_1, an effective match can be achieved, as the constant conductance circles through A_1, A_2, A_3, etc., all intersect circle 2.

We conclude, therefore, that for double stub matching to be effective, Stub 1 must be connected at a minimum distance l_1 from the termination, where the value of l_1 depends on the value of the load conductance.

Alternatively, by keeping the Stub 1 connected right at the load, the same result can be achieved by altering the distance between the stubs. If Stub 1 is connected at the load, then an effective match can be achieved by altering the distance between the stubs until the circle 2 cuts the constant conductance circle through the load admittance point.

Example 10.3 (Double stub matching: graphical approach).

Problem. The reflection coefficient at a load terminating a 50 Ω lossless line is $0.667\angle90°$. We need to match this load to the line using a double stub matching network comprising two stubs, spaced 0.375λ apart. The stub nearest to the load is 0.1λ away from it. Determine the possible combinations of stub lengths which are required to match the load to the line. Carry out two designs: one using open-circuit and one using short-circuit stubs. You may assume all line sections and stubs are 50 Ω.

Solution. Once again, since we are adding components in parallel, we will use the Smith Chart in admittance mode.

The stubs are $0.375\lambda = \frac{3}{8}\lambda$ apart. Therefore, we start by drawing the unit conductance circle (circle 1) and the same circle rotated through $\frac{3}{8}\lambda$ toward the load (circle 2) in Figure 10.15.

Next, we locate the load reflection coefficient $= 0.667\angle90°$ as the impedance point "A" on the Smith Chart, and draw the constant VSWR ($=5$) circle through this point. The VSWR $= 5$ circle is designated as circle 3 in Figure 10.15. We obtain the load admittance as point B by rotating the point A around circle 3 through $180°$ in either direction. From this point onward, all coordinates on the Smith Chart represent normalized admittance ($g + jb$).

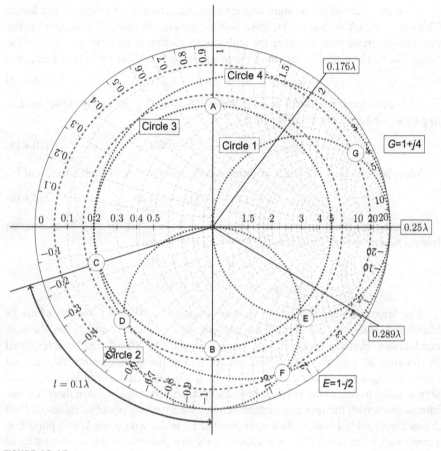

FIGURE 10.15

Double stub matching, Example 10.3.

We are told that the distance to Stub 1 is 0.1λ from the load. We therefore locate point C by rotating the load admittance 0.1λ toward the generator in Figure 10.15. We note that point C lies on the $g = 0.2$ constant conductance circle (circle 4), which means that we can move back and forth along this circle by adjusting the susceptance of Stub 1.

We see that the $g = 0.2$ constant conductance circle (circle 4) intersects the translated unit conductance circle (circle 2) at points D and F. Since there are two points of intersection, there must be two values of Stub 1 that can provide a match. The normalized susceptance corresponding to point D is found by looking at which constant susceptance circles points C and D lie on and subtracting one from the other. This gives us:

$$b_{\text{Stub1D}} = (-0.40) - (-0.15) = -0.25 \tag{10.4.13}$$

We locate the -0.25 constant susceptance circle on the perimeter of the Smith Chart and read off the value of 0.461λ on the "wavelengths toward generator" scale. For a short-circuit stub, we trace the stub length starting at the $y = \infty$ point on the Smith Chart. The electrical length of the short-circuit Stub 1 for point D is, therefore,

$$l_{\text{Stub1Ds}} = 0.461 - 0.25 = 0.211\lambda \qquad (10.4.14)$$

The open-circuit Stub 1 for point D is $\pm 0.25\lambda$ away (we obviously chose $\pm 0.25\lambda$ to give us a positive electrical length):

$$l_{\text{Stub1Do}} = 0.211 + 0.25 = 0.461\lambda \qquad (10.4.15)$$

We apply the same approach to determine the susceptance of Stub 1 for point F:

$$b_{\text{Stub1F}} = (-1.6) - (-0.15) = -1.45 \qquad (10.4.16)$$

Locating the point representing susceptance $b = -1.45$ on the "wavelengths toward generator scale" gives us the following stub lengths:

$$l_{\text{Stub1Fs}} = 0.346 - 0.25 = 0.096\lambda$$

$$l_{\text{Stub1Fo}} = 0.096 + 0.25 = 0.346\lambda$$

The length of line separating the two stubs ($d = 0.375\lambda$) has the effect of translating points D and F into the corresponding points E and G on the unit conductance circle in Figure 10.15. The susceptance of Stub 2 can be determined by looking at which constant susceptance circle these two points lie on, and remembering that the susceptance of Stub 2 must have equal magnitude but opposite sign in order to cancel the residual line susceptance. Once again, since there are two intersections with the unit conductance circle, we have two possible values of Stub 2 that can provide a match. Note that point D is paired with point E and point F is paired with point G, so there are actually only two possible solutions consisting of the stub pairs D, E and F, G.

For points E and G we can read off the residual line susceptances as:

$$b_{\text{E}} = -2.00$$

$$b_{\text{G}} = 4.00$$

Locating the opposite polarity susceptance points on the "wavelengths toward generator" scale on the outer boundary of the Smith Chart gives the following electrical lengths for Stub 2 using short-circuit stubs:

$$l_{\text{Stub2Gs}} = 0.289 - 0.25 = 0.039\lambda$$

$$l_{\text{Stub2Es}} = 0.176 + 0.25 = 0.426\lambda$$

The equivalent open-circuit stubs are $\pm 0.25\lambda$ in length, therefore,

$$l_{\text{Stub2Go}} = 0.039 + 0.25 = 0.289\lambda$$

$$l_{\text{Stub2Eo}} = 0.426 - 0.25 = 0.176\lambda$$

So, we have four possible double stub matching network solutions, which are summarized in Table 10.2.

Table 10.2 Double Stub Matching: Graphical Solutions

	Normalized Susceptance	Length (s/c Stub)	Length (o/c Stub)
Solution 1: (D, E)	Stub 1 = −0.25 Stub 2 = 2.00	Stub 1 = 0.211λ Stub 2 = 0.426λ	Stub 1 = 0.461λ Stub 2 = 0.176λ
Solution 2: (F, G)	Stub 1 = −1.45 Stub 2 = −4.00	Stub 1 = 0.096λ Stub 2 = 0.039λ	Stub 1 = 0.346λ Stub 2 = 0.289λ

Notwithstanding the arbitrary assignment of "Solution 1" and "Solution 2" categories, and rounding errors, the reader can see that these results are basically the same as those obtained by numerical calculation in Example 10.2.

For clarity, a schematic representation of the two double stub matching network solutions implemented in microstrip implementation is shown in Figure 10.16.

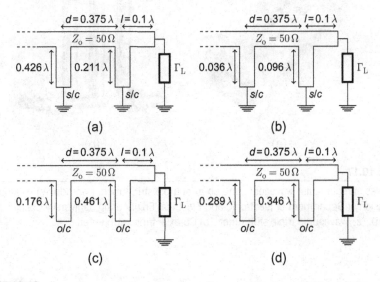

FIGURE 10.16

Four possible double stub matching networks for the load $\Gamma_L = 0.667 < 90°$ implemented in microstrip. (a) Solution 1 (points D, E) s/c stubs. (b) Solution 2 (points F, G) s/c stubs. (c) Solution 1 (points D, E) o/c stubs. (d) Solution 2 (points F, G) o/c stubs.

10.5 **TRIPLE STUB MATCHING**

It can be seen from Figure 10.12 that double stub matching is not possible for certain values of load impedance and stub placements. One solution is to add a third stub. Triple stub matching is rarely used as a design technique in fixed media, such as microstrip, but commercial "Triple Stub Tuners" are available as standard components in waveguide or co-axial media. These tuners are frequently implemented with three stubs spaced at unequal intervals, which ensures that any value of passive load can be matched. Short-circuit stubs are mainly used in co-axial or waveguide, for ease of adjustment.

We will not go into design techniques for triple stub tuners here. The interested reader is referred to Collin [3] for more description and design methodology.

Examples of waveguide and co-axial triple stub tuners are shown in Figure 10.17.

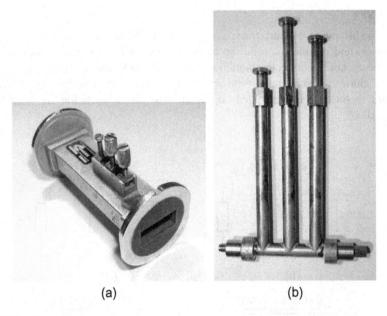

(a)   (b)

FIGURE 10.17

Examples of rectangular waveguide and co-axial triple stub tuners (reproduced by kind permission of Department of Electronic and Electrical Engineering, University College London). (a) Waveguide triple stub tuner. (b) Co-axial triple stub tuner.

10.6 **QUARTER-WAVE TRANSFORMER MATCHING**

Consider a situation where we have connected a length of transmission line of length $\lambda/4$ and characteristic impedance Z_1 in front of the load, as shown in Figure 10.18.

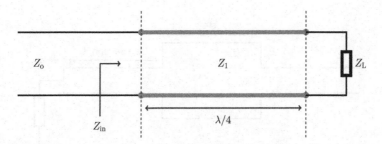

FIGURE 10.18

Quarter-wave transformer.

From equation (2.7.15), the input impedance of any lossless line section, of characteristic impedance Z_1, is given by:

$$Z_{in} = Z_1 \left(\frac{Z_L + jZ_1 \tan(\beta l)}{Z_1 + jZ_L \tan(\beta l)} \right) \tag{10.6.1}$$

But in Figure 10.18 we have set $l = \lambda/4$, so equation (10.6.1) now becomes:

$$Z_{in} = Z_1 \cdot \frac{Z_L + jZ_1 \tan\left(\frac{2\pi}{\lambda} \cdot \frac{\lambda}{4}\right)}{Z_1 + jZ_L \tan\left(\frac{2\pi}{\lambda} \cdot \frac{\lambda}{4}\right)}$$

$$= Z_1 \cdot \frac{Z_L + jZ_1 \tan\left(\frac{\pi}{2}\right)}{Z_1 + jZ_L \tan\left(\frac{\pi}{2}\right)} \tag{10.6.2}$$

Since $\tan(\pi/2) = \infty$ equation (10.6.2) reduces to:

$$Z_{in} = \frac{Z_1^2}{Z_L} \tag{10.6.3}$$

Equation (10.6.3) tells us that a quarter-wave line section can act as a matching network between a load, Z_L, and a required input impedance, Z_{in}, provided that the line section has a characteristic impedance given by:

$$Z_1 = \sqrt{Z_L Z_{in}} \tag{10.6.4}$$

Such a line section is called a *quarter-wave transformer* (QWT).

Equation (10.6.4) implies that, since the characteristic impedance of a lossless QWT must be real, this QWT can only be used to match purely resistive loads, $(Z_L = R_L + j0)$. To match complex loads, we would need a transmission line with a complex characteristic impedance, that is the line would have to be lossy. This is generally undesirable. One alternative is to transform the complex load into a real quantity by inserting a transmission line section between the load and the QWT, as

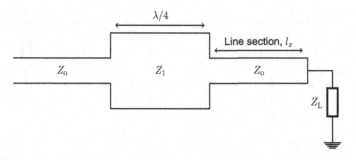

FIGURE 10.19

Quarter-wave transformer with a complex load.

shown, in microstrip form, in Figure 10.19. For convenience we will assume that the added line section has the system characteristic impedance, Z_0.

The effect of adding the extra line section, l_x, in Figure 10.19 can be demonstrated by considering the input impedance, Z_{in} of any line section terminated by a load Z_L. Z_{in} is given by equation (2.7.15), which can be expressed in normalized terms as:

$$z_{in} = \frac{z_L + jt}{1 + jz_L t} \tag{10.6.5}$$

where again

$$t = \tan\left(2\pi \frac{l_x}{\lambda}\right)$$

Since we need the input impedance of the added line section in Figure 10.19 to be purely real, we set $\mathrm{Im}(z_{in}) = 0$ in equation (10.6.5), which results in:

$$\frac{(x_L + t)(1 - x_L t) - r_L^2 t}{(1 - x_L)^2 + r_L^2 t} = 0 \tag{10.6.6}$$

where $z_L = r_L + jx_L$.

Equation (10.6.6) implies:

$$(x_L + t)(1 - x_L t) - r_L^2 t = 0 \tag{10.6.7}$$

Rearranging equation (10.6.7), and noting that $|z_L|^2 = r_L^2 + x_L^2$, results in the following quadratic in t:

$$x_L t^2 - t(1 - |z_L|^2) - x_L = 0 \tag{10.6.8}$$

We can now determine t as follows:

$$t = \frac{(1 - |z_L|^2) \pm \sqrt{(1 - |z_L|^2)^2 + 4x_L^2}}{2x_L} \qquad (10.6.9)$$

The two solutions embodied in equation (10.6.9) represent the two intersections of the constant VSWR circle through z_L with the real axis. One of these intersections gives $\text{Re}(z_{in}) < z_o$ and the other gives $\text{Re}(z_{in}) > z_o$. With a given value of t, we determine the electrical length of the line section, l_x, from:

$$\frac{l_x}{\lambda} = \frac{1}{2\pi} \tan^{-1}(t)$$

Figure 10.20 shows how such a matching network can be designed using a Smith Chart. We plot the normalized load impedance, z_L, on the chart at point "A." We then draw the constant VSWR circle through the point A, and we note that this circle crosses the real axis of the chart at two points, B and C. At both these points the input impedance of the line section is purely real.

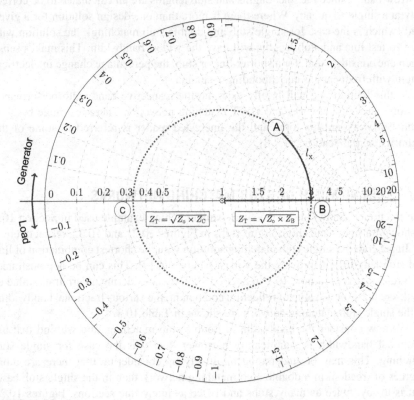

FIGURE 10.20

Quarter-wave transformer with complex load: graphical solution.

We note that at point B $Z_p > Z_o$ and at point C $Z_p < Z_o$, the lengths of series line required to achieve this are given by the distances AB and AC = AB + 0.25λ in Figure 10.20. Once the length of series line l_x has been chosen, the QWT is then used to transform Z_B or Z_C into Z_o. Choosing point B will require a transformer with a characteristic impedance greater than Z_o and choosing point C will result in a transformer with Z_T less than Z_o. Whichever point is chosen, Z_T is computed using equation (10.6.4).

If the QWT is being implemented in microstrip medium, the lower value of Z_T is generally preferred because low impedance lines (i.e., thicker traces) are easier to reproduce accurately.

10.7 BANDWIDTH OF DISTRIBUTED ELEMENT MATCHING NETWORKS

The distributed matching networks described in this chapter are all inherently "narrow band," since the line lengths and stub lengths are all calculated to be correct only at a single frequency. Where there is more than one design solution for a given load (which is the case for single stub and double stub matching), the solution with the shortest line and stub lengths will have the widest bandwidth. This makes sense when one considers that, for a shorter line or stub, the percentage change in electrical length with frequency is correspondingly smaller.

In this section, we will briefly examine the respective bandwidth performance of stub matching networks and QWT matching networks. There is a huge body of published work on this topic and the interested reader is referred to some of the authoritative publications [1,7,8].

10.7.1 BANDWIDTH OF STUB MATCHING NETWORKS

The fractional bandwidth for the single stub matching example used in Section 10.3 is shown for open- and short-circuit stubs in Figures 10.21 and 10.22, respectively.

In general, the single stub matching solution with the shortest combination of line and stub lengths will provide the widest bandwidth [9]. This can be demonstrated, with reference to Example 10.1 in Section 10.3, by considering an arbitrary value of $|\Gamma_{in}|$, say 0.2. At this value of reflection coefficient, the various fractional bandwidths of the single stub matching network are shown in Table 10.3.

We now consider the bandwidth of double stub matching, and we find that the fractional bandwidth is significantly narrower than was the case for single stub matching. This may be understood intuitively by considering that there are more degrees of freedom in a double stub matching network than in the single stub case, that is to say, twice as many stubs and twice as many line sections. Figures 10.23 and 10.24 show fractional bandwidth for typical open- and short-circuit double stub matching networks, respectively.

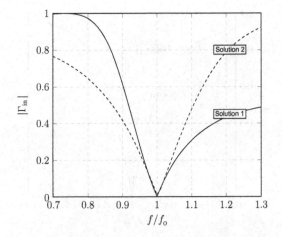

FIGURE 10.21

Single stub matching network bandwidth: open-circuit stubs.

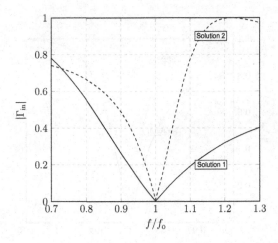

FIGURE 10.22

Single stub matching network bandwidth: short-circuit stubs.

Table 10.3 Fractional Bandwidth of a Typical Single Stub Matching Network

	Open-Circuit Stub (%)	Short-Circuit Stub (%)
Solution 1	9	19
Solution 2	8	6

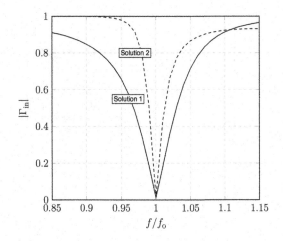

FIGURE 10.23

Double stub matching network bandwidth: open-circuit stubs.

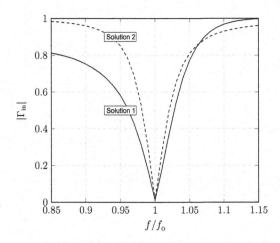

FIGURE 10.24

Double stub matching network bandwidth: short-circuit stubs.

We can compare the fractional bandwidths of the double stub solutions of Figures 10.23 and 10.24 by adopting a similar approach to that taken with the single stub solutions, only this time, because the bandwidth is narrower, we will choose a higher value of $|\Gamma_{in}| = 0.3$. The results are shown in Table 10.4.

Table 10.4 Fractional Bandwidth of a Typical Double Stub Matching Network

	Open-Circuit Stubs (%)	Short-Circuit Stubs (%)
Solution 1	3	4
Solution 2	1	2

10.7.2 BANDWIDTH OF QUARTER-WAVE TRANSFORMERS

The QWT is also inherently narrow band, since, by definition, it is only exactly a quarter wavelength at a single frequency. It turns out, however, that the closer the load, Z_L, is to the characteristic impedance of the main line, Z_0, the wider the fractional bandwidth of the quarter-wavelength transformer will be. This is illustrated for several purely real values of load impedance Z_L in Figure 10.25. It turns out that the bandwidth of a QWT increases according to the following quantity:

$$\frac{\sqrt{Z_L Z_0}}{|Z_L - Z_0|} \tag{10.7.1}$$

In other words, the bandwidth will increase as Z_L approaches Z_0 from either direction; either $Z_L > Z_0$ or $Z_L < Z_0$. This is shown in Figure 10.25, where $Z_L = 100\ \Omega$ results in the same bandwidth curve as $Z_L = 25\ \Omega$, since, in both cases, the parameter given in equation (10.7.1) has the same value (i.e., $\sqrt{2}$). Similarly, $Z_L = 150\ \Omega$ results in the same bandwidth curve as $Z_L = 16.67\ \Omega$ and $Z_L = 75\ \Omega$ results in the same bandwidth curve as $Z_L = 37.5\ \Omega$, and so on.

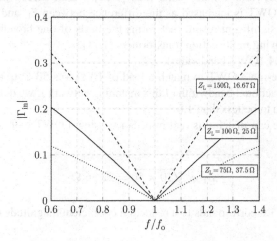

FIGURE 10.25

Quarter-wave transformer bandwidth.

A closed form expression for the fractional bandwidth of a single section transformer has been derived [2] as follows:

$$\frac{\Delta f}{f_0} = 2 - \frac{4}{\pi} \cos^{-1}\left[\frac{\Gamma_m}{\sqrt{(1-\Gamma_m^2)}} \frac{2\sqrt{Z_L Z_0}}{|Z_L - Z_0|}\right] \tag{10.7.2}$$

where Γ_m is the maximum reflection coefficient that we can accept as defining a "good" match.

We can see from equation (10.7.2) that the fractional bandwidth of the QWT increases as Z_L becomes closer to Z_0, as outlined above. For a value of $|\Gamma_{in}| = 0.1$, the fractional bandwidth of the QWT shown in Figure 10.25 with various loads is summarized in Table 10.5.

Table 10.5 Fractional Bandwidth of a Typical QWT ($|\Gamma_{in}| = 0.1$)

Load	Fractional Bandwidth (%)
$Z_L = 150\,\Omega$ or $16.67\,\Omega$	22
$Z_L = 100\,\Omega$ or $25\,\Omega$	36
$Z_L = 75\,\Omega$ or $37.5\,\Omega$	66

As an exercise, the reader may wish to verify that the bandwidths shown in Table 10.5 are in accordance with equation (10.7.2).

In general, we can say that QWTs offer wider fractional bandwidths than stub matching networks for the same load [9,10]. Furthermore, the fact that the bandwidth of a QWT is increased as the difference between Z_L and Z_0 is reduced is an extremely significant result, that forms the basis of one broadband matching technique employing multisection transformers [1,11].

Example 10.4 (QWT bandwidth).

Problem. Design a QWT to match a load of 10 Ω to a 50 Ω transmission line. Determine the fractional bandwidth of this matching network if we define the VSWR for a good match to be less than 1.6.

Solution. We can match this real impedance using a QWT with a characteristic impedance of:

$$Z_T = \sqrt{10 \times 50} = 22.36\,\Omega$$

A VSWR of 1.6 corresponds to a reflection coefficient magnitude of:

$$\Gamma_m = \frac{VSWR - 1}{VSWR + 1} = \frac{1.6 - 1}{1.6 + 1} = 0.231$$

The fractional bandwidth can now be computed from equation (10.7.2) as:

$$\frac{\Delta f}{f_0} = 2 - \frac{4}{\pi}\cos^{-1}\left[\frac{0.231}{\sqrt{(1 - 0.231^2)}} \frac{2\sqrt{10 \times 50}}{|10 - 50|}\right]$$

$$\frac{\Delta f}{f_0} = 0.342$$

The fractional bandwidth of this transformer is therefore 34.2%.

10.8 SUMMARY

In this chapter, we presented four basic types of matching network that can be implemented with distributed elements (i.e., transmission line sections), namely:

1. single line matching;
2. single stub matching;
3. double stub matching; and
4. QWT.

Both numerical and graphical design approaches were presented for stub matching and QWT networks.

These networks are usually classified as "narrow band" (i.e., having fractional bandwidths of less than 20% depending on the quality of match required). The bandwidth of the QWT, however, is strongly dependent on the difference between the load and Z_0, and can be very wide when this difference is small.

Single line matching is the simplest approach but is only applicable for a limited range of loads. Single stub matching can be used to match any load and is very easily implemented in planar media such as microstrip. This is consequently a very widely used narrow band matching technique.

In cases where there is a limitation on the physical placement of the stubs with respect to the load, double stub matching is a more appropriate technique, as the first stub can be located at any distance from the load, subject to a certain minimum distance that depends on the load conductance. The double stub technique is often used in variable stub tuners, as a wide range of loads can be matched simply by adjusting the stub lengths whilst maintaining a fixed stub spacing and distance from the load. This idea is extended further with the "triple stub tuner," which was not discussed in any detail in this chapter.

In general, a double stub matching network provides a narrower fractional bandwidth than an equivalent single stub matching network for a given load.

In the case of a QWT, matching is achieved by altering the characteristic impedance of a fixed quarter-wavelength line section. The basic QWT can only match purely resistive loads, but this can be extended to complex loads by inserting an additional section of line between the load and the transformer to translate the complex load into a real impedance at the input of the transformer. The QWT is widely used as a matching network in its own right, and is also an important building block in broadband multisection transformers [1,11].

10.9 TAKEAWAYS

1. Distributed element matching refers to the use of transmission line sections, as opposed to lumped elements, such as resistances, capacitors, and inductors to produce the necessary impedances and admittances in the matching network. The distributed element approach has implementation advantages in certain transmission line media, such as microstrip.

2. As with lumped element matching networks, distributed element matching networks can be designed either analytically (using a computer) or graphically using the Smith Chart.

3. The simplest distributed element matching network technique is known as *single line matching*, which consists of a particular length of series transmission line between the source and load. The range of load impedances that can be matched using this technique is strictly limited, meaning that this technique is rarely used in practice.

4. Single stub matching is a simple and versatile technique that uses a series length of transmission line plus a parallel short-circuit or open-circuit stub at a specified distance from the load. Single stub matching can be used to match any complex load and is therefore widely used in practice. There are always two stub matching solutions for any given matching problem, which differ in terms of their bandwidth.

5. The main limitation of single stub matching network is that the designer has no flexibility in the choice of stub location (relative to the load), as the length of the transmission line section between the stub and the load forms an integral part of the matching network. This limitation can be overcome by means of a *double stub* matching network, at the expense of reduced bandwidth.

6. A QWT ($\lambda/4$) consists of a single series length of transmission line having a length of one quarter wavelength at the center frequency of interest. The basic QWT can only match purely resistive loads, but this can be extended to complex loads by inserting an additional section of line between the load and the transformer to translate the complex load into a real impedance at the input of the transformer.

7. Bandwidth may be the main consideration in the decision whether to employ a single stub, double stub, or QWT matching network. In general, a double stub matching network provides a narrower fractional bandwidth than an equivalent single stub matching network for a given load. The bandwidth of the QWT is strongly dependent on the difference between the load and source impedances, and can be very wide when this difference is small.

TUTORIAL PROBLEMS

Problem 10.1 (Impedance transformation using a line section). Using a Smith Chart, determine the input admittance of a transmission line of characteristic impedance 75 Ω and electrical length 0.123λ terminated with a load $Z_L = (15 + j30)$ Ω.

Problem 10.2 (Single line matching). Determine whether it is possible to use the single line matching approach to match a load $Z_L = (41 + j49)$ Ω to a $Z_o = 50$ Ω system. If so, calculate the length of $Z_o = 30$ Ω line required.

Problem 10.3 (Single stub matching). A co-axial transmission line of characteristic impedance 70 Ω is terminated in a load impedance $Z_L = 84 + j85.75$ Ω. Design a single stub matching network, using a short-circuit stub, to match this load to the line. Assume that the system characteristic impedance is the same throughout.

Problem 10.4 (Single stub matching). A load consists of an inductor having a series resistance of 10 Ω and an unknown inductance. This is matched to a 50 Ω co-axial transmission line system using a variable single stub tuner consisting of a variable length short-circuit stub connected to the load via a variable length transmission line section. At a frequency of 318 MHz, we match the load, initially, with the stub located 0.346λ from the load and with stub susceptance $b_s = -140$ Ω. We now add another inductor of 0.5 nH in series with the load inductor and find that the stub has to be adjusted and moved to a new location 0.035λ from the load to achieve a match. Determine the inductance of the original inductor. *Hint*: use the Smith Chart.

Problem 10.5 (Double stub matching). Calculate the required electrical lengths of short-circuit Stubs 1 and 2 in the double stub matching network shown in Figure 10.26. The load, y_L, is normalised and system characteristic impedance is 50 Ω throughout.

Problem 10.6 (Forbidden regions). Can a load $Z_L = (17 - j7)$ Ω be matched to a 50 Ω system using a double stub tuner with a stub separation of 0.1λ?

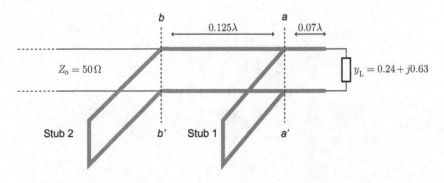

FIGURE 10.26

Figure for Problem 10.5.

Problem 10.7 (QWT). Design a QWT to match a load of $Z_L = (60 - j50)$ Ω to a 50 Ω transmission line. Explain why equation (10.7.2) could not be used to give an accurate determination of the percentage bandwidth in this case.

REFERENCES

[1] G. Matthaei, E. Jones, L. Young, Microwave Filters Impedance Matching Networks and Coupling Structures, McGraw-Hill, New York, NY, 1961.

[2] D. Pozar, Microwave Engineering, second ed., John Wiley and Sons, New York, NY, 1998.

[3] R. Collin, Foundations for Microwave Engineering, second ed., John Wiley and Sons, New York, NY, 2005, ISBN 9788126515288.

[4] R. Thomas, A Practical Introduction to Impedance Matching, Artech House Inc., Boston, MA, 1976.

[5] R. Carson, High Frequency Amplifiers, John Wiley and Sons, New York, NY, 1975, ISBN 0471137057.

[6] C. Bowick, RF Circuit Design, Newnes Elsevier, Burlington, MA, 2008.

[7] H. Carlin, Gain-bandwidth limitations on equalizers and matching networks, Proc. IRE 42 (11) (1954) 1676-1685, ISSN 0096-8390, http://dx.doi.org/10.1109/JRPROC.1954.274663.

[8] H. Carlin, A. Giordano, Network Theory: An Introduction to Reciprocal and Non-Reciprocal Circuits, Prentice-Hall Inc., Englewood Cliffs, NJ, 1964.

[9] C. Davidson, Transmission Lines for Communications, Macmillan Press Ltd, London, 1978.

[10] G. French, E. Fooks, Double section matching transformers (correspondence), IEEE Trans. Microw. Theory Tech. 17 (9) (1969) 719, ISSN 0018-9480, http://dx.doi.org/10.1109/TMTT.1969.1127042.

[11] R. Levy, J. Helszajn, Specific equations for one and two section quarter-wave matching networks for stub-resistor loads, IEEE Trans. Microw. Theory Tech. 30 (1) (1982) 55-63, ISSN 0018-9480, http://dx.doi.org/10.1109/TMTT.1982.1131017.

Microwave circuit design

Microwave semiconductor materials and diodes

CHAPTER OUTLINE

INTENDED LEARNING OUTCOMES

- *Knowledge*
  - Be aware of the various types of compound semiconductors that are used at microwave frequencies.
  - Be acquainted with the basic fabrication processes for microwave semiconductor devices, such as photolithography and molecular beam epitaxy (MBE).

Microwave Active Circuit Analysis and Design. http://dx.doi.org/10.1016/B978-0-12-407823-9.00011-1

- Be familiar with the various types of microwave diode, such as Tunnel, IMPATT, TRAPATT, and Gunn diodes, their equivalent circuits and applications.
- Understand the concept of *dynamic negative resistance* as it applies to microwave two-terminal device.
- Understand the operating principle and application of the varactor diode and the effect of doping profile on electrical characteristics.
- *Skills*
  - Be able to work out the doping profile of a varactor needed to achieve a particular *C-V* characteristic.
  - Be able to design basic microwave negative resistance diode circuits using the load line concept.
  - Be able to determine the oscillation frequency of a Gunn diode of specific geometry and design a simple Gunn diode oscillator.
  - Be able to design a PIN diode attenuator.

11.1 INTRODUCTION

The reader familiar with low-frequency active devices, fabricated in silicon (Si) or germanium (Ge) material, will be struck by the much wider range of semiconductor materials and exotic semiconductor devices that are encountered at microwave frequencies.

Operation at microwave frequencies exposes the limitations of the traditional semiconductor materials Si and Ge. Over the past few decades, a great deal of research has gone into creating and refining compound semiconductor materials, such as gallium arsenide (GaAs), indium phosphide (InP), and gallium nitride (GaN), that are more expensive to produce but have performance benefits that outweigh the additional cost at microwave frequencies. These materials will be introduced briefly here in the context of diode fabrication, although they are also used to fabricate transistors, to be discussed in the next chapter.

At low frequencies, the overwhelming majority of diodes are based on the familiar *pn*-junction or the Schottky barrier junction. The relatively high junction capacitance of the conventional *pn*-junction renders it unsuitable for many RF applications, although this property can be put to deliberate use in the case of the varactor diode. The Schottky barrier is also used at lower frequencies but is more widely used at higher frequencies due to its fast switching properties. In addition to these familiar varieties, there is a proliferation of novel two-terminal device types available at microwave frequencies, with names like Gunn, PIN, and IMPATT, which have no analogs at lower frequencies. All of these devices, and their applications, will be introduced in this chapter although varactors, tunnel diodes, and Gunn diodes will be discussed in more detail than the others due to their more widespread use.

11.2 CHOICE OF MICROWAVE SEMICONDUCTOR MATERIALS

A complete understanding of the physics of semiconductor materials would require a book in itself, and there are indeed many well-known texts in this area that the interested reader may consult, such as Watson [1], Roy and Mitra [2], Sze [3], and Yngvesson [4] to name just a few. In this section, our purpose is simply to explain the various materials that are commonly used to fabricate microwave active devices today and their relative strengths and weaknesses, so that the reader can acquire a basic understanding of why a particular material is used in a particular application.

As briefly touched on in Chapter 1, semiconductors are materials that have an electrical conductivity that lies somewhere between that of conductors and insulators. It is the existence of these materials, and the fact that their electrical properties can be altered by adding small quantities of other elements (known as *dopants*), that makes all of modern electronics possible. A pure sample of semiconductor material is referred to as an *intrinsic* semiconductor and has equal numbers of negative charge carriers ("electrons") and positive carriers ("holes"). A so-called *extrinsic* semiconductor is an intrinsic semiconductor to which a dopant has been added for the purpose of changing its electrical properties.

The electrical conductivity of an elemental solid is largely determined by something called the *Band-gap*, which is defined as the range of energies where electron states cannot exist within that solid. Alternatively the Band-gap may be defined as the energy difference (in eV) between the top of the valence band and the bottom of the conduction band [3]. This definition applies to insulators and semiconductors. In the case of conductors, such as metals, the valence and conduction bands overlap, so there is no Band-gap as such. In general, if the Band-gap is greater than around 4-5 eV at room temperature then the material is an insulator. If the Band-gap is less than around 4 eV, but still nonzero then the material is a semiconductor. This is shown in Figure 11.1.

The concentration of electrons or holes in a semiconductor are defined in relation to the *Fermi level*, E_f, which is also shown in Figure 11.1. The Fermi level is defined as the energy level that would have a 50% probability of being occupied at any given time [3]. The Fermi level is usually found half way between the valence and conduction bands. An alternative way of understanding the Fermi level is to think of it as the highest occupied molecular orbital in the material when the temperature is reduced to absolute zero [5].

The process of doping produces two categories of semiconductors: the negative charge conductor (*n*-type) and the positive charge conductor (*p*-type). *n*-type material is produced by adding *Pentavalent* impurities (i.e., elements having five valence electrons), such as antimony, arsenic, or phosphorous. *p*-type material is produced by adding *Trivalent* impurities (i.e., elements having three valence electrons), such as boron, aluminum, or gallium.

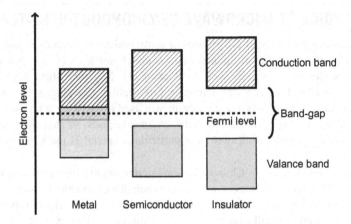

FIGURE 11.1

Electronic band structures of metals, semiconductors, and insulators.

Table 11.1 Groups 2-6 of the Periodic Table

II	III	IV	V	VI
	Bo	C	N	O
	Al	Si	P	S
Zn	Ga	Ge	As	Se
Cd	In	Sn	Sb	Te
Hg	Tl	Pb	Bi	Po

Table 11.1 shows part of groups 2 (II) to 6 (VI) of the periodic table which is the region containing the semiconducting elements. The most well known of these are Si and Ge. Certain elements in Groups III and V, or Groups II and VI, or Groups IV and VI of the periodic table may be combined to form what are called *Compound Semiconductors*. These are particularly important at RF and microwave frequencies due to their superior electrical properties at high frequencies. Perhaps the most important example of a compound semiconductor material is GaAs which first started emerging in the 1960s and is still the material of choice for high-volume commercial microwave applications. GaAs and the more recently developed materials, InP and GaN, are referred to as "three to five compounds" after the periodic table groups from which their constituent elements originate.

The last few decades have seen a further expansion in the availability of different compound semiconductor materials that have beneficial properties at microwave frequencies, such as silicon carbide (SiC), indium arsenide (InAs), gallium phosphide (GaP), and indium selenide (InSe). The advantage of compound semiconductors is that they provide the device engineer with a wide range of energy gaps and mobilities, so that materials can be engineered with properties that meet specific requirements. Some of these compound semiconductors are called *wide band-gap* semiconductors [6]. Combinations of these materials, such as GaN on SiC, are also being used in specialist applications [7]. Active devices fabricated with these semiconductor materials offer a wide range of performance advantages, such as low-noise figure, high power handling, and high-frequency operation up to THz frequencies [7–9].

In addition to three to five compound semiconductors, there is increasing use of silicon germanium (SiGe) as a compound semiconductor, as it offers high-frequency device performance at relatively low cost and allows CMOS logic and RF heterojunction bipolar transistors (HBTs) to be integrated on the same chip.

A summary of some important semiconductor materials, plus diamond as a reference, is listed in Table 11.2 [6,10]. Desirable material properties include a large band-gap (E_g), high thermal conductivity (κ), high critical electric field for breakdown (E_c), and high power density (W mm^{-1}).

A wide band-gap generally translates into an ability to support high internal electric fields before electronic breakdown occurs. This is of particular interest for high power microwave devices [6]. Table 11.2 shows that GaN and SiC have energy band-gaps about two to three times those in the conventional semiconductors such as Si and GaAs. These materials can operate at much higher voltages, frequencies, and temperature than conventional semiconductor materials

The thermal conductance of the material is important as this indicates the ease with which dissipated power can be extracted from the device. Poor thermal conductivity results in device operation at elevated temperature with degraded performance. Compound semiconductors such as GaAs and InP are, in general, poor thermal conductors and this introduces complexity in the design of high power devices. Diamond and SiC are excellent thermal conductors and are often used as heat sink materials.

The critical electric field for electronic breakdown parameter, E_c, is an indication of the strength of the electric fields that can be supported internally in the device before breakdown. High electric fields permit large terminal RF signal voltages to be supported and this is necessary for the generation of high RF power. The wide band-gap materials in Table 11.2, such as SiC and GaN, have a value of critical field, E_c, that is typically an order of magnitude greater than the other materials listed.

Silicon and silicon-germanium

Si is the dominant material for low-frequency electronics and is also used for active devices operating up to a few GHz. The widespread use of Si across the electronics industry results in the ready availability of Si fabs, used for huge production volumes. This results in relatively low cost of Si device fabrication as compared with other semiconductor materials. There is therefore a built-in bias toward the use of Si, if it is applicable to the frequency of operation.

Table 11.2 Material Properties of Some Specified Materials

Materials	Ge	Si	GaAs	InP	SiC	GaN	Diamond
Band-gap (E_g)	0.67	1.12	1.43	1.34	3.2	3.4	5.6
Dielectric constant (ε_r)	16	11.9	13.1	12.5	10	9.5	5.5
Intrinsic resistivity, ρ ($\Omega \cdot$ cm)	47	2.3×10^5	10^8	8.6×10^7	10^{12}	10^{10}	
Electron mobility	3900	1350	8500	4600	1140	1250	7700
Hole mobility	1900	480	400	150	50	850	
Thermal conductivity (κ)	0.6	1.5	0.46	0.68	3.7	1.3	20-30
Breakdown field (E_c)	1×10^5	3×10^5	6.0×10^5	5.0×10^5	3.5×10^6	2×10^6	5×10^6
Power density (W mm^{-1})	N/A	1.5	0.67	0.54	4	5-12	20-30

Unlike the other compound semiconductors mentioned in this chapter, SiGe is not fabricated as a bulk semiconductor material, but as the base region of a transistor in an otherwise Si wafer. This means that SiGe manufacturing is very much the same as conventional Si IC manufacturing, except for the addition of an extra epitaxial reactor to implant the Ge into the Si lattice. The addition of Ge allows higher dopant concentrations in the base region of the transistor because a band-gap now exists between the base and the emitter. Higher doping concentration in the base region means that the base can be made narrower which speeds up the transit time.

The development of SiGe technology is intimately linked to the development of the HBT, which will be discussed later in Chapter 12. The first SiGe HBT was reported by IBM in 1987 [11] and since then SiGe devices have become commonplace in microwave active devices up to the low tens or even hundreds of GHz. IBM has recently announced SiGe integrated circuits operating at 300 GHz [12].

Gallium arsenide

GaAs was the first compound semiconductor to extend the frequency of operation of active devices beyond what was possible with Si. GaAs was also the material within which the *transferred electron effect* was first discovered [13,14], which enabled the production of negative resistance Gunn diodes, to be discussed later in this chapter. As Si does not exhibit the transferred electron effect, devices such as Gunn diodes do not have any Si counterparts.

The first GaAs diode was reported in 1958 [15] and GaAs transistors started to appear in the 1960s [16]. Due to the higher cost of the material in the early days of its introduction, GaAs was restricted to high performance military and space-based systems for the first decade or so. GaAs components first started to make a significant impact in the commercial wireless markets in the early 1990s [17]. These days, GaAs devices are commonly deployed in consumer applications operating above a few GHz.

Key advantages of GaAs over Si are [18,19]:

- GaAs has higher saturated electron drift velocity and low field mobility than Si. This leads to faster devices.
- Si has higher substrate loss at microwave frequencies. GaAs has a much higher resistivity than Si (see Table 11.2), to the extent that it is often referred to as a "semi-insulator." This facilitates devices with low parasitics and good inter-device isolation.

GaAs is a versatile material that also has other advantages over Si that are not of direct relevance to a book on microwave circuit design, such as light-emitting properties and photovoltaic properties.

Indium phosphide

InP has superior electron velocity with respect to both Si and GaAs. InP has had an established presence for some time as a common material for optoelectronics devices like laser diodes. It is also used as a substrate for epitaxial Indium GaAs-based optoelectronic devices. In terms of high-frequency active device properties, InP surpasses both Si and GaAs with submillimeter wave MMICs being now routinely fabricated in InP [20,21].

Silicon carbide

SiC is an extremely hard material that has been widely used as an abrasive since the 1830s. SiC also happens to be a wide band-gap semiconductor material, making it applicable for short wavelength optoelectronic, high temperature, radiation resistant, and high-power/high-frequency applications. Electronic devices made from SiC can operate at extremely high temperatures without suffering from intrinsic conduction effects because of the wide energy band-gap. The very wide band-gap of the material also allows SiC to emit and detect short wavelength light which makes it suitable for the fabrication of blue light emitting diodes and nearly "solar blind" ultraviolet photodetectors.

SiC can withstand a voltage gradient (or electric field) over eight times greater than Si or GaAs without undergoing avalanche breakdown. This high breakdown electric field enables the fabrication of very high-voltage, high-power devices. Additionally, it allows the devices to be placed very close together, providing high device packing density for integrated circuits. At room temperature, SiC has a higher thermal conductivity than any metal. This property enables SiC devices to operate at extremely high power levels.

SiC devices can operate at extremely high frequencies because of the high saturated electron drift velocity, and SiC power Metal Semiconductor Field Effect Transistors (MESFETs) have been reported with multi-octave to decade bandwidths [22].

Gallium nitride

GaN is another wide band-gap semiconductor that has been commonly used in bright light-emitting diodes since the 1990s. More recently GaN has been used to manufacture microwave active devices and MMICs offering high power density, high voltage operation, higher reliability, and very wideband performance.

By comparison to some other compound semiconductors, GaN has a higher resistance to ionizing radiation, making it a suitable material for solar cell arrays for spacecraft and some military applications [23]. The high temperature and high operating voltage characteristics of GaN transistors finds them increasingly used as power amplifiers at microwave frequencies [24].

11.3 MICROWAVE SEMICONDUCTOR FABRICATION TECHNOLOGY

There are many excellent books on the main techniques of semiconductor fabrication [25,26], so we will not go into too much detail here. We feel it is useful, however, to outline the main methods of fabrication, and their advantages and disadvantages, as they relate specifically to the manufacture and performance of microwave semiconductor devices.

Active electronic devices are usually fabricated in an *Epitaxial* layer which is grown on top of the base substrate. Both low- and high-resistivity substrates may be used, depending upon the type of device to be fabricated. Low-resistivity substrates are generally used for vertical devices, such as diodes and bipolar transistors, where the current flow must pass vertically through the substrate material. High-resistivity substrates are used for surface-oriented devices, such as field-effect transistors, where the current flow is parallel to the substrate. For the latter devices, good DC and RF performance generally require that current flow be confined to the epitaxial layer and blocked from the substrate.

11.3.1 PHOTOLITHOGRAPHY

Photo-lithography is a process whereby light is used to transfer a geometric pattern from a photomask to a light-sensitive chemical (the *photo-resist*) that has been applied as a thin coating on the semiconductor substrate. A series of chemical treatments then either engraves the exposure pattern into, or enables deposition of a new material in the desired pattern upon, the material underneath the photo-resist. The basic steps in photo-lithography are shown in Figure 11.2, and can be summarized as follows:

1. *Cleaning and preparation*: The wafer surface is cleaned using a solvent to remove any organic contaminants.
   The wafer is initially heated to a temperature of around 150 °C for 10 min to drive off any moisture that may be present on the wafer surface. A liquid or gaseous "adhesion promoter," is applied to promote adhesion of the photo-resist to the wafer. The surface layer of silicon dioxide on the wafer reacts with the adhesion promoter to form a water-repellent layer that prevents the aqueous developer from penetrating between the photo-resist layer and the wafer's surface. The wafer is then covered and placed over a hot plate to dry at 120 °C.
2. *Photo-resist application*: A liquid solution of photo-resist is dispensed onto the wafer and the wafer is spun rapidly to produce a uniformly thick layer. The photo-resist layer deposited by spin coating is between 0.5 and 2.5 μm thick. The photo-resist coated wafer is then prebaked to drive off excess photo-resist solvent, typically at a temperature of between 90 and 100 °C for 30-60 s.
3. *Photo-resist exposure*: After prebaking, the photo-resist is exposed to a pattern of intense light via a *mask* that defines the circuit pattern required on the surface

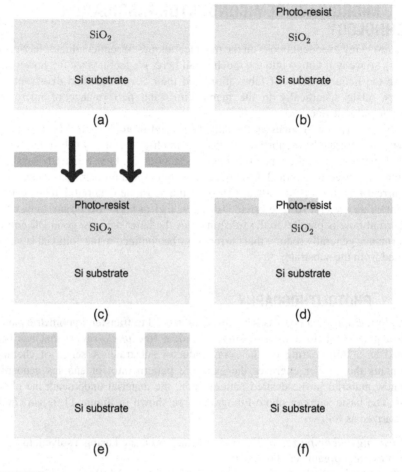

FIGURE 11.2

Simplified photo-lithography process in semiconductor microfabrication. (a) Cleaning and preparation. (b) Photo-resist application. (c) Photo-resist exposure. (d) Photo-resist developing. (e) Etching. (f) Photo-resist removal.

of the wafer. The exposure to light causes a chemical change in the photo-resist that allows the exposed areas to be removed by a suitable solvent, called the *developer*.

4. *Photo-resist developing*: The developer is applied uniformly across the wafer, using the same spinning process that was used to apply the photo-resist earlier. The developer removes the photo-resist in those areas that have been exposed to light, leaving behind photo-resist in the unexposed areas. The resulting wafer is then "hard-baked," typically at between 120 and 180 °C for 20-30 min. The hard

bake solidifies the remaining photo-resist, to make it able to survive the subsequent etching processes to follow.

5. *Etching*: A chemical etching agent is applied to remove the uppermost layer of the substrate in the areas that are not protected by photo-resist. In semiconductor fabrication, dry etching techniques are generally used, as they can be made *anisotropic*, and avoid significant undercutting of the photo-resist pattern. This is essential when the width of the features to be defined is similar to or less than the thickness of the material being etched (i.e., when the aspect ratio approaches unity).

6. *Photo-resist removal*: After the etching process, the photo-resist is no longer needed and must be removed from the substrate. This usually requires a liquid "resist stripper," which chemically alters the resist so that it no longer adheres to the substrate.

The above key process steps will usually be accompanied by a number of subsidiary processes, such as *wafer annealing*, which is a high-temperature furnace process that relieves stress in the crystal lattice, activates any ion-implanted dopants, and reduces structural defects and stress.

11.3.2 MOLECULAR BEAM EPITAXY

Molecular beam epitaxy (MBE) is a precision process that involves firing molecular beams of different semiconductor elements at a sample so as to build up thin layers of different materials. Typically, each element is delivered in a separately controlled beam, so the choice of elements and their relative concentrations can be adjusted for any given layer, thereby defining the precise composition and electrical and optical characteristics of that layer. The MBE process was invented in the late 1960s at Bell Telephone Laboratories by J.R. Arthur and Alfred Y. Cho [27].

MBE enables fabrication of compound semiconductor materials with great precision (<0.01 nm) and purity ($>99.99999\%$). The MBE process needs to take place in conditions of very high vacuum (basic pressure 10-13 bar), so an MBE machine will consist of an ultra-high-vacuum vessel containing the "guns" (or *effusion cells*) that fire molecular beams at the *substrate*. MBE also enables semiconductor materials to be layered one on top of the other, rather like a many-layered cake, to form complex semiconductor devices, such as transistors, with performance characteristics that simply cannot be obtained by the conventional photo-lithographic processes.

Each gas beam can be turned on and off rapidly with a shutter or a valve within 0.2 s. MBE offers tremendous control over layer thickness, composition, and purity. The growth rate is sufficiently slow that layers only a few atoms thick can be produced reliably. Thicker layers are obtained with longer deposition times. A typical MBE machine is shown in Figure 11.3.

FIGURE 11.3

MBE equipment at UCL.

Reproduced by kind permission of Department of Electronic and
Electrical Engineering, University College London.

11.4 THE *pn*-JUNCTION

Given the fundamental importance of the *pn*-junction in all electronic devices, it is worth briefly reviewing what the reader may already be familiar with. We will not go into too much detail on the semiconductor physics, but will just touch on the behavior of the *pn*-junction particularly as it relates to voltage dependence of the junction capacitance, which is widely employed in microwave circuitry.

Consider a piece of semiconductor material, such as Si, having one side doped as *n*-type and the other side doped as *p*-type. Electrons in the *n* region and holes in the *p* region are called *majority carriers* because they are in the majority in the respective region. The other charge carriers in each respective region are called *minority* carriers.

As a thought experiment, we may consider this as two separate pieces of semiconductor, one *n*-type and the other *p*-type, being brought into contact so that continuity of the crystal lattice is maintained at the junction. This is a useful abstraction, although *pn*-junctions are not, of course, fabricated this way in practice.

When the two materials are brought into contact, the Fermi levels have to become aligned so as to be the same throughout the crystal. Surplus electrons from the *n*-region will diffuse into the *p*-region leaving a region of net positive charge in the *n*-region, near the junction. Similarly, surplus holes from the *p*-region will diffuse into

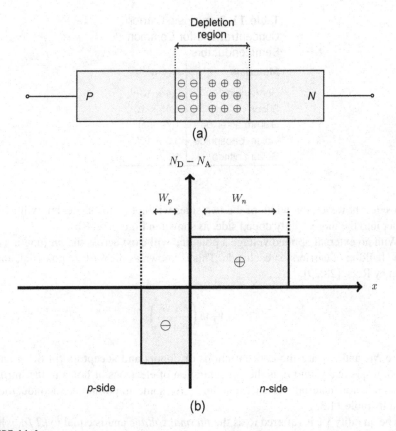

FIGURE 11.4

Simplified representation of a *pn*-junction. (a) *pn*-junction with depletion region.
(b) *pn*-junction charge density profile.

the *n*-region leaving a region of net negative charge in the *p*-region, near the junction. As electrons diffuse from the *n*-region into the *p*-region and holes diffuse from the *p*-region into the *n*-region, the ionized donors and acceptors that created the electrons and holes are left behind as the donors and acceptors are bound in place by their bonds with the atoms in the semiconductor crystal. The region immediately either side of the junction will now have been depleted of majority carriers, and is therefore referred to as the *depletion region*, also referred to as the *space charge* region, as shown in Figure 11.4. The physical width of the depletion region in a typical Si diode ranges from a fraction of a micrometer to tens of micrometers depending on device geometry, doping profile, and external bias.

The potential difference that now exists in the depletion region will prevent the further flow of majority carriers, so the junction will be in equilibrium, with equal amounts of charge existing on both sides. Since the doping levels are not the same on

Table 11.3 Intrinsic Carrier Concentrations for Common Semiconductors

Materials	$n_i(cm^{-3})$
Germanium	2.4×10^{13}
Silicon	1.45×10^{10}
Gallium arsenide	1.79×10^{6}
Indium phosphide	3.3×10^{7}
Gallium nitride	10^{10}

both sides; however, the width of the two sides will not be the same, but will extend deeper into the more lightly doped side, as shown in Figure 11.4(b).

With no external applied voltage a potential will exist across the pn-junction due to the buildup of carriers on each side. This is known as the *built-in* potential, and is given by Refs. [28,29]:

$$V_o = V_T \ln \left(\frac{N_A N_D}{n_i^2} \right) \tag{11.4.1}$$

where N_D and N_A are the concentrations of donors and acceptors (in the n and p sides), respectively, and n_i is the concentration of electrons or holes in the intrinsic semiconductor material. As an example, n_i for some important semiconductors is listed in Table 11.3.

The quantity V_T is referred to as the *thermal voltage* and is equal to kT/q, where k is Boltzmann's constant ($1.3806488 \times 10^{-23}$ J K^{-1}), T is the absolute temperature of the pn-junction, and q is the electron charge ($q = 1.60217657 \times 10^{-19}$ coulombs). At "room temperature" (i.e., around 300 K) V_T is approximately equal to 25.85 mV.

For a Si diode at room temperature equation (11.4.1) gives a value of V_o in the range 0.6-0.8 V. It should be noted, however, that this voltage is not measurable at the external terminals of the device, since the metal-semiconductor contact voltages at the points where the external connection is made to the semiconductor crystal exactly cancel out the pn-junction built in potential [29].

If we denote the width of the depletion region on the p side by W_p and on the n side by W_n, as per Figure 11.4(b), we can state the *charge equality condition* as [28]:

$$qW_p A N_A = qW_n A N_D \tag{11.4.2}$$

where A is the cross-sectional area of the junction. Equation (11.4.2) can be rearranged to give:

$$\frac{W_n}{W_p} = \frac{N_A}{N_D} \tag{11.4.3}$$

In other words, the ratio or depletion region widths on the *p* and *n* sides is the inverse of the ratio of the respective doping levels. Standard semiconductor physics textbooks [28,30] give the total width of the depletion region as[1]:

$$W = W_n + W_p = \sqrt{\frac{2\varepsilon_r\varepsilon_0 V_o}{q} \cdot \left(\frac{1}{N_A} + \frac{1}{N_D}\right)} \qquad (11.4.4)$$

When a forward bias is applied externally, the depletion region shrinks as negative charge carriers are repelled from the negative terminal toward the junction and holes are repelled from the positive terminal toward the junction. This reduces the energy required for charge carriers to cross the depletion region. As the applied voltage increases, current starts to flow across the junction once the applied voltage reaches the *barrier potential* [1]. In this forward biased mode, the current through the diode, *I*, as a function of applied voltage, *V*, is defined by the *Shockley diode equation* [28]:

$$I = I_S\left(e^{\frac{qV}{kT}} - 1\right) \qquad (11.4.5)$$

where I_o is the reverse *saturation current* of the *pn*-junction. When any forward bias voltage significantly greater than V_T is applied, the exponential term in equation (11.4.5) becomes much greater than unity and the current increases exponentially with applied voltage, as shown in Figure 11.5.

When *V* is negative (i.e., under *reverse bias*) the depletion region widens, thereby preventing the flow of majority carriers. Only the small *saturation current*, I_o, will flow under these conditions. This is reflected in equation (11.4.5) as the exponential term will approach zero with a negative value of *V* and the total current, *I*, will reduce to $-I_o$.

The saturation current, I_o, is made up of electron-hole pairs being produced in the depletion region. The saturation current remains more-or-less constant as voltage is increased (at a given temperature). At a high enough value of applied voltage, however, the current will suddenly increase. The diode is then operating in what is called *reverse breakdown* mode. There is nothing inherently destructive about reverse breakdown, provided that we ensure the current is limited by an external resistor. Provided we do not exceed a certain maximum current, the diode will return to normal operation once the reverse bias voltage is reduced to below the breakdown voltage.

Although the reverse biased *pn*-junction appears as a very high resistance to any external circuit, it will present an appreciable capacitance due to the structure of the reverse biased junction, where the nonconducting depletion region is analogous to a

---

[1]For interested readers, a full derivation of equation (11.4.4) based on solving the Poisson equation in one dimension is presented in Pierret's book [30].

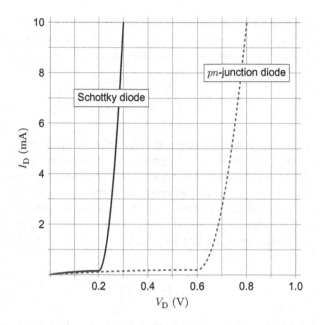

FIGURE 11.5

Schottky diode vs. *pn*-junction forward bias characteristics (Si).

dielectric insulator between two parallel plates (the respective *n* and *p* regions). The fact that the width of the depletion region is a function of the applied voltage, is put to good use in the form of the *Varactor*, which we elaborate on in Section 11.6.

11.5 SCHOTTKY DIODES

Unlike the *pn*-junction diode, the Schottky diode makes use of a metal-semiconductor junction to perform the rectifying function. There is no *pn*-junction, as such.

The main advantage of such devices is that they have very fast switching times due to their small capacitance and the fact that they are *majority carrier* devices. Schottky diodes have a very short reverse *recovery time*, which is defined as the time needed for the diode to switch from the conducting to the nonconducting state. For *pn*-junctions, the reverse recovery time is between 5 and 100 ns. Schottky diodes, by contrast, have typical switching times of less than 1ns. For this reason, Schottky diodes are widely used in RF circuits as mixers and detectors.

The forward bias characteristic of a Si Schottky junction versus a Si *pn*-junction is shown in Figure 11.5.

The Schottky barrier is extremely important in the context of microwave electronics, as it forms the basis of the MESFET which will be introduced in Chapter 12.

11.6 VARACTOR DIODES

As explained in Section 11.4, the reverse biased *pn*-junction will exhibit a capacitance which will be a function of the reverse bias voltage. A varactor (which is a shortened form of *variable reactor*) is simply a *pn*-junction diode that has been engineered to maximize the value and range of junction capacitance with the goal of applying the device as a voltage controlled capacitor in various tuned circuits such as filters and oscillators. Varactors are widely applied as electronic tuning devices in microwave systems.

Suppose a *pn*-junction is reversed biased (so almost no current flows). If the reverse bias is increased then the two parts of the depletion layer will widen by an amount ΔW_p and ΔW_n on the respective sides, as shown in Figure 11.6.

We can understand the variation of junction capacitance with applied voltage by considering the total charge, Q, stored in the depletion area in Figure 11.6 in terms of the size of the depletion region [28]. We define the equilibrium values of the depletion widths (i.e., with no external bias) as W_{n0} and W_{p0}. The equilibrium value of $W_0 = W_{n0} + W_{p0}$ is given in equation (11.4.4). Since we are dealing with a nonequilibrium situation (i.e., an external voltage V has been applied), we use a modified version of equation (11.4.4) with bias voltage, V, as follows:

$$W = \sqrt{\frac{2\varepsilon_r\varepsilon_0(V_0 - V)}{q} \cdot \left(\frac{1}{N_A} + \frac{1}{N_D}\right)} \qquad (11.6.1)$$

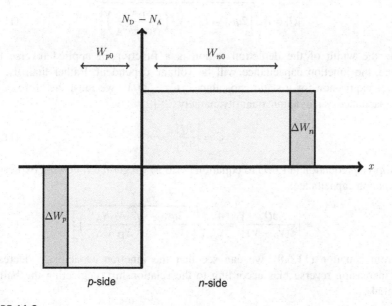

FIGURE 11.6

Varactor depletion layer.

According to equation (11.6.1), the width of the depletion region is increased for reverse bias and decreased for forward bias. Since the amount of charge stored on each side of the junction varies with W, variations in the applied voltage will result in corresponding variations in the charge, which results in the capacitive behavior. For a junction having a cross-sectional area A, the amount of stored charge, Q, can be written in terms of the doping concentration and the depletion region widths as follows:

$$|Q| = qAW_{n0}N_D = qAW_{p0}N_A \tag{11.6.2}$$

Relating the total width of the depletion region W to the widths of the individual regions W_{n0} and W_{p0} we have:

$$W_{n0} = \frac{N_A}{N_A + N_D} W \tag{11.6.3}$$

$$W_{p0} = \frac{N_D}{N_A + N_D} W \tag{11.6.4}$$

From equations (11.6.2) and (11.6.3) we can write the charge on either side of the junction as:

$$|Q| = Aq \frac{N_A N_D}{N_A + N_D} W \tag{11.6.5}$$

Replacing W in equation (11.6.5) by equation (11.6.1) we get:

$$|Q| = A \sqrt{\left[2q\varepsilon_0\varepsilon_r(V_0 - V) \left(\frac{N_D N_A}{N_D + N_A} \right) \right]} \tag{11.6.6}$$

As the width of the depletion region is a function of applied reverse bias voltage, the junction capacitance will be voltage dependent. Rather than use the familiar expression for a static capacitance, $C = Q/V$, we must therefore define the capacitance as a dynamic quantity, namely [28]:

$$C = \left| \frac{dQ}{dV} \right| \tag{11.6.7}$$

Applying equation (11.6.7) to equation (11.6.5) we get the following expression for junction capacitance:

$$C_j = \left| \frac{dQ}{d(V_0 - V)} \right| = \frac{A}{2} \cdot \sqrt{\left[\frac{2q\varepsilon_0\varepsilon_r}{(V_0 - V)} \left(\frac{N_D N_A}{N_D + N_A} \right) \right]} \tag{11.6.8}$$

From equation (11.6.8), we can see that the junction capacitance decreases with increasing reverse bias according to the relationship (neglecting the built-in potential):

$$C_j \propto \frac{1}{|V|^{1/2}} \tag{11.6.9}$$

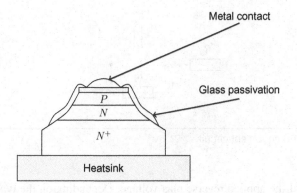

FIGURE 11.7

Varactor diode construction.

Note that this relationship refers to an *abrupt* (i.e., step like) doping profile. Other doping profiles can be engineered and lead to a more general relationship:

$$C_j \propto \frac{1}{|V|^n} \tag{11.6.10}$$

By varying the doping profile of the junction it is possible to change the way capacitance varies with voltage. This relationship between doping profile and C-V profile also means that measuring C_j as a function of V is a useful way of investigating the doping profile of a given junction.

Varactors find wide application in RF and microwave systems, with the primary application being as electronic tuning elements in oscillators and variable filters. Varactor tuned microwave oscillators started to appear in the late 1960s and early 1970s [31,32] and have been the primary tuning element in microwave oscillators ever since. The current state of the art sees varactors being applied in oscillators and frequency multipliers up to THz frequencies [33,34].

Figure 11.7 is a simplified cross section of a typical discrete varactor diode fabricated using a *mesa* structure, as opposed to the planar structure used for MMIC. The mesa structure is a "table shaped" structure, as shown in Figure 11.7, which avoids the high field regions at the edges, which tends to occur in planar structures. Most discrete varactors are manufactured in this format.

The varactor shown in Figure 11.7 resembles a conventional *pn*-junction diode formed on top of a low-resistance substrate layer consisting of highly doped $N+$ material.

A simplified equivalent circuit for a varactor is shown in Figure 11.8. The capacitor C_j is the variable junction capacitance we are primarily interested in. The series resistance, R_S, models the resistance of the semiconductor in the areas outside the depletion region, plus the parasitic resistance of the lead and package elements. The resistor R_j represents the junction leakage resistance in reverse bias. Like C_j, R_j

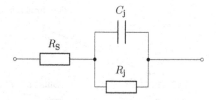

FIGURE 11.8

Varactor diode AC equivalent circuit.

is a function of the applied reverse bias voltage. Depending on the type of package and the frequency range, the model may need to include some inductive elements, which we have omitted for the time being.

Typical component values in Figure 11.8 are

$$R_S = 0.4 \text{ to } 0.8 \ \Omega \tag{11.6.11}$$

$$C_j = 1 \text{ to } 6 \text{ pF} \tag{11.6.12}$$

$$R_j > 10 \text{ M}\Omega \tag{11.6.13}$$

An important characteristic of any varactor diode is its Q factor. This is particularly important in tuned oscillator applications as a high Q varactor will result in a higher Q tank circuit which will in turn reduce the phase noise produced by the oscillator. Varactor Q is also very important in tuned filter applications as higher varactor Q will result in a sharper frequency response.

Considering the equivalent circuit of Figure 11.8, we can write an approximate expression for varactor Q by ignoring R_j and considering the varactor as a series RC circuit, that is,

$$Q_v = \frac{1}{\omega_0 C_j R_S} \tag{11.6.14}$$

Aside from the obvious observation that Q can be increased by reducing the series resistance, R_S, equation (11.6.14) also reveals that there is a trade-off between capacitance and Q. Although in many applications we are inclined to select a varactor with the highest capacitance, we need to take into account the effect that the reduction in Q will have on the circuit.

By careful control of the quantity of dopant used in the manufacturing process, it is possible to create specific doping profiles for the *pn*-junction which gives rise to specific *C-V* characteristics for the varactor. Terms such as *Abrupt* and *Hyperabrupt* refer to the doping profile of the *pn*-junction, and consequently different *C-V* characteristics.

For an abrupt varactor the doping concentration is held constant. The abrupt varactor exhibits an inverse square law *C-V* function. This provides for an inverse fourth law frequency dependence (due to the fact that resonant frequency is proportional to $1/\sqrt{C}$).

A hyperabrupt varactor provides a C-V curve that has an inverse square law curve over at least some of the characteristics. This provides a roughly linear frequency variation over a limited range. In addition to this, the hyperabrupt junction gives a much greater capacitance change for the given voltage change. The advantages of the hyperabrupt varactor comes at a cost, however, as these devices have a substantially lower Q when compared to abrupt varactor diodes.

The frequency response of Si varactors is constrained by minority carrier storage effects which limits their use at microwave frequencies. For this reason, microwave Schottky barrier varactors fabricated in GaAs [35] or other compound semiconductors [8] have become more widely used.

11.7 PIN DIODES

The acronym PIN stands for "P-Intrinsic-N," and refers to the fact that the PIN diode is a *pn*-junction device that has a region of very lightly doped, or *intrinsic*, semiconductor material located between the *p*-type and *n*-type regions. The addition of the intrinsic region results in unique characteristics that are very useful in microwave applications.

Under reverse bias conditions the intrinsic region results in very high values of breakdown voltage, whereas the device capacitance is greatly reduced by the increased separation between the *p* and the *n* regions.

Under forward bias conditions the conductivity of the intrinsic region is controlled or modulated by the injection of charge from the end regions. The diode therefore acts as a *bias-current controlled resistor* with excellent linearity. PIN diodes are used extensively in microwave circuits as voltage controlled switches and voltage variable attenuators in such applications as amplitude modulators, phase shifters, and limiters.

Figure 11.9(a) is a diagrammatic representation of the PIN diode structure.

Equivalent circuit parameters of a typical PIN diode are shown in Table 11.4 [36].

Table 11.4 Typical PIN Diode Parameters

Parameter	Value
C_j	0.1-1 pF
R_f	0.4-1 Ω
R_s	0.5-4 Ω
L_{int}	$\approx$0.3 nH
C_p	0.08-0.18 pF
τ	0.5-5 μS

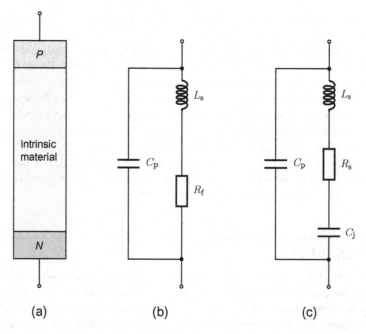

FIGURE 11.9

PIN diode structure and equivalent circuits. (a) PIN diode construction. (b) PIN diode equivalent circuit: forward bias. (c) PIN diode equivalent circuit: reverse bias.

The equivalent circuit model of a PIN diode under forward and reverse bias conditions is shown in Figure 11.9(b) and (c), respectively [36]. More complex equivalent circuits than these may be used, depending on the frequency range and the degree of accuracy required in the design [1].

The admittance of the reverse bias equivalent circuit shown in Figure 11.9(c) is as follows [1]:

$$Y_r = \frac{1}{R_s}\left[\frac{(f/f_{co})}{(f/f_{co})^2 + (1 - (f/f_r)^2)^2}\right] + j\omega\left[\frac{C_j(1 - (f/f_r)^2)}{(f/f_{co})^2 + (1 - (f/f_r)^2)^2} + C_c\right] \quad (11.7.1)$$

where

$$\omega = 2\pi f$$

$$f_{co} = \frac{1}{2\pi R_s C_j}$$

$$f_r = \frac{1}{2\pi \sqrt{L_s C_j}}$$

f_{co} is the cut-off frequency of the diode and f_r is the reverse biased series resonant frequency of the diode. In practice, f_{co} is normally much higher than f_r. For example, using typical values from Table 11.4 gives $f_{co} \approx 128$ GHz and $f_r \approx 12$ GHz. We can therefore make the following approximation in relation to equation (11.7.1) [1]:

$$Y_r = \frac{1}{R_s} \left[\frac{(f/f_{co})}{1 - (f/f_r)^2} \right]^2 + j\omega \left[\frac{C_j}{1 - (f/f_r)^2} + C_c \right] \tag{11.7.2}$$

PIN diodes are generally selected so as to minimize the variation of device performance with frequency. In other words, devices are selected such that $(f/f_r)^2 \ll 1$ over the frequency range of operation. This leads to the further approximation:

$$Y_r \approx G_r + j\omega C_t \tag{11.7.3}$$

where $G_r = (1/R_s)(f/f_{co})^2$ and $C_t = C_j + C_c$. The approximation equation (11.7.3) is very widely used, to the extent that manufacturers often specify C_t rather than C_j and C_c individually. When $f > f_r$ equation (11.7.2) indicates that the sign of the susceptance component becomes negative, meaning that the reverse biased PIN diode becomes inductive at higher frequencies.

The admittance of the forward bias equivalent circuit shown in Figure 11.9(b) is as follows [1]:

$$Y_f = \left[\frac{R_f}{R_f^2 + (\omega L_s)^2} \right] + j \left[\omega C_c - \frac{\omega L_s}{R_f^2 + (\omega L_s)^2} \right] \tag{11.7.4}$$

We can derive two approximations from equation (11.7.4) depending on the frequency range of operation. At low frequencies where $(\omega C_c) \ll 1/(\omega L_s)$, the impedance of the diode under forward bias conditions is:

$$Z_f = \frac{1}{Y_f} \approx R_f + j\omega L_s \tag{11.7.5}$$

At high frequencies, that is, $R_f^2 \ll (\omega L_s)^2$, we can use the following approximation:

$$Y_f \approx \left(\frac{R_f}{\omega L_s} \right)^2 + j \left(\omega C_c - \frac{1}{\omega L_s} \right) \tag{11.7.6}$$

The above approximations can be employed in the design of a PIN diode attenuators. Figure 11.10 shows the schematic diagram of a simple shunt-connected PIN diode attenuator, comprising a PIN diode, a bias network, such as a choke inductor and decoupling capacitor, and two DC blocking capacitors, C_1 and C_2. At microwave frequencies, the PIN diode under forward bias appears essentially as a pure linear resistor, R_{rf}, whose value can be controlled by the DC bias. At low frequencies, the PIN diode behaves as an ordinary *P-N* junction diode.

Depending on the type of transmission line in use, diodes can be mounted in either a shunt or series configuration. In the case of waveguide, it is more convenient

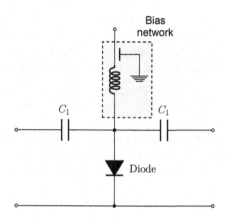

FIGURE 11.10

PIN diode attenuator schematic.

to mount the diode on a pillar inside the waveguide, that is, in shunt configuration. In the case of planar transmission lines, such as microstrip, striplines, and coplanar waveguide, a series configuration may be more convenient, but shunt configuration can also be used whereby the diode is mounted in a hole through the substrate.

We will proceed to derive equations for the insertion loss of a simple PIN diode switch by considering a shunt mounted PIN diode embedded in a transmission line system, of characteristic impedance Z_0. We will employ the PIN diode equivalent circuit models for forward and reverse bias shown in Figure 11.9(b) and (c). The essential requirements for such a switch are that it provides minimum attenuation in the "on" state (reverse biased condition) and maximum isolation in the "off" state (forward biased condition). If we represent the total shunt admittance (diode plus associated matching circuitry) as Y, then these conditions ideally correspond to $Y = 0$ and $Y = \infty$, respectively.

We first start with the "on" condition shown in equivalent circuit form, ignoring biasing and decoupling elements, in Figure 11.11.

In order to minimize the loss at the design frequency, f_0, we need to add an inductance L_p to resonate out the net capacitive reactance of the PIN diode, C_t. The value of this inductance is given by:

$$L_p = \frac{1}{(2\pi f_0)^2 C_t} \tag{11.7.7}$$

It can be shown that the insertion loss arising from the presence of any shunt admittance, $Y = G + jB$, in a transmission line of characteristic impedance, Z_0, is given by:

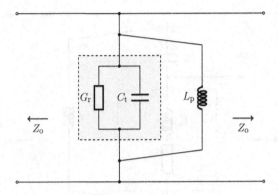

FIGURE 11.11

PIN diode attenuator equivalent circuit: reverse bias ("on") state.

$$\alpha_{LY}(\text{dB}) = 10 \log_{10}\left[\left(1 + \frac{GZ_0}{2}\right)^2 + \left(\frac{BZ_0}{2}\right)^2\right] \tag{11.7.8}$$

From equation (11.7.8) we can see that α_{LY} is at a minimum when $B = 0$, that is at resonance. With reference to Figure 11.11, at resonance we have $G = G_r$ and $B = \omega C - 1/\omega L$, so the insertion loss of the circuit in Figure 11.11 becomes:

$$\alpha_{\text{on}}(\text{dB}) = 10 \log_{10}\left(1 + \frac{G_r Z_0}{2}\right)^2 \tag{11.7.9}$$

When the diode is forward biased, it will act as a short circuit across the transmission line, resulting in a high insertion loss. The switch will therefore be in the "off" state. In order to minimize the shunt impedance across the transmission line in the forward biased case, we need to add a series capacitor, C_s to "tune out" the effects of the series inductance L_s at f_0. The value of C_s is given by:

$$C_s = \frac{1}{(2\pi f_0)^2 L_s} \tag{11.7.10}$$

The complete circuit is shown in Figure 11.12.

From equation (11.7.8) we can derive the insertion loss α_H resulting from the presence of a shunt impedance $Z = R + jX$, as follows:

$$\alpha_{LZ}(\text{dB}) = 10 \log_{10}\left[\left(1 + \frac{RZ_0}{2(R^2 + X^2)}\right)^2 + \left(\frac{XZ_0}{2(R^2 + X^2)}\right)^2\right] \tag{11.7.11}$$

It can be shown that α_{LZ} is at a maximum when $X = 0$, that is at resonance. With reference to Figure 11.12, at resonance we have $R = R_f$ and $X = \omega L_s - 1/\omega C_s$, so the insertion loss becomes:

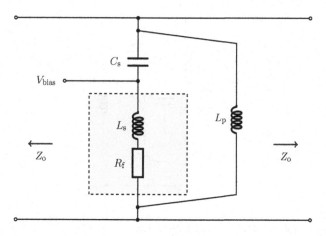

FIGURE 11.12

PIN diode attenuator equivalent circuit: forward bias ("off") state.

$$\alpha_{\text{off}}(\text{dB}) = 10 \log_{10} \left(1 + \frac{Z_0}{2R_f} \right)^2 \qquad (11.7.12)$$

From equations (11.7.9) and (11.7.12) we can see that, for a good quality PIN diode switch, that is one that has a low value of α_{on} and a high value of α_{off}, we need a high value of G_f in the reverse biased state and a low value of R_f in the forward biased state.

PIN diodes, in discrete or MMIC form, are routinely used in circuits similar to Figure 11.10 as attenuators and switches up to hundreds of GHz [37].

11.8 TUNNEL DIODES

The Tunnel diode is basically a very highly doped *pn*-junction (around 10^{19} to 10^{20} cm^{-3}) that makes use of a quantum mechanical effect called *tunneling*. This type of diode is also known as an *Esaki diode* [38], after the inventor, Leo Esaki, who discovered the effect in 1957, a discovery for which he was awarded the Nobel Prize in Physics in 1973. As a consequence of the very high doping, a tunnel diode will have a very narrow depletion region, typically less than 10 nm.

We will not delve into the physics of tunneling, which is well covered in standard texts [3,28]. The important point about the tunneling mechanism, from the engineering point of view, is that it gives rise to a region of *negative resistance* in the *I-V* characteristics, shown as region "*B*" in Figure 11.13. In region "*B*," an increase in forward voltage will result in a decrease in forward current, and vice versa. This is equivalent to saying that the device exhibits negative resistance in this

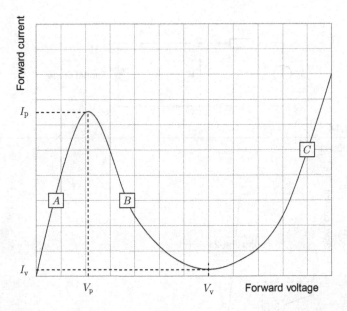

FIGURE 11.13

Tunnel diode *I-V* characteristics.

region although, strictly speaking, we should call this negative dynamic resistance, as it refers to the negative slope of the *V-I* characteristics, not a physical "negative" resistor, which does not exist, of course.

Region "*A*" in Figure 11.13 is actually the region where tunneling occurs. Region "*C*" is the region of normal *pn*-junction behavior. In this sense, region "*B*" can be considered as the region of transition between region "*A*," where the *I-V* characteristic is linear, and region "*C*" where the *I-V* characteristic obeys equation (11.4.5). With reference to Figure 11.13 we can see that, as the bias voltage is increased from zero, the current increases linearly along curve "*A*" until a peak current is reached, at the bias voltage V_p. This corresponds to the *n*-side conduction band becoming aligned with the *p*-side valance band in the device. At this point tunneling stops, at a current level called the *peak tunneling current*, I_p in Figure 11.13. I_p is also known as the "Esaki current."

We can analyze the circuit behavior of a tunnel diode with DC bias with the aid of Figure 11.14, from which, by inspection, we can write:

$$V_S = V_D - I_D R \qquad (11.8.1)$$

The current through the diode is then given by:

$$I_D = \frac{V_D}{R} - \frac{V_S}{R} \qquad (11.8.2)$$

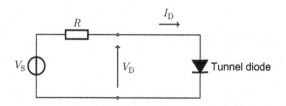

FIGURE 11.14

Tunnel diode circuit.

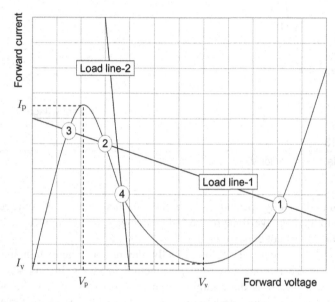

FIGURE 11.15

Tunnel diode characteristic with a load line.

Equation (11.8.2) is in the form of a straight line current/voltage graph with slope $(-1/R)$ and an intercept on the current axis of $(I_D = V_D/R)$. This is called a *load line*. As the voltage across the series combination of resistor plus diode increases, the load line is raised with its x-axis intercept at the applied voltage. Figure 11.15 shows two possible load lines for the circuit in Figure 11.14, depending on the chosen value of R.

With higher values of R, the load line will be the shallower load line-1 in Figure 11.15 that intersects the diode characteristic at three points, 1, 2, and 3, meaning that the circuit has three possible operating points. Point 2 is an unstable operating point, as any perturbations in bias voltage will cause the diode to jump from point 2 to either point 1 or point 3 on the load line. The circuit will therefore settle at either point 1 or point 3 depending on the history. It is in this mode that tunnel diodes are used as switched or memory devices. This mode is of little interest to the microwave circuit designer, however.

If the value of R, is reduced the load line will be much steeper, resembling load line-2 in Figure 11.15. In this case, the circuit has only one operating point, point 4. The total differential resistance is negative (because $R < |R_d|$). In this mode, the diode can be made to oscillate at a microwave frequency dependent on the external L and C components.

Because the negative resistance phenomenon in tunnel diodes relies on a tunneling phenomenon, the currents generated are necessarily quite small and the amount of RF power generated by a tunnel diode oscillator is quite low. Consequently, tunnel diodes are only suitable for low power applications and even in this arena they have been largely superseded by transistors.

11.9 GUNN DIODES

The Gunn diode is the exception to all the devices covered in this chapter as it constructed of only one type of semiconductor material. There is no *pn*-junction of any kind. The common structure is a sandwich of variously doped *n*-type material in the form $n^+ - n - n^+$, as shown in Figure 11.16. The Gunn diode is named after its inventor, J.B. Gunn who first published a paper in 1962 on what later became known as the *Gunn effect* in GaAs [13,39]. The effect was later discovered in other compound semiconductor materials such as InP and GaN [14].

The band structure of certain compound semiconductors, such as GaAs and InP, has two local minima in the conduction band: one where the electrons have a low effective mass and a high mobility and a second local minimum at a higher energy level where electrons have a higher effective mass and a lower mobility [1,28,40]. At low electric fields, nearly all of the conduction electrons will occupy the lower energy minimum with the lesser effective mass and high mobility. As the electric field is increased, the electrons will have enough energy to occupy energies in the second minima with its associated higher effective mass and lower mobility. As the field is increased further, the proportion of electrons with the lower mobility increases and the drift-velocity will continue to decrease until all the electrons share the lower mobility and the drift-velocity will level off. The drop in electron mobility with

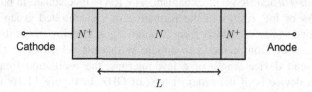

FIGURE 11.16

Gunn diode construction.

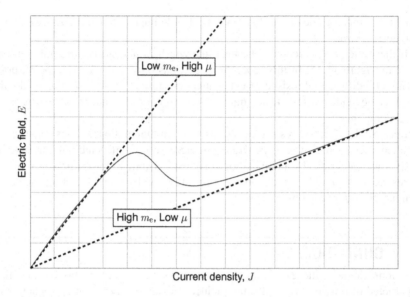

FIGURE 11.17

Electric field/current density characteristic for GaAs or InP.

increasing electric field means that a sample of this material will exhibit a decrease in current with increasing applied voltage, that is to say a negative differential resistance. At higher voltages, the normal increase of current with voltage relation resumes once the bulk of the carriers are kicked into the higher energy-mass region. Therefore, the negative resistance only occurs over a limited range of voltages. This phenomenon is shown in Figure 11.17.

The existence of a drop in mobility with increasing electric field and resulting possibility of negative resistance was actually predicted a year earlier by theoretical physicists Ridley and Watkins, and by Hilsum who published a paper on the effect in 1961 [41]. This phenomenon is therefore also sometimes referred to as the *Ridley-Watkins-Hilsum* (RWH) mechanism. The RWH mechanism in bulk material, such as GaAs or InP, results in the formation of concentrated groupings of free electrons called *domains*, which form around regions of low mobility and move through the device from cathode to anode. With drift-velocities of the order of 10^7 cm s^{-1} and device length of a few microns, the oscillation frequency of a typical Gunn device is of the order of tens of GHz. In Figure 11.16, the electron domains originate at the N^+ cathode region, move through the N region until they reach the N^+ anode region where the domain is transformed into a pulse of current flow.

The known time for formation of domains within the bulk material is given by Ref. [3]:

$$\tau_R = \frac{\varepsilon_0 \varepsilon_r}{q n_0 |\mu_-|} \tag{11.9.1}$$

where μ_- is the negative differential mobility in the material (typically 2000 cm^2 V^{-1} S^{-1} for GaAs). The sample must be long enough to allow the domain to fully form before it reaches the opposite electrode, so we can say that, for a sample of physical length L, as shown in Figure 11.16, we have the requirement:

$$\frac{L}{v_d} > \tau_R \tag{11.9.2}$$

where v_d is electron domain drift velocity. Combining equations (11.9.1) and (11.9.2) we have the following requirement for Gunn oscillation to take place:

$$n_0 L > \frac{\varepsilon_0 \varepsilon_r v_d}{q n_0 |\mu_-|} \tag{11.9.3}$$

The product of the carrier concentration and device length, $n_0 L$, is an important figure of merit for a Gunn device and sets constraints on the physical size and doping level of the bulk semiconductor sample. It turns out that for GaAs and InP we need $n_0 L$ to be greater than 10^{12} cm^{-2} for there to be enough time for domains to form in the material and Gunn oscillation to take place [3]. The corresponding frequency of oscillation is given by:

$$f = \frac{v_d}{L} \tag{11.9.4}$$

where v_d is the electron drift velocity in the semiconductor material, and is a function of temperature, doping, and applied electric field.

The DC current voltage characteristic of a Gunn diode is shown in Figure 11.18. Although superficially similar to the *I-V* characteristic of the tunnel diode shown in Figure 11.13, it is important to remember that these two devices are based on totally different operating principles. One consequence of this is that Gunn diode oscillators can deliver much higher RF signal powers than tunnel diode oscillators.

11.9.1 GUNN DIODE OSCILLATORS

A Gunn diode can be used to construct an RF oscillator simply by applying a suitable direct current through the device. In effect, the negative differential resistance created by the diode will negate the positive resistance of an actual load and thus create a "zero" resistance circuit which will sustain oscillations as long as DC bias remains applied. The oscillation frequency is determined partly by the physical properties of the Gunn device but largely by the characteristics of an external resonator. The resonator can take the form of a waveguide, microwave cavity, or Yitrium Iron Garnet (YIG) sphere, as discussed in more detail in Chapter 15.

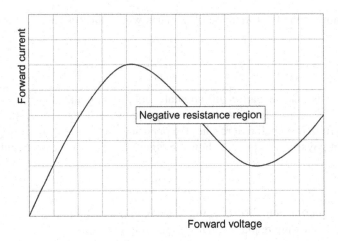

FIGURE 11.18

Gunn diode *I-V* characteristics.

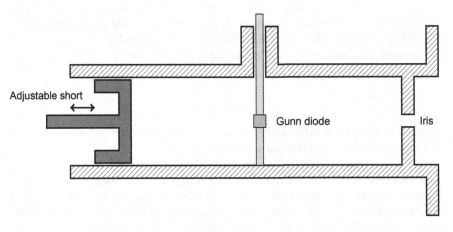

FIGURE 11.19

GUNN diode waveguide oscillator.

Figure 11.19 is a diagrammatic representation of a simple Gunn diode oscillator implemented in waveguide. Such simple waveguide Gunn oscillators have a long history and are still used today in such everyday microwave applications as radar speed guns, traffic light radars and intruder alarms, and in experimental oscillators operating in the hundreds of GHz [42]. The waveguide Gunn oscillator consists of a Gunn diode mounted on a metal pillar which ensures correct positioning of the diode within the waveguide cavity and allows the external bias current to be

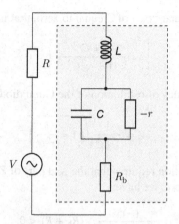

FIGURE 11.20

GUNN diode AC equivalent circuit.

applied. At one end of the waveguide cavity is an adjustable short circuit in the form of a metal plunger. The other end of the cavity (the output end) is closed off by a metal diaphragm containing an iris of the right diameter to allow the RF energy to escape. The distance between the Gunn diode and the sliding short needs to be $\lambda/4$ for coherent oscillation to take place. Adjustment of the short position therefore represents a crude tuning mechanism. An adjustable metal screw is sometimes inserted into the waveguide cavity as an additional impedance matching device.

Although waveguide-based oscillators are still widely used, contemporary Gunn diode oscillators are also implemented in planar form, suitable for integration into MMICs with resonators implemented as planar structures and where tuning is implemented electronically with varactors or other devices [43,44].

The AC equivalent circuit of a Gunn diode is shown in Figure 11.20, where $-r$ represents the dynamic negative resistance of the device at a particular bias point. The same load line design methodology we introduced in the case of the tunnel diode can be applied to Gunn diode circuit design. The DC source, V, and external load resistor, R, are selected to give a load line that biases the device in the negative-resistance region.

Inductance L arises from the wire leads, C is the effective capacitance of the device, and R_b is the bulk resistance of the device. The AC equivalent circuit of Figure 11.20 can thus be analyzed by writing the total impedance of the Gunn diode plus load as follows:

$$Z = j\omega L + \left(\frac{(-r)}{1 - j\omega Cr} \right) + (R_b + R) \qquad (11.9.5)$$

We now set the imaginary part of Z equal to zero, that is,

$$\left[\omega L - \frac{\omega C r^2}{1 + \omega^2 C^2 r^2}\right] = 0 \tag{11.9.6}$$

which defines the frequency of oscillation of the Gunn diode, ω_0, as:

$$\omega_0 = \frac{1}{\sqrt{LC}}\sqrt{\left(1 - \frac{L}{r^2 C}\right)} \tag{11.9.7}$$

The oscillation condition requires that the real part of Z be negative at ω_0. From equation (11.9.5), therefore, we have:

$$\frac{(-r)}{1 + \omega_0^2 C^2 r^2} + (R_b + R) < 0 \tag{11.9.8}$$

Substituting equation (11.9.7) into equation (11.9.8) we obtain the condition for oscillation of the circuit in Figure 11.20, given that r must be negative, as:

$$\frac{R_b + R}{r} < \frac{L}{r^2 C} < 1 \tag{11.9.9}$$

The requirement that $(R_b + R)/r < 1$ is equivalent to stating that the negative slope of the circuit load line must be greater than the slope of the negative-resistance curve, shown in Figure 11.18.

GaAs Gunn diodes are commonly reported operating in the submillimeter wave frequency range up to 200 GHz [45]. GaN Gunn devices generally have higher operating frequencies and power capabilities than GaAs Gunn diodes, making them suitable for operation in the THz frequency range [46,47]. The Gunn effect has been observed in other more "exotic" materials at submillimeter wave and THz frequencies [48–50], which suggests the continued importance of the Gunn diode as a microwave signal source.

11.10 THE IMPATT DIODE FAMILY

We use the term "family" here to describe a variety of different microwave diodes whose operation are based on avalanche phenomena. The archetype diode of this type is the IMPATT, which stands for IMPact Avalanche and Transit Time [28]. An IMPATT diode is basically a *pn*-junction diode which is operated in reverse bias mode, up to breakdown, at which point an avalanche of electron-hole pairs is produced in the high field region of the depletion layer by impact ionization. The IMPATT diode is also known as a Read diode, after its inventor W.T. Read who, together with Ralph L. Johnston at Bell Laboratories, proposed that an avalanche diode that exhibited significant transit time delay might exhibit a negative resistance characteristic [4].

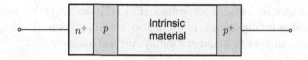

FIGURE 11.21

IMPATT diode construction.

The frequency of operation of an IMPATT depends on device thickness, similar to the case of the Gunn diode. IMPATT diodes are typically applied at frequencies between a few GHz and a few hundred GHz, but experimental devices have recently been reported operating in the THz range [51].

The main advantage of IMPATT devices, compared to the other two terminal devices covered in this chapter, is their ability to generate high signal powers [19]. This is achieved, however, at the cost of higher phase noise.

Figure 11.21 is a diagrammatic representation of the $n^+ - p - i - p^+$ structure of an IMPATT diode. The device essentially consists of two regions: the $n^+ - p$ region within which avalanche multiplication occurs and an *intrinsic* region through which the generated holes drift toward the p^+ contact. A similar device could be constructed with the opposite polarities, that is $p^+ - n - i - n^+$, in which case electrons would drift through the intrinsic region toward an n^+ contact. The operating principle of the IMPATT diode is rather complicated and so we will refer the interested reader to other sources for this [3,28].

A TRAPATT diode is similar to an IMPATT, having a structure p^+nn^+ or n^+pp^+. The acronym "TRAPATT" stands for Trapped Plasma Avalanche Triggered Transit. The main difference in terms of performance is that the TRAPATT has a much higher DC to RF conversion efficiency when compared to the IMPATT (40-60% [2], compared with 15% [10]).

Other diodes in this family, having similar properties, include such devices as the "BARRITT" diode (which stands for BARRier Injection Triggered Transit) [19] and the MITATT diode (which stands for Mixed Tunneling and Avalanche Transit Time) [51]. What all these devices have in common is their application in high power microwave oscillators.

11.11 TAKEAWAYS

1. In addition to devices fabricated using the traditional semiconductor materials, Si and Ge, the microwave circuit designer needs to be familiar with devices made from a range of different compound semiconductor, such as GaAs, InP, and GaN, which have advantageous properties at microwave frequencies.

2. Varactor diodes take advantage of the voltage dependant capacitance of a reverse biased *pn*-junction or Schottky barrier. A varactor is simply a conventional diode that has been designed to maximize this junction capacitance,

and have a specific capacitance-voltage (C-V) characteristic. Varactor diodes are widely used as tuning devices in oscillators and filters, as well as having other applications, such as in frequency multipliers and parametric amplifiers.

3. The C-V characteristic of a varactor depends on the doping profile. Conversely, the doping profile of a varactor can be determined from measurements of the C-V characteristic. Doping profiles are classified as either "abrupt" or "hyperabrupt." For an abrupt profile, the doping concentration is held constant, which results in an inverse square root C-V characteristic. A hyperabrupt profile, on the other hand, has an inverse square law C-V characteristic, meaning that such a device, when combined with an inductor in a tuned circuit, will provide a linear voltage-frequency characteristic.

4. A PIN diode consists of a *pn*-junction device that has a region of very lightly doped, or *intrinsic*, semiconductor material located between the *p*-type and *n*-type regions. The PIN diode behaves as a high-frequency voltage controlled resistance that can be used as an attenuator or a switch when inserted in a transmission line system.

5. There are basically three distinct mechanisms of negative resistance generation in microwave diodes, namely (a) tunneling mechanisms (as seen in tunnel diodes), (b) the "Transferred Electron" mechanism (as seen in Gunn diodes), and (c) avalanche mechanisms (as seen in the IMPATT diode family).

6. The transferred electron mechanism, otherwise known as Ridley-Watkins-Hilsum (RWH) mechanism, is observed within bulk samples of certain compound semiconductors such as GaAs and InP, and results in a negative dynamic resistance over a certain portion of the I-V characteristic. Gunn diodes operate on this principle. The oscillation frequency of a Gunn diode can be calculated based on device geometry and the drift velocity of electron domains within the sample.

7. Circuit design using negative resistance two terminal devices is carried out based on the "load line" concept.

REFERENCES

[1] H. Watson, M. Barber, Microwave Semiconductor Devices and Their Circuit Applications, McGraw-Hill, 1969.

[2] S. Roy, M. Mitra, Microwave Semiconductor Devices, PHI Learning, 2003. ISBN 9788120324183.

[3] S. Sze, Semiconductor Devices, Physics and Technology, Wiley, 1985.

[4] S. Yngvesson, Microwave Semiconductor Devices, Springer US, 1991. ISBN 9780792391562.

[5] D. Morgan, J. Howes, Solid State Electronic Devices, Wykeham Science Series, Wykeham, 1972. ISBN 9780851092409.

[6] R. Trew, High-frequency solid-state electronic devices, IEEE Trans. Electron. Dev. 52 (5) (2005) 638-649. ISSN 0018-9383, http://dx.doi.org/10.1109/TED.2005.845862.

[7] R. Pengelly, S. Wood, J. Milligan, S. Sheppard, W. Pribble, A review of GaN on SiC high electron-mobility power transistors and MMICs, IEEE Trans. Microw. Theory Tech. 60 (6) (2012) 1764-1783. ISSN 0018-9480, http://dx.doi.org/10.1109/TMTT.2012.2187535.

[8] T. Vu, Compound Semiconductor Integrated Circuits, Selected Topics in Electronics and Systems, World Scientific, 2003. ISBN 9789812796844.

[9] L. Samoska, An overview of solid-state integrated circuit amplifiers in the submillimeter-wave and THz regime, IEEE Trans. Terahertz Sci. Technol. 1 (1) (2011) 9-24. ISSN 2156-342X, http://dx.doi.org/10.1109/TTHZ.2011.2159558.

[10] R. Ludwig, G. Bogdanov, RF Circuit Design, second ed., Pearson Education Inc., Upper Saddle River, NJ, 2009.

[11] S. Iyer, G. Patton, S. Delage, S. Tiwari, J. Stork, Silicon-germanium base heterojunction bipolar transistors by molecular beam epitaxy, in: 1987 International Electron Devices Meeting, vol. 33, 1987, pp. 874-876. http://dx.doi.org/10.1109/IEDM.1987.191578.

[12] P.-Y. Chiang, Z. Wang, O. Momeni, P. Heydari, A 300GHz frequency synthesizer with 7.9% locking range in 90 nm SiGe BiCMOS, in: 2014 IEEE International Solid-State Circuits Conference Digest of Technical Papers (ISSCC), ISSN 0193-6530, 2014, pp. 260-261. http://dx.doi.org/10.1109/ISSCC.2014.6757426.

[13] J. Gunn, Instabilities of current in III-V semiconductors, IBM J. Res. Dev. 8 (2) (1964) 141-159. ISSN 0018-8646, http://dx.doi.org/10.1147/rd.82.0141.

[14] H. Kroemer, Theory of the Gunn effect, Proc. IEEE 52 (12) (1964) 1736-1736. ISSN 0018-9219, http://dx.doi.org/10.1109/PROC.1964.3476.

[15] D. Jenny, A gallium arsenide microwave diode, Proc. IRE 46 (4) (1958) 717-722. ISSN 0096-8390, http://dx.doi.org/10.1109/JRPROC.1958.286772.

[16] C. Maed, Schottky barrier gate field effect transistor, Proc. IEEE 54 (2) (1966) 307-308.

[17] M. Golio, B. Newgard, The history and future of GaAs devices in commercial wireless products, in: 2001 IEEE Emerging Technologies Symposium on Broadband Communications for the Internet Era Symposium Digest, 2001, pp. 63-69. http://dx.doi.org/10.1109/ETS.2001.979423.

[18] Y. Yung, A tutorial on GaAs vs silicon, in: Proceedings of Fifth Annual IEEE International ASIC Conference and Exhibit, 1992, pp. 281-287. http://dx.doi.org/10.1109/ASIC.1992.270234.

[19] H. Thim, Active Microwave Semiconductor Devices, in: Fourth European Microwave Conference, 1974, pp. 1-15. http://dx.doi.org/10.1109/EUMA.1974.332002.

[20] M. Rodwell, M. Le, B. Brar, InP bipolar ICs: scaling roadmaps, frequency limits, manufacturable technologies, Proc. IEEE 96 (2) (2008) 271-286. ISSN 0018-9219, http://dx.doi.org/10.1109/JPROC.2007.911058.

[21] G. Raghavan, M. Sokolich, W. Stanchina, Indium phosphide ICs unleash the high-frequency spectrum, IEEE Spectr. 37 (10) (2000) 47-52. ISSN 0018-9235, http://dx.doi.org/10.1109/6.873917.

[22] R. Sadler, S. Allen, W. Pribble, T. Alcorn, J. Sumakeris, J. Palmour, SiC MESFET hybrid amplifier with 30-W output power at 10 GHz, in: Proceedings of 2000 IEEE/Cornell Conference on High Performance Devices, ISSN 1529-3068, 2000, pp. 173-177. http://dx.doi.org/10.1109/CORNEL.2000.902535.

[23] A. Lidow, A. Nakata, M. Rearwin, J. Strydom, A. Zafrani, Single-event and radiation effect on enhancement mode gallium nitride FETs, in: 2014 IEEE Radiation Effects Data Workshop (REDW), 2014, pp. 1-7. http://dx.doi.org/10.1109/REDW.2014.7004594.

[24] V. Paidi, S. Xiex, R. Coffie, B. Moran, S. Heikman, S. Keller, A. Chini, S. DenBaars, U. Mishra, S. Long, M. Rodwell, High linearity and high efficiency of class-B power amplifiers in GaN HEMT technology, IEEE Trans. Microw. Theory Tech. 51 (2) (2003) 643-652. ISSN 0018-9480, http://dx.doi.org/10.1109/TMTT.2002.807682.

[25] D. Gupta, Semiconductor Fabrication: Technology and Metrology, in ASTM STP 990, ASTM Committee F-1 on Electronics and Semiconductor Equipment and Materials Institute, 1989. ISBN 9780803112735.

[26] M. Henini, Molecular Beam Epitaxy: From Research to Mass Production, Elsevier Science, 2012. ISBN 9780123918598.

[27] A. Cho, J. Arthur, Molecular beam epitaxy, Prog. Solid State Chem. 10 (Part 3) (1975) 157-191. ISSN 0079-6786, http://dx.doi.org/10.1016/0079-6786(75)90005-9, http://www.sciencedirect.com/science/article/pii/0079678675900059.

[28] B. Streetman, S. Banerjee, Solid State Electronic Devices, Pearson Education, Limited, 2014. ISBN 9780133356038.

[29] A. Sedra, K. Smith, Microelectronic Circuits, Oxford University Press, 1998. ISBN 9780195116632.

[30] R. Pierret, Semiconductor Device Fundamentals, Addison-Wesley, 1996. ISBN 9780201543933.

[31] I. Kuru, Frequency modulation of the Gunn oscillator, Proc. IEEE 53 (10) (1965) 1642-1643.

[32] C. Aitchison, B. Newton, Varactor-tuned X-Band Gunn oscillator using lumped thin-film circuits, Electron. Lett. 7 (4) (1971) 93-94.

[33] T. Crowe, T. Grein, R. Zimmermann, P. Zimmermann, Progress toward solid-state local oscillators at 1 THz, IEEE Microw. Guided Wave Lett. 6 (5) (1996) 207-208. ISSN 1051-8207, http://dx.doi.org/10.1109/75.491507.

[34] G. Chattopadhyay, Technology, capabilities, and performance of low power terahertz sources, IEEE Trans. Terahertz Sci. Technol. 1 (1) (2011) 33-53. ISSN 2156-342X, http://dx.doi.org/10.1109/TTHZ.2011.2159561.

[35] Howes and Tebbenham, Design of microwave oscillators, in: Microwave Circuit and System Design, Leeds University Summer School, 1983.

[36] I. Bahl, P. Bhartia, Microwave Solid State Circuit Design, John Wiley and Sons, New York, 1988.

[37] W. Hongtao, G. Xuebang, W. Hongjiang, W. Bihua, L. Yanan, W-band GaAs PIN diode SPST switch MMIC, in: 2012 International Conference on Computational Problem-Solving (ICCP), 2012, pp. 93-95. http://dx.doi.org/10.1109/ICCPS.2012.6384329.

[38] L. Esaki, Discovery of the tunnel diode, IEEE Trans. Electron Devices 23 (7) (1976) 644-647. ISSN 0018-9383, http://dx.doi.org/10.1109/T-ED.1976.18466.

[39] R. van Zyl, W. Perold, R. Botha, The Gunn-diode: fundamentals and fabrication, in: Proceedings of the 1998 South African Symposium on Communications and Signal Processing, COMSIG '98, 1998, pp. 407-412, http://dx.doi.org/10.1109/COMSIG.1998.736992.

[40] M. Shur, GaAs Devices and Circuits, Microdevices: Physics and Fabrication Technologies, Springer, 1987. ISBN 9780306421921.

[41] B. Ridley, T. Watkins, The Possibility of Negative Resistance Effects in Semiconductors, Proc. Phys. Soc. 78 (2) (1961) 293. http://stacks.iop.org/0370-1328/78/i=2/a=315.

[42] C. Balocco, S. Kasjoo, L. Zhang, Y. Alimi, S. Winnerl, A. Song, Planar terahertz nanodevices, in: 2011 European Microwave Integrated Circuits Conference (EuMIC), 2011, pp. 585-588.

[43] H. Scheiber, K. Lubke, D. Grutzmacher, C. Diskus, H. Thim, MIMIC-compatible GaAs and InP field effect controlled transferred electron (FECTED) oscillators, IEEE Trans. Microw. Theory Tech. 37 (12) (1989) 2093-2098. ISSN 0018-9480, http://dx.doi.org/10.1109/22.44127.

[44] A. Springer, C. Diskus, K. Lubke, H. Thim, A 60-GHz MMIC-compatible TED-oscillator, IEEE Microw. Guided Wave Lett. 5 (4) (1995) 114-116. ISSN 1051-8207, http://dx.doi.org/10.1109/75.372809.

[45] F. Amira, C. Mitchell, N. Farrington, M. Missousa, Advanced Gunn diode as high power terahertz source for a millimetre wave high power multiplier, in: Proc. SPIE 7485, Millimetre Wave and Terahertz Sensors and Technology II, vol. 74850I-1, 2009. http://dx.doi.org/10.1117/12.830296.

[46] E. Alekseev, D. Pavlidis, GaN Gunn diodes for THz signal generation, in: 2000 IEEE MTT-S International Microwave Symposium Digest, vol. 3, ISSN 0149-645X, 2000, pp. 1905-1908. http://dx.doi.org/10.1109/MWSYM.2000.862354.

[47] Z. Gribnikov, R. Bashirov, V. Mitin, Negative effective mass mechanism of negative differential drift velocity and terahertz generation, IEEE J. Sel. Top. Quantum Electron. 7 (4) (2001) 630-640. ISSN 1077-260X, http://dx.doi.org/10.1109/2944.974235.

[48] H. Eisele, R. Kamoua, Submillimeter-wave InP Gunn devices, IEEE Trans. Microw. Theory Tech. 52 (10) (2004) 2371-2378. ISSN 0018-9480, http://dx.doi.org/10.1109/TMTT.2004.835974.

[49] M. Dragoman, D. Dragoman, TeraHertz Gunn amplification in semiconductor carbon nanotubes, in: CAS 2004: Proceedings of 2004 International Semiconductor Conference, vol. 1, 2004, pp. 88. http://dx.doi.org/10.1109/SMICND.2004.1402810.

[50] R. Parida, N. Agrawala, G. Dash, A. Panda, Characteristics of a GaN-based Gunn diode for THz signal generation, J. Semicond. 33 (8) (2012) 084001. http://stacks.iop.org/1674-4926/33/i=8/a=084001.

[51] T. Wu, Performance of GaN Schottky contact MITATT diode at terahertz frequency, Electron. Lett. 44 (14) (2008) 883-884. ISSN 0013-5194, http://dx.doi.org/10.1049/el:20080568.

Microwave transistors and MMICs

CHAPTER OUTLINE

Microwave Active Circuit Analysis and Design. http://dx.doi.org/10.1016/B978-0-12-407823-9.00012-3
© 2016 Elsevier Ltd. All rights reserved.

INTENDED LEARNING OUTCOMES

- *Knowledge*
  - Be familiar with the specific types of transistors used at microwave frequencies and their various limitations.
  - Understand the origins of equivalent circuit models for microwave bipolar and field-effect transistors.
  - Understand the multilayered construction of MMICs and how passive components can be constructed in MMIC form.
- *Skills*
  - Be able to distinguish between a BJT, HBT, MESFET, and HEMT transistor types, and be able to explain the basic operation of these devices and their relative performance characteristics.
  - Be able to draw the equivalent circuit of a BJT/HBT and a MESFET/HEMT and be able to explain the physical origin of equivalent circuit components, including parasitics.
  - Be able to estimate the f_T and f_{max} of a given transistor type, with given equivalent circuit component values.
  - Be able to identify the main passive component types used in contemporary MMIC design and their equivalent circuits.

12.1 INTRODUCTION

Although incremental advances in technology have enabled "conventional" silicon transistors to operate well into the GHz region, the demand for higher and higher frequency operation has driven innovations in materials, architectures, processes, and geometries that have resulted in completely novel devices that only inhabit the higher frequency domains. Possibly the most significant distinction between microwave transistors and their lower frequency counterparts is in the area of materials. Whereas low-frequency devices are fabricated almost exclusively in silicon, the use of relatively costly compound semiconductors, such as gallium arsenide (GaAs) and indium phosphide (InP), becomes economical at microwave frequencies due to their performance advantages over silicon. The drive to higher frequencies has also engendered highly sophisticated material configurations, such as the heterojunction, that have no low-frequency counterparts [1,2]. In this chapter, we will give a brief overview of the main types of microwave transistor in use today, and how they can be modeled and applied in practical circuits. We will not dwell on the details of semiconductor physics or device fabrication as there are plenty of excellent texts on this subject for the interested reader to consult [3–7].

As is the case at lower frequencies, microwave transistors may be broadly divided into two categories: the bipolar junction transistors (BJTs) and the field-effect transistors (FETs). The latter category, at lower frequencies, contains the junction FET (JFET) and the metal oxide FET (MOSFET), both of which have structural characteristics that normally limit their high-frequency operation. In the 1970s, a new type of FET device, the GaAs metal semiconductor FET (MESFET) came on the scene, employing a direct connection between the gate metal and the semiconductor material, pushing the frequency of operation well into the tens of GHz. GaAs MESFET technology developed rapidly during the 1970s and 1980s while microwave bipolar transistor technology progressed more gradually [8]. In the intervening decades, however, bipolar device technology has caught up and it is now common to find variants of the BJT operating at hundreds of GHz [9].

Although low-frequency BJTs all employ the same basic structure, at microwave frequencies there are a number of specialist adaptations to the basic BJT construction designed to achieve higher frequency performance, lower noise figure, or higher power handling capacity. The main factors limiting the conventional BJT at high frequencies are high inter-electrode capacitance, carrier transit times, and lead inductance. These limitations can be overcome with special geometries, process control and packaging, and by the use of compound semiconductor materials. High precision manufacturing technologies, such as molecular beam epitaxy (MBE), as discussed in Chapter 11, have given rise to the Heterojunction bipolar transistor (HBT) [6,10]. Finally, the high frequency technology of the bipolar world is mirrored in the FET world in the form of the high electron mobility transistor (HEMT) [6,7,11], all of which will be briefly described in this chapter.

12.2 MICROWAVE BIPOLAR JUNCTION TRANSISTORS

The BJT is so named because both types of charge carrier (electrons and holes) take part in the current-carrying mechanism.

The BJT is the earliest transistor structure, being invented by Shockley, Bardeen, and Brittain at Bell Labs in 1948 [12,13], for which they were awarded the Nobel Prize in Physics in 1956. The invention of the BJT has had a profound impact on the modern world. Since the BJTs are constructed from both p-type and n-type materials ("bipolar"); they come in two configurations: npn and pnp, depending on the arrangement of the respective n and p doped materials. Microwave bipolar transistors are almost always of the npn type, because of the higher mobility of electrons, when compared to holes.

12.2.1 BJT CONSTRUCTION

The basic construction of a planar epitaxial BJT is shown in Figure 12.1. The device consists of three differently doped semiconductor regions, such as the emitter region, the base region, and the collector region. The term *epitaxial* refers to the fact that the device is constructed by the deposition of successive crystalline layers on top of a crystalline substrate. The three regions in Figure 12.1 are, respectively, n-type, p-type, and n-type in an npn transistor and p-type, n-type, and p-type in a pnp transistor. Metal contacts connect the respective semiconductor region to the external circuit.

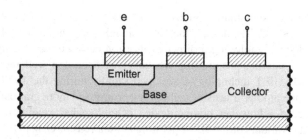

FIGURE 12.1

Conceptual BJT cross-section (not to scale).

The base region is physically located between the emitter and the collector and is made from lightly doped, high resistivity material. The base surrounds the emitter region and the collector surrounds the base region, meaning that electrons injected into the base region from the emitter (in the case of an *npn* device) are very likely to be swept across the thin base region and "collected" by the collector. The rate at which this happens is dictated by the bias applied to the base-emitter junction, hence the transistor action. Figure 12.1 shows that the collector-base junction has a considerably larger surface area than the emitter-base junction. This results in the equivalent c-b junction capacitance being much larger than the equivalent b-e junction capacitance, which has implications for high-frequency operation, as will be explained.

Figure 12.1 also suggests that the BJT is not a symmetrical device. This means that the collector and emitter cannot be interchanged in circuit operation. The lack of symmetry is not just due to geometry but also a result of the doping ratios of the emitter and the collector. The emitter is heavily doped, while the collector is lightly doped, allowing a large reverse bias voltage to be applied before the c-b junction breaks down. The c-b junction is reverse biased in normal operation. The reason the emitter is heavily doped is to make it a better reservoir of carriers, in other words to increase the emitter injection efficiency which is defined as the ratio of carriers injected by the emitter to those injected by the base. For high current gain, most of the carriers injected into the e-b junction must come from the emitter.

12.2.2 THE BJT EQUIVALENT CIRCUIT

Our starting point for understanding how BJT behave at microwave frequencies will be to review BJT behavior at low frequencies, with which we will assume the reader already has some familiarity.

For small signal currents and voltages, the BJT may be represented by two alternative equivalent circuits called the *T-model* and the *π-model*, respectively, as shown in Figure 12.2. The resistances used in these models, r_e and r_π are not the same. Their physical meaning can be defined as follows:

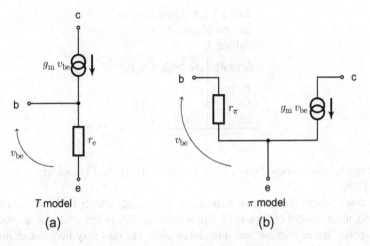

FIGURE 12.2

BJT equivalent circuit models. (a) T-model. (b) π-model.

- r_π is the small signal active mode input resistance seen looking into the base, with emitter grounded.
- r_e is the small signal active mode input resistance seen looking into the emitter, with the base grounded.

The fact that these two resistances are not the same means that the transistor is not a reciprocal device. As it happens, r_π is much greater than r_e, and it can be shown that r_e and r_π are related by [14]:

$$r_\pi = (1 + \beta)r_e \qquad (12.2.1)$$

where β is the forward current gain of the transistor, which is the dimensionless ratio of emitter current to base current. It typically has a value from tens to hundreds. The collector current is related to the base-emitter voltage, v_{be}, via the all-important quantity g_m, which is the transconductance of the transistor, having the units of Siemens (S). The transconductance, g_m, is directly proportional to collector current and, as a rule of thumb, its value is around 40 mS per milliamp of collector current.

The inverse of g_m has the dimensions of resistance, that is, $1/g_m = r_e$. The resistance r_e, as shown in Figure 12.2(a), is known as the *dynamic resistance* of the base-emitter junction, and its value is a function of both current (hence the adjective "dynamic"), and the temperature of the device. A good approximation of r_e is given by:

$$r_e \approx \frac{25}{I_e \ (\text{mA})} \ \Omega \qquad (12.2.2)$$

For more detailed explanation and derivation of these terms the reader may consult some of the general texts on this subject [14–16].

Table 12.1 Approximate Values of g_m and r_e Values I_e

I_e (mA)	g_m (mS)	r_e (Ω)
1	40	25
10	400	2.5
25	1000	1

By application of equation (12.2.2) we get some typical values of r_e as shown in Table 12.1.

In cases where the emitter terminal is grounded, which is the most common scenario, the π-model of Figure 12.2(b) is more useful. In any case, the π-model and the T-model are equivalent and interchangeable, but one may be more appropriate than the other, depending on the design task.

High-frequency BJT performance

At higher frequencies, the gain of the transistor drops off with frequency due to charge storage effects within the various regions of the device. These effects can be modeled by adding parasitic capacitors to the low-frequency π-model of Figure 12.2(b). The resulting equivalent circuit, commonly referred to as the *hybrid-π model* is shown in Figure 12.3, which represents the intrinsic device alone, without package parasitics.

The elements that have been added in Figure 12.3 are as follows:

1. *Base-emitter capacitance (C_π)*: includes the base-emitter diffusion capacitance and the base-emitter junction capacitance. In microwave transistors, this combined capacitance ranges from a few femto-Farads (fF) to a few pico-Farads (pF).

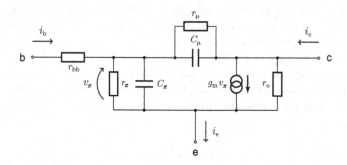

FIGURE 12.3

BJT hybrid-π model equivalent circuit.

2. *Collector-base feedback capacitance (C_μ)*: representing the junction capacitance across the reverse biased collector-base junction. Because the junction is reverse biased, under normal operating conditions there is no diffusion capacitance in this case. Although this capacitor is usually small, < 1 pF, its influence is multiplied by the transistor gain which amplifies the voltage variation at the base connection and applies it to the collector connection, exaggerating the influence of this capacitor through the Miller effect, which was discussed in Chapter 5.

3. *Intrinsic base-collector resistance (r_μ)*: representing the resistance between the collector and the base when the junction is under reverse bias. It is thus usually a high value (hundreds of kΩ to several MΩ).

4. *Base spreading resistance (r_{bb})*: representing the resistance between the base metal contact and the active region of the semiconductor material making up the base region of BJT. For a microwave transistor its value ranges from a few ohms to tens of ohms.

5. *Intrinsic base-emitter resistance (r_π)*: the internal resistance of the forward biased base-emitter junction, as described in Section 12.2.1. Its value is a function of the current gain of the device and the operating current.

6. *Output resistance (r_o)*: representing the bulk resistance of the material between the collector and the emitter. It is usually quite high, in the order of 100 kΩ. r_o is also a dynamic resistance as it is a function of the variation of the collector current with the transistor bias voltages. This phenomenon is referred to as the *Early effect* [15].

A very important figure of merit for both BJTs and FETs is the current gain cut-off frequency, f_T, which indicates the frequency at which the short-circuit current gain drops to unity (0 dB) in the common emitter configuration. The cut-off frequency is important not only because it relates different transistor parameters together but also because it is a quantity that is easily measurable and consequently is always reported in manufacturer's data sheets. f_T for a BJT can be calculated from the circuit shown in Figure 12.3. If we apply a short-circuit between the collector and the emitter, so as to measure i_c, we find that the base-emitter voltage, v_π, is also applied across C_μ. Hence we can write:

$$i_b = \frac{v_\pi (1 + j\omega r_\pi C_\pi)}{r_\pi} \tag{12.2.3}$$

$$i_c = (g_m - j\omega C_\mu)v_\pi \tag{12.2.4}$$

For the frequency range for which this model is valid, ωC_μ is much less than g_m and, therefore, we can calculate the short-circuit current gain as follows:

$$\frac{i_c}{i_b} \approx \frac{g_m r_\pi}{1 + j\omega(C_\mu + C_\pi)r_\pi} \tag{12.2.5}$$

If $\omega_T = 2\pi f_T$ is the frequency for which the current gain is unity, that is,

$$\left| \frac{i_c}{i_b} \right| = 1 \tag{12.2.6}$$

then, at $\omega = \omega_T$ we can write:

$$\frac{g_m r_\pi}{\sqrt{1 + \omega_T^2 (C_\mu + C_\pi)^2 r_\pi^2}} = 1 \qquad (12.2.7)$$

Hence,

$$1 + \omega_T^2 (C_\mu + C_\pi)^2 r_\pi^2 = (g_m r_\pi)^2 \qquad (12.2.8)$$

Which yields ω_T as follows:

$$\omega_T^2 = \frac{(g_m r_\pi)^2 - 1}{(C_\mu + C_\pi)^2 r_\pi^2} \qquad (12.2.9)$$

Since $(g_m r_\pi)^2 \gg 1$ we can write:

$$\omega_T \approx \frac{g_m}{C_\mu + C_\pi} \qquad (12.2.10)$$

Thus, the unity-gain bandwidth, f_T, of a BJT can be approximated, in terms of the hybrid-π model parameters as follows:

$$\boxed{f_T \approx \frac{g_m}{2\pi (C_\mu + C_\pi)}} \qquad (12.2.11)$$

At lower frequencies, a further approximation of equation (12.2.11) is often made by noting that C_π is usually much larger than C_μ, resulting in:

$$f_T \approx \frac{g_m}{2\pi C_\pi} \qquad (12.2.12)$$

At microwave frequencies, however, we usually cannot ignore C_μ, so the approximation equation (12.2.11) is seldom used. Furthermore, it should be noted that f_T is also a function of the transistor bias voltages and currents, as both C_π and C_μ are functions of these [14–16].

f_T is also known as the *gain bandwidth product*, a term which emphasizes the trade-off between current gain, which is proportional to g_m, and bandwidth, which is proportional to $1/(C_\pi + C_\mu)$. For microwave transistors, f_T, is in the range of tens to hundreds of GHz depending on the device technology. Since g_m is proportional to collector current, i_c, the gain bandwidth product has to be specified at a particular value of collector current.

Though common emitter current gain is equal to unity at f_T, by definition, there may still be considerable power gain at f_T due to different input and output matching conditions. Thus, f_T does not necessarily represent the highest useful frequency of operation of a transistor. Another figure of merit, the *maximum frequency of oscillation*, (f_{max}), is often used. f_{max} is the frequency at which common emitter power gain is equal to unity and is related to f_T as follows [17]:

$$f_{\max} = \sqrt{\frac{f_T}{8\pi r_{bb} C_\mu}} \qquad (12.2.13)$$

We note, from equation (12.2.13) that unlike f_T, $f_{\max}$ is a function of r_{bb}. The value of r_{bb} therefore has to be carefully controlled to ensure high-frequency operation. These two figures of merit, f_T and $f_{\max}$, are sometimes referred to collectively as the *characteristic frequencies* [18]. Although $f_{\max}$ will generally be higher than f_T, we should not always assume that this is the case. Since f_T is a measure of voltage gain whereas $f_{\max}$ is a measure of power gain, a transistor that still has a voltage gain greater than unity above $f_{\max}$ will have $f_T > f_{\max}$ [18].

The high-frequency model presented in Figure 12.3 becomes increasingly less accurate as the frequency extends into the microwave regime. One way of improving the accuracy of this model at microwave frequencies is to introduce additional parasitic elements, such as the parasitic capacitances, inductances, and resistances associated with terminal connectors or packaging. Although these may have values of only a few femto-Farads, pico-Henries, and a few ohms, they will have appreciable impedance at microwave frequencies. Figure 12.4 shows the addition of package parasitics to the intrinsic BJT equivalent circuit. Obtaining the exact values of these parasitic elements is an extremely difficult task. Approximate values are normally

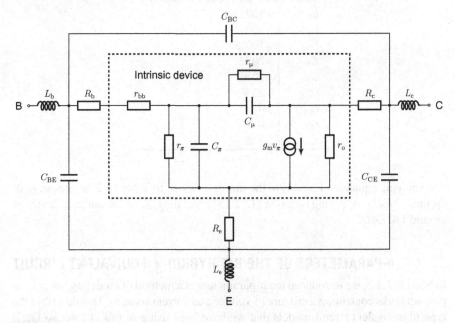

FIGURE 12.4

BJT hybrid-π model equivalent circuit with package parasitics.

obtained by taking extensive measurements of a transistor's S-parameters and then carrying out de-embedding and parameter extraction assuming a particular extended circuit model, an example of which is given in Figure 12.4. The extraction of equivalent circuit parameters from measured S-parameter data is a specialist field in which a substantial body of work has been undertaken [19–21].

By way of illustration, hybrid-π parameters for a typical microwave BJT are shown in Tables 12.2 and 12.3 [21].

Model parameters for a typical microwave BJT

Table 12.2 Intrinsic Parameters

g_m	80 mS(@$I_C = 2$ mA)
r_{bb}	38 Ω
r_π	1.1 kΩ
r_μ	270 kΩ
C_π	130 fF
C_μ	8 fF
r_o	5 MΩ

Table 12.3 Extrinsic Parameters

L_e	>100 pH
L_c	>100 pH
L_b	>100 pH
C_{BC}	>20 fF
C_{BE}	>20 fF
C_{CE}	>20 fF
r_e	3.4 Ω
r_c	2.0 Ω
r_b	3.8 Ω

Applying equation (12.2.11) to the intrinsic device in Table 12.2, we get an f_T of around 92 GHz. Applying equation (12.2.13) gives the f_{max} of the intrinsic device as around 110 GHz.

12.2.3 *h*-PARAMETERS OF THE BJT HYBRID-π EQUIVALENT CIRCUIT

In Section 5.1.3, we introduced the *h*-parameters as a method of modeling linear two-port networks containing a mixture of voltage and current sources. This describes the type of equivalent circuit models that we have been using in this chapter so far. It therefore makes sense to be able to relate the *h*-parameter framework to the hybrid-π model parameters. Keep in mind that at microwave frequencies we will mostly be working with Y-parameters or S-parameters, but we can easily obtain these from a given set of h-parameters using the standard conversion formulae in Appendix A.

Recapping the definition of the h-parameters from Section 5.1.3 we have, for the common emitter configuration,

$$\left.\begin{array}{l} h_{ie} = \dfrac{v_{be}}{i_b}\bigg|_{v_{ce}=0} \\[12pt] h_{re} = \dfrac{v_{be}}{v_{ce}}\bigg|_{i_b=0} \\[12pt] h_{fe} = \dfrac{i_c}{i_b}\bigg|_{v_{ce}=0} \\[12pt] h_{oe} = \dfrac{i_c}{v_{ce}}\bigg|_{i_b=0} \end{array}\right\} \qquad (12.2.14)$$

In other words, we measure the parameters h_{ie} and h_{fe} with the output short circuited, whereas the parameters h_{re} and h_{oe} are measured with the input open circuited.

Applying a short circuit at the output of Figure 12.3 we can therefore write:

$$h_{ie} = \frac{v_{be}}{i_b} = r_{bb} + \frac{r_\pi}{1 + j\omega r_\pi (C_\mu + C_\pi)} \qquad (12.2.15)$$

where r_x is the parallel combination of r_π and r_μ, that is,

$$r_x = \frac{r_\pi r_\mu}{r_\pi + r_\mu} \qquad (12.2.16)$$

Since, in general, $r_\pi \gg r_\mu$, we can write:

$$h_{ie} \approx r_{bb} + \frac{r_\pi}{1 + j\omega r_\pi (C_\mu + C_\pi)} \qquad (12.2.17)$$

It is also usually the case that $C_\pi \gg C_\mu$, so equation (12.2.17) can be further simplified to:

$$\boxed{h_{ie} = r_{bb} + \frac{r_\pi}{1 + j\omega r_\pi C_\pi}} \qquad (12.2.18)$$

With the output short circuited and applying the same approximations with regard to r_μ and C_μ as above, we can determine h_{fe} by establishing the relationship between v_{be} and i_b in Figure 12.3 as follows:

$$v_{be} = \frac{i_b r_\pi}{1 + j\omega r_\pi (C_\mu + C_\pi)} \approx \frac{i_b r_\pi}{1 + j\omega r_\pi C_\pi} \qquad (12.2.19)$$

Noting that $i_c = g_m i_b$ we can write:

$$h_{fe} = \frac{i_c}{i_b} = \frac{g_m v_{be}}{i_b} \qquad (12.2.20)$$

Combining equations (12.2.19) and (12.2.20) gives:

$$h_{fe} = \frac{g_m r_\pi}{1 + j\omega r_\pi C_\pi}$$

(12.2.21)

To compute the parameters h_{re} and h_{oe} we now apply an open circuit to the input. Applying a hypothetical voltage generator to this input we can write:

$$h_{re} = \frac{v_{be}}{v_{ce}} = \frac{\left(\dfrac{r_\pi}{1 + j\omega r_\pi C_\pi}\right)}{\left(\dfrac{r_\pi}{1 + j\omega r_\pi C_\pi}\right) + \left(\dfrac{r_\mu}{1 + j\omega r_\mu C_\mu}\right)}$$

(12.2.22)

Applying the same approximations with regard to r_μ and C_μ as above, we can approximate equation (12.2.22) as follows:

$$h_{re} \approx \left(\frac{r_\pi}{r_\mu}\right) \cdot \frac{1 + j\omega r_\mu C_\mu}{1 + j\omega r_\pi C_\pi}$$

(12.2.23)

Finally, we determine the parameter h_{oe} by reference to Figure 12.3 as follows:

$$h_{oe} = \frac{i_c}{v_{ce}} = \frac{1}{r_o} + \frac{1}{\left(\dfrac{r_\mu}{1 + j\omega r_\mu C_\mu}\right) + \left(\dfrac{r_\pi}{1 + j\omega r_\pi C_\pi}\right)} + g_m \left(\frac{r_\pi}{r_\mu}\right) \cdot \frac{1 + j\omega r_\mu C_\mu}{1 + j\omega r_\pi C_\pi}$$

(12.2.24)

The last term is a result of the presence of the current generator. This term can be written as $g_m h_{re}$. Applying this, and the approximations with regard to r_μ and C_μ, we can rewrite equation (12.2.24) as follows:

$$h_{oe} \approx \frac{1}{r_o} + \frac{1}{r_\mu} + j\omega C_\mu + g_m \left(\frac{r_\pi}{r_\mu}\right) + \frac{g_m \left(1 - \dfrac{r_\pi C_\pi}{r_\mu C_\mu}\right) \dfrac{C_\mu}{C_\pi} j\omega}{j\omega + \dfrac{1}{r_\pi C_\pi}}$$

(12.2.25)

Since $r_\pi r_\mu \gg r_\pi C_\pi$ we can make the approximation:

$$\left(1 - \frac{r_\pi C_\pi}{r_\mu C_\mu}\right) \approx 1$$

(12.2.26)

Equation (12.2.25) can now be approximated as:

$$h_{oe} \approx \frac{1}{r_o} + \frac{1}{r_\mu} + j\omega C_\mu + g_m \left[\frac{r_\pi}{r_\mu} + \frac{j\omega C_\mu}{\frac{1}{r_\pi} + j\omega C_\pi}\right]$$

(12.2.27)

Note that all the above equations only consider the intrinsic hybrid-π model of Figure 12.3. The above closed form approach is rarely applied to the more complete model of Figure 12.4 as the resulting equations tend to be cumbersome. The approach usually adopted with more complex equivalent circuits is based on computer simulation.

12.3 HETEROJUNCTION BIPOLAR TRANSISTOR

The heterojunction bipolar transistor (HBT) is a type of BJT that uses a different type of semiconductor material for the emitter and base regions, creating a *heterojunction*. The main benefit of the HBT is higher frequency performance, which is a function of the type of semiconductor material used and the geometry of the device [22]. For example, HBTs fabricated using GaAs/aluminum gallium arsenide (GaAs/AlGaAs) compound semiconductor material can have f_T well into the hundreds of GHz. Even higher frequencies can be obtained with InP/indium gallium arsenide (InGaAs) double heterojunction devices [9,23]. HBTs find application either as oscillators or as low-noise amplifiers [6,24]. Another benefit of HBT is high efficiency and high power density, which makes them highly suitable for microwave power amplifier applications [25].

There are two versions of HBT, the single heterojunction bipolar transistor (SHBT) and the double heterojunction bipolar transistor (DHBT) as shown in Figure 12.5 [26].

In both cases, the HBT structure is typically formed by creating the emitter from a wide-bandgap material such as AlInAs (1.45 eV) or InP (1.35 eV) and the base from a narrower-bandgap compound, GaInAs (0.75 eV), for instance. In the case of an SHBT the same material is used for both collector and base whereas a third material is used as the collector in the case of a DHBT.

This mix of materials has several advantages. First, the bandgap of GaInAs, being narrower than that of silicon and GaAs, produces InP HBTs that have a very low turn-on voltage and are therefore ideal for low-voltage applications. Second, the valence

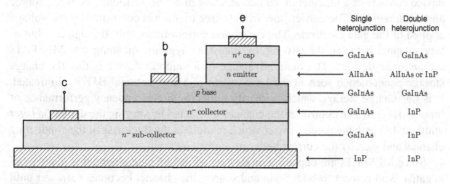

FIGURE 12.5

Simplified conceptual HBT cross-section with materials.

band offset, which blocks base-emitter hole flow, allows base doping to be one to two orders of magnitude higher in an HBT than in a single-material (homojunction) device. The effect is to lower base resistance, increasing the maximum operating frequency (f_{max}) and permitting smaller device dimensions.

12.4 MICROWAVE FIELD-EFFECT TRANSISTORS

12.4.1 THE METAL SEMICONDUCTOR FET

FETs differ from bipolar transistors in that the controlled current is comprised of only one carrier type (electrons or holes). The most common types of FET used at low frequencies are the JFET and the MOSFET, these are limited in their maximum operating frequency due to their relatively large gate capacitance. To operate at microwave frequencies, a way must be found to reduce this gate capacitance.

One solution to this problem is to place the gate electrode directly on top of the semiconductor, as opposed to having an insulating layer separating the gate contact from the semiconductor material (as in the "O" in MOSFET) or *pn*-junction in the case of a JFET. The resulting device is referred to as the MESFET [13].

The main benefit of using a metal-semiconductor contact in the gate of a FET is the fast recovery time resulting from the small amount of stored charge in the region depleted of carriers below the gate. This makes the MESFET structure more suitable for high-frequency operation. The first MESFET, fabricated in GaAs material, was reported by Mead in 1966 [27]. Microwave MESFETs are usually fabricated from compound semiconductor technologies such as GaAs, InP, or GaN [28,29].

Good accounts of the historical development of the GaAs MESFET or its mode of operation can be found in various references [5,30,31].

12.4.2 MESFET CONSTRUCTION

A simplified illustration of the structure of a MESFET is shown in Figure 12.6. The device consists of a channel of semiconducting material positioned between source and drain regions. The carrier flow from source to drain is controlled by the voltage applied to the gate electrode. The controlled current flows only through the thin *n*-type channel between the two highly doped n^+ regions, meaning that MESFETs are *unipolar* devices. The main advantage of a unipolar device is that the charge storage phenomenon seen in the base region of a conventional BJT is eliminated. It is this charge storage which primarily limits the high-frequency performance of bipolar devices. The control of the channel is affected by varying the depletion layer width underneath the metal contact which modulates the thickness of the conducting channel and thereby the current between source and drain.

Since MESFETs are depletion mode devices, as the gate potential is made more negative with respect to both drain and source, the channel becomes narrower until eventually the gate is sufficiently negative to close the channel completely with the result that the drain-source current falls to zero.

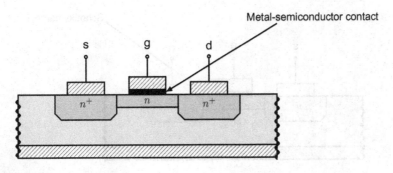

FIGURE 12.6

Conceptual MESFET cross-section (not to scale).

The maximum practical operating frequency of a MESFET is dependent mainly on the width of the gate electrode which defines the channel length. The smaller the channel length, the lower the associated capacitances, and the higher the maximum frequency of operation. Very narrow gate devices are difficult to manufacture by conventional photolithographic processes but can be manufactured by e-beam techniques, albeit at considerably higher unit costs. Aside from the increased maximum operating frequency, reduced gate length also tends to result in improved noise figure. The gate length is therefore one of the most important parameters of a MESFET.

12.4.3 HIGH ELECTRON MOBILITY TRANSISTORS CONSTRUCTION

The HEMT is basically the heterojunction approach applied to the MESFET topology. This means that the channel region of the FET is constructed of two materials with different band gaps instead of a doped region of single material in the case for the simple MESFET, an approach known as *modulation doping* [32]. The heterojunction approach results in higher electron mobility in the channel, allowing the device to respond to rapid changes in the gate voltage. In equivalent circuit terms, this translates as a reduction in C_{gs}.

The first HEMT was demonstrated by Mimura in 1980 [33,34]. HEMTs are capable of operating at very high frequencies, well into the submillimeter wave region (several hundreds of GHz), coupled with very low-noise performance. The HEMT is also referred to as the MOdulation-Doped FET (MODFET).

Unlike the conventional MESFET structure, the HEMT consists of two different material layers, typically undoped GaAs and doped AlGaAs. That is to say, it is a *heterostructure*. A simplified illustration of the structure of a HEMT is shown in Figure 12.7.

In the case of conventional MESFETs, the free electrons generated in the doped semiconductor must share the same space with the donors simultaneously generated. The ensuing electron scattering by donors reduces the electron mobility and velocity.

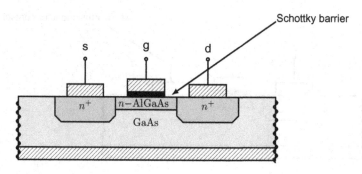

FIGURE 12.7

Conceptual HEMT cross-section (not to scale).

By contrast, in the undoped semiconductor material of the HEMT, electrons can travel unhindered by donors, with consequently greater mobility and velocity [35].

12.5 MESFET AND HEMT EQUIVALENT CIRCUIT

The representation of the MESFET by an equivalent circuit of passive components and controlled current sources has been the subject of a large amount of research and published literature.

Since MESFETs are only used at microwave and millimeter wave frequencies, that is, tens of GHz upward, the devices are characterized by S-parameter measurements rather than Y- or h-parameters. An equivalent circuit having identical S-parameters with an actual device at all frequencies would, in general, contain an infinite number of elements. Thus, the question of accuracy arises since any practical equivalent circuit will contain fewer elements than are required to represent the device exactly.

As with the BJT equivalent circuit discussed in Section 12.2.2, the components in a MESFET equivalent circuit can be classified as either *intrinsic* or *extrinsic*, depending on whether they are inherent to the structure of the device or are due to parasitic elements. The behavior of the MESFET S-parameters with frequency is generally more complex if the device is packaged, making it necessary to include more extrinsic components in the equivalent circuit. A microwave equivalent circuit should ideally model the channel as a distributed RC network [36], but a simpler lumped element circuit has been shown to be capable of describing the MESFET quite accurately [36–38].

The first equivalent circuit derived specifically for the GaAs MESFET by considering both the intrinsic and the extrinsic elements was proposed by Wolf [39] and is shown in Figure 12.8.

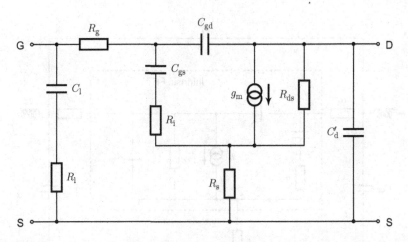

FIGURE 12.8

MESFET equivalent circuit due to Wolf [39].

The physical meaning of the various components in Figure 12.8 are as follows:

1. C_{gs} and C_{gd} correspond to the gate-source and gate-drain junction capacitances, respectively.
2. C_l and C_d' correspond to the extrinsic gate-source and drain-source capacitances.
3. R_i is the resistance of the ohmic channel between the source and the gate.
4. g_m represents the steady-state transconductance of the device, which is frequency dependent (see below).
5. R_{ds} is the channel resistance between the drain and the source.

Aside from the various extrinsic elements, the reader will notice that the main difference between the MESFET equivalent circuit of Figure 12.8 and the BJT hybrid-π equivalent circuit of Figure 12.3 is that C_{gs} effectively blocks the DC path between gate and source in the MESFET, whereas there is a DC path, via r_π in the case of the BJT. This is to be expected if one understands the differences in device construction and operation. There is also no equivalent of r_μ in the case of the MESFET, which is another way of saying that most models assume the gate-drain resistance to be infinite, which is pretty close to reality.

There have been numerous developments of the MESFET equivalent circuit since Wolf's original. The main trend in later models is the addition of an intrinsic capacitance, C_d, between drain and source terminals of the device [40]. R_g and R_i also tend to be combined together in later models to form one intrinsic component [37,41,42].

The circuit of Figure 12.9, or some variant thereof, is the most widely used today and is applied, with different component values, to both MESFETS and HEMTs. In Figure 12.9, we have clearly delineated the intrinsic FET, whose parameters are bias dependent, and extrinsic, bias independent, inductances and resistances that model

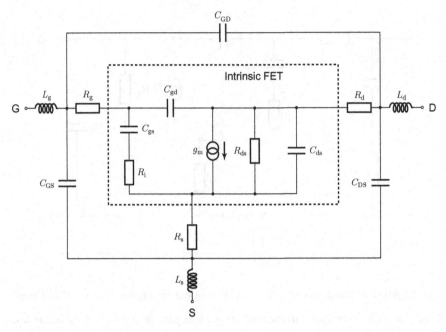

FIGURE 12.9

MESFET or HEMT equivalent circuit with package parasitics.

the gate, source, and drain lead parasitics. This is similar to the approach adopted in the case of the BJT in Figure 12.4.

For high-frequency modeling, it is also important to note that the transconductance g_m is modeled as a complex quantity in order to account for the time τ taken by the carriers to travel across the gate region. The frequency dependence of g_m can thus be modeled as:

$$g_m = g_{m_0}\, e^{-j\omega\tau} \tag{12.5.1}$$

For frequencies in the tens of GHz region, more complex models than that of Figure 12.9 may be required. Such models account for coupling capacitances across the device terminals (known as geometric capacitances) and for transmission line effects on the metal terminals.

By way of illustration, the equivalent circuit parameters for a typical microwave MESFET, calculated from S-parameter measurements and with reference to Figure 12.9 are shown in Tables 12.4 and 12.5 [43–45].

Model parameters for a typical microwave MESFET

Table 12.4 Intrinsic Parameters

g_m	45 mS
τ	6 pS
R_i	0.5 Ω
R_{ds}	1.1 kΩ
C_{gs}	0.55 pF
C_{gd}	0.04 pF
C_{ds}	0.15 pF

Table 12.5 Extrinsic Parameters

L_s	10 pH
L_d	460 pH
L_g	340 pH
C_{GD}	70 fF
C_{GS}	70 fF
C_{SD}	70 fF
R_s	3.4 Ω
R_d	2.0 Ω
R_g	3.8 Ω

Note that the extrinsic parameters values shown in Table 12.5 are the same as those given for the BJT in Table 12.3, since package parasitics will be the same for the same type of package, irrespective of the nature of the intrinsic device inside the package.

The MESFET electrical behavior is described by two models; a small signal model and a large signal one. The large signal model defines the relation between voltages and currents and can be used to extract the small signal parameters at a given bias point. Monolithic microwave integrated circuit (MMIC) foundries provide large signal model parameters based on one or more of the many models available in the literature and implemented in different microwave CAD packages. Foundries also provide MESFET transfer characteristics (I_{ds} versus V_{ds} for different values of V_{gs}) and power transfer characteristics data. Designers are also provided with small signal model parameters (usually scalable with respect to device size/geometry). These parameters are provided at particular bias points (V_{ds} and I_{ds}) and if they were to be used the designer is restricted to such bias points.

For low-noise designs, noise models are of great importance. The intrinsic MESFET noise is modeled by two (gate and drain) correlated noise sources. Resistive elements generate thermal noise that adds to the intrinsic noise. Again, there are several models used to describe MESFET noise [46,47] and either one or more of such model parameters will be made available by device manufacturers.

We discuss the general subject of noise in amplifiers in Chapter 14. The details of noise generation in BJT and FET devices is outside the scope of this book, but the interested reader is referred to some of the specialist texts on this subject, such as Pengelly [5] and Razavi [48].

12.5.1 MESFET EQUIVALENT CIRCUIT PARAMETER EXTRACTION

As we are operating in the GHz frequency range with MESFETs, the h-parameters used to model the BJT are not suitable. Ideally, we would attempt to relate the model parameters directly to the S-parameters, but this is quite difficult. As a half-way house we will use the Y-parameters, which are immittance parameters, and can therefore be related to terminal voltages and currents in the equivalent circuit, but can also be readily converted into S-parameters [49].

We can obtain the Y-parameters of the intrinsic FET in Figure 12.9, after de-embedding the device from the extrinsic parasitics [50–52] by straightforward circuit analysis. Recall the definition of Y-parameters from Section 5.1.1 as follows:

$$\left.\begin{aligned} Y_{11} &= \frac{i_1}{v_1}\bigg|_{v_2=0} \\[2mm] Y_{12} &= \frac{i_1}{v_2}\bigg|_{v_1=0} \\[2mm] Y_{21} &= \frac{i_2}{v_1}\bigg|_{v_2=0} \\[2mm] Y_{22} &= \frac{i_2}{v_2}\bigg|_{v_1=0} \end{aligned}\right\} \tag{12.5.2}$$

By inspection of the intrinsic equivalent circuit in Figure 12.9 and applying the definitions of equation (12.5.2), we can write:

$$Y_{11} = \frac{\omega^2 C_{gs}^2 R_i^2}{1 + \omega^2 C_{gs}^2 R_i^2} + j\omega \left(\frac{C_{ds}}{1 + \omega^2 C_{gs}^2 R_i^2} + C_{dg} \right) \tag{12.5.3}$$

$$Y_{12} = -j\omega C_{dg} \tag{12.5.4}$$

$$Y_{21} = \frac{g_{m0}\, e^{-j\omega\tau}}{1 + j\omega C_{gs} R_i} - j\omega C_{gd} \tag{12.5.5}$$

$$Y_{22} = \frac{1}{R_{ds}} + j\omega(C_{ds} + C_{gd}) \tag{12.5.6}$$

Taking the Y-parameter expressions (equations 12.5.3–12.5.6), and separating them into real and imaginary parts, the various intrinsic equivalent circuit elements can be found analytically as follows [43,53]:

$$C_{dg} = \frac{-\text{Im}(Y_{21})}{\omega} \tag{12.5.7}$$

$$C_{ds} = \frac{\text{Im}(Y_{22}) - \omega C_{dg}}{\omega} \tag{12.5.8}$$

$$C_{ds} = \frac{\mathrm{Im}(Y_{11}) + \mathrm{Im}(Y_{12})}{\omega} \left[1 + \frac{\mathrm{Re}(Y_{12})^2}{(\mathrm{Im}(Y_{11}) + \mathrm{Im}(Y_{12}))^2} \right] \qquad (12.5.9)$$

$$R_i = \frac{\mathrm{Re}(Y_{11})}{\mathrm{Re}(Y_{11})^2 + (\mathrm{Im}(Y_{11}) + \mathrm{Im}(Y_{12}))^2} \qquad (12.5.10)$$

$$g_m = \sqrt{(1 + \omega^2 R_i^2 C_{gs}^2)\mathrm{Re}(Y_{21})^2 + (\mathrm{Im}(Y_{22}) + \omega C_{dg})^2} \qquad (12.5.11)$$

$$G_{ds} = \mathrm{Re}(Y_{22}) \qquad (12.5.12)$$

$$\tau = \frac{1}{\omega} \arctan \left[\frac{-\mathrm{Im}(Y_{21}) - \omega R_i C_{gs}\mathrm{Re}(Y_{21}) - \omega C_{gd}}{\mathrm{Re}(Y_{21}) - \omega R_i C_{gs}\mathrm{Im}(Y_{21}) - \omega^2 R_i C_{gs} C_{dg}} \right] \qquad (12.5.13)$$

where arctan is the arctangent or inverse tangent function.

Analysis of the intrinsic equivalent circuit shown in Figure 12.9 yields the following expression for the gain-bandwidth product of the MESFET, which is similar in form to that for a BJT given in equation (12.2.12):

$$f_T \approx \frac{g_m}{2\pi(C_{gs} + C_{gd})} \qquad (12.5.14)$$

As with the case of BJTs, f_T does not necessarily represent the highest useful frequency of operation of a MESFET as there may be useful power gain that can be extracted even above f_T. We therefore need a definition of f_{max}, the maximum oscillation frequency of a MESFET, which is as follows [54]:

$$f_{max} = \frac{f_T}{2} \sqrt{\frac{R_{ds}}{R_i + R_g + R_s}} \qquad (12.5.15)$$

12.6 MONOLITHIC MICROWAVE INTEGRATED CIRCUITS

MMICs are ICs, containing active, passive, and interconnect components and designed to operate at frequencies from hundreds of MHz to hundreds of GHz. Most of today's MMICs are fabricated on III-V compound substrates such as GaAs, InP, and GaN [55–58], although silicon and SiGe MMICs are also becoming commonplace, especially where complex mixed signal systems need to be integrated on the same chip [59].

From the mid-1970s, the growth in military and commercial demand for reliable high-frequency circuits led to a large investment in GaAs foundries mainly aimed at developing MMIC processes [60]. Recently, MMICs are widely used in applications ranging from specialist phased arrays and multi-Gbit per second optical communications components to mobile phones and home satellite receivers [61]. The upper frequency performance of today's MMICs approaches the THz frequency range [62].

The main features of MMICs will be discussed in this section. The discussion will concentrate on aspects of interest to circuit designers, such as MMIC element models and design methodologies. The key types of active devices used in MMICs are MESFETs, HEMTs, and HBTs. Other active devices, such as Gunn diodes, PIN diodes, and varactors, as well as opto-electronic devices are also to be found in specialist MMICs. Opto-electronic integrated circuits (OEICs) are, as the name suggests, MMICs with opto-electronic devices integrated in them and are increasingly used in high-speed optical communications systems [63].

MMICs are the components of choice for most of today's high-frequency applications. They offer several advantages over their discrete or hybrid counterparts, such as reduced size, low cost, and high reproducibility and repeatable performance.

The main disadvantages of MMICs are shared with all other ICs in that it is difficult (if not impossible) to tune the performance once the IC is fabricated. Most MMICs are fabricated using foundry processes. The key to successful MMIC design is to have well-characterized devices and foundry models. The following sections will describe the basic structure of typical MMICs and the associated active and passive device models. This is followed by a brief discussion of the design procedures of MMICs and examples of wideband MMIC amplifier designs.

12.7 MMIC TECHNOLOGIES

As stated above, MMICs can be fabricated using various material systems and device technologies. GaAs MMICs are the oldest and most commonly used to date. Although MMIC technologies differ in the details of implementation, they share common characteristics in terms of general device geometries and their multilayered structure. In this section, we use a typical GaAs MMIC process to illustrate the multilayer structure of MMICs and the type of elements used in them.

A GaAs MMIC is composed of several layers, all structures are built on a semi-insulating GaAs substrate. The semi-insulating nature of GaAs allows the easy integration of various passive components. A typical process can include eight or more layers of ion implanted GaAs, dielectric insulator layer(s) and metallization layer(s). Detailed description of MMIC processes can be found in two excellent books by Soares [29], and Robertson and Lucyszyn [64].

A cross section of a typical GaAs MESFET MMIC is shown in Figure 12.10. The GaAs wafer (2-4 in. in diameter) is usually "thinned" to a height ranging from 100 to 300 μm, depending on the process used. Many processing steps are involved in the creation of the MMIC, as described in Section 11.3. In this illustrative example, the various layers can be briefly described as follows:

1. *The active layer*: usually two levels of doping (sublayers) define the active device; n active sublayer (doping density $\approx 10^{17}$ cm^{-3}) and n^+ low resistance contact sublayer (doping density $\approx 10^{18}$ cm^{-3}). The layers are produced, in most cases, by ion implantation. For HEMT and HBT MMICs, however, MBE is normally used. The MESFET regions are defined through processes of etching to isolate the gate, drain, and source regions.

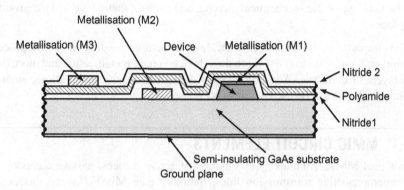

FIGURE 12.10

Typical MMIC cross-sectional view (not to scale).

2. *Ohmic contact layer (M1 in Figure 12.10)*: made of a high conductivity metal alloy, such as AuGe-Ni-Au, is commonly used for contacts to the n^+ layers of MESFETs drain and source terminals.

3. *Schottky (gate) metallization*: is the most critical layer (and manufacturing process) of the MMIC. This is the layer that defines the gates of the MESFETs. The gate metallization is applied after the gate region is recessed to provide the appropriate pinch-off voltage. Metallization is applied in three (or more) metal layers (Ti, Pt, Au). Two types of gates are commonly used; the "T" (or *mushroom*) type and the *Vertical* type. For commercially available foundries, gate widths range from 100 nm to 1 μm. The total metal thickness can be up to few μm.

4. *Second metal layer (M2 in Figure 12.10)*: Alloy metals are used to form contacts to the ohmic contact layer and to form the lowest layer of metal-insulator-metal (MIM) capacitors.

5. *Dielectric layers*: usually silicon nitride (SiN) or polymide, are used for passivation of exposed semiconductor layers and as a dielectric layer for MIM capacitors. Three of these layers are shown in Figure 12.10. The dielectric constant of such layers can be as high as seven, and with layer thickness ranging from 0.1 to 0.3 μm, capacitance values from 100 to 700 pF mm^{-2} can be manufactured, with break down voltages in excess of 50 V.

6. *Third metal layer (M3 in Figure 12.10)*: few microns thick (low resistance) metal used to form top layers of MIM capacitors, interdigital capacitors, spiral inductors, transmission lines, and other interconnect components such as air bridges.

7. *Ground plane*: In many MMICs, the underside of the thinned wafer is metalized to form a ground plane. This is particularly important when microstrip transmission lines are used. It is also important to create connections to ground for various circuit functions by means of *via holes*, which are "drilled" through

the GaAs substrate by chemical etching, and connect metal layers to the ground plane.

It is important to note that the MMIC layer structure may differ from one process to another. Some foundries use more than three layers of metallization and more than two layers of dielectric. Additional layers of high resistivity metal alloys, such as Nichrome (NiCr) can be used to construct resistors.

12.8 MMIC CIRCUIT ELEMENTS

A practical MMIC contains a combination of active devices, passive components, and interconnecting transmission lines. Schottky gate MESFETs and diodes are the main active devices used in GaAs MMICs. Different types of passive devices such as resistors, capacitors, and inductors are also used together with a variety of interconnecting transmission lines. The parasitics that are inevitably associated with three-dimensional MMIC structures means that these passive elements are all, in reality, complex impedances that are designed in such a way as to emphasize one particular property (capacitance in the case of capacitors, inductance in the case of inductors, or resistance in the case of resistors). The presence of significant parasitics does mean, however, that the key to a good MMIC design lies in having accurate DC and RF models of all the MMIC elements and basing the MMIC simulation on layout parameters and dimensions.

MMIC foundries provide users with layout-based models, obtained from comprehensive sets of measurements of the different MMIC elements. In the following sections, the main elements used in GaAs MMICs are described and their models outlined. We have necessarily summarized what has become a huge area of research, so we feel obliged to refer the interested reader to other texts that focus specifically and comprehensively on MMIC technology such as Robertson and Lucyszyn [65], Ladbrooke [46], Razavi [48], Voinigescu [66], and Goyal [47].

12.8.1 MMIC MESFETS

The MESFET is built on the two active sublayers of the GaAs substrate. The operational characteristics of a given MESFET are strongly dependent on its geometry and size. A cross section of a typical MESFET is shown in Figure 12.11.

For a given MMIC process, while the frequency behavior of the MESFET is dependent on the narrowness of the gate line (the so-called gate length) which determines the channel width, the gain of the MESFET depends on its active area, which is defined by the total width (longer dimension) of the gate. Several gates can be "cascaded" to increase the gain. One of the most common gate cascading geometries is known as the "Π-geometry," where several gate lines, also known as gate "fingers," are connected together. This is shown in Figure 12.12.

The device shown in Figure 12.12 can be viewed as a cascade of four identical MESFETs; each having a single gate finger. The Π-geometry is used to reduce the overall lateral size of the MESFET.

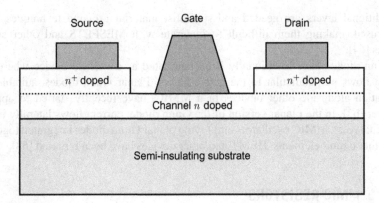

FIGURE 12.11

Typical MESFET structure.

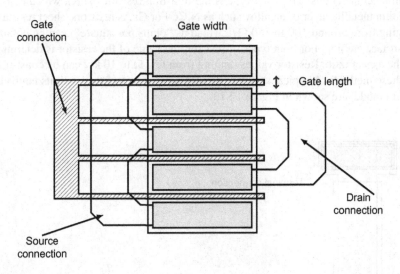

FIGURE 12.12

Typical MESFET layout (four-finger Π-structure).

12.8.2 MMIC DIODES

Schottky diodes are rarely used in MMICs. Occasionally, they are useful either as nonlinear elements in very high-frequency mixers and switches or as bias elements. Diodes can be easily constructed using a Schottky junction similar to that of the MESFET gate, or alternatively the MESFET itself can be used as a diode by shorting any two of its three terminals and using it in a forward bias configuration (positive V_{gs}). PIN diodes have been implemented in MMICs, however, but are quite rare as additional layers are needed and specialist material growing techniques have to be used, making them difficult to integrate with MESFETS and other active devices [64].

Gunn diodes have traditionally been fabricated as a vertical device where the current flows perpendicular to the active layers. Planar Gunn diodes, suitable for integration alongside other devices in a MMIC, have recently started to appear, however [67]. In the planar version of the Gunn diode, current flows laterally in the epitaxial layers. MMIC oscillators employing planar Gunn diodes integrated together with other circuit elements, HEMT and other devices have been reported [68].

12.8.3 MMIC RESISTORS

Two types of resistors are available in MMICs, the first is the *implant resistor*, which is constructed by defining an area on the ion implanted semiconductor having a specific resistivity, defined by the level of doping, and using ohmic contacts for terminal connections. The other type is the thin-film resistor (TFR), which consists of a thin metallic layer of an alloy such as NiCr. For GaAs resistors, sheet resistance ranging from around 100 to 350 $\Omega/\square$ (read as "ohms per square") can be obtained by surface etching. For both types of resistor, the value of the resistor is determined by the aspect ratio. Resistor values ranging from few Ω to 10 kΩ can be constructed by these methods. A typical structure of such a resistor, together with its equivalent circuit model are shown in Figure 12.13.

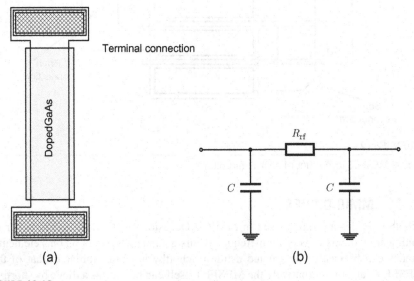

FIGURE 12.13

GaAs resistor and equivalent circuit. (a) GaAs resistor. (b) Equivalent circuit.

Care must be taken when using ion implanted GaAs resistors, as such resistors exhibit nonlinear behavior at high voltages, due to velocity saturation (the same effect that makes the MESFET nonlinear). The alternative is to use alloy metal resistors such as NiCr. The sheet resistance obtained for NiCr, however, is around 50 $\Omega/\square$, resulting in smaller resistor values, up to few hundred Ω. Since NiCr resistors are made of metal they are always linear. There is a limitation, however, associated with the maximum current, and therefore maximum voltage that can be applied across them, before they "melt"! NiCr resistors are best modeled as distributed elements as they behave as lossy transmission lines. Microwave CAD packages have appropriate models for such elements.

The area of the ohmic contact pads is particularly important in determining resistor tolerance, as contact resistance is difficult to control precisely and must be minimized by making pads large. A practical limit is imposed by higher parasitic capacitance of large pads and constraints on resistor's physical size.

Figure 12.14 is a photomicrograph of part of a 4.8 Gbit s^{-1} optical receiver GaAs MMIC containing several implant resistors.

Implant resistors

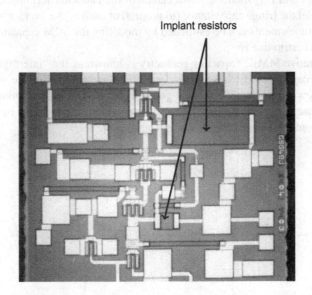

FIGURE 12.14

Photomicrograph of an MMIC showing various implant resistors (reproduced with kind permission of Department of Electronic and Electrical Engineering, University College London).

12.8.4 MMIC CAPACITORS

The most commonly used type of MMIC capacitors is the MIM capacitor, also referred to as "overlay" capacitor. It is simply constructed by using two metallic layers with a dielectric layer, such as SiN, "sandwiched" in between. Capacitance values ranging from hundreds of fF to 100 pF can be achieved by this type of capacitor construction.

The primary capacitance, C, in Figure 12.15 is given by:

$$C \approx \frac{\varepsilon_0 \varepsilon_r A}{t} \tag{12.8.1}$$

where $\varepsilon_0 \varepsilon_r$ is the permittivity of the insulator, t is its thickness, and A is the capacitor area.

A typical equivalent circuit model for such capacitors is shown in Figure 12.15. The parallel conductance, G, is associated with leakage currents and dielectric losses. This element is often ignored in practice. The series resistance, R, can be calculated from the resistivity of the metal used to fabricate the capacitor, together with its geometry. L_{m3} and L_{m2} model the inductance of the capacitor terminals, while C_{m3} and C_{m2} model the fringe capacitance (to ground) of each of the overlay metal plates. These parasitic elements can be estimated by modeling the MIM capacitor as a short length of microstrip line [64].

An alternative MMIC capacitor geometry is known as the "interdigital" or "interdigitated" capacitor, which is a planar structure formed in one of the metallization layers, as shown in Figure 12.16. This geometry is used when particularly accurate values of capacitance are required. The capacitance can be controlled by controlling the finger geometry (length, width, and spacing) [46].

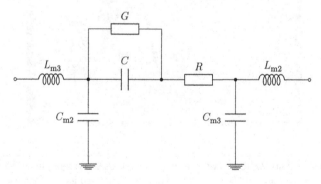

FIGURE 12.15

MIM capacitor equivalent circuit.

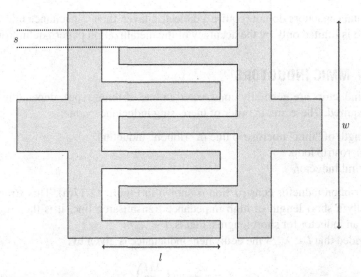

FIGURE 12.16

Interdigital capacitor layout.

The following approximate expression for the primary capacitance, C, of an interdigital capacitor having $N > 4$ fingers is given in Robertson and Lucyszyn's book [64]:

$$C = \frac{l(\varepsilon_r + 1)}{w}[(N - 3)A_1 + A_2] \qquad (12.8.2)$$

where A_1 and A_2 are given by $8.85 \times 10^{-12} \times w$ and $9.92 \times 10^{-12} \times w$, respectively, with w being the total width of the structure in centimeter as shown in Figure 12.16. Equation (12.8.2) is an approximation that is valid for cases where the substrate thickness, h, is large in relation to the strip spacing, s (i.e., $h/s > 100$). There is actually a very wide field of research into complex transmission structures such as this and the above equation is specific to the type of structure in Figure 12.16 implemented in GaAs. Equation (12.8.2) will not apply to other geometries and the parameters A_1 and A_2 will be different for other substrate materials.

The equivalent circuit of Figure 12.15 can also be used to describe the interdigital capacitor, with appropriate values. Note that MMIC capacitors are not reciprocal components, because of the different characteristics associated with metal plates constructed on two different metallization layers in the case of MIM capacitors, or nonsymmetrical geometry in the case of interdigital capacitors.

The disadvantage of interdigital capacitors, compared with MIM capacitors, is that they take up a considerable area. For example, a 0.5 pF interdigital capacitor will measure approximately 400×400 μm^2. For capacitances greater than around 1 pF, capacitor area becomes significant and the resulting distributed effects become serious, limiting the maximum usable operating frequency. On the other hand, since interdigital capacitors do not require a dielectric layer, their capacitance tolerance is good and is limited only by the accuracy of the metallization pattern definition.

12.8.5 MMIC INDUCTORS

MMIC inductors are generally constructed in one of three types, depending on the value required. These are, in order of increasing inductance value:

1. A length of "thin" microstrip line (a "ribbon" inductor).
2. A microstrip loop.
3. A spiral inductor.

The ribbon inductor construction is shown in Figure 12.17(a). This structure is essentially a short length of high impedance transmission line, it is therefore only valid as an inductor for short lengths, that is, $l < \lambda_g/4$.

Provided that $l < \lambda_g/4$ the equivalent inductance is given by:

$$L = \frac{Z_0}{2\pi f} \sin\left(\frac{2\pi l}{\lambda_g}\right)$$

(12.8.3)

There will inevitably be parasitic elements that will have to be taken into account in the design of such inductors. The dominant parasitics in this case will be the shunt capacitances shown in Figure 12.17(b), whose values are given by:

$$C = \frac{1}{2\pi f Z_0} \tan\left(\frac{\pi l}{\lambda_g}\right)$$

(12.8.4)

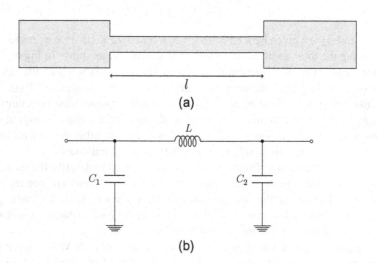

(a)

(b)

FIGURE 12.17

MMIC ribbon inductor. (a) MMIC ribbon inductor. (b) MMIC ribbon inductor equivalent circuit.

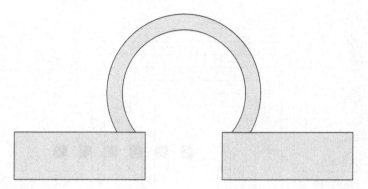

FIGURE 12.18

Basic MMIC loop inductor.

If slightly larger inductances are required, then a microstrip loop inductor can be used, as shown in Figure 12.18. This is essentially just a longer version of the ribbon inductor, so the equivalent circuit is the same as that shown in Figure 12.17(b), and the values can also be calculated from equations (12.8.3) and (12.8.4).

Figure 12.19 is a photomicrograph of the very first MMIC, an 8-12 GHz low-noise amplifier manufactured by Plessey in the UK 1974 [69,70], that nicely illustrates the use of both loop inductors and interdigital capacitors.

The above two inductor types are suitable for implementing low values of inductance, up to a few nH. The spiral inductor is used to implement higher values. Whereas ribbon inductors can be realized using a single layer of metal, the spiral

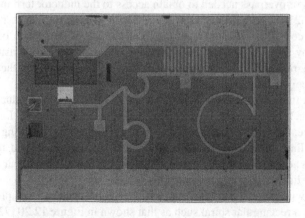

FIGURE 12.19

Photomicrograph of the very first MMIC, illustrating the use of loop inductors and interdigital capacitors [69,70].

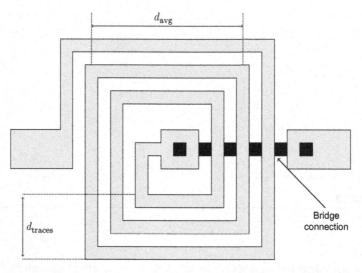

FIGURE 12.20

MMIC spiral inductor.

inductor is a multilayer structure that requires a "bridge" connection between the center back to the outside circuit. This bridge connection must pass under or over the main inductor spiral and is usually implemented as a dielectric underpass or an "airbridge" [71], as shown in Figure 12.20.

The main drawback of spiral inductors is that they exhibit high-frequency resonance behavior, due to the relatively high parasitic capacitances associated with the underpass or overpass needed to obtain access to the inductor terminals.

Figure 12.21 shows a typical equivalent circuit of a spiral inductor. The inductance of the coil is represented by L_{prime} and the series resistance of the line is represented by R_S. The parallel capacitance C_{fb} represents the parasitic coupling between the parallel lines. The shunt elements C_{m1} and C_{m2} represent the capacitance between the coil and the ground planes.

There is a trade-off between inductance and Q for such spiral inductors. Higher inductance can be achieved by more turns, but this increases the total line length. As a consequence, the series resistance R_S as well as the parasitic coupling capacitance between the lines both increase. In addition, the inductor is limited to operate at frequencies where the total length is less than a quarter of the signal wavelength, otherwise the line will act as a resonator.

The following expression has demonstrated good accuracy for predicting the inductance of a rectangular spiral such as that shown in Figure 12.20 [72,73]:

$$L = \frac{2\mu_0 n^2 d_{avg}}{\pi} \left[\ln\left(\frac{2.067}{\rho}\right) + 0.178 + 0.125\rho^2 \right] \tag{12.8.5}$$

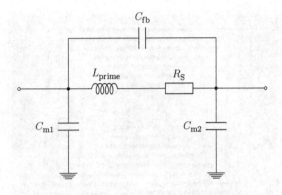

FIGURE 12.21

MMIC spiral inductors equivalent circuit.

where n is the number of turns in the spiral, μ_o is the permeability of free space, d_{avg} represents the average diameter of the spiral, as shown in Figure 12.20, and ρ represents the percentage of the inductor area that is filled by metal traces and is defined according to Figure 12.20 as [73]:

$$\rho = \frac{d_{traces}}{d_{avg}} \qquad (12.8.6)$$

Equation (12.8.5) is based on a current sheet approximation of the spiral structure, and is valid only for square spirals. Interested readers are referred to the often cited paper by Mohan [72] which also presents a range of expressions that are valid for other planar inductor geometries, such as circular spirals. Figure 12.22 is a photomicrograph of a large MMIC spiral inductor [70].

For readers who wish to know more, a very detailed discussion of MMIC spiral inductors can be found in Razavi's book [48].

12.8.6 MMIC TRANSMISSION LINES

Transmission lines on MMICs are implemented as microstrip lines using one of the higher (nonohmic) layer metals (Figure 12.23). For MMICs operating at frequencies higher than 20 GHz, Coplanar waveguides are typically used in place of microstrip lines. Transmission line models for each of these two categories are widely available and almost all of the modern microwave simulators include appropriate models not only for "straight runs" of lines but also for common discontinuities such as bends, coupled lines, semi-circular lines, "T," and cross junctions. Other shapes and structures of transmission lines may be used in MMICs as required. For such structures, detailed microwave modeling will be required for accurate prediction of circuit behavior. For more detailed explanation on design and modelling of MMIC transmission line structures, the reader is referred to the book by Garg et al. [74].

FIGURE 12.22

Photomicrograph of a large MMIC spiral inductor [70].

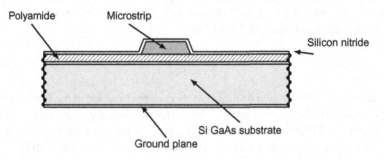

FIGURE 12.23

MMIC microstrip line.

12.8.7 VIA HOLES, BOND PADS, AND OTHER STRUCTURES

Interconnection between different metallization layers (including connection to ground plane) is achieved by means of cylindrical via holes that pass vertically through the various layers. Via holes are generally modeled as a series LC combination. Appropriate, measurement-based, models need to be carefully chosen to incorporate via hole parasitics into the overall design in order to prevent undesirable resonance effects.

The MMIC is connected to the outside world using bond wires that are connected to bond pads constructed on the uppermost metal layer of the MMIC. Bond pads are modeled by a simple capacitance (values in the tens of fF). Different sizes of bond pads may be used for RF and DC connections. Bond wires are modeled as inductors with inductance values directly proportional to length. The inductance of a bond wire can be approximately determined from equation (1.3.1) and typical values will be in the order of hundreds of pH. For MMICs operating at frequencies exceeding 20 GHz multiple parallel bond wires or ribbon bonding can be used to reduce the effective inductance.

Many other passive structures are found in MMICs, such as air bridges, transmission line filters, and even antennas. Recent developments have included the incorporation of microelectromechanical systems (MEMS) that are created on the same die, alongside conventional electronic devices. For a survey of MEMS and their uses in microwave systems, the interested reader is referred to Rebeiz's book [75]. A nice example of how MEMS are incorporated with an MMIC is given by Kim [76].

12.9 MMIC APPLICATION EXAMPLE

In this section, we will briefly describe the design and implementation of a wideband distributed amplifier (DA), operating from tens of MHz to hundreds of GHz, as a way of illustrating the capabilities of contemporary MMIC technology [9]. The DA is a particular broadband amplifier topology that will be discussed in more detail in Chapter 13.

This example uses an InP DHBT process with 250 nm emitter width and where the gain of the transistor is dependent on the number of emitter fingers and, of course, the transistor bias. The unity-gain cut-off frequency of these transistors, f_T, is close to 350 GHz and the maximum frequency of oscillation, f_{max}, is close to 650 GHz. The process is described in detail in Ref. [77].

The process has four metallization layers (M1-M4) and four dielectric layers. The dielectric consists of a 1 μm layer of the organic compound benzocyclobutene (BCB). A schematic cross-section view of the back-end is shown in Figure 12.24.

The thickness of the M1, M2, and M3 layers is $\sim$ 1 μm, whereas M4 is 3 μm thick to support higher current densities. The process includes TFR and MIM capacitors. The InP substrate thickness is 100 μm. The topmost layer, M4, is utilized for signal connections and the second metal layer, M2, is used as a ground plane. This allows low inductive connections (between emitter and ground) when the transistors are used in common emitter configuration which is highly important for achieving high gain when operating up to frequencies close to the f_{max} of the transistors. The first metal layer, M1, is shielded from the RF signal by M2 and can be used for DC connections without disturbing the RF signals on M4.

A photomicrograph of a three-stage DA MMIC is shown in Figure 12.25. The die size is 0.86 mm × 0.37 mm. In this circuit, each of the amplifying stages has two DHBT transistors configured as a *cascade* gain cell, that is, an input transistor in common emitter configuration, the output of which is connected to a transistor in common base configuration. The common emitter devices are visible just at the top of the "L"-shaped transmission lines and the common-base devices can be seen on the lower right of the same lines.

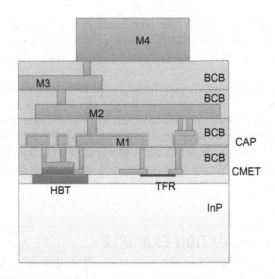

FIGURE 12.24

Schematic cross-section view of the multilayer interconnect back-end process.

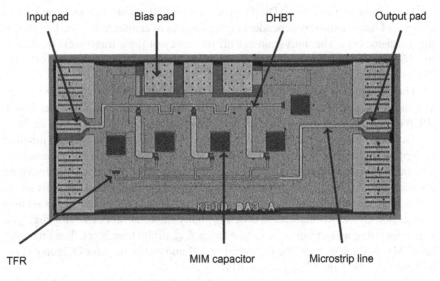

FIGURE 12.25

Photomicrograph of a three-stage DA MMIC. Circuit size: 0.86 mm × 0.37 mm.

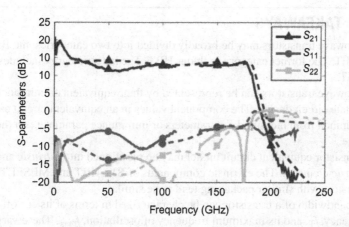

FIGURE 12.26

Simulated and measured frequency response of the three-stage MMIC DA in Figure 12.25.

A high impedance microstrip line ($Z_0 = 70 \ \Omega$) was used at the output of each gain cell to improve the bandwidth. The connections from the input transmission line to the common emitter devices were designed to be as short as possible, since any inductance here reduces the gain and the bandwidth. The two devices in each gain cell are connected with an "L"-shaped low impedance, low-loss microstrip line.

Figure 12.26 shows the measured (solid-line) and simulated (dotted-line) frequency responses of the three-stage DA MMIC shown in Figure 12.25. The amplifier demonstrates more than 10 dB gain from 70 kHz up to 180 GHz. At 180 GHz and above, the gain drops sharply. Due to the very wide bandwidths involved, the amplifier had to be measured in three separate frequency bands, using three different network analyzers:

  (i) between 70 kHz and 115 GHz;
 (ii) 130-220 GHz; and
(iii) 220-300 GHz.

Unfortunately, no measurements could be carried out between 115 and 130 GHz due to nonavailability of test equipment in this band.

The amplifier briefly described here was a member of a family of three amplifiers that were fabricated using the same process, as detailed in Ref. [9]. One of these amplifiers, at the time of publication (2014), was the widest band amplifier reported, with an average gain of 16 dB over a bandwidth of 235 GHz. Such results would have been inconceivable before the advent of the advanced microwave transistors, such as the HBT and MMIC integration technologies described in this chapter.

12.10 TAKEAWAYS

1. Microwave transistors may be broadly divided into two categories: the BJT and the FET. The former category includes HBT. The latter category includes HEMT.

2. Microwave transistors can be represented by their equivalent circuit models to facilitate circuit design. The component values in an equivalent circuit can be determined from measured S-parameters or immittance parameters on the device.

3. A transistor equivalent circuit model may be partitioned into *intrinsic* and *extrinsic* elements. The extrinsic components of BJT/HBT and MESFET/HEMT transistors with similar packaging tend to be similar.

4. The bandwidth of a transistor can be characterized in terms of its cut-off frequency, f_T, and its maximum frequency of oscillation, f_{max}. These vary by transistor type and geometry, and can be calculated based on equivalent circuit component values.

5. There are a number of different techniques for fabricating lumped elements, such as resistors, inductors, and capacitors in MMIC format. The choice of technique depends on the value required and other parameters, such as power handling requirements (in the case of resistors) or dielectric loss requirements (in the case of capacitors).

REFERENCES

[1] E. Kasper, D. Kissinger, P. Russer, R. Weigel, High speeds in a single chip, IEEE Microw. Mag. 10 (7) (2009) 28-33. ISSN 1527-3342, http://dx.doi.org/10.1109/MMM.2009.934691.

[2] H. Cooke, Microwave transistors: theory and design, Proc. IEEE 59 (8) (1971) 1163-1181. ISSN 0018-9219, http://dx.doi.org/10.1109/PROC.1971.8362.

[3] S. Roy, M. Mitra, Microwave Semiconductor Devices, PHI Learning, 2003. ISBN 9788120324183.

[4] M. Rodwell, High-Speed Integrated Circuit Technology: Towards 100 GHz Logic, Selected Topics in Electronics and Systems, World Scientific Publishing Company Incorporated, 2001. ISBN 9789812810014.

[5] R. Pengelly, Microwave Field-Effect Transistors: Theory, Design, and Applications, Electronic & Electrical Engineering Research Studies: Electronic Devices and Systems Series, Research Studies Press, 1986.

[6] F. Ali, A. Gupta, HEMTs and HBTs: Devices, Fabrication, and Circuits, Artech House Antennas and Propagation Library, Artech House, 1991. ISBN 9780890064016.

[7] J. Golio, Microwave MESFETs and HEMTs, Artech House, Dedham, MA, 1991. ISBN 0-89006-426-1.

[8] F. Schwierz, Microwave transistors: the last 20 years, in: Proceedings of the 2000 Third IEEE International Caracas Conference on Devices, Circuits and Systems, 2000, pp. D28/1-D28/7. http://dx.doi.org/10.1109/ICCDCS.2000.869833.

[9] K. Eriksson, I. Darwazeh, H. Zirath, InP DHBT wideband amplifiers with up to 235 GHz bandwidth, in: 2014 IEEE MTT-S International Microwave Symposium (IMS), 2014, pp. 1-4. http://dx.doi.org/10.1109/MWSYM.2014.6848436.

[10] S. Iyer, G. Patton, S.S. Delage, S. Tiwari, J. Stork, Silicon-germanium base heterojunction bipolar transistors by molecular beam epitaxy, in: 1987 International Electron Devices Meeting, vol. 33, 1987, pp. 874-876. http://dx.doi.org/10.1109/IEDM.1987.191578.

[11] P. Solomon, A comparison of semiconductor devices for high-speed logic, Proc. IEEE 70 (5) (1982) 489-509. ISSN 0018-9219, http://dx.doi.org/10.1109/PROC.1982.12333.

[12] W. Shockley, The path to the conception of the junction transistor, IEEE Trans. Electron Devices 23 (7) (1976) 597-620. ISSN 0018-9383, http://dx.doi.org/10.1109/T-ED.1976.18463.

[13] G. Haddad, R. Trew, Microwave solid-state active devices, IEEE Trans. Microw. Theory Tech. 50 (3) (2002) 760-779. ISSN 0018-9480, http://dx.doi.org/10.1109/22.989960.

[14] I. Darwazeh, L. Moura, Introduction to Linear Circuit Analysis and Modelling, Newnes, 2005. ISBN 0750659327.

[15] A. Sedra, K. Smith, Microelectronic Circuits, Oxford University Press, 1998. ISBN 9780195116632.

[16] M. Halkias, Integrated Electronics, McGraw-Hill Electrical and Electronic Engineering Series, Tata McGraw-Hill Publishing Company, 1972. ISBN 9780074622452.

[17] A. Phillips, Transistor Engineering and Introduction to Integrated Se Conductor Circuits, McGraw-Hill Series in Solid State Engineering, McGraw-Hill, 1962.

[18] F. Schwierz, J. Liou, Modern Microwave Transistors: Theory, Design, and Performance, Wiley, 2003. ISBN 9780471417781.

[19] Q. Cai, J. Gerber, U. Rohde, T. Daniel, HBT high-frequency modeling and integrated parameter extraction, IEEE Trans. Microw. Theory Tech. 45 (12) (1997) 2493-2502. ISSN 0018-9480, http://dx.doi.org/10.1109/22.643865.

[20] A. Alt, D. Marti, C. Bolognesi, Transistor modeling: robust small-signal equivalent circuit extraction in various HEMT technologies, IEEE Microw. Mag. 14 (4) (2013) 83-101. ISSN 1527-3342, http://dx.doi.org/10.1109/MMM.2013.2248593.

[21] A. Oudir, M. Mahdouani, R. Bourguiga, Direct extraction method of HBT equivalent-circuit elements relying exclusively on S-parameters measured at normal bias conditions, IEEE Trans. Microw. Theory Tech. 59 (8) (2011) 1973-1982. ISSN 0018-9480, http://dx.doi.org/10.1109/TMTT.2011.2158441.

[22] G. Freeman, B. Jagannathan, S.-J. Jeng, J.-S. Rieh, A. Stricker, D. Ahlgren, S. Subbanna, Transistor design and application considerations for >200-GHz SiGe HBTs, IEEE Trans. Electron Devices 50 (3) (2003) 645-655. ISSN 0018-9383, http://dx.doi.org/10.1109/TED.2003.810467.

[23] M. Urteaga, M. Seo, J. Hacker, Z. Griffith, A. Young, R. Pierson, P. Rowell, A. Skalare, M. Rodwell, InP HBT integrated circuit technology for terahertz frequencies, in: 2010 IEEE Compound Semiconductor Integrated Circuit Symposium (CSICS), ISSN 1550-8781, 2010, pp. 1-4. http://dx.doi.org/10.1109/CSICS.2010.5619675.

[24] J. Higgins, GaAs heterojunction bipolar transistors: a second generation microwave power amplifier transistor, Microw. J. 34 (1991) 176.

[25] V. Radisic, D. Scott, S. Wang, A. Cavus, A. Gutierrez-Aitken, W. Deal, 235 GHz amplifier using 150 nm InP HBT high power density transistor, IEEE Microwave Wireless Compon. Lett. 21 (6) (2011) 335-337. ISSN 1531-1309, http://dx.doi.org/10.1109/LMWC.2011.2139196.

[26] G. Raghavan, M. Sokolich, W. Stanchina, Indium phosphide ICs unleash the high-frequency spectrum, IEEE Spectr. 37 (10) (2000) 47-52. ISSN 0018-9235, http://dx.doi.org/10.1109/6.873917.

[27] C. Maed, Schottky barrier gate field effect transistor, Proc. IEEE 54 (2) (1966) 307-308.

[28] S. Mohammad, A. Salvador, H. Morkoc, Emerging gallium nitride based devices, Proc. IEEE 83 (10) (1995) 1306-1355. ISSN 0018-9219, http://dx.doi.org/10.1109/5.469300.

[29] R. Soares, Applications of GaAs MESFETs, Artech House Inc., Dedham, MA, 1983.

[30] Howes and Morgan, Microwave Solid State Devices, Device Circuit Interactions, John Wiley and Sons, London, 1976.

[31] C. Liechti, Microwave Field-Effect Transistors—1976, IEEE Trans. Microw. Theory Tech. 24 (6) (1976) 279-300. ISSN 0018-9480, http://dx.doi.org/10.1109/TMTT.1976.1128845.

[32] L. Esaki, R. Tsu, Superlattice and negative differential conductivity in semiconductors, IBM J. Res. Dev. 14 (1) (1970) 61-65. ISSN 0018-8646, http://dx.doi.org/10.1147/rd.141.0061.

[33] T. Mimura, S. Hiyamizu, T. Fujii, K. Nanbu, A new field-effect transistor with selectively doped GaAs/n-Al$_x$ Ga$_{1-x}$ As heterojunctions, Jpn J. Appl. Phys. 19 (5) (1980) L225. http://stacks.iop.org/1347-4065/19/i=5/a=L225.

[34] T. Mimura, The early history of the high electron mobility transistor (HEMT), IEEE Trans. Microw. Theory Tech. 50 (3) (2002) 780-782. ISSN 0018-9480.

[35] H. Morkoc, P. Solomon, The HEMT: a superfast transistor, IEEE Spectr. 21 (2) (1984) 28-35. ISSN 0018-9235, http://dx.doi.org/10.1109/MSPEC.1984.6370174.

[36] S. Liao, Microwave Circuit Analysis and Amplifier Design, Prentice Hall, Inc., Englewood Cliffs, 1987.

[37] R. Dawson, Equivalent circuit of the Schottky-barrier gate field-effect transistor at microwave frequencies, IEEE Trans. Microw. Theory Tech. 23 (1975) 499-501.

[38] G. Vendelin, Design of Amplifiers and Oscillators by the S-Parameter Method, John Wiley and Sons, New York, 1982.

[39] P. Wolf, Microwave properties of Schottky barrier field-effect transistors, IBM J. Res. Dev. (1970) 125-141.

[40] G. Vendelin, M. Omori, Circuit model for the GaAs MESFET valid to 12 GHz, Electron. Lett. 11 (3) (1975) 60-61.

[41] C. Liechti, R. Tillman, Design and performance of microwave amplifiers with GaAs Schottky-gate field-effect transistors, IEEE Trans. Microw. Theory Tech. 22 (5) (1974) 510-517. ISSN 0018-9480, http://dx.doi.org/10.1109/TMTT.1974.1128271.

[42] M. Minasian, Simplified GaAs MESFET model to 10 GHz, Electron. Lett. 13 (18) (1977) 549-551.

[43] E. Arnold, M. Golio, M. Miller, B. Beckwith, Direct extraction of GaAs MESFET intrinsic element and parasitic inductance values, in: IEEE MTT-S International Microwave

Symposium Digest, vol. 1, 1990, pp. 359-362. http://dx.doi.org/10.1109/MWSYM.1990. 99594.

[44] U. Iqbal, M. Ahmed, N. Memon, An efficient small signal parameters estimation technique for submicron GaAs MESFET's, in: Proceedings of the IEEE Symposium on Emerging Technologies, 2005, pp. 312-317. http://dx.doi.org/10.1109/ICET.2005. 1558900.

[45] B. Ooi, J. Ma, An improved but reliable model for MESFET parasitic capacitance extraction, in: 2003 IEEE MTT-S International Microwave Symposium Digest, vol. 1, ISSN 0149-645X, 2003, pp. A53-A56. http://dx.doi.org/10.1109/MWSYM.2003.1211032.

[46] P. Ladbrooke, MMIC Design: GaAs FETs and HEMTs, Microwave Library, Artech House, 1989. ISBN 9780890063149.

[47] R. Goyal, Monolithic Microwave Integrated Circuits: Technology and Design, Artech House Microwave Library, Artech House, 1989. ISBN 9780890063095.

[48] B. Razavi, RF Microelectronics, second ed., Prentice Hall, Englewood Cliffs, 2011.

[49] P. Aaen, J. Plá, J. Wood, Modeling and Characterization of RF and Microwave Power FETs, The Cambridge RF and Microwave Engineering Series, Cambridge University Press, 2007. ISBN 9781139468121.

[50] T. Gonzalez, D. Pardo, Monte Carlo determination of the intrinsic small-signal equivalent circuit of MESFET's, IEEE Trans. Electron Devices 42 (4) (1995) 605-611. ISSN 0018-9383, http://dx.doi.org/10.1109/16.372061.

[51] G. Dambrine, A. Cappy, F. Heliodore, E. Playez, A new method for determining the FET small-signal equivalent circuit, IEEE Trans. Microw. Theory Tech. 36 (7) (1988) 1151-1159. ISSN 0018-9480, http://dx.doi.org/10.1109/22.3650.

[52] B.-L. Ooi, M.-S. Leong, P.-S. Kooi, A novel approach for determining the GaAs MESFET small-signal equivalent-circuit elements, IEEE Trans. Microw. Theory Tech. 45 (12) (1997) 2084-2088. ISSN 0018-9480, http://dx.doi.org/10.1109/22.643741.

[53] M. Kameche, M. Feham, Simple techniques for determining the small-signal equivalent circuit of MESFETs, Int. J. Infrared Millimeter Waves 27 (5) (2006) 687-705. ISSN 0195-9271, http://dx.doi.org/10.1007/s10762-006-9112-9, http://dx.doi.org/ 10.1007/s10762-006-9112-9.

[54] G. Vendelin, A. Pavio, U. Rohde, Microwave Circuit Design Using Linear and Nonlinear Techniques, Wiley, 2005. ISBN 9780471715825.

[55] M. Micovic, A. Kurdoghlian, H. Moyer, P. Hashimoto, A. Schmitz, I. Milosavljevic, P. Willadsen, W.-S. Wong, J. Duvall, M. Hu, M. Wetzel, D. Chow, GaN MMIC technology for microwave and millimeter-wave applications, in: IEEE Compound Semiconductor Integrated Circuit Symposium, CSIC '05, 2005. http://dx.doi.org/10.1109/CSICS.2005. 1531801.

[56] J. Magarshack, Microwave integrated circuits on GaAs, in: 12th European Microwave Conference, 1982, pp. 5-15. http://dx.doi.org/10.1109/EUMA.1982.333136.

[57] R. Smith, S. Sheppard, Y.-F. Wu, S. Heikman, S. Wood, W. Pribble, J. Milligan, AlGaN/GaN-on-SiC HEMT technology status, in: IEEE Compound Semiconductor Integrated Circuits Symposium, CSIC '08, ISSN 1550-8781, 2008, pp. 1-4. http://dx.doi.org/ 10.1109/CSICS.2008.57.

[58] M. Barsky, M. Biedenbender, X. Mei, P.-H. Liu, R. Lai, Advanced InP and GaAs HEMT MMIC technologies for MMW commercial products, in: 65th Annual Device Research

Conference, ISSN 1548-3770, 2007, pp. 147-148. http://dx.doi.org/10.1109/DRC.2007. 4373691.

[59] J. Yu, F. Zhao, J. Cali, F. Dai, D. Ma, X. Geng, Y. Jin, Y. Yao, X. Jin, J. Irwin, R. Jaeger, An X-band radar transceiver MMIC with bandwidth reduction in 0.13 μm SiGe technology, IEEE J. Solid State Circuits 49 (9) (2014) 1905-1915. ISSN 0018-9200, http://dx.doi.org/10.1109/JSSC.2014.2315650.

[60] R. Van Tuyl, The early days of GaAs ICs, in: 2010 IEEE Compound Semiconductor Integrated Circuit Symposium (CSICS), ISSN 1550-8781, 2010, pp. 1-4. http://dx.doi.org/10.1109/CSICS.2010.5619700.

[61] C. Mahle, H. Huang, MMIC's in communications, IEEE Commun. Mag. 23 (9) (1985) 8-16. ISSN 0163-6804, http://dx.doi.org/10.1109/MCOM.1985.1092639.

[62] L. Samoska, An Overview of Solid-State Integrated Circuit Amplifiers in the Submillimeter-Wave and THz Regime, IEEE Trans. Terahertz Sci. Technol. 1 (1) (2011) 9-24. ISSN 2156-342X, http://dx.doi.org/10.1109/TTHZ.2011.2159558.

[63] S. Forrest, Optoelectronic integrated circuits, Proc. IEEE 75 (11) (1987) 1488-1497. ISSN 0018-9219, http://dx.doi.org/10.1109/PROC.1987.13910.

[64] I. Robertson, S. Lucyszyn, Institution of Electrical Engineers, RFIC and MMIC Design and Technology, IEE Circuits, Devices and Systems Series, Institution of Engineering and Technology, 2001. ISBN 9780852967867.

[65] I. Robertson, MMIC Design (IEE Circuits and Systems), Institution of Engineering and Technology, 1995. ISBN 0-85296-816-7.

[66] S. Voinigescu, High-Frequency Integrated Circuits, High-frequency Integrated Circuits, Cambridge University Press, 2013. ISBN 9780521873024.

[67] A. Khalid, N. Pilgrim, G. Dunn, M. Holland, C. Stanley, I. Thayne, D. Cumming, A planar Gunn diode operating above 100 GHz, IEEE Electron Device Lett. 28 (10) (2007) 849-851. ISSN 0741-3106, http://dx.doi.org/10.1109/LED.2007.904218.

[68] V. Papageorgiou, A. Khalid, C. Li, D. Cumming, Cofabrication of planar Gunn diode and HEMT on InP substrate, IEEE Trans. Electron Devices 61 (8) (2014) 2779-2784. ISSN 0018-9383, http://dx.doi.org/10.1109/TED.2014.2331368.

[69] R. Pengelly, Early GaAs FET monolithic microwave integrated circuit developments for radar applications at Plessey, UK, in: 2008 IEEE MTT-S International Microwave Symposium Digest, ISSN 0149-645X, 2008, pp. 827-830. http://dx.doi.org/10.1109/MWSYM.2008.4632960.

[70] J. Turner, D. Smith, Plessey GaAs lives on, Euro III-Vs Rev. 3 (6) (1990) 20-22, ISSN 0959-3527, http://dx.doi.org/10.1016/0959-3527(90)90174-R, GaAs IC Symposium and Device Processing. http://www.sciencedirect.com/science/article/pii/095935279090174R.

[71] I. Bahl, Lumped Elements for RF and Microwave Circuits, Artech House Microwave Library, Artech House, 2003. ISBN 9781580536615.

[72] S. Mohan, M. del Mar Hershenson, S. Boyd, T. Lee, Simple accurate expressions for planar spiral inductances, IEEE J. Solid State Circuits 34 (10) (1999) 1419-1424. ISSN 0018-9200, http://dx.doi.org/10.1109/4.792620.

[73] R. Bunch, D. Sanderson, S. Raman, Quality factor and inductance in differential IC implementations, IEEE Microw. Mag. 3 (2) (2002) 82-92. ISSN 1527-3342, http://dx.doi.org/10.1109/MMW.2002.1004055.

[74] R. Garg, I. Bahl, M. Bozzi, Microstrip Lines and Slotlines, third ed., Microwave & RF, Artech House, 2013. ISBN 9781608075355.

[75] G. Rebeiz, RF MEMS: Theory, Design, and Technology, Wiley, 2004. ISBN 9780471462880.

[76] Y. Kim, N.-G. Kim, J.-M. Kim, S. Lee, Y. Kwon, Y.-K. Kim, 60-GHz Full MEMS antenna platform mechanically driven by magnetic actuator, IEEE Trans. Ind. Electron. 58 (10) (2011) 4830-4836. ISSN 0278-0046, http://dx.doi.org/10.1109/TIE.2011.2114317.

[77] J. Hacker, M. Seo, A. Young, Z. Griffith, M. Urteaga, T. Reed, M. Rodwell, THz MMICs based on InP HBT Technology, in: 2010 IEEE MTT-S International Microwave Symposium Digest (MTT), ISSN 0149-645X, 2010, p. 1. http://dx.doi.org/10.1109/MWSYM.2010.5518124.

Microwave amplifier design

13

CHAPTER OUTLINE

INTENDED LEARNING OUTCOMES

- *Knowledge*
  - Understand the operation of a single-stage microwave transistor amplifier.
  - Be familiar with the unilateral approximation, its application and limitations.
  - Understand the benefits of adding feedback in a transistor amplifier and be able to apply feedback design techniques to a single transistor amplifier. Be introduced to the principle of distributed amplification and it's advantages over the traditional cascaded amplifier approach.
- *Skills*
  - Be able to design a single-stage microwave transistor amplifier for maximum available gain.

Microwave Active Circuit Analysis and Design. http://dx.doi.org/10.1016/B978-0-12-407823-9.00013-5

- Be able to apply constant gain circles to design a single-stage transistor amplifier to meet a particular gain specification.
- Be able to design a feedback network to unilateralize a transistor.
- Be able to design a single-stage broadband transistor amplifier using resistive shunt/series feedback.
- Be able to design a basic distributed amplifier (DA).

13.1 INTRODUCTION

Whether characterized as low noise, high power, wideband, or otherwise, all electronic amplifiers have in common the defining characteristic of providing finite positive *Power Gain* at the frequency, or range of frequencies, of interest [1]. The superficially simple concept of power gain, that is, that the signal power delivered to the load exceeds the signal power supplied to the input of the amplifier, is actually a little tricky to define in practice, as there is more than one way to define "power in" and "power out." This results in a number of different definitions of power gain as explained in Chapter 7.

In this chapter, we look in a bit more detail at the actual processes of designing an amplifier. We shall elaborate on precisely how to design an amplifier to meet specified gain and bandwidth characteristics, starting with simple single-stage amplifiers.

We investigate the special case of an active device for which $S_{12} \approx 0$. Assuming that $S_{12} = 0$ simplifies the design considerably, but the error involved in this assumption needs to be carefully considered, as we will explain.

We will look at the use of feedback to achieve a specific performance specification, and we will show how the application of feedback can be used to achieve unilaterality or otherwise improve the parameters of the basic transistor.

We then move on to discuss multistage and broadband amplifiers, employing various design techniques such as feedback, lossy interstage matching, and the distributed amplifier (DA) topology.

13.2 SINGLE-STAGE AMPLIFIER DESIGN

The simplest possible single-stage amplifier consists of a single transistor, a bipolar junction transistor (BJT) in common emitter configuration, for instance, connected to a source and load both having the system characteristic impedance, Z_0 as shown in Figure 13.1.

If we consider the definition of the S-parameters for any two-port, as outlined in Chapter 6, we can write the relationship between the incident and the reflected power waves at the input and output ports of the simple amplifier in Figure 13.1 as follows:

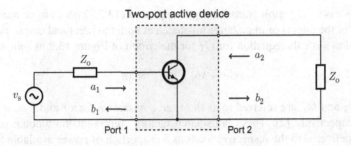

FIGURE 13.1

The simplest possible single transistor amplifier.

$$b_1 = S_{11}a_1 + S_{12}a_2 \qquad (13.2.1)$$
$$b_2 = S_{21}a_1 + S_{22}a_2 \qquad (13.2.2)$$

One working definition of the power gain of this circuit is the ratio of the power entering at port 1 to the power emerging from port 2, with no power being incident at port 2. If we therefore set $a_2 = 0$ in equation (13.2.2) we can write the power gain of Figure 13.1 as:

$$G = \frac{|b_2|^2}{|a_1|^2} = |S_{21}|^2 \qquad (13.2.3)$$

This gain parameter is actually referred to as the *Transducer Power Gain*, or simply *Transducer Gain*, and is represented by G_T. We have already mentioned that the power gain of a two-port can be defined in a number of different ways. We will soon show that all of these definitions are essentially derived from the transducer power gain.

We will now extend the simple case above to account for the more realistic situation of arbitrary source and load terminations, Z_S and Z_L (reflection coefficients Γ_S and Γ_L), as shown in Figure 13.2.

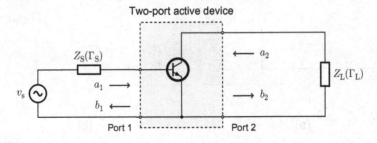

FIGURE 13.2

Simple single transistor amplifier with arbitrary terminations.

In this case, the simple transducer gain equation (13.2.3) needs to be modified to account for the effects of impedance mismatch at both the input and output ports. We now need to write the equation for G_T for the circuit of Figure 13.2 as follows:

$$G_T = M_S \times |S_{21}|^2 \times M_L \tag{13.2.4}$$

where M_S and M_L are referred to as the *source mismatch factor* and *load mismatch factor*, respectively [2]. These "mismatch factors" represent the amount of power actually delivered to the respective loads as a proportion of power available from the respective sources, in other words:

$$M_S = \frac{P_{in}}{P_{AVS}} \tag{13.2.5}$$

and

$$M_L = \frac{P_L}{P_{AVN}} \tag{13.2.6}$$

where P_{AVS} is the power available from the source and P_{AVN} is the power available from the amplifier output port. Since P_{AVS} and P_{AVN} represent maximum power available from the respective sources (under conjugately matched conditions), we can therefore deduce that M_S and M_L can never be greater than unity.

In order to express the mismatch factors in terms of termination impedances, we need to consider the power flow at the input and output of the amplifier separately, as shown in Figure 13.3.

We will first consider the situation at the input port, as shown in Figure 13.3(a). The "power available" from the source is defined as the power delivered into a conjugately matched load. If we consider the source in Figure 13.3(a) as being

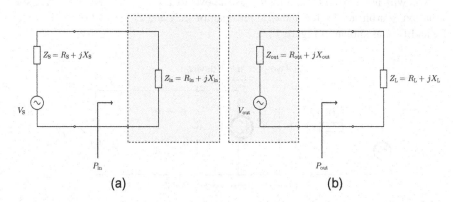

(a)   (b)

FIGURE 13.3

Mismatch factors. (a) Input port matching. (b) Output port matching.

conjugately matched to the input of the transistor, that is, $Z_S = Z_{in}^*$, then the average power available from the source is:

$$P_{AVS} = \frac{1}{2} \frac{|V_S|^2}{R_S} \qquad (13.2.7)$$

The power delivered to the input of the transistor in Figure 13.3(a) is given by:

$$P_{in} = \frac{1}{2} \left| \frac{V_S}{Z_S + Z_{in}} \right|^2 R_{in} \qquad (13.2.8)$$

By substituting equation (13.2.7) into equation (13.2.8), with some manipulation we get:

$$P_{in} = \frac{1}{8} \frac{|V_S|^2}{R_S} \left(\frac{4R_S R_{in}}{|Z_S + Z_{in}|^2} \right) = P_{AVS} \left(\frac{4R_S R_{in}}{|Z_S + Z_{in}|^2} \right) \qquad (13.2.9)$$

Applying equation (13.2.9) to equation (13.2.5) we have:

$$M_S = \frac{4R_S R_{in}}{|Z_S + Z_{in}|^2} \qquad (13.2.10)$$

We shall now convert equation (13.2.10) to reflection coefficient form by applying equation (2.2.16). First, we write R_S in terms of Γ_S as follows:

$$R_S = \frac{1}{2}(Z_S + Z_S^*) = \frac{Z_o}{2} \left[\frac{1 + \Gamma_S}{1 - \Gamma_S} + \frac{1 + \Gamma_S^*}{1 - \Gamma_S^*} \right] = Z_o \frac{1 - |\Gamma_S|^2}{|1 - \Gamma_S|^2} \qquad (13.2.11)$$

By similar reasoning we can express the real part of Z_{in} in terms of Γ_{in}:

$$R_{in} = Z_o \frac{1 - |\Gamma_{in}|^2}{|1 - \Gamma_{in}|^2} \qquad (13.2.12)$$

By substituting equations (13.2.11) and (13.2.12) into equation (13.2.10), and replacing Z_S and Z_{in} in equation (13.2.10) by their reflection coefficient equivalents, we obtain the following expression for M_S in reflection coefficient terms:

$$M_S = \frac{(1 - |\Gamma_{in}|^2)(1 - |\Gamma_S|^2)}{|1 - \Gamma_{in}\Gamma_S|^2} \qquad (13.2.13)$$

A similar analysis to the above, applied to the output port of Figure 13.3(b) results in a corresponding expression for the load mismatch factor, M_L:

$$M_L = \frac{(1 - |\Gamma_{out}|^2)(1 - |\Gamma_L|^2)}{|1 - \Gamma_{out}\Gamma_L|^2} \qquad (13.2.14)$$

From equation (13.2.14) we can deduce that for $M_L = 1$ the load must be conjugately matched to the output of the transistor, that is, $\Gamma_L = \Gamma_{out}^*$. A quick look at equation (13.2.13) will similarly reveal that $M_S = 1$ when the source is conjugately matched to the input of the transistor, that is, when $\Gamma_S = \Gamma_{in}^*$. This supports the intuitive understanding that the maximum available gain will be obtained from the transistor when we conjugately match the input and output ports.

For any linear, nonunilateral, two-port, Γ_{in} will be dependent on the load termination and Γ_{out} will be dependent on the source termination, these quantities being related by equations (6.2.5) and (6.2.7). This means that for a nonunilateral device the mismatch factors M_S and M_L are interdependent. The ports cannot be conjugately matched independently as changing the value of Γ_L will alter the value of Γ_{in} and vice versa. In this case, we must employ a technique referred to as "simultaneous conjugate matching" which was discussed in Chapter 7.

The interrelationship between the input and output ports of a nonunilateral two-port means that if we combine equations (13.2.13) and (13.2.14) together in a single gain expression, as in equation (13.2.4), we can eliminate one of either Γ_{in} or Γ_{out}. We therefore have a choice of two possible expressions for the transducer power gain that we obtain by substituting equations (13.2.13) and (13.2.14) into equation (13.2.4) and applying equations (6.2.5) and (6.2.7), as follows:

$$G_T = \frac{(1 - |\Gamma_S|^2)}{|1 - \Gamma_{in}\Gamma_S|^2}|S_{21}|^2\frac{(1 - |\Gamma_L|^2)}{|1 - S_{22}\Gamma_L|^2} \qquad (13.2.15)$$

or

$$G_T = \frac{(1 - |\Gamma_S|^2)}{|1 - S_{11}\Gamma_S|^2}|S_{21}|^2\frac{(1 - |\Gamma_L|^2)}{|1 - \Gamma_{out}\Gamma_L|^2} \qquad (13.2.16)$$

The above equivalent expressions for transducer power gain, equations (13.2.15) and (13.2.16), were derived in Chapter 7 and it was observed that they both consist of $|S_{21}|^2$ "sandwiched" between the source mismatch factor and the load mismatch factor, the latter two accounting for the degree of mismatch at the respective ports. The reader can easily confirm this by setting $\Gamma_S = 0$ and $\Gamma_L = 0$ in either equation (13.2.15) or (13.2.16), which will result in $G_T = |S_{21}|^2$ in both cases.

We have so far demonstrated that the gain performance of a transistor is a function of the source and load terminations, Γ_S and Γ_L, and that these terminations can be optimized to extract maximum gain from the transistor. In the real world, we do not always have control over the sources and loads with which we are presented. We need a way to translate a given source and load to the terminating values we need at the transistor ports in order to achieve our specific performance requirements.

In order to construct a practical amplifier, we need to add input and output matching networks in order to optimize gain, as well as bias networks to supply DC supply power to the transistor, as shown in Figure 13.4.

In Chapter 7, it was shown that certain values of input and output terminations can result in instability and that areas of stable operation can be mapped out on a Smith Chart representing the load or source plane using *Stability Circles*. The concept of *Stability Criteria* were also defined in Chapter 7 as a short-cut to determining stability without plotting these circles.

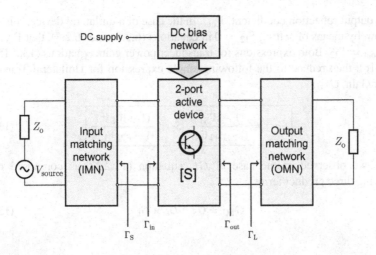

FIGURE 13.4

Single active device with input and output matching network.

13.2.1 THE UNILATERAL APPROXIMATION

The parameter S_{12} represents internal feedback from the output to the input of a two-port device. In terms of the operation of practical amplifier circuits, it is desirable to have S_{12} as small as possible for two reasons:

1. The smaller the value of S_{12} the greater is the degree of isolation between the output and input of a stage. This is an important consideration when stages are to be cascaded.
2. The smaller the value of S_{12} the greater is the degree of stability of a given stage.

However, there is a third reason why a small value of S_{12} is desirable and it relates to the process of designing amplifier stages. If S_{12} is small enough we can make an approximation by setting $S_{12} = 0$ and the design process can be greatly simplified. This so-called *Unilateral Approximation* also leads to greater conceptual simplicity when studying some of the factors which affect amplifier performance.

Of course, with practical transistors, $S_{12} \neq 0$ and any design procedure which assumes unilaterality will yield approximate results. The degree of error introduced by such an approximation can be quantified and a judgment can be made as to its acceptability in a given circumstance. Conditions close to unilaterality can be achieved by the application of feedback to a transistor, as will be discussed in Section 13.3. Unilateral design techniques can be safely applied in such cases.

13.2.2 UNILATERAL GAIN

We recall that the two alternative expressions for transducer power gain, given by equations (13.2.15) and (13.2.16), contain either the input reflection coefficient, Γ_{in}, or the output reflection coefficient, Γ_{out}. In the case of a unilateral device, it is trivial to show, by means of setting $S_{12} = 0$ in equations (6.2.5) and (6.2.7), that $\Gamma_{in} = S_{11}$ and $\Gamma_{out} = S_{22}$. Both expressions for transducer power gain, equations (13.2.15) and (13.2.16), then reduce to the following single expression for Unilateral Transducer Power Gain, G_{TU}:

$$G_{TU} = \frac{(1 - |\Gamma_S|^2)}{|1 - S_{11}\Gamma_S|^2}|S_{21}|^2\frac{(1 - |\Gamma_L|^2)}{|1 - S_{22}\Gamma_L|^2} \qquad (13.2.17)$$

As we observed in the case of G_T, equation (13.2.17) is composed of the following three product terms:

$$G_{TU} = G_S \times G_o \times G_L \qquad (13.2.18)$$

where

$$G_S = \frac{|1 - |\Gamma_S|^2|}{|1 - S_{11}\Gamma_S|^2} \qquad (13.2.19)$$

$$G_o = |S_{21}|^2 \qquad (13.2.20)$$

$$G_L = \frac{|1 - |\Gamma_L|^2|}{|1 - S_{22}\Gamma_L|^2} \qquad (13.2.21)$$

This time the mismatch factors, G_S and G_L, are independent of each other, meaning that the input and output ports of the device can be matched independently. G_S and G_L result from equations (13.2.13) and (13.2.14) when we set $\Gamma_{in} = S_{11}$ and $\Gamma_{out} = S_{22}$. G_o simply represents the transducer gain of the active device when terminated in the system characteristic impedance, Z_o.

It is a simple matter to show that, for a unilateral device, maximum gain is obtained when we set:

$$\Gamma_S = S_{11}^* \qquad (13.2.22)$$

and

$$\Gamma_L = S_{22}^* \qquad (13.2.23)$$

Equation (13.2.17) then gives the value of the maximum unilateral gain as:

$$G_{TU_{max}} = \frac{1}{|1 - |S_{11}|^2|}|S_{21}|^2\frac{1}{|1 - |S_{22}|^2|} \qquad (13.2.24)$$

A further useful consequence of assuming unilaterality is that Rollett's stability factor tends to infinity and the stability criteria equation (7.4.29) simply reduce to:

$$|S_{11}| < 1$$
$$|S_{22}| < 1 \qquad (13.2.25)$$

13.2.3 CIRCLES OF CONSTANT UNILATERAL GAIN

It was shown in the previous section that the unilateral gain is made up of three independent gain blocks, which means that the input and output ports of the device can be matched independently.

Referring to the relationships equations (13.2.17) to (13.2.21), for a given device, $G_o = |S_{21}|^2$ is fixed so we need only consider the values of G_S and G_L, which depend on the values of Γ_S and Γ_L, respectively.

Considering firstly equation (13.2.19), we note that G_S is maximum when the input port is conjugately matched, that is, when $\Gamma_S = S_{11}^*$. In this case, we have $G_S = G_{S_{max}}$, where

$$G_{S_{max}} = \frac{1}{1 - |S_{11}|^2} \qquad (13.2.26)$$

We will define the normalized unilateral gain parameter, g_{Su} as:

$$g_{Su} = \frac{G_S}{G_{S_{max}}} = \frac{|1 - |\Gamma_S|^2|}{|1 - S_{11}\Gamma_S|^2}(1 - |S_{11}|^2) \qquad (13.2.27)$$

Once again we note that the equation of a circle on the Γ_S plane is of the form:

$$|\Gamma_S - C_{gu}|^2 = |\gamma_{gu}|^2 \qquad (13.2.28)$$

where C_{gu} is the center and γ_{gu} is the radius of the constant unilateral gain circle. We can rearrange equation (13.2.27) to be in the form of equation (13.2.28), as follows:

$$\left|\Gamma_S - \frac{g_{Su}S_{11}^*}{1 - |S_{11}|^2(1 - g_{Su})}\right|^2 = \left|\frac{\sqrt{1 - g_{Su}}(1 - |S_{11}|^2)}{1 - |S_{11}|^2(1 - g_{Su})}\right|^2 \qquad (13.2.29)$$

By comparing equation (13.2.29) with equation (13.2.28), we determine the centers of the source plane constant unilateral gain circles as follows:

$$\boxed{C_{Su} = \frac{g_{Su}S_{11}^*}{1 - |S_{11}|^2(1 - g_{Su})}} \qquad (13.2.30)$$

The radii of these circles are given by:

$$\boxed{\gamma_{Su} = \frac{\sqrt{1 - g_{Su}}(1 - |S_{11}|^2)}{1 - |S_{11}|^2(1 - g_{Su})}} \qquad (13.2.31)$$

We can carry out a similar analysis starting with equation (13.2.21) for G_L, which will lead us to the centers of the load plane constant unilateral gain circles as follows:

$$\boxed{C_{Lu} = \frac{g_{Lu}S_{22}^*}{1 - |S_{22}|^2(1 - g_{Lu})}} \qquad (13.2.32)$$

The radii of these circles are given by:

$$\gamma_{Lu} = \frac{\sqrt{1 - g_{Lu}}(1 - |S_{22}|^2)}{1 - |S_{22}|^2(1 - g_{Lu})}$$

(13.2.33)

where the normalized load plane unilateral gain parameter, g_{Lu}, is defined as:

$$g_{Lu} = \frac{G_L}{G_{L_{max}}} = \frac{|1 - |\Gamma_L|^2|}{|1 - S_{22}\Gamma_L|^2}(1 - |S_{22}|^2)$$

(13.2.34)

and

$$G_{L_{max}} = \frac{1}{1 - |S_{22}|^2}$$

(13.2.35)

By inspection of equations (13.2.30)–(13.2.33) we can make the following observations about constant unilateral gain circles:

1. The angle of C_{Su} will always be equal to $\pm\angle S_{22}$, which means that the centers of the constant unilateral gain circles on the source plane will always lie on the line drawn between S_{11}^* and the origin. Similarly, the centers of the constant unilateral gain circles on the load plane will always lie on the line drawn between S_{22}^* and the origin.
2. Maximum unilateral gain occurs when $g_{Su} = 1$ (i.e., when $G_S = G_{S_{max}}$). The source plane constant unilateral gain circle then reduces to a single point at S_{11}^* which corresponds to a conjugate match at the input port. The same applies to the output port.
3. When $g_{Su} = 1 - |S_{11}|^2$ or $G_S = 1$, we have $|C_{Su}| = |\gamma_{Su}|$. This indicates that the constant unilateral gain circle passes through the origin of the input reflection coefficient plane. This represents the boundary of the input network for providing gain, any values beyond this circle represent loss.

13.2.4 ERROR INVOLVED IN THE UNILATERAL APPROXIMATION

Making the assumption of unilaterality can result in a simplified design procedure. It is important, however, to be able to estimate the discrepancy between the results obtained by such a procedure and those obtained by using the unapproximated design equations introduced in the previous sections. A unilateral figure of merit has been proposed by, and named after, Mason [3] which is defined by:

$$U = \frac{|S_{11}S_{12}S_{21}S_{22}|}{(1 - |S_{11}|^2)(1 - |S_{22}|^2)}$$

(13.2.36)

The quantity "U" can be used to determine the maximum error associated with using G_{TU} instead of G_T in the design procedure, by using the following relationship:

$$\frac{1}{(1+U)^2} \leq \frac{G_T}{G_{TU}} \leq \frac{1}{(1-U)^2} \tag{13.2.37}$$

The decision whether to apply the unilateral approximation is, of course, a matter of judgment. Typically, we can be comfortable using the approximation if the gain error, defined by the ratio $\frac{G_T}{G_{TU}}$ in equation (13.2.37), is within $\pm 10\%$, that is, if $U < 0.05$. The interested reader is referred to an extensive treatment of the errors associated with assuming unilaterality published by Scanlan and Young [4].

Example 13.1 (The unilateral approximation).

Problem. What would be the error associated with assuming that the transistor in Example 7.4 were unilateral?

Solution. We calculated the Transducer Power Gain in Example 7.4 as $G_T = 12.810$ (11.1 dB). We now calculate the Unilateral Transducer Power Gain, G_{TU}, using equation (13.2.17).

$$G_{TU} = \frac{(1-|\Gamma_S|^2)}{|1-S_{11}\Gamma_S|^2}|S_{21}|^2\frac{(1-|\Gamma_L|^2)}{|1-S_{22}\Gamma_L|^2}$$
$$= \frac{0.889}{0.721}9.090\frac{0.960}{0.802}$$
$$= 13.40 \text{ (11.28 dB)}$$

We can see that, in this case, the gain error involved in applying the unilateral approximation is 7.5%.

13.3 SINGLE-STAGE FEEDBACK AMPLIFIER DESIGN

The application of feedback to a two-port active device adds extra degrees of freedom in amplifier design, and allows us to achieve performance characteristics that would not be possible without feedback. We can apply the three-port design techniques set out in Chapter 8 to determine the effect of a given feedback termination on the S-parameters of the resulting "reduced" two-port. We should recall that there are two types of feedback that can be applied to a single transistor, namely series feedback and shunt feedback, as shown in Figure 13.5.

One application of feedback is to "unilateralize" the active device, that is, we choose feedback terminations that will result in $S_{12} = 0$ for the reduced two-port in Figure 13.5. A unilateral two-port has the advantages of being inherently stable and having its input and output ports isolated from each other, so that input and output matching networks can be independently designed.

It should be noted that equation (8.6.6) applies equally to either shunt or series feedback, depending on whether shunt or series three-port S-parameters are being used. In practice, shunt feedback is more often used to achieve unilateralization

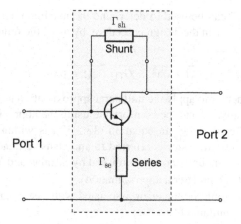

FIGURE 13.5

Application of shunt and series feedback to a transistor.

[5,6]. The design procedure for a single-stage feedback amplifier based on the unilateralization technique is illustrated in the following example.

Example 13.2 (Feedback amplifier design).

Problem. Design a 1.8 GHz amplifier based on the unilateralized BFP93 transistor in Example 8.6 using open-circuit stub matching networks at the input and output to match the amplifier to a 50 Ω system. Evaluate whether it is appropriate to use the unilateral approximation in the design.

Solution. The S-parameters of the BFR93 at bias conditions $V_{cc} = 5$ V and $I_c = 10$ mA were given in Example 8.6 as:

$$\begin{bmatrix} S_{11} & S_{12} \\ S_{21} & S_{22} \end{bmatrix} = \begin{bmatrix} 0.382\angle161.7 & 0.202\angle67 \\ 2.053\angle59.3 & 0.275\angle-53.5 \end{bmatrix}$$

The optimum passive shunt feedback termination that will minimize $|S_{12}|$ was calculated in Example 8.6 as $\Gamma_3 = 1\angle41.4°$. This is equivalent to a reactance of 2.646 Ω. At 1.8 GHz, this translates to an inductance of 23.4 nH. We also need to include a DC blocking capacitor to isolate the base and collector bias voltages, however. This capacitor should have a reactance of no more than one-tenth of the reactance of the feedback inductor, in order to avoid unduly influencing the designed feedback impedance. If we choose a capacitive reactance of around 1 Ω at 1.8 GHz, it means we need a capacitor value of around 88 pF.

Applying this shunt feedback network to the transistor results in a reduced two-port with the following S-parameters at 1.8 GHz:

$$\begin{bmatrix} S_{11} & S_{12} \\ S_{21} & S_{22} \end{bmatrix} = \begin{bmatrix} 0.108\angle136.2° & 0.026\angle-8.0° \\ 2.969\angle70.0° & 0.904\angle-18.6° \end{bmatrix}$$

We can apply equation (13.2.36) to determine U for the "unilateralized" transistor, giving:

$$U = \frac{|S_{11}S_{12}S_{21}S_{22}|}{(1 - |S_{11}|^2)(1 - |S_{22}|^2)}$$

$$= \frac{0.108 \times 0.026 \times 2.969 \times 0.904}{(1 - 0.01166) \times (1 - 0.8172)}$$

$$= \frac{0.0075}{0.1807}$$

$$= 0.042$$

Applying condition (13.2.37) indicates that the expected error in applying the unilateral approximation is around 4%. We therefore conclude that using the unilateral approximation is justified in this case.

We can calculate the maximum Unilateral Transducer Power Gain, $G_{TU_{max}}$, using equation (13.2.24).

$$G_{TU_{max}} = \frac{1}{|1 - |S_{11}|^2|} |S_{21}|^2 \frac{1}{|1 - |S_{22}|^2|}$$

$$= \frac{1}{(1 - 0.01166)} 8.815 \frac{1}{(1 - 0.8172)}$$

$$= 48.8 \ (16.9 \ \text{dB})$$

Since we have unilateralized the transistor, the input and output reflection coefficients, Γ_{in} and Γ_{out}, shown in Figure 13.6 reduce to S_{11} and S_{22}, respectively. The input and output matching networks therefore simply need to match S_{11} and S_{22} to 50 Ω. Using the relevant formulas from Section 10.3.1, or the graphical techniques outlines in Section 10.3.2, we obtain the following solutions:

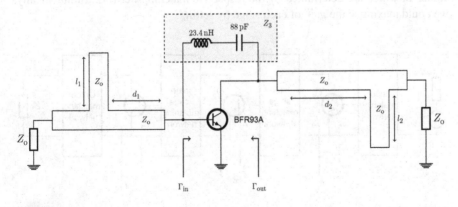

FIGURE 13.6

Transistor feedback amplifier with input and output stub matching networks.

Solution	Input Matching Network	Output Matching Network
1	$d_1 = 0.364\,\lambda$ $l_1 = 0.034\,\lambda$	$d_2 = 0.479\,\lambda$ $l_2 = 0.213\,\lambda$
2	$d_1 = 0.131\,\lambda$ $l_1 = 0.466\,\lambda$	$d_2 = 0.409\,\lambda$ $l_2 = 0.287\,\lambda$

13.4 MULTISTAGE AMPLIFIERS

The gain of the single-stage amplifiers so far discussed is ultimately limited by the parameters of the transistor used. The conventional approach to increasing the gain above that of a single-stage is to cascade multiple stages. In order to achieve maximum power transfer between the stages, interstage matching networks are required, as shown in Figure 13.7, that uses a BJT for illustration.

Ideally, with perfect lossless matching networks, the gain of a multistage amplifier is the product of the gains of each stage. The overall numeric gain of the amplifier in Figure 13.7 is, therefore,

$$G_{\text{tot}} = G_{Q_1} \cdot G_{Q_2} \cdot G_{Q_3} \tag{13.4.1}$$

If the gains are expressed in dB then equation (13.4.1) can be written as:

$$G_{\text{tot(dB)}} = G_{Q_1\text{(dB)}} + G_{Q_2\text{(dB)}} + G_{Q_3\text{(dB)}} \tag{13.4.2}$$

In Chapter 7, we showed that the power gain of an amplifier is a function of the source and load terminations. In the context of Figure 13.7, this means that the individual stage gains in equations (13.4.1) and (13.4.2) will depend on the source and load terminations presented by the respective interstage matching networks (or input and output matching networks in the case of the first and last stages). Each transistor, Q_n, in Figure 13.7 sees a source and load termination Γ_{S_n} and Γ_{L_n}, the values of which are determined by the respective matching network. Simplistically, we could maximize the gain of each stage by setting:

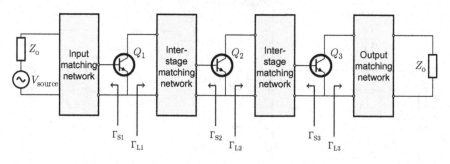

FIGURE 13.7

Conceptual multistage amplifier.

$$\Gamma_{S_n} = \Gamma_{ms_n} \tag{13.4.3}$$

and

$$\Gamma_{L_n} = \Gamma_{ml_n} \tag{13.4.4}$$

where Γ_{ms_n} and Γ_{ml_n} are the optimum terminations for that particular transistor, defined by equations (7.5.24) and (7.5.25). This assumes, of course, that the device in question is unconditionally stable, as defined in Section 7.4.3.

The design of the interstage matching networks is therefore critically important to the performance of the overall amplifier and, consequently, a large body of work exists on this subject. In addition to fulfilling the gain specification, the design of the interstage matching circuitry must take into account several other requirements, such as:

- *Impedance matching*: As a general rule we seek to achieve maximum power transfer between each stage of a multistage amplifier. Exceptions to this rule are when we need to achieve minimum noise figure (as explained in Chapter 14) or when doing so would lead to instability of the overall amplifier.
- *DC isolation*: It is often the case that the output port of one stage is at a different DC potential to the input of the next, due to different bias requirements at base and collector (or gate and drain) terminals of a transistor. Each stage will therefore need to have DC isolation from the preceding and following stages. This is usually accomplished by means of a DC blocking capacitor. Depending on the capacitor value, its reactance may have to be incorporated in the AC analysis.
- *Frequency response*: Attempting to satisfy the above two requirements will inevitably impact on a third-important consideration, namely the frequency response of the overall amplifier. We often seek the maximum gain over the widest possible bandwidth and one approach is to eliminate any interstage coupling capacitors, which we would otherwise use for DC blocking. Such "direct coupled" amplifiers are more challenging to design.

A key figure of merit which is particularly important for multistage amplifiers is the *Gain-Bandwidth Product* or "GBP," which is defined as the product of the amplifier's 3 dB bandwidth and the nominal gain (i.e., the gain at the center of the passband). Since gain is a dimensionless ratio (not dB), the GBP is expressed in units of Hz. For the purposes of this analysis, we shall assume that the amplifier has a first-order frequency response, described by:

$$A(\omega) = \frac{A_0}{\sqrt{1 + \left(\frac{\omega}{\omega_c}\right)^2}} \tag{13.4.5}$$

where A_0 is the nominal or "passband" gain of the amplifier and ω_c is the corner frequency of the first-order response, that is, the 3 dB frequency. We can show that the

GBP is approximately constant, in other words gain and bandwidth can be "traded-off" against each other. The proof of constant GBP for an amplifier is as follows, starting with the definition of GBP at a frequency, ω:

$$GBP = A(\omega) \cdot \omega \tag{13.4.6}$$

$$= \frac{A_0}{\sqrt{1 + \left(\frac{\omega}{\omega_c}\right)^2}} \cdot \omega \approx \frac{A_0}{\sqrt{\left(\frac{\omega}{\omega_c}\right)^2}} \cdot \omega \tag{13.4.7}$$

$$= A_0 \cdot \omega_c \tag{13.4.8}$$

Since both A_0 and ω_c are constants, it follows that $A_0 \cdot \omega_c$ must be a constant. For transistors, the current GBP is the same as the transition frequency, f_T, that was defined for BJTs and field-effect transistors (FETs) in equations (12.2.12) and (12.5.14), respectively. Transistors are usually operated at frequencies well below their f_T in order to obtain useful power gain. In bipolar transistors, f_T varies with collector current, I_c, reaching a maximum at a particular value of I_c.

So, when designing multistage amplifiers, we encounter the problem of *bandwidth shrinkage*. This is due to the fact that the gain versus frequency transfer functions are multiplied, increasing the overall gain but also increasing the steepness of the gain roll-off. In simple terms, when two identical amplifier stages, each having a 20 dB per decade roll-off characteristic, are cascaded the resultant cascade will have a 40 dB per decade roll-off characteristic. The −3 dB bandwidth of the cascade will therefore be less than that of the individual stages. We therefore need to be aware that, as we increase the gain by adding more stages, we will inevitably be sacrificing bandwidth.

Consider a multistage amplifier, such as that shown in Figure 13.7, generalized to n identical stages, each having a voltage gain A and a 3 dB bandwidth B.

Let us again assume a "first-order" frequency dependence for the individual stages, so that the frequency dependence of amplifier gain magnitude for each stage is represented by:

$$A(f) = \frac{A_0}{\sqrt{1 + \left(\frac{f}{B}\right)^2}} \tag{13.4.9}$$

For the overall amplifier consisting of n such stages in cascade we have, from an extension of equation (13.4.1) to n stages:

$$A_t(f) = \left[\frac{A_0}{\sqrt{1 + \left(\frac{f}{B}\right)^2}} \right]^n \tag{13.4.10}$$

With an amplifier having a first-order response, the overall 3 dB bandwidth of the amplifier is the frequency at which the voltage gain falls to $1/\sqrt{2}$ of its nominal value, A_t, that is, where $A_t(f) = A_t/\sqrt{2}$. At this frequency, which we will call B_t, and based on equation (13.4.10) we can write:

$$\frac{A_t}{\sqrt{2}} = \left[\frac{A_o}{\sqrt{1 + \left(\dfrac{B_t}{B}\right)^2}} \right]^n \tag{13.4.11}$$

Given that $A_t = A_o^n$, we can derive the bandwidth of the overall amplifier in terms of the individual stage bandwidths from equation (13.4.11) as:

$$B_t = B\sqrt{2^{1/n} - 1} \tag{13.4.12}$$

A cascade of n identical amplifier stages will therefore have a bandwidth less than that of an individual stage by a factor of $\sqrt{2^{1/n} - 1}$. The GBP for each individual stage can thus be written as:

$$\text{GBP}_S = A_o \times B = \left(\frac{B_t}{\sqrt{2^{1/n} - 1}} \right) \cdot (A_t)^{1/n} \tag{13.4.13}$$

An interesting question arising from equation (13.4.13) is, what is the optimum number of stages which will be required to achieve the maximum overall GBP specification for a cascade of identical amplifier stages? Jindal [7] answered this question by differentiating equation (13.4.13) with respect to n and obtaining the optimum number of stages, n_{opt}, as a function of the overall gain, A_t, as follows:

$$n_{\text{opt}} = \frac{\ln(2)}{-\ln\left[-\dfrac{\ln(2)}{2\ln(A_t)} \right]} \tag{13.4.14}$$

If $A_t \gg \sqrt{2}$ we can approximate equation (13.4.14) to:

$$n_{\text{opt}} = 2\ln(A_t) \tag{13.4.15}$$

Jindal [7] gives the example of a multistage amplifier having a gain $A_t = 100$, for which equation (13.4.15) gives $n_{\text{opt}} = 9.2$. If we required the overall amplifier to have a bandwidth of 880 MHz, we find that the optimum value for the GBP for each individual stage, from equation (13.4.13) is 5.2 GHz.

Example 13.3 (Multistage amplifier design).

Problem. You are required to design a broadband multistage amplifier comprising a number of identical stages and having a total GBP of 20 GHz. The overall amplifier gain should be 30 dB. Determine the number of stages required, and the necessary gain and bandwidth of the individual stages.

Solution. First, we need to convert 30 dB into a numerical voltage gain:

$$30 = 20 \log_{10}(A_t)$$
$$A_t = 10^{(30/20)} = 31.63$$

The optimum number of stages is then given by equation (13.4.15):

$$n_{opt} = 2 \ln(31.63) \approx 7 \tag{13.4.16}$$

Using equation (13.4.13) we calculate the GBP of each stage as:

$$\text{GBP}_S = \left(\frac{20}{\sqrt{2^{1/7} - 1}} \right) \cdot (31.63)^{1/7}$$
$$\approx 102 \text{ GHz}$$

Given that the gain of each stage is $31.63^{1/7} = 1.638$ (i.e., 4.3 dB), we can calculate the required bandwidth of each stage as:

$$\text{BW}_S = \frac{102}{1.638} \approx 62 \text{ GHz} \tag{13.4.17}$$

13.4.1 FUNDAMENTAL LIMITS ON THE BANDWIDTH OF INTERSTAGE MATCHING NETWORKS

All matching networks are essentially filter structures, and interstage matching network design is closely related to passive filter design, which we have not covered in this book but is very well covered elsewhere [8,9]. When we consider the design of broadband interstage matching networks, it is worth considering whether there are any fundamental limits on the achievable match and what they might be. Theoretical limits of bandwidth that can be achieved when matching an arbitrary source to different types of load were first proposed by Bode and Fano in the 1930s, who found that the integral of return loss is bound by a constant [10,11]. The so-called *Bode-Fano Criteria* set a theoretical limit on the minimum reflection magnitude, in other words the degree of matching, for a lossless matching network terminated in an arbitrary load [8]. The criterion provides the theoretical upper limit of performance that can be obtained. This gives us a benchmark and allows us to trade off between reflection coefficient, bandwidth, and network complexity.

The Bode-Fano criteria for simple *RC* and *RL* loads matched with a passive, lossless matching network are shown in Figure 13.8.

If it were possible to build a matching network having the perfectly rectangular reflection coefficient response of Figure 13.9(a), the Bode-Fano limits for the four simple load circuits shown in Figure 13.8 would be:

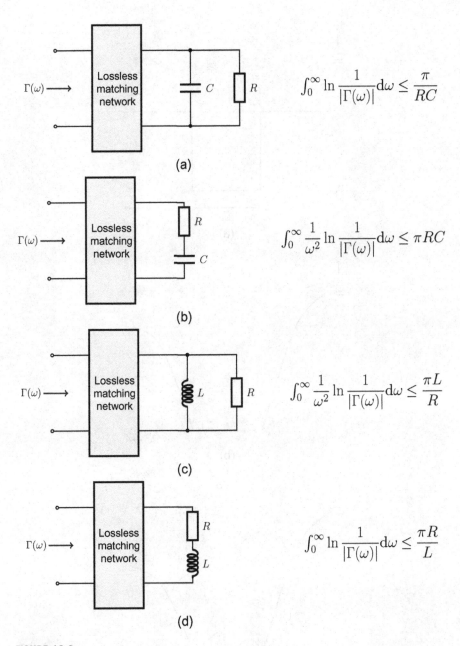

FIGURE 13.8

Bode-Fano criteria for four types of simple load impedance. (a) Matching parallel *RC* loads. (b) Matching series *RC* loads. (c) Matching parallel *LR* loads. (d) Matching series *LR* loads.

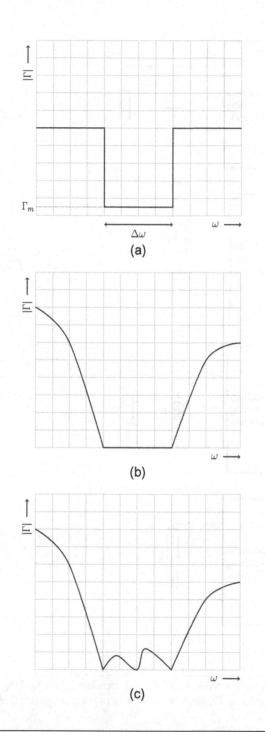

FIGURE 13.9

Bode-Fano matching illustration. (a) Idealized matching network response. (b) Nonrealizable matching network response. (c) Realizable matching network response.

$$\frac{\Delta\omega}{\omega_0}\frac{1}{\Gamma_{\text{avg}}} \leq \frac{\pi}{R(\omega_0 C)} \qquad (13.4.18)$$

$$\frac{\Delta\omega}{\omega_0}\frac{1}{\Gamma_{\text{avg}}} \leq \frac{\pi(\omega_0 L)}{R} \qquad (13.4.19)$$

$$\frac{\Delta\omega}{\omega_0}\frac{1}{\Gamma_{\text{avg}}} \leq \frac{\pi R}{(\omega_0 L)} \qquad (13.4.20)$$

$$\frac{\Delta\omega}{\omega_0}\frac{1}{\Gamma_{\text{avg}}} \leq \pi R(\omega_0 C) \qquad (13.4.21)$$

where $\Delta\omega$ is the width of the passband, Γ_{avg} is the average absolute value of the reflection coefficient looking into the matching network in the passband, and ω_0 is the center frequency of the passband. The quantity $\Delta\omega/\omega_0$ is the fractional bandwidth of the matching network. The way to interpret equation (13.4.18), for example, is that the area under the $\ln(1/\Gamma)$ can never exceed π/RC. Any increases in the bandwidth of the matching network can therefore only be achieved at the expense of less power transfer, in other words a poorer quality of match in the passband.

The Bode-Fano limits can be written in terms of reactance/susceptance for a generalized load impedance/admittance, respectively, as follows:

$$\frac{\Delta\omega}{\omega_0}\frac{1}{\Gamma_{\text{avg}}} \leq \frac{\pi G}{B} \qquad (13.4.22)$$

$$\frac{\Delta\omega}{\omega_0}\frac{1}{\Gamma_{\text{avg}}} \leq \frac{\pi R}{X} \qquad (13.4.23)$$

where G is the load conductance, B is the load susceptance, R is the load resistance, and X is the load reactance.

The load that was initially considered by Bode was the simple parallel RC circuit shown in Figure 13.8(a). In this case, the match performance limit can be described in terms of load-Q, as follows:

$$\frac{\Delta\omega}{\omega_0}\frac{1}{\Gamma_{\text{avg}}} \leq \frac{\pi}{Q} \qquad (13.4.24)$$

It turns out that equation (13.4.24) applies equally to the other prototype loads in Figure 13.8. equation (13.4.24) suggests that the more reactive energy that is stored in a load, the narrower the bandwidth of a match. The higher the Q is, the narrower the bandwidth of the match for the same average in-band reflection coefficient. Accordingly, it will be much harder to design the matching network to achieve a specified matching bandwidth. Only when the load is purely resistive can a match over all frequencies be found.

The ideal situation is to maintain $1/\Gamma$ constant over the frequency range of interest, $\Delta\omega$, and have it equal to one outside this range, as shown in Figure 13.9(a). A matching profile like that of Figure 13.9(a) is unrealizable as it would require an infinite number of elements in the matching network [8]. Even a more realistic response like that shown in Figure 13.9(b) is unrealizable as it is not possible in practice to achieve a perfectly flat passband response. Any realizable matching network will have a response resembling the one shown in Figure 13.9(c), where neither the passband or stopband response are "ideal."

The implications of the Bode-Fano criterion are as follows:

1. For a given load (e.g., a fixed RC product) broader bandwidth ($\Delta\omega$) can be achieved only at the expense of higher reflection coefficient in the passband.
2. The reflection coefficient in the passband cannot be zero unless $\Delta\omega = 0$. Thus a perfect match can be achieved only at a spot frequency.
3. As R and/or C increases, the quality of the match ($\Delta\omega$ and/or $1/\Gamma_m$) must decrease. Thus higher-Q circuits are intrinsically harder to match than are lower-Q circuits.

The results obtained above for the simple high-pass/low-pass prototype loads in Figure 13.8 can be extended to the two-element bandpass cases by well-known lowpass to bandpass transformations [8].

13.5 BROADBAND AMPLIFIERS

In the previous section, we looked at the essential trade-off between gain and bandwidth in multistage amplifiers, that is, as we cascade more and more stages to achieve higher gain, the overall bandwidth of the amplifier inevitably "shrinks." In the ongoing quest for wider and wider bandwidths, a number of strategies have been developed to overcome the limitations of the conventional multistage amplifier approach exemplified by Figure 13.7.

Three broadband amplifier topologies are particularly well suited to monolithic microwave integrated circuit (MMIC) implementation, so we will focus on these as follows [12]:

1. *Lossy matched amplifier (LMA)*: Where lossy elements (i.e., resistors) are used to achieve good input, output, and interstage matching across a wide range of frequencies.
2. *Feedback amplifiers*: Where negative feedback is applied to extend the bandwidth. Both shunt and series feedback configurations have been used in this context but shunt feedback usually yields the best results.
3. *Distributed amplifiers*: Which employ the concept of "additive amplification" and offer ultra-broadband operation that can extend from DC up to the cut-off frequency of the active devices.

These three approaches will be discussed in turn in the following sections.

13.5.1 LOSSY MATCHED AMPLIFIERS

An Lossy Matched Amplifier (LMA) uses resistors within its matching networks to enable flat gain to be achieved over a broad bandwidth.

The most common topology employs resistors in series with high impedance stubs on both the input and output, as shown in Figure 13.10(a) [13], where an FET has been used for example and the matching networks consist of a mixture of resistors and transmission line stubs. At low frequencies, the stubs have little reactance, so the resistors form the major part of the load on the transistor, lowering its gain. At higher frequencies, the reactance of the stubs increases (tending to infinity when the stubs are a quarter-wavelength long), and the resistors have less effect on the transistor gain. Hence, the matching networks can introduce a positive gain slope to compensate the transistor's gain roll-off, without resorting to mismatching.

Lossy match amplifiers have moderate gain and acceptable gain flatness. The use of resistors also enhances the stability of the amplifier at low frequencies. The downside of the lossy matching approach, compared with reactively matched amplifiers, are lower gain, lower output power, and higher noise figure, as the resistors add thermal noise.

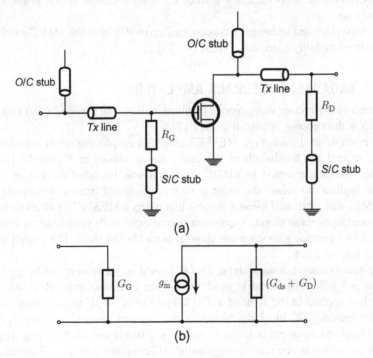

FIGURE 13.10

Lossy matching of an FET amplifier. (a) Basic lossy matching topology. (b) Low-frequency model.

Niclas [14,15] uses the low-frequency model of a lossy match FET amplifier, shown in Figure 13.10(b), to illustrate lossy matching. From Figure 13.10(b) it can be seen that:

$$S_{11} = \frac{1 - G_G Z_0}{1 + G_G Z_0} \tag{13.5.1}$$

$$S_{22} = \frac{1 - (G_{ds} + G_D)Z_0}{1 + (G_{ds} + G_D)Z_0} \tag{13.5.2}$$

and

$$\text{Gain} \approx \left[\frac{g_m Z_0}{2}(1 + S_{11})(1 + S_{22}) \right] \tag{13.5.3}$$

where G_{ds} is the drain-source conductance ($1/R_{ds}$) and G_D is the lossy match drain loading conductance. From equation (13.5.3), it can be seen that if $G_G = (G_{ds} + G_D) = 1/Z_0$ (i.e., the gate loading resistor is 50 Ω) and the drain loading resistor in parallel with R_{ds} gives 50 Ω, then $S_{11} = 0$, $S_{22} = 0$ and the gain (in dB) is equal to $20\log((g_m Z_0)/2)$, which gives typically 8 dB for an 800 μm metal semiconductor FET (MESFET). This low-frequency model clearly shows how the gate-width of the FET determines the low-frequency gain of the lossy match amplifier in the ideally matched case.

The lossy matched technique has been used up to 90 GHz with MMIC employing high electron mobility transistors (HEMTs) [16].

13.5.2 BROADBAND FEEDBACK AMPLIFIERS

It is generally understood that negative feedback will extend the bandwidth of a given amplifier at the expense of overall gain [2,17].

An example of a multistage MESFET amplifier employing shunt voltage feedback to extend the bandwidth of the first stage is shown in Figure 13.11. This amplifier was implemented in MMIC form and was intended for use in optical receiver applications where the input is current generated from a photodiode [18]. The MMIC was fabricated using a process that offers a MESFET f_T of 20 GHz. The circuit comprises three stages, a common source input with voltage shunt feedback followed by a cascade gain stage and then an output buffer stage. The overall MMIC area is 2 mm by 3 mm.

The first common source stage, Q_1, is biased by providing a voltage to the gate via a 5 kΩ resistor and is used to provide moderate gain. Resistive shunt feedback is applied in the form of a 220 Ω resistor. A 30 pF metal-insulator-metal capacitor provides DC blocking, to isolate the gate and drain bias. The purpose of this feedback arrangement is solely to extend the bandwidth of the stage. The first stage is connected to the second stage using AC coupling with an LC circuit. This LC combination is used as a filter to shape the frequency response of the amplifier. The second stage is a cascade stage, made of two transistors, Q_2 and Q_3. Q_2 is a common source MESFET loaded by the common gate MESFET, Q_3. The cascade

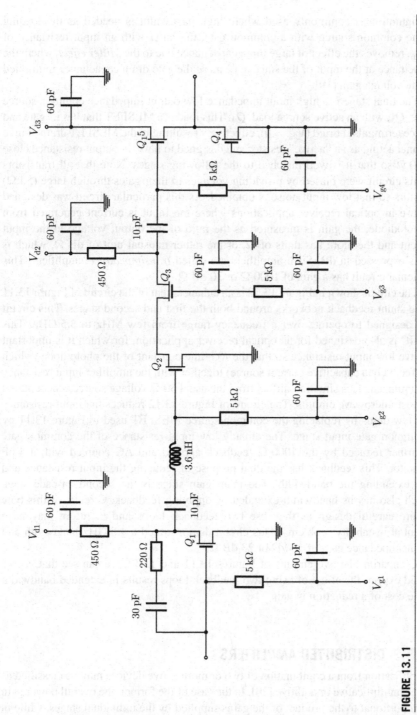

FIGURE 13.11

Common source (2.4 Gbit s⁻¹) MMIC feedback amplifier.

configuration is commonly used where high bandwidth is needed as the loading of the common source with a common gate, the latter with an input resistance of $1/g_m$, removes the effect of large input capacitance due to the *Miller effect*, where the capacitance at the input of the stage appears as the gate drain capacitance multiplied by the voltage gain [19].

The final stage is a high input impedance-low output impedance common source stage, Q_5, with an active source load, Q_4. This load is a MESFET that has its gate and source terminals shorted ($V_{gs} = 0$), effectively resulting in the MESFET drain-source channel acting as an "active" resistor and designed to have an output resistance close to 50 Ω so that it is well matched to the following stages. Note that all transistors in this circuit were biased by providing voltages to their gates through large (5 kΩ) resistors so that low input noise is obtained. As this particular circuit was designed for use in optical receiver applications where the input is current generated from a photodiode, the gain is measured as the ratio of the output voltage to the input current and therefore has units of Ω, or the rather unusual unit of dB Ω, which is ohms expressed in dB. Such amplifiers are called *transimpedance* amplifiers. This particular circuit has a gain of 1000 Ω or 30 dB Ω.

The circuit shown in Figure 13.12 is an enhancement of the circuit of Figure 13.11 using shunt feedback networks around both the first and second stages. This circuit was designed to operate over a frequency range from few MHz to 3.5 GHz. This MMIC is also designed for an optical receiver application, for which it is important to have low input resistance so that the *RC* time constant of the photodiode (which is effectively a capacitive current source) together with the amplifier input resistance is minimized. This situation differs from the usual 50 Ω voltage sources encountered in most microwave circuits. The circuit of Figure 13.12 reduces the input resistance to a few ohms by replacing the common source MESFET used in Figure 13.11 by a common gate input stage. The already low input resistance of the common gate is further reduced by the 1000 Ω feedback applied and AC coupled with a 4 pF capacitor. This feedback has the dual purpose of reducing the input resistance and also extending the bandwidth. The main gain stage is the second cascade stage which also has its bandwidth extended by applying feedback. Circuits of this type require careful design as they use two feedback loops and therefore may have potential instability problems. This circuit demonstrated a 4.8 GHz bandwidth and a transimpedance gain of 180 Ω or 23 dB Ω.

Comparing the two circuits of Figures 13.11 and 13.12, we can see that, as we would expect, the effect of using two feedback loops results in extended bandwidth at the cost of a reduction in gain.

13.5.3 DISTRIBUTED AMPLIFIERS

Amplification from a combination of two or more active devices may be classified as either multiplicative or additive [20]. In the case of the former, the overall power gain is proportional to the product of the gains supplied by the individual stages, while in

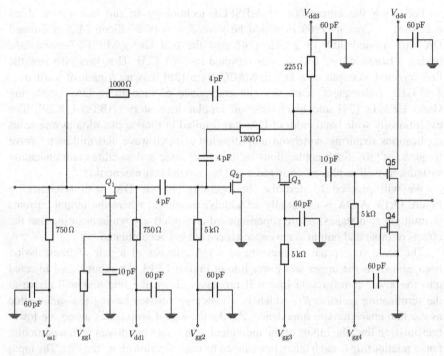

FIGURE 13.12

Common gate (5 Gbit s^{-1}) MMIC feedback amplifier.

the latter case it is proportional to the sum of the powers contributed by the individual active devices.

The vast majority of multistage amplifiers, and all those discussed in this chapter so far, make use of the multiplicative process through cascading. In this section, we will introduce an entirely different multistage amplifier architecture called the *Distributed Amplifier (DA)* [21] in which the output powers of the individual stages are combined additively. While in most practical applications this approach produces less gain per device than the multiplicative approach, it yields significant bandwidth benefits.

The concept of distributed amplification was first proposed by Percival in 1936, who introduced this novel idea in a patent specification relating to the design of thermionic valve circuits [21]. The term "distributed amplifier" was actually coined by Ginzton et al., in their 1948 paper [22] on such circuits. Prior to the advent of the transistor in 1948, vacuum tubes were employed as the active devices in DAs. One of the earliest reports of a DA implemented using transistors was made by Enloe and Rogers [23], who reported two simple DA circuits based on BJT technology.

Following the emergence of MESFET technology in the late 1960s, Jutzi published a paper in 1969 in which he gave details of a silicon MESFET-based DA with a bandwidth of 2 GHz [24], and the first GaAs MESFET-based DA, having a bandwidth of 6 GHz, was reported in 1981 [25]. That year also saw the first reported example of a DA in MMIC form [26], having a gain of 9 dB over 1-13 GHz. Subsequent years saw the emergence of monolithic DAs employing GaAs HEMTs [27] and heterojunction bipolar transistors (HBTs) [28,29]. The exceptionally wide bandwidth of DAs has resulted in their application in numerous applications requiring wideband amplification at microwave and millimeter wave frequencies [30]. Such applications include fiber optic and satellite communication systems, as well as phased array radar and broadband instrumentation.

We will proceed to describe the operation of the DA, with reference to Figure 13.13. A DA is essentially an additive amplifier, where the output currents of multiple gain stages, A, are superimposed constructively while ensuring that the effects of input and output shunt capacitances are not accumulated.

The basic conceptual architecture of a DA consists of a pair of transmission lines, shown as the upper and lower line in Figure 13.13. An input signal injected into the lower transmission line will propagate down the line and will arrive at the terminating resistor, R_1, which is a matching resistor having the same value as the line characteristic impedance, Z_0. As the signal propagates along the lower transmission line, the inputs of the individual gain stages are driven with a particular phase relationship to each other, determined by their position along the line. The input signal is amplified by each gain stage, and the stage outputs are combined coherently (i.e., with their phase relationships preserved) in the upper transmission line. The propagation characteristics of the upper and lower transmission lines in Figure 13.13 must be designed to be equal so as to ensure that output signals from each individual gain stage sum in phase.

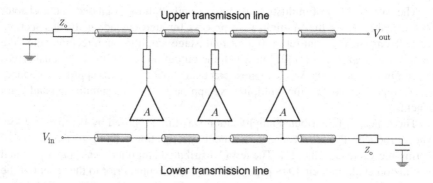

FIGURE 13.13

Conceptual DA architecture.

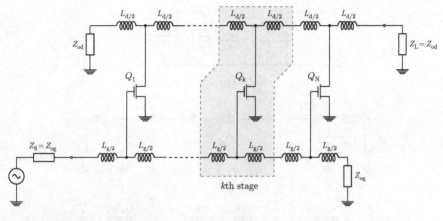

FIGURE 13.14

Basic AC circuit schematic of an *N*-stage MESFET DA.

The gain stages in Figure 13.13 can be implemented as single transistor amplifiers, as shown in Figure 13.14, which shows the basic AC circuit of a DA with *N* identical transistor amplifier stages. The active devices are shown in Figure 13.14 are MESFETs or HEMTs, but the following description applies equally to bipolar transistors (BJT or HBT).

A single DA section is comprised of four inductors plus an active device as indicated in the shaded section of Figure 13.14. The simplified small signal equivalent circuit of each transistor is shown in Figure 13.15(a), and is based on the intrinsic part of the equivalent circuit in Figure 12.9 of Chapter 12, with the omission of the input resistance, R_i, the output drain-source resistance, R_{ds}, and the gate-drain feedback capacitance, C_{gd}. The omission of these components do not materially affect our analysis of distributed amplified operation. The important equivalent circuit parameters are C_{gs}, the intrinsic gate-source capacitance, and C_{ds}, the drain-source capacitance.

Figure 13.15(b) and (c) shows the simplified small-signal equivalent circuit models for the input and output portions of an MESFET DA, respectively. The current source in parallel with C_{ds} in Figure 13.15(c) represents the intrinsic current that is generated by an MESFET through its gain mechanism. g_m is the complex transconductance which, when multiplied by the input voltage v_{gs} developed across C_{gs}, gives the value of the drain current produced internally by the MESFET. L_g and L_d are the total inductances present between the input base and output collector terminals of adjacent transistors, respectively.

We demonstrated in Chapter 2 that an ideal uniform lossless transmission line can be represented by an equivalent electrical model consisting of a distributed total series inductance ΔL and a distributed shunt capacitance ΔC [31], as shown in Figure 13.16. The quantities ΔL and ΔC are termed "distributed" because they

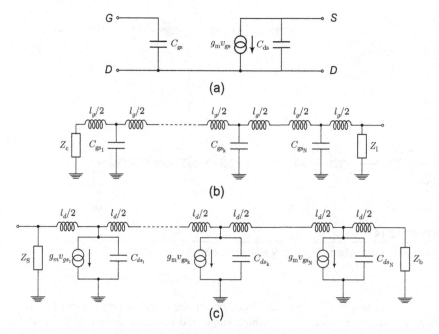

(a)

(b)

(c)

FIGURE 13.15

Simplified small-signal equivalent circuit models of MESFET-based DA artificial transmission lines. (a) Individual MESFET equivalent circuit. (b) Gate-line equivalent circuit. (c) Drain-line equivalent circuit.

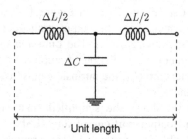

Unit length

FIGURE 13.16

Unit length element of an ideal uniform lossless real transmission line.

are defined per unit length of transmission line. If we compare Figure 13.15 with Figure 13.16 we notice that the input and output portions of a DA each resemble the equivalent circuit model of a lossless transmission line. Consequently, a direct analogy can be established between the two circuit portions of a DA and a transmission line where the transistor parasitic capacitances, C_π and C_{ce}, are incorporated into what is, in effect, an *artificial transmission line* (ATL). A DA can essentially be viewed as a pair of transmission lines, coupled via the active devices and thus possess amplifying properties.

When field effect devices are utilized in a DA, the input and output ATLs are known as the gate-line and the drain-line, respectively. Conversely, when bipolar devices are employed, the input and output ATLs are referred to as the base line and the collector line, respectively.

An important consequence of the transistor input and output capacitances being incorporated into an ATL is that they are not lumped together at the input and output ports, as they would be if the transistors were simply connected in parallel. A parallel connection would increase the total capacitance and thereby reduce the bandwidth of the amplifier, whereas the distributed approach actually makes use of these capacitances. Consider an ideal real transmission line (RTL) that has an inherently wide bandwidth and possesses low-pass filtering characteristics. The upper bandwidth cut-off frequency of a lossless RTL is determined solely by the values of ΔL and ΔC. Due to their transmission line-like nature, DAs possess the key attributes of RTLs. In particular, an ATL cut-off frequency depends only upon the amount of inductance and capacitance present per DA section and not upon the actual number of sections employed. This is in contrast to a conventional amplifier in which the upper cut-off frequency decreases with the number of cascaded gain stages.

Ideally, the gain of a DA can be made larger by increasing the number of amplifier sections, while bandwidth is preserved since it is fixed by the cut-off characteristics of the ATLs. By virtue of the independence of the amplifier properties that set bandwidth and gain, the GBP of a DA may even exceed f_T of the active devices themselves.

Having ignored R_i and R_{ds} in the MESFET equivalent circuits, the gain of the amplifier is given by [32]:

$$G = \frac{1}{4}N^2 g_m^2 Z_{og} Z_{od} \tag{13.5.4}$$

where N is the total number of stages and Z_{og} and Z_{od} are the characteristic impedances of the gate and drain ATLs, respectively, and are therefore the values of the terminating impedances employed in Figure 13.14. These characteristic impedances are given by:

$$Z_{og} = \sqrt{\frac{L_g}{C_{gs}}} \tag{13.5.5}$$

$$Z_{od} = \sqrt{\frac{L_d}{C_{ds}}} \tag{13.5.6}$$

Equation (13.5.4) would seem to suggest that it is possible to increase the amplifier gain by simply increasing the number of stages. The losses introduced by R_i and R_{ds}, however, which we have ignored in this analysis so far, result in an optimum number of stages to maximize the gain for a given active device [32]. In theory, the gain of a DA should remain flat up to the f_T of the active devices, provided that the cut-off frequency of the ATLs is made much higher than the device f_T. The cut-off frequencies of the gate and drain ATLs are given by:

$$f_{cg} = \left(\frac{1}{\pi \sqrt{L_g C_{gs}}} \right) \tag{13.5.7}$$

$$f_{cd} = \left(\frac{1}{\pi \sqrt{L_d C_{ds}}} \right) \tag{13.5.8}$$

A consequence of the equations (13.5.5) to (13.5.8) is that once the characteristic impedance of the ATL has been chosen then the cut-off frequencies cannot be chosen independently. Up till now we have considered the use of matching networks to interconnect circuits having different characteristic impedances. In the case of DAs, which are by their nature extremely broadband, the characteristic impedance of the ATLs that make up the DA are constrained by the need to match directly to preceding and subsequent circuits.

In this chapter, we have focused on conventional and broadband amplifiers, without regard to their noise performance. Consequently we have used resistors to effect the necessary matching, such as in the case of lossy matched and DAs. The disadvantage of using resistors is that they inevitably add thermal noise. In the next chapter, we will specifically address the issue of noise in electronic circuits and low-noise amplifier design.

13.6 TAKEAWAYS

1. In order to obtain the optimum gain from a microwave transistor, matching networks are required at the input and output ports of the device to match the required source and load terminations to the system characteristic impedance. The source and load terminations presented to the device are constrained by requirements of gain and stability, and may be otherwise constrained (as in the case of low-noise amplifiers).

2. Single-stage transistor amplifier design can be greatly simplified by assuming that the transistor is unilateral, that is, $S_{12} = 0$. The validity of this assumption can be quantified using a parameter called the unilateral figure of merit (U), which can be calculated from the device S-parameters and is invariant under lossless, reciprocal embeddings.

3. Feedback can be used to change the S-parameters of a given transistor so as to simplify the amplifier design and allow more degrees of freedom. For example, we can apply shunt feedback to reduce S_{12} to a minimum, allowing the unilateral assumption to be applied to the subsequent design.

4. If more gain is needed, this can be obtained by cascading single-stage amplifiers to form a multistage amplifier. This requires interstage matching networks to

ensure that the correct terminating impedances are presented to each stage, ensuring optimum power transfer between stages. The *Bode-Fano criteria* set the fundamental limits on the degree of match that can be achieved over a particular bandwidth.

5. There is an inevitable trade-off between gain and bandwidth in multistage amplifiers, but bandwidth can be optimized by the use of feedback or by such strategies as lossy interstage matching.

6. The *Distributed Amplifier (DA)* topology overcomes the limitations of conventional multistage amplifiers by adopting an additive rather than a multiplicative approach to combining the outputs of multiple stages.

REFERENCES

[1] R. Carson, High Frequency Amplifiers, John Wiley and Sons, New York, 1975, ISBN 0471137057.

[2] G. Gonzalez, Microwave Transistor Amplifiers, Analysis and Design, second ed., Prentice Hall Inc., Englewood Cliffs, NJ, 1997.

[3] S. Mason, Some properties of three-terminal devices, IRE Trans. Circuit Theory 4 (4) (1957) 330-332, ISSN 0096-2007, http://dx.doi.org/10.1109/TCT.1957.1086413.

[4] S. Scanlan, G. Young, Error considerations in the design of microwave transistor amplifiers, IEEE Trans. Microw. Theory Tech. 28 (1980) 1163-1168.

[5] M. Gupta, Power gain in feedback amplifiers, a classic revisited, IEEE Trans. Microw. Theory Tech. 40 (5) (1992) 864-879, ISSN 0018-9480, http://dx.doi.org/10.1109/22.137392.

[6] S. Mason, Power gain in feedback amplifiers, Trans. IRE Prof. Group Circuit Theory 1 (2) (1954) 20-25, ISSN 0197-6389, http://dx.doi.org/10.1109/TCT.1954.1083579.

[7] R. Jindal, Gigahertz-band high-gain low-noise AGC amplifiers in fine-line NMOS, IEEE J. Solid State Circuits 22 (4) (1987) 512-521, ISSN 0018-9200, http://dx.doi.org/10.1109/JSSC.1987.1052765.

[8] D. Pozar, Microwave Engineering, second ed., John Wiley and Sons Inc., New York, 1998.

[9] G. Matthaei, E. Jones, L. Young, Microwave Filters Impedance Matching Networks and Coupling Structures, McGraw-Hill, New York, 1961.

[10] R. Fano, Theoretical limitations on the broad-band matching of arbitrary impedances, J. Franklin Inst. 249 (1960) 57-83.

[11] H. Bode, Network Analysis and Feedback Amplifier Design, Bell Telephone Laboratories Series, D. Van Nostrand Company, Inc., New York, 1950.

[12] B. Virdee, A. Virdee, B. Banyamin, Broadband Microwave Amplifiers, Artech House Microwave Library, Artech House, Norwood, MA, 2004, ISBN 9781580538930.

[13] I. Robertson, S. Lucyszyn, Institution of Electrical Engineers, RFIC and MMIC Design and Technology, IEE Circuits, Devices and Systems Series, Institution of Engineering and Technology, 2001, ISBN 9780852967867.

[14] K. Niclas, On design and performance of lossy match GaAs MESFET amplifiers, IEEE Trans. Microw. Theory Tech. 30 (11) (1982) 1900-1907, ISSN 0018-9480, http://dx.doi.org/10.1109/TMTT.1982.1131341.

[15] K. Niclas, W. Wilser, R. Gold, W. Hitchens, The matched feedback amplifier: ultrawide-band microwave amplification with GaAs MESFET's, IEEE Trans. Microw. Theory Tech. 28 (4) (1980) 285-294, ISSN 0018-9480, http://dx.doi.org/10.1109/TMTT.1980.1130067.

[16] Y. Inoue, M. Sato, T. Ohki, K. Makiyama, T. Takahashi, H. Shigematsu, T. Hirose, A 90-GHz InP-HEMT lossy match amplifier with a 20-dB gain using a broadband matching technique, IEEE J. Solid State Circuits 40 (10) (2005) 2098-2103, ISSN 0018-9200, http://dx.doi.org/10.1109/JSSC.2005.854616.

[17] M. Halkias, Integrated Electronics, McGraw-Hill Electrical and Electronic Engineering Series, Tata McGraw-Hill Publishing Company, New Delhi, 1972, ISBN 9780074622452.

[18] I. Darwazeh, P. Lane, W. Marnane, P. Moreira, L. Watkins, M. Capstick, J. O'Reilly, GaAs MMIC optical receiver with embedded signal processing, IEE Proc. G Circuits Devices Syst. 139 (2) (1992) 241-243, ISSN 0956-3768.

[19] L. Moura, Error analysis in Miller's theorems, IEEE Trans. Circuits Syst. I Fundam. Theory Appl. 48 (2) (2001) 241-249, ISSN 1057-7122, http://dx.doi.org/10.1109/81.904891.

[20] K. Niclas, R. Pereira, The matrix amplifier: a high-gain module for multioctave frequency bands, IEEE Trans. Microw. Theory Tech. 35 (3) (1987) 296-306, ISSN 0018-9480, http://dx.doi.org/10.1109/TMTT.1987.1133642.

[21] T. Wong, Fundamentals of Distributed Amplification, Artech House, Norwood, MA, 1993, ISBN 0-89006-615-9.

[22] E. Ginzton, W. Hewlett, J. Jasberg, J. Noe, Distributed amplification, Proc. IRE 36 (8) (1948) 956-969, ISSN 0096-8390, http://dx.doi.org/10.1109/JRPROC.1948.231624.

[23] L. Enloe, P. Rogers, Wideband transistor distributed amplifiers, in: 1959 IEEE International Solid-State Circuits Conference—Digest of Technical Papers, vol. II, 1959, pp. 44-45, http://dx.doi.org/10.1109/ISSCC.1959.1157032.

[24] W. Jutzi, A MESFET distributed amplifier with 2 GHz bandwidth, Proc. IEEE 57 (6) (1969) 1195-1196, ISSN 0018-9219, http://dx.doi.org/10.1109/PROC.1969.7188.

[25] J. Archer, F. Petz, H. Weidlich, GaAs FET distributed amplifier, Electron. Lett. 17 (13) (1981) 433-433, ISSN 0013-5194, http://dx.doi.org/10.1049/el:19810303.

[26] Y. Ayasli, J. Vorhaus, R. Mozzi, L. Reynolds, Monolithic GaAs travelling-wave amplifier, Electron. Lett. 17 (12) (1981) 413-414, ISSN 0013-5194, http://dx.doi.org/10.1049/el:19810287.

[27] S. Bandy, C. Nishimoto, C. Yuen, R. Larue, M. Day, J. Eckstein, Z. Tan, C. Webb, G. Zdasiuk, A 2-20 GHz high-gain monolithic HEMT distributed amplifier, IEEE Trans. Electron Devices 34 (12) (1987) 2603-2609, ISSN 0018-9383, http://dx.doi.org/10.1109/T-ED.1987.23360.

[28] B. Nelson, C. Perry, R. Dixit, B. Allen, M. Kim, A. Oki, J. Camou, D. Umemoto, High-linearity, low DC power GaAs HBT broadband amplifiers to 11 GHz, in: 11th Annual Gallium Arsenide Integrated Circuit (GaAs IC) Symposium, Technical Digest, 1989, pp. 79-82, http://dx.doi.org/10.1109/GAAS.1989.69298.

[29] B. Nelson, D. Umemoto, C. Perry, R. Dixit, B. Allen, M. Kim, A. Oki, High-linearity, low DC power monolithic GaAs HBT broadband amplifiers to 11 GHz, in: IEEE 1990 Microwave and Millimeter-Wave Monolithic Circuits Symposium, Digest of Papers, 1990, pp. 15-18, http://dx.doi.org/10.1109/MCS.1990.110928.

[30] L. Samoska, An overview of solid-state integrated circuit amplifiers in the submillimeter-wave and THz regime, IEEE Trans. Terahertz Sci. Technol. 1 (1) (2011) 9-24, ISSN 2156-342X, http://dx.doi.org/10.1109/TTHZ.2011.2159558.

[31] J. Karakash, Transmission Lines and Filter Networks, Macmillan, New York, 1950.

[32] J. Beyer, S. Prasad, R. Becker, J. Nordman, G. Hohenwarter, MESFET distributed amplifier design guidelines, IEEE Trans. Microw. Theory Tech. 32 (3) (1984) 268-275, ISSN 0018-9480, http://dx.doi.org/10.1109/TMTT.1984.1132664.

Low-noise amplifier design

14

CHAPTER OUTLINE

INTENDED LEARNING OUTCOMES

- *Knowledge*
  - Understand the most important sources of electrical noise, such as thermal noise, shot noise, and flicker noise, and their characteristics.
  - Know the definition of noise factor and noise figure for a two-port network.
  - Understand the relationship between noise factor and effective noise temperature.
  - Understand the relationship between noise figure and source termination for a single-stage and multistage amplifiers.
  - Understand the basic principles of noise figure measurement and transistor noise characterization.

Microwave Active Circuit Analysis and Design. http://dx.doi.org/10.1016/B978-0-12-407823-9.00014-7

- *Skills*
  - Be able to design a single-stage microwave transistor amplifier having the minimum possible noise figure.
  - Be able to design a single-stage microwave transistor amplifier having a specified noise figure and gain.
  - Be able to calculate the overall noise figure of a receiver chain.
  - Be able to design a single-stage microwave transistor amplifier having the minimum noise measure, and thereby design a multistage amplifier having the minimum possible noise figure.
  - Be able to design a single-stage microwave transistor amplifier with a specified noise measure and gain, and thereby design a multistage amplifier having a specified noise figure.

14.1 INTRODUCTION

So far we have largely ignored the presence and effects of electrical noise in the circuits we have been designing. Noise is always present in any electronic or microwave system, however, and can often be the deciding factor in assessing system performance. In this chapter and Chapter 16, we will specifically address the problem of how to design amplifiers (this chapter) and oscillators (Chapter 16) for optimum noise performance.

The term "electrical noise" refers to unwanted signals that corrupt, mask, or interfere with the desired signal which is being processed by an electronic circuit. Noise adds to or modifies the wanted signal, resulting in degraded circuit and system performance. Noise sources are sometimes classified into two categories, intrinsic and extrinsic. Intrinsic noise sources are those that are inherent to the electronic devices in question and arise from fundamental physical effects. Examples of such noise sources are thermal (or Johnson) noise, electronic shot-noise, and $1/f$ noise. Extrinsic noise sources are those that arise outside the circuit in question or from interactions between the circuit and the surrounding environment. In this chapter, we will address the question of how noise can be minimized by careful design. We will therefore be concerning ourselves exclusively with intrinsic noise sources. It is important to emphasize that intrinsic noise does not arise from faulty contacts or any other spurious effects caused by poor construction of the circuit, but is an inherent property of the components themselves [1].

We start by briefly introducing the three types of noise and their properties. What is special about electronic noise is that it results from random processes and is therefore best described statistically. In this book, we are mainly concerned with design aspects, so we will not delve into the detailed mathematics of noise theory, which is well covered elsewhere [2,3]. Our focus is on modeling the noise performance of microwave amplifiers so as to arrive at a set of design tools.

We will then explore the relationship between the noise figure of a given amplifier stage and the value of the input termination presented to the device, and we will show that the input matching network can be designed specifically to minimize the amount of noise added by the amplifier.

Since the input termination which minimizes the noise figure is in general not equal to that which maximizes the gain, there will usually be a trade-off between gain and noise performance. We will introduce *circles of constant noise figure* as a means of evaluating the effect of differing source terminations and making an effective trade-off between gain and noise performance. This trade-off becomes more complex when we consider multistage amplifiers, where we are faced with the problem of deciding which combination of gain/noise figure in each stage will optimize the noise performance of the overall amplifier. In this chapter, we will introduce the concept of *noise measure* as a more relevant measure of performance for a low-noise stage when stages are to be cascaded.

The chapter then concludes with a brief discussion of the measurement of amplifier noise figure.

14.2 TYPES OF ELECTRICAL NOISE

14.2.1 THERMAL NOISE

The most common form of intrinsic electrical noise in circuits is *thermal noise*, which is generated by the random thermal motion of electrons within any conducting or semi-conducting material. This thermal motion would cease to exist if the material is "properly frozen", that is taken down to absolute zero (0 K). Thermal noise is also known as *Johnson noise* after J.B. Johnson who first observed the phenomenon in 1927 [4].

Thermal noise is characterized as "white noise" due to the fact that its power spectral density is flat across the full frequency spectrum. Of course, this does not mean that there is energy being generated at all frequencies up to infinity (as this would imply infinite energy), but simply that the power spectral density is flat across the frequency range of interest. The amplitude distribution of the signal is described by a Gaussian probability density function. This means that it is not possible to predict exactly the instantaneous noise voltage (or current) by means of a closed form equation. Having said this, knowledge of the noise statistics and its physical origins allows us to calculate the noise power as an averaged quantity. This is usually calculated as the equivalent mean square (MS) value of voltage or current measured in units of V^2 or A^2, respectively.

The MS value of thermal noise voltage and current in a resistor R (in Ω) in a bandwidth Δf (in Hz) and at an absolute operating temperature T_o (in K) are given, respectively, by the two equations below:

$$\overline{|v_{nt}|^2} = 4k_B T_o R \Delta f \quad \text{(in units of } V^2) \tag{14.2.1}$$

$$\overline{|i_{nt}|^2} = \frac{4k_B T_o \Delta f}{R} \quad \text{(in units of } A^2) \tag{14.2.2}$$

where k_B is Boltzmann's constant, which has a value of $1.3806488 \times 10^{-23}$ J K^{-1}.

$\overline{|v_{nt}|^2}$ and $\overline{|i_{nt}|^2}$ are in fact equal to the variances of the Gaussian distributions that describe the noise voltage and current, respectively. It is convenient for circuit designers to express noise in units of volts or amperes. These are expressed as root mean square (RMS) values. The RMS voltage, v_{n_t}, and the corresponding current, i_{n_t}, due to thermal noise in a resistance R (in Ω) may simply be obtained by taking the square roots of the quantities in equations (14.2.1) and (14.2.2), giving:

$$v_{n_t} = \sqrt{4k_B T_0 R \Delta f} \tag{14.2.3}$$

$$i_{n_t} = \sqrt{\frac{4k_B T_0 \Delta f}{R}} \tag{14.2.4}$$

To find the thermal noise power generated by an arbitrary resistor R, we can apply one or both of equations (14.2.3) and (14.2.4). We then have the noise power, P_{n_t}, generated by the resistor R as:

$$P_{n_t} = v_{n_t} \cdot i_{n_t} \tag{14.2.5}$$

$$= \frac{v_{n_t}^2}{R} = i_{n_t}^2 R \tag{14.2.6}$$

$$= 4k_B T_0 \Delta f \tag{14.2.7}$$

One way of understanding equation (14.2.5) is to think of P_{n_t} as the power dissipated in the noise generating resistor itself when it is terminated by a short circuit. Note that this power is directly proportional to the bandwidth and the absolute temperature, but is independent of the resistor value. What we are primarily interested in is the amount of noise power that will be transferred to an external circuit. According to the maximum power transfer theorem, the maximum noise power will be extracted from the resistor, R, when the equivalent resistance of the external circuit is also equal to R, as shown in Figure 14.1.

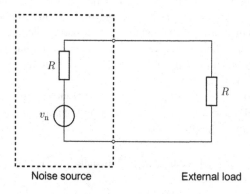

Noise source   External load

FIGURE 14.1

Maximum noise power extraction from a resistor R.

The noise voltage across the external load resistor in Figure 14.1, is $v_n/2$, where v_n is defined by equation (14.2.3). The maximum available noise power from R is therefore given by:

$$P_{n_t}(\text{max}) = k_B T_0 \Delta f \qquad (14.2.8)$$

It is important to keep in mind that the load resistor in Figure 14.1 is also a source of thermal noise, and that each one of the two participating resistors generates and dissipates noise in both itself and in the other resistor. This does not alter the validity of equation (14.2.8), since, as the two resistors are physically separate entities, their noise voltages are not correlated and so do not add constructively.

We can express the noise power in dBm as follows:

$$P_{n_t}(\text{dBm}) = 10\log_{10}(k_B T_0 \Delta f \times 1000) \qquad (14.2.9)$$

where the factor of 1000 in equation (14.2.9) is present because dBm is a ratio of the power to 1 mW. We can separate out the bandwidth element of equation (14.2.9) from the constant elements as follows:

$$P_{n_t}(\text{dBm}) = 10\log_{10}(k_B T_0 \times 1000) + 10\log_{10}(\Delta f) \qquad (14.2.10)$$

If we take T_0 to be room temperature (290 K), equation (14.2.10) can be written in a compact form as:

$$P_{n_t}(\text{dBm}) \approx -174 + 10\log_{10}(\Delta f) \qquad (14.2.11)$$

If we take the bandwidth to be 1 Hz, equation (14.2.11) gives us the *Thermal Noise Floor* as -174 dBm at room temperature. For a bandwidth of 1 MHz the thermal noise floor becomes approximately -114 dBm, and for a bandwidth of 1 GHz the thermal noise floor is approximately -84 dBm.

14.2.2 SHOT NOISE

Shot noise in electronic devices arises from the discrete nature of electric current and relates to the arrival of charge carriers at a particular place, that is, when electrons cross some type of physical "gap," such as a *pn* or Schottky junction. Unlike thermal noise, shot noise is characterized by the Poisson distribution [3], which describes the occurrence of independent and discrete random events. There are numerous examples of Poisson distributions outside of the field of electronics, such as arrival of people at a bus stop, traffic flow, and data flow patterns on the Internet.

When the number of events is sufficiently high, as in the case of the flow of electrons in a circuit with "normal" operating currents, the Poisson distribution resembles the Gaussian distribution. For most practical cases, therefore, we usually assume that the shot noise and thermal noise have the same distribution. This makes our circuit analysis and design more straightforward. In other words, we simply add the shot noise component to the thermal noise component. Furthermore, shot noise, just like thermal noise, can be characterized as "white noise" due to its flat power spectral density.

As the shot noise has its physical origin in electrons crossing a junction, it is normally expressed in terms of electron flow, in other words, current. The RMS value of the shot noise current is given by [2]:

$$i_{n_s} = \sqrt{2Iq\Delta f} \qquad (14.2.12)$$

where I is the DC current, q is the electron charge, and Δf is the bandwidth in Hz.

In all active circuits where semiconductor devices are biased, shot noise exists and has to be accounted for by designers. We note from equation (14.2.12) that shot noise is not a function of temperature, unlike thermal noise. We should also note that conductors and resistors do not exhibit shot noise because there is no "gap" as such.

14.2.3 FLICKER NOISE

In addition to thermal noise, which is present in every conductor having a finite resistance, semiconductor devices also exhibit a particular type of noise called *flicker noise* or $1/f$ noise, after its frequency characteristic which falls off steadily as frequency increases from zero.

Because of its spectral characteristics flicker noise is sometimes referred to as "Pink" noise (as opposed to thermal and shot noise which have a "white" spectrum). It was first noticed as an excess low-frequency noise in vacuum tubes and then, much later, in semiconductors [5]. Models of $1/f$ noise, based on detailed physical mechanisms, were developed by Bernamont [6] in 1937 for vacuum tubes and by McWhorter [7] in 1955 for semiconductors.

The $1/f$ characteristic of flicker noise implies that the noise power tends to infinity as the frequency approaches zero. Various researchers have tried to investigate whether indeed there is a cut off frequency below which flicker noise deviates from the $1/f$ characteristic, and have found none. Additionally, unlike other types of noise, $1/f$ noise has been described as a nonstationary random process [5], in other words its statistics vary with time.

The *flicker noise corner frequency*, f_c, defines the boundary between flicker noise dominant and thermal noise dominant regions in the frequency domain. In fact, $1/f$ noise has spectral characteristics that can be described as comprising a number of $1/f^{\alpha}$ curves with various cut-off frequencies depending upon the value of the integer α. This corner frequency and the actual spectral density of $1/f$ noise depend on the type of material used to construct a semiconductor device, the device geometry and the bias. Generally, both the $1/f$ noise spectral density and the corner frequency increase with bias current. The corner frequencies range from tens of Hz to tens of kHz [8].

14.3 NOISE FACTOR, NOISE FIGURE, AND NOISE TEMPERATURE

We now introduce an important figure of merit for low-noise amplifiers namely the *Noise Factor, F*, which is a measure of the degradation of signal-to-noise ratio (SNR), caused by any two-port network. The noise factor of a two-port network is calculated as a simple ratio of input SNR to output SNR, as follows:

$$F = \frac{\text{SNR}_{\text{in}}}{\text{SNR}_{\text{out}}} \tag{14.3.1}$$

For any real-world device or circuit in which internal noise will be generated as described in the previous sections, the input SNR will never be less than the output SNR. The noise factor, F, for such a device can therefore never be less than 1.

The SNR quantities in equation (14.3.1) are numerical power ratios, so the noise factor itself is a dimensionless ratio. The noise factor is most often presented in the form of the *Noise Figure*, which is simply the noise factor expressed in dB as follows:

$$F_{\text{dB}} = 10 \log_{10}(F) = 10 \log_{10}\left(\frac{\text{SNR}_{\text{in}}}{\text{SNR}_{\text{out}}}\right) \tag{14.3.2}$$

At this point we would like to draw the reader's attention to the fact that we are using the term "noise factor" to refer to the dimensionless ratio in equation (14.3.1) and the term "noise figure" to refer to the dB representation in equation (14.3.2). This distinction is often blurred, even in some text books, which may lead to confusion. From now on in this book we will adopt the convention of using "noise figure" as an umbrella term to refer to both quantities but we will make the distinction between "noise factor" and "noise figure" wherever necessary.

So, a perfect, noise-free amplifier will have a noise factor of 1 ($F_{\text{dB}} = 0$ dB), and any real amplifier will have a noise factor greater than 1 ($F_{\text{dB}} > 0$ dB).

These formulae are valid at a specified operating temperature, T_0, since, according to equation (14.2.8), any changes in temperature will affect the noise power. We can therefore define something called the *effective noise temperature* of any device or circuit as being the absolute temperature at which a perfect resistor, of equal resistance to the device or circuit, would generate the same noise power as that device or circuit at room temperature. We can also define the effective input noise temperature of an amplifier or other two-port network as the source noise temperature that would result in the same output noise power, when connected to an ideal "noise-free" network or amplifier, as that of the actual network or amplifier connected to a noise-free source.

The noise temperature, T_e, of a device in terms of it's noise factor is as follows:

$$T_e = T_0(F - 1) \tag{14.3.3}$$

where T_0 is the actual operating temperature (in K).

Noise temperature is useful in calculating the overall noise figure of a system that includes both one-port and two-port elements, such as a receiver with an antenna connected to its input. This is because strictly speaking, a one-port device like an antenna cannot be defined in terms of noise factor or noise figure, which is only defined for two-port devices (refer to equation 14.3.1). We can, however, refer to the effective noise temperature of the antenna to describe the amount of noise it introduces to the system.

Another reason for using noise temperature as a figure of merit is that it provides greater resolution at very small values of noise factor (where $F \approx 1$). For this reason extremely low-noise amplifiers may be characterized by their effective noise temperature.

The relationship between noise figure in dB and noise temperature is defined by:

$$F = \left(1 + \frac{T_e}{T_0}\right) \tag{14.3.4}$$

or in dB terms:

$$F_{dB} = 10 \, \log_{10}\left(1 + \frac{T_e}{T_0}\right) \tag{14.3.5}$$

The behavior of equation (14.3.5) is illustrated by Table 14.1, which shows some typical values of noise figure, noise factor, and noise temperature and by Figure 14.2, which shows a plot of noise figure in dB versus noise temperature.

Table 14.1 Noise Factor, Noise Figure (dB), and Noise Temperature

Noise Factor	Noise Figure (dB)	Noise Temperature (K)
1	0	0 (absolute zero)
1.26	1	75.1
2	3	290
10	10	2610
100	20	28,710

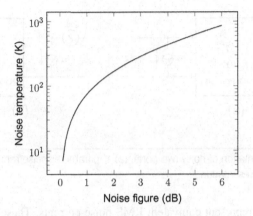

FIGURE 14.2

Noise temperature versus noise figure (dB).

14.4 REPRESENTATION OF NOISE IN ACTIVE TWO-PORT NETWORKS

As a prelude to discussing design methodologies for low-noise amplifiers, we need to understand how to predict the effect that an external circuit will have on the noise figure of a particular two-port active device. There is a large body of work on this subject going back many decades and some of it is highly theoretical. In this section, we aim to give the reader a general understanding of the underlying mechanisms, so as to facilitate the design of low-noise active circuits.

So far in this chapter we have discussed how all passive components such as resistors, contain thermal noise sources and that, in addition to thermal noise, active devices contain additional sources of shot noise and $1/f$ noise. The analysis of amplifier noise is further complicated by the fact that the amplifier will be amplifying its own internal noise sources along with the wanted signal. The approach that is generally taken to analyzing amplifier noise is therefore to refer all the noise sources back to the input.

One of the seminal papers in this field was published by Rothe and Dahlke in 1956 [9] in which they represented a noisy two-port by a noise-free two-port with external noise sources at the input and output, as shown in Figure 14.3(a).

This noisy two-port, which could be an amplifier, for example, may be represented in terms of its Y-parameters, introduced in Chapter 5, as follows:

$$i_1 = Y_{11}v_1 + Y_{12}v_2 + i_{n1}$$
$$i_2 = Y_{21}v_1 + Y_{22}v_2 + i_{n2}$$

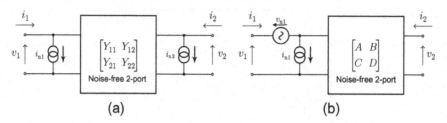

FIGURE 14.3

Immittance representation of noisy two-ports. (a) Y-parameter noise representation.
(b) $ABCD$ noise representation.

where i_{n1} and i_{n2} represent equivalent RMS noise currents. These currents, i_{n1} and i_{n2}, represent the RMS noise currents that would appear to originate at the input and output of the noiseless two-port, respectively, if the terminals were left open.

The analysis is simplified if we represent the noisy two-port in terms of its chain matrix ($ABCD$ matrix) [10,11] so that both noise sources may be located at the input port. Reconfiguration of Figure 14.3(a), based on this approach, is shown in Figure 14.3(b) and the alternative set of equations is:

$$i_1 = AV_2 + BI_2 + i_{n1}$$
$$v_1 = CV_2 + DI_2 + v_{n1} \tag{14.4.1}$$

where v_{n1} is the equivalent input RMS noise voltage generator, that is that noise voltage that would appear to originate at the input of the noiseless two-port.

This representation combines the effect of all internal noise sources into two generators at the input. The reason why we need to use two separate noise sources, i_{n1} and v_{n1}, as shown in Figure 14.3(b), is because the internal noise generating processes contain a combination of both voltage and current noise components. If voltage noise is the dominant mechanism then the noise output can be minimized by connecting an open-circuit source termination. If current noise is the dominant mechanism then the noise output can be minimized by connecting a short-circuit source termination. When both noise sources are present then the output noise is minimized at some intermediate complex value of source impedance.

Recalling equations (14.2.1) and (14.2.2) we can define equivalent noise resistance and conductance parameters in terms of the mean-square (MS) values of the equivalent noise voltage and current sources v_{n1} and i_{n1}, as follows [12]:

$$R_n = \frac{\overline{|v_{n1}|^2}}{4k_B T_o \Delta f} \tag{14.4.2}$$

$$G_n = \frac{\overline{|i_{n1}|^2}}{4k_B T_o \Delta f} \tag{14.4.3}$$

The equivalent noise sources v_{n1} and i_{n1} describe complex processes that are not fully correlated with each other. We can therefore define a complex correlation coefficient, ρ, that is a function of the particular circuit configuration. We can also define a correlation admittance, Y_{cor}, that relates v_{n1} and i_{n1} as follows [9]:

$$Y_{cor} = \rho \sqrt{\frac{\overline{|i_{n1}|^2}}{\overline{|v_{n1}|^2}}} \qquad (14.4.4)$$

The noise-free two-port of Figure 14.3(b) has the same SNR at its input and output. Therefore, the noise figure of the overall two-port can be derived by considering the input noise network alone (v_{n1} and i_{n1}).

Consider the input noise network in Figure 14.3(b) connected to a source of internal admittance $Y_S = G_S + jB_S$ and a noise current i_{ns} which is uncorrelated with v_{n1} or i_{n1}. Given equation (14.4.4), we can replace the noise voltage source, v_{en1}, with an equivalent current source $Y_{cor}v_{n1}$, as shown in Figure 14.4.

The mean-square short-circuit output current of this network is given by:

$$\overline{|i_{ntot}|^2} = \overline{|i_{ns} + i_u + v_{n1}(Y_S + Y_{cor})|^2} \qquad (14.4.5)$$

Since the components of the right-hand side of equation (14.4.5) are uncorrelated with each other, the mean-square value of i_{ntot} is equal to the sum of the mean-square values of the components. We can therefore rewrite equation (14.4.5) as:

$$\overline{|i_{ntot}|^2} = \overline{|i_{ns}|^2} + \overline{|i_u|^2} + \overline{|e_{n1}|^2}|Y_S + Y_{cor}|^2 \qquad (14.4.6)$$

We defined the noise factor of a two-port network in equation (14.3.1). Another definition of the noise factor is the ratio of the total output noise factor per unit bandwidth to the total input noise power per unit bandwidth [13]. Using this definition, the noise factor of the circuit of Figure 14.3(b) may be written as:

$$F = \frac{\overline{|i_{ntot}|^2}}{\overline{|i_{ns}|^2}} \qquad (14.4.7)$$

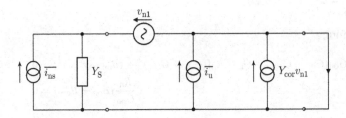

FIGURE 14.4

Input noise equivalent model.

Applying equation (14.4.6) this becomes:

$$F = 1 + \frac{\overline{|i_u|^2}}{|i_{ns}|^2} + \frac{\overline{|e_{n1}|^2}}{|i_{ns}|^2}|Y_S + Y_{cor}|^2 \tag{14.4.8}$$

The various voltage and current components of equation (14.4.8) may all be defined in terms of equivalent noise resistances and conductances as follows:

$$\overline{|i_{ns}|^2} = 4k_B T_0 G_S \Delta f \tag{14.4.9}$$

$$\overline{|i_u|^2} = 4k_B T_0 G_u \Delta f \tag{14.4.10}$$

$$\overline{|e_{n1}|^2} = 4k_B T_0 R_n \Delta f \tag{14.4.11}$$

Substituting these definitions (14.4.9)–(14.4.11) into equation (14.4.8) results in:

$$F = 1 + \frac{G_u}{G_S} + \frac{R_n}{G_S}|Y_S + Y_{cor}|^2 \tag{14.4.12}$$

or

$$F = 1 + \frac{G_u}{G_S} + \frac{R_n}{G_S}[(G_S + G_{cor})^2 + (B_S + B_{cor})^2] \tag{14.4.13}$$

The noise factor of the two-port is therefore an explicit function of the source admittance and depends upon four parameters, G_u, R_n, G_{cor}, and B_{cor}.

These parameters are in general determined by D.C. bias and operating frequency. Equation (14.4.13) represents a quasi-elliptic paraboloid [14] in the three-dimensional space bounded by the F, G_S, and B_S axes shown in Figure 14.5. This figure shows that the noise factor is a minimum at a particular value of source admittance, Y_{on}. The value of $Y_{on}(= G_{on} + jB_{on})$ and the associated minimum noise figure, F_{min}, may be found by differentiating equation (14.4.13) with respect to G_s and B_s and setting the derivatives simultaneously equal to zero:

$$\frac{\partial F}{\partial G_S} = 2\frac{R_n}{G_S}(G_S + G_{cor}) - \frac{G_u}{G_S^2} - \frac{R_n}{G_S^2}[(G_S + G_{cor})^2 + (B_S + B_{cor})^2] \tag{14.4.14}$$

$$\frac{\partial F}{\partial B_S} = 2\frac{R_n}{G_S}(B_S + B_{cor}) \tag{14.4.15}$$

Therefore,

$$2R_n G_{on}(G_{on} + G_{cor}) - G_u - R_n[(G_{on} + G_{cor})^2 + (B_{on} + B_{cor})^2] = 0 \tag{14.4.16}$$

$$2\frac{R_n}{G_{on}}(B_{on} + B_{cor}) = 0 \tag{14.4.17}$$

Equation (14.4.17) gives:

$$B_{on} = -B_{cor} \tag{14.4.18}$$

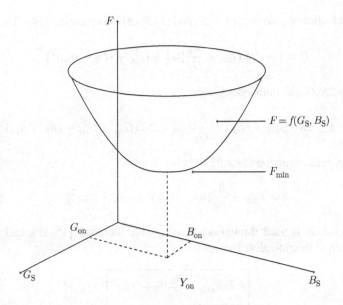

FIGURE 14.5

Noise factor, F, as a function of Y_S.

Substituting equation (14.4.18) into equation (14.4.14) gives:

$$2R_n G_{on}(G_{on} + G_{cor}) - R_n(G_{on} + G_{cor})^2 - G_u = 0 \qquad (14.4.19)$$

which, after some manipulation, gives us the value of G_{on} as:

$$G_{on} = \sqrt{\frac{R_n G_{cor}^2 + G_u}{R_n}} \qquad (14.4.20)$$

In order to find the value of the minimum noise figure, we substitute equations (14.4.18) and (14.4.20) into equation (14.4.13) which, after some further manipulation, results in:

$$F_{min} = 1 + 2R_n(G_{on} + G_{cor}) \qquad (14.4.21)$$

From the point of view of the circuit designer, it is more useful to have the noise figure of a device in terms of G_{on} and B_{on} as these are more readily measured than G_{cor} and B_{cor}. To express F in terms of G_{on} and B_{on} we must rearrange equation (14.4.20) to give:

$$G_{cor} = \sqrt{\frac{G_{on}^2 R_n - G_u}{R_n}} \qquad (14.4.22)$$

Then substitute equations (14.4.22) and (14.4.18) into equation (14.4.13) to give:

$$F = 1 + 2R_n G_{cor} + \frac{R_n}{G_S}[G_S^2 + G_{on}^2 + (B_S - B_{on})^2]$$

Adding $2R_n G_{on}$ to both sides yields:

$$F = 1 + 2R_n(G_{cor} + G_{on}) + \frac{R_n}{G_S}[G_S^2 - 2G_S G_{on} + G_{on}^2 + (B_S - B_{on})^2]$$

By employing equation (14.4.21) this becomes:

$$F = F_{min} + \frac{R_n}{G_S}[(G_S - G_{on})^2 + (B_S - B_{on})^2] \tag{14.4.23}$$

This equation is used throughout the available literature [15–19], and is more often written in its equivalent form:

$$\boxed{F = F_{min} + \frac{R_n}{G_S}|Y_S - Y_{on}|^2} \tag{14.4.24}$$

where Y_{on} is the optimum source termination ($Y_{on} = G_{on} + jB_{on}$). Let us define the excess noise figure using equation (14.4.8):

$$(F - 1) = \frac{\overline{|i_{n1} Y_S e_{n1}|^2}}{4k_B T_0 \Delta f} \tag{14.4.25}$$

Penfield [20] has shown that this equation can be expanded to yield:

$$(F - 1) = \frac{R_n}{G_S}\left[|Y_S|^2 + \frac{G_n}{R_S} + 2\sqrt{\frac{G_n}{R_S}}((\rho)G_S - Im(\rho)R_S)\right] \tag{14.4.26}$$

where $Re(\rho)$ and $Im(\rho)$ refer to the real and imaginary parts of ρ. Equation (14.4.26) may be rearranged to give:

$$(F - 1) = (F - 1)_{min} + \frac{R_n}{G_S}(|Y_S| + |Y_{on}| - 2(G_{on}G_S + B_{on}B_S)) \tag{14.4.27}$$

Comparing equations (14.4.25) and (14.4.26) yields:

$$|Y_{on}|^2 = \frac{G_n}{R_n} \tag{14.4.28}$$

$$G_{on} = -\sqrt{\frac{G_n}{R_n}}Re(\rho) \tag{14.4.29}$$

$$B_{on} = \sqrt{\frac{G_n}{R_n}}Im(\rho) \tag{14.4.30}$$

In the microwave frequency range, we are more accustomed to working with reflection coefficients than with impedances or admittances. Equation (14.4.24) may be translated into the source reflection coefficient plane by using the relationships:

$$Y_S = \frac{1}{Z_0} \frac{(1 - \Gamma_S)}{(1 + \Gamma_S)}$$

$$Y_{on} = \frac{1}{Z_0} \frac{(1 - \Gamma_{on})}{(1 + \Gamma_{on})} \tag{14.4.31}$$

This leads to the equation:

$$\boxed{F = F_{min} + 4r_n \frac{|\Gamma_S - \Gamma_{on}|^2}{|1 + \Gamma_{on}|^2(1 - |\Gamma_S|^2)}} \tag{14.4.32}$$

where r_n is the normalized equivalent input noise resistance, which is defined as:

$$r_n = \frac{R_n}{Z_0} \tag{14.4.33}$$

The four scalar parameters, F_{min}, $|\Gamma_{on}|$, $\angle\Gamma_{on}$, and R_n, are known as the *Noise Parameters* and are often specified in manufacturer's data sheets for a given microwave transistor, alongside the *S*-parameters. The interested reader is referred to Hartmann [21], who gives a good overview of the above theory with many relevant references.

14.5 SINGLE-STAGE LOW-NOISE AMPLIFIER DESIGN

In this section, we will build on the two-port noise analysis of the previous sections to set out a design methodology for low-noise microwave transistor amplifiers.

We rely on the noise parameters that are usually provided by the device manufacturer, but can be measured if necessary, using techniques set out in Section 14.7.2. There are two real and one complex parameters we need for this purpose, being the parameters used in equation (14.4.32), namely:

- The minimum noise figure: F_{min} in dB.
- The equivalent noise resistance: R_n in Ω.
- The complex optimum source termination: Γ_{on} (which is dimensionless).

As a reminder, we will use the symbol R_n to represent the ohmic value of equivalent noise resistance and the symbol r_n to represent the normalized value.

14.5.1 CIRCLES OF CONSTANT NOISE FIGURE

In Section 14.4, we showed that the noise factor of a given device can be minimized at a particular value of source termination. This concept is embodied in equation (14.4.24) (for admittances) and equation (14.4.32) (for reflection coefficients).

A graphical representation of the effect of variations in Γ_s on the noise factor of an amplifier provides a means of assessing the "trade-off" between noise figure and gain, when plotted on the same axes.

It can be shown that loci of constant noise factor obtained from equation (14.4.32) are circles in the source reflection coefficient plane [19]. We will not cover the admittance plane approach here, but this is well documented elsewhere [22]. We will focus instead on the reflection coefficient-based approach, which is more widely used today.

We start by considering equation (14.4.32):

$$F = F_{min} + 4r_n \frac{|\Gamma_S - \Gamma_{on}|^2}{|1 + \Gamma_{on}|^2(1 - |\Gamma_S|^2)} \qquad (14.4.32)$$

Rearranging equation (14.4.32) above gives:

$$\frac{(F - F_{min})|1 + \Gamma_{on}|^2}{4r_n} = \frac{|\Gamma_S - \Gamma_{on}|^2}{(1 - |\Gamma_S|^2)} \qquad (14.5.1)$$

which can be rearranged as:

$$N_i(1 - |\Gamma_S|^2) = |\Gamma_S|^2 + |\Gamma_{on}|^2 - \Gamma_S^* \Gamma_{on} - \Gamma_S \Gamma_{on}^* \qquad (14.5.2)$$

where[1]

$$N_i = \frac{(F - F_{min})|1 + \Gamma_{on}|^2}{4r_n} \qquad (14.5.3)$$

Rearranging equation (14.5.2) leads to:

$$|\Gamma_S|^2 - \Gamma_S \frac{\Gamma_{on}^*}{(1 + N_i)} - \Gamma_S^* \frac{\Gamma_{on}}{(1 + N_i)} = \frac{N_i - |\Gamma_{on}|^2}{(1 + N_i)} \qquad (14.5.4)$$

adding $\frac{|\Gamma_{on}|^2}{(1+N_i)^2}$ to both sides leads to:

$$|\Gamma_S|^2 + \frac{|\Gamma_{on}|^2}{(1 + N_i)^2} - \Gamma_S \frac{\Gamma_{on}^*}{(1 + N_i)} - \Gamma_S^* \frac{\Gamma_{on}}{(1 + N_i)} = \frac{(N_i - |\Gamma_{on}|^2)(1 + N_i) + |\Gamma_{on}|^2}{(1 + N_i)^2} \qquad (14.5.5)$$

Equation (14.5.5) is of the form of a circle in the Γ_s plane, that is to say, it is of the form:

$$|\Gamma_S - C_{Sn}| = \gamma_{Sn}^2 \qquad (14.5.6)$$

where the center is given by:

$$C_{Sn} = \frac{\Gamma_{on}}{1 + N_i} \qquad (14.5.7)$$

and the radius is given by:

---

[1] The noise figure parameter, N_i, has been defined by Pengelly [15].

$$\gamma_{\text{Sn}} = \frac{\sqrt{N_i^2 + N_i(1 - |\Gamma_{\text{on}}|^2)}}{1 + N_i} \qquad (14.5.8)$$

Notwithstanding our distinction between noise factor and noise figure, the constant noise factor circles defined here are often referred to as *constant noise figure circles*, which reflects the fact that they are often labeled with the noise figure value in dB.

We will now illustrate the use of constant noise figure circles with an example.

Example 14.1 (Noise figure circles).

Problem. Draw constant noise figure circles for $F = 1.4$ dB, $F = 2$ dB, and $F = 3$ dB on the source plane for the Avago ATF-34143 Low-Noise HEMT operating at 10 GHz, and hence, or otherwise, determine the lowest possible noise figure commensurate with the maximum gain available from this device. The S-parameters and noise parameters of the device with bias conditions $V_{\text{DS}} = 3$ V, $I_{\text{DS}} = 40$ mA are as follows:

S-parameters:

$$\begin{bmatrix} S_{11} & S_{12} \\ S_{21} & S_{22} \end{bmatrix} = \begin{bmatrix} 0.760\angle 28 & 0.144\angle -84 \\ 1.647\angle -84 & 0.410\angle 23 \end{bmatrix}$$

Noise parameters:

$$F_{\text{min}} = 1.22 \text{ dB}$$
$$\Gamma_{\text{on}} = 0.61\angle -39°$$
$$R_n = 25 \ \Omega$$

Solution. First, we need to investigate the stability of the device, for which we will use the Edwards-Sinsky stability criteria defined by equations (7.4.36) and (7.4.37), that is,

$$\mu_1 = \frac{1 - |S_{11}|^2}{|S_{22} - \Delta S_{11}^*| + |S_{12}S_{21}|} = \frac{0.4224}{0.3134} = 1.1887$$

$$\mu_2 = \frac{1 - |S_{22}|^2}{|S_{11} - \Delta S_{22}^*| + |S_{12}S_{21}|} = \frac{0.8319}{0.8246} = 1.0435$$

Since both μ_1 and μ_2 are greater than 1 we conclude that the device is unconditionally stable, so we are free to choose any terminating impedances lying within the $|\Gamma| = 1$ boundary of the source and load plane Smith Charts. Maximum available gain (MAG) occurs when the source and load are simultaneously conjugately matched. The necessary terminating reflection coefficients are determined using equations (7.5.24) and (7.5.25) as follows:

$$\Gamma_{\text{ms}} = C_1^* \left[\frac{B_1 - \sqrt{B_1^2 - 4|C_1|^2}}{2|C_1|^2} \right]$$

$$= 0.5601\angle - 34° \left[\frac{1.1413 - \sqrt{1.1413^2 - 4 \times 0.5601^2}}{2 \times 0.5601^2} \right]$$

$$= 0.8233\angle - 34°$$

$$\Gamma_{ml} = C_2^* \left[\frac{B_2 - \sqrt{B_2^2 - 4|C_2|^2}}{2|C_2|^2} \right]$$

$$= 0.1182\angle - 97° \left[\frac{0.3223 - \sqrt{0.3223^2 - 4 \times 0.1182^2}}{2 \times 0.1182^2} \right]$$

$$= 0.4364\angle - 97°$$

where

$$B_1 = 1 + |S_{11}|^2 - |S_{22}|^2 - |\Delta|^2 = 1.1413$$
$$C_1 = S_{11} - \Delta S_{22}^* = 0.5601\angle 34°$$
$$B_2 = 1 - |S_{11}|^2 + |S_{22}|^2 - |\Delta|^2 = 0.3223$$
$$C_2 = S_{22} - \Delta S_{22}^* = 0.1182\angle 97°$$

With the above conjugate terminations the MAG of the device, from equation (7.5.26) is:

$$\text{MAG} = \frac{|S_{21}|}{|S_{12}|} \left[K - \sqrt{K^2 - 1} \right]$$

$$= \frac{1.647}{0.144} \left[1.1016 - \sqrt{1.1016^2 - 1} \right]$$

$$= 7.3151$$

which is equal to 8.6 dB.

Where K is the Rollett stability factor, calculated using equation (7.4.27).

In order to draw the constant noise figure circles for $F = 1.4$ dB, $F = 2$ dB, and $F = 3$ dB on the Γ_S plane, the first step is to calculate the parameter N_i, as defined by equation (14.5.3) for the various values of noise figure. For example, for $F = 1.4$ dB we have:

$$N_i = \frac{(F - F_{min})|1 + \Gamma_{on}|^2}{4r_n}$$

$$= \frac{(10^{(1.4/10)} - 10^{(1.22/10)})|1 + 0.61\angle - 39|^2}{4 \times 25/50}$$

$$= \frac{0.0560 \times 2.3202}{2} = 0.0650$$

Similarly, we calculate N_i for $F = 2$ dB and $F = 3$ dB to be 0.3023 and 0.7783, respectively. Employing equations (14.5.7) and (14.5.8) we can now calculate the centers and radii of the three noise figure circles as follows.

1.4 dB noise figure circle:

$$C_{Sn_{1.4}} = \frac{0.61\angle - 39}{1 + 0.0650} = 0.5730\angle - 39$$

$$\gamma_{Sn_{1.4}} = \frac{\sqrt{(0.0042 + 0.0650 \times 0.6279)}}{1 + 0.0650} = 0.1993$$

2 dB noise figure circle:

$$C_{Sn_{2.0}} = \frac{0.61\angle - 39}{1 + 0.3023} = 0.4684\angle - 39$$

$$\gamma_{Sn_{2.0}} = \frac{\sqrt{(0.0914 + 0.3023 \times 0.6279)}}{1 + 0.3023} = 0.4072$$

3 dB noise figure circle:

$$C_{Sn_{3.0}} = \frac{0.61\angle - 39}{1 + 0.7783} = 0.3430\angle - 39$$

$$\gamma_{Sn_{3.0}} = \frac{\sqrt{(0.6058 + 0.7783 \times 0.6279)}}{1 + 0.7783} = 0.5883$$

The above noise figure circles are plotted on the Smith Chart in Figure 14.6, together with the optimum source termination, Γ_{on}, which is basically the center of the noise figure circle of zero radius (i.e., when we set $F = F_{min}$ in equation (14.5.8) we get $\gamma_{Sn} = 0$). We have also plotted the optimum source termination for MAG, Γ_{ms}, on Figure 14.6 and we can see that this lies between the $F = 1.4$ dB and $F = 2$ dB noise figure circles, indicating that the noise figure of the device, when simultaneously conjugately matched for maximum gain, will have a noise figure between 1.4 and 2 dB.

We can calculate the exact noise figure of the simultaneously conjugately matched device by employing equation (14.4.32). If the input port is matched with Γ_{ms}, the noise figure will be:

$$
\begin{aligned}
F &= F_{min} + 4r_n \frac{|\Gamma_{ms} - \Gamma_{on}|^2}{|1 + \Gamma_{on}|^2(1 - |\Gamma_{ms}|^2)} \\
&= 1.3243 + 4 \times \frac{25}{50} \times \frac{|0.8233\angle - 34° - 0.61\angle - 39°|^2}{|1 + 0.61\angle - 39°|^2(1 - |0.8233\angle - 34°|^2)} \\
&= 1.3243 + 2.0 \times \frac{0.0489}{2.3202 \times 0.3222} \\
&= 1.3243 + 2.0 \times 0.0654 = 1.4551
\end{aligned}
$$

which is equal to 1.62 dB.

Which corresponds with our assessment based on the noise figure circles in Figure 14.6.

In many cases, the amplifier designer will be required to design an amplifier to meet a specified gain and noise figure requirement. This involves a trade-off between

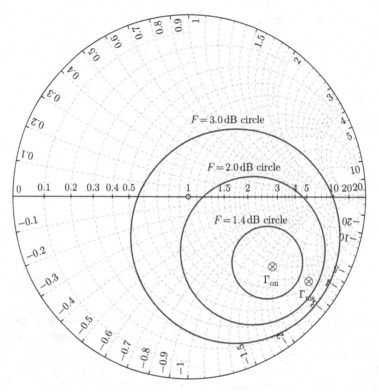

FIGURE 14.6

Constant noise figure circles for Avago ATF-34143 at 10 GHz ($V_{DS} = 3$ V, $I_{DS} = 40$ mA).

noise figure and gain, as these optima generally do not occur with the same source termination. Furthermore, the active device may be potentially unstable, which will further restrict the choice of source and load termination. We can illustrate this rather more complicated scenario by means of another example, as follows.

Example 14.2 (Low-noise design for specified gain).

Problem. You are required to design an 18 GHz low-noise amplifier having a gain of at least 10 dB and a noise figure of less than 2 dB, using the BFU730F Silicon-Germanium BJT from NXP. The S-parameters and noise parameters of the device with bias conditions $V_C = 2.0$ V, $I_C = 10$ mA are as follows.

S-parameters:

$$\begin{bmatrix} S_{11} & S_{12} \\ S_{21} & S_{22} \end{bmatrix} = \begin{bmatrix} 0.691\angle 63° & 0.178\angle -20° \\ 2.108\angle -55° & 0.218\angle 97° \end{bmatrix}$$

Noise parameters:

$$F_{min} = 1.79 \text{ dB}$$
$$\Gamma_{on} = 0.667\angle307°$$
$$R_n = 28.6 \ \Omega$$

Solution. First, we need to investigate the stability of the device, for which we will use the Edwards-Sinsky stability criteria of equations (7.4.36) and (7.4.37), that is,

$$\mu_1 = \frac{1 - |S_{11}|^2}{|S_{22} - \Delta S_{11}^*| + |S_{12}S_{21}|} = \frac{0.4224}{0.3134} = 0.885$$

$$\mu_2 = \frac{1 - |S_{22}|^2}{|S_{11} - \Delta S_{22}^*| + |S_{12}S_{21}|} = \frac{0.8319}{0.8246} = 0.962$$

Since both μ_1 and μ_2 are less than 1 we conclude that the device is potentially unstable. We therefore need to plot stability circles in order to determine the acceptable range of source terminations. Since we need to focus on matching the input port to achieve the desired noise specification, we first use equations (7.4.14) and (7.4.15) to calculate the center and radius of the source plane stability circle, as follows:

$$C_{SS} = \frac{C_1^*}{|S_{11}|^2 - |\Delta|^2} = \frac{0.6149\angle - 62°}{0.2492} = 2.4680\angle - 69°$$

$$r_{SS} = \left| \frac{|S_{12}S_{21}|}{|S_{11}|^2 - |\Delta|^2} \right| = \frac{0.3752}{0.2492} = 1.5060$$

Once again, we calculate the parameter N_i, as defined by equation (14.5.3), for various values of noise figure circle (say, $F = 2$ dB, $F = 3$ dB, and $F = 5$ dB). We then calculate the respective noise figure circle centers and radii using equations (14.5.7) and (14.5.8). The resulting calculations are summarized in the following table.

F (dB)	N_i	$|C_{Sn}|$	$\angle C_{Sn}$	γ_{Sn}
2	0.0735	0.6213	−53°	0.2002
3	0.4766	0.4517	−53°	0.4749
5	1.6231	0.2543	−53°	0.7168

The above noise figure circles are plotted on the source plane Smith Chart, together with the stability circle as shown in Figure 14.7.

We now check the gain available from the device when terminated for minimum noise figure, that is, when the source termination is $\Gamma_{on} = 0.667\angle307°$. For this we use equation (7.5.13) for available power gain:

$$
\begin{aligned}
G_A &= \frac{|S_{21}|^2(1 - |\Gamma_{on}|^2)}{|1 - S_{11}\Gamma_{on}|^2 - |S_{22} - \Delta\Gamma_{on}|^2} \\
&= \frac{2.108^2 \times (1 - 0.667^2)}{|1 - 0.691\angle 63° \times 0.667\angle 307°|^2 - |0.218\angle 97° - 0.4778\angle 120° \times 0.667\angle 307°|^2} \\
&= \frac{2.467}{0.276} = 8.943(9.5 \text{ dB})
\end{aligned}
$$

If we set the source termination to obtain minimum noise figure, therefore, we will not be able to achieve the required gain specification. In order to determine a range of source terminations that can achieve the lowest noise figure consistent with 10 dB of gain we should draw the 10 dB constant available gain circle on the source plane and see where this circle intersects with the noise figure circles. The available gain circle is calculated by applying equations (7.5.41) and (7.5.42). First, we need to calculate the normalized gain parameter g_a as defined by equation (7.5.37):

$$
g_a = \frac{G_A}{|S_{21}|^2} = \frac{10^{(10/10)}}{2.108^2} = 2.250
$$

The centers and radii of the 10 dB constant gain circle on the source reflection coefficient plane are now calculated using equations (7.5.41) and (7.5.42) as follows:

$$
\begin{aligned}
C_{gS} &= \frac{g_a C_1^*}{1 + g_a D_1} \\
&= \frac{2.250 \times 0.6149\angle - 69.2°}{1 + 2.250 \times 0.2492} \\
&= \frac{1.384\angle - 69.2°}{1.561} \\
&= 0.8860\angle - 69.2°
\end{aligned}
$$

$$
\begin{aligned}
\gamma_{gS} &= \frac{\sqrt{1 - 2K|S_{12}S_{21}|g_a + |S_{12}S_{21}|^2 g_a^2}}{1 + g_a D_1} \\
&= \frac{0.361}{1.561} = 0.231
\end{aligned}
$$

where

$$
\Delta = S_{11}S_{22} - S_{12}S_{21} = 0.478\angle 120°
$$

$$
K = \frac{1 - |S_{11}|^2 - |S_{22}|^2 + |\Delta|^2}{2|S_{21}S_{12}|} = 0.9372
$$

$$
C_1 = S_{11} - \Delta S_{22}^* = 0.6149\angle 69.2°
$$

$$
D_1 = (|S_{11}|^2 - |\Delta|^2) = 0.2492
$$

We can see from Figure 14.7 that there is a region where the 10 dB constant gain circle overlaps the $F = 2$ dB constant noise figure circle. Any source termination

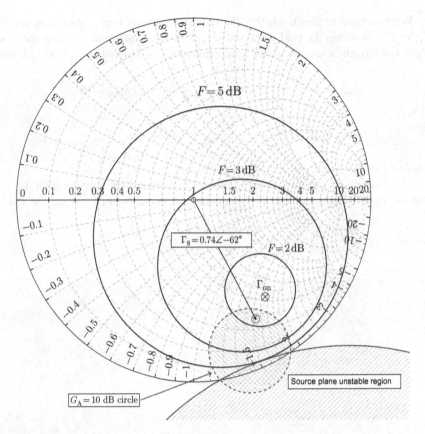

FIGURE 14.7

Constant noise figure circles, Γ_{on}, and stability and gain circles on the source plane for the NXP BFU730F at 18 GHz ($V_C = 2.0$ V, $I_C = 10$ mA).

lying within this region will have a gain greater than 10 dB and a noise figure less than 2 dB. We therefore choose a source termination of $\Gamma_S = 0.74\angle - 62°$ as indicated in Figure 14.7, and we can be confident that this source termination will result in $F < 2$ dB and $G_A > 10$ dB. With this source termination, the output reflection coefficient of the transistor can be calculated using equation (6.2.7), as follows:

$$\Gamma_{out} = S_{22} + \frac{S_{12}S_{21}\Gamma_S}{1 - S_{11}\Gamma_S}$$

$$= 0.218\angle 97° + \frac{0.178\angle - 20° \times 2.108\angle - 55° \times 0.74\angle - 62°}{1 - 0.691\angle 63° \times 0.74\angle - 62°}$$

$$= 0.218\angle 97° + \frac{0.2775\angle - 137°}{0.4892\angle - 1°}$$

$$= 0.4694\angle - 157.7°$$

We now need to check whether the required value of load termination, set by $\Gamma_L = \Gamma_{out}^*$ is within the load plane stable region. The center and radius of the load plane stability circle are calculated using equations (7.4.10) and (7.4.12) as follows:

$$C_{SL} = \frac{C_2^*}{|S_{22}|^2 - |\Delta|^2} = \frac{0.2152\angle - 164°}{-0.1808} = 1.1903\angle - 16.3°$$

$$r_{SL} = \left| \frac{|S_{12}S_{21}|}{|S_{22}|^2 - |\Delta|^2} \right| = \frac{0.3752}{0.1808} = 2.0753$$

where $C_2 = S_{22} - \Delta S_{11}^* = 0.2152\angle - 164°$.

The load plane stability circle is plotted in Figure 14.8 together with $\Gamma_L = 0.4694\angle 157.7°$. Since the load plane stability circle encloses the origin, the stable

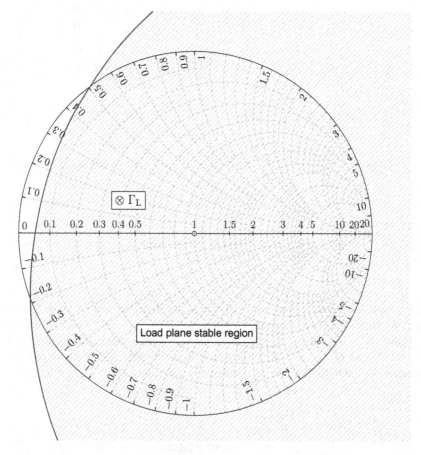

FIGURE 14.8

Load plane stability circles for the NXP BFU730F at 18 GHz ($V_C = 2.0$ V, $I_C = 10$ mA) with designated input match.

region is represented by the interior the circle. This means that the stable region encompasses most of the load plane Smith Chart except for a small sliver on the left-hand side, as shown in Figure 14.8. Our chosen value of Γ_L is therefore comfortably inside the stable region of the load plane.

14.5.2 NOISE FACTOR OF PASSIVE TWO-PORTS

Although this book focuses on active circuit design, we will digress briefly to talk about the noise factor of passive circuits. This is because the microwave designer will often be required to calculate the noise factor of a number of active and passive circuits in combination, as will be illustrated in the next section, so some familiarity with the noise factor of passive circuits is desirable.

By definition, a passive two-port has a gain, G, that is less than unity. Passive circuits are therefore usually characterized by their *attenuation*, which is defined by:

$$A = \frac{1}{G} \tag{14.5.9}$$

The equivalent noise temperature of a passive two-port having an attenuation, A, and at operating temperature, T_o, can be shown to be [23]:

$$T_e = (A - 1)T_o \tag{14.5.10}$$

Thus, we can write the output noise temperature of such a passive two-port as:

$$
\begin{aligned}
T_{out} &= G(T_{in} + T_e) \\
&= \frac{(T_{in} + T_e)}{A} \\
&= \frac{T_{in}}{A} + \frac{(A - 1)T_o}{A}
\end{aligned}
$$

which reduces to:

$$T_{out} = \frac{T_{in}}{A} - \frac{T_o}{A} + T_o \tag{14.5.11}$$

The meaning of equation (14.5.11) is as follows: as the attenuation, A, approaches unity (i.e., the lossless case), we find that T_{out} approaches T_{in}. In other words, the noise passes through a lossless device unaltered, and the device will generate no internal noise of its own. This makes sense from a physical point of view, since no loss means no internal resistive elements inside the two-port to generate thermal noise.

Let us now consider the case where the attenuation, A, becomes very large. In this case, the input noise is completely absorbed by the two-port. The noise at the device output now consists of noise that is entirely generated inside the two-port.

The output noise temperature will therefore become $T_{out} = T_o$, that is, equal to the physical temperature of the device.

We can now determine the noise factor of a passive two-port by combining equation (14.3.4) on page 482 and equation (14.5.10) to give:

$$F = 1 + \frac{(A-1)T_o}{T_o} = 1 + (A-1) = A \qquad (14.5.12)$$

In other words, for any passive two-port device, the noise factor, F, is equal to the attenuation of the device, A.

14.6 MULTISTAGE LOW-NOISE AMPLIFIER DESIGN

In Section 13.4, we explained how a required gain and bandwidth can be obtained by cascading several single stages. In the context of this chapter, cascading stages in this way raises the question of the relationship between the noise factor of a multistage amplifier and the noise factors of the individual stages.

One might intuitively expect that a minimum noise factor multistage amplifier could be constructed by simply cascading a number of individual stages each optimized for minimum noise factor. It turns out, however, that this approach does not result in the lowest overall noise factor for the cascade, due to the trade-off between noise factor and gain inherent in single-stage amplifier design, as outlined in the previous sections.

Figure 14.9 shows a cascade of single-stage amplifiers, with the nth stage having a noise factor F_n and available gain G_n.

It turns out that the overall noise factor of the multistage amplifier in Figure 14.9 depends not only on the noise factor of the individual stages but also on the gain of the individual stages. This is embodied in the so-called *Friis noise formula* which is named after its originator, Harald Friis [24], and can be stated as follows:

$$F = F_1 + \frac{F_2 - 1}{G_1} + \frac{F_3 - 1}{G_1 G_2} + \cdots + \frac{F_n - 1}{G_1 G_2 G_3 \ldots G_{n-1}} \qquad (14.6.1)$$

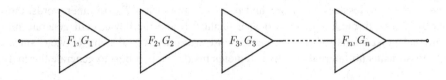

FIGURE 14.9

Cascaded amplifiers.

where

F = noise factor of the cascade;

F_n = noise factor of the nth stage; and

G_n = gain of the nth stage.

We can deduce the following by studying equation (14.6.1):

(i) The noise factor of the first stage is much more important than the noise factors of subsequent stages, as these are divided by the gain of the preceding stages. This suggests that the first-stage noise factor should be made as small as possible.

(ii) In order to make subsequent stage noise factors insignificant, the first-stage gain should be as high as possible.

It will be apparent that the requirements represented by statements (i) and (ii) above are mutually incompatible since, as was illustrated in Example 14.1, minimum noise figure and maximum gain do not normally occur at the same source termination. A compromise must therefore be struck between the gain and noise figure of each stage which results in the minimum overall noise figure for the cascade. This can be done by selecting a suitable source termination which is in general not equal to Γ_{ms} or Γ_{on}.

Consider the case of two stages that are to be cascaded. Let their noise figures be F_1 and F_2 and their available gains be G_{A1} and G_{A2}. There are two possible arrangements as shown in Figure 14.10.

If stage 1 is placed first, as in Figure 14.10(a), the overall noise factor of the cascade, from equation (14.6.1), will be:

$$F_{12} = F_1 + \frac{F_2 - 1}{G_{A1}} \tag{14.6.2}$$

On the other hand, if stage 2 is placed first, as in Figure 14.10(b), the overall noise factor of the cascade, from equation (14.6.1) will be:

$$F_{21} = F_2 + \frac{F_1 - 1}{G_{A2}} \tag{14.6.3}$$

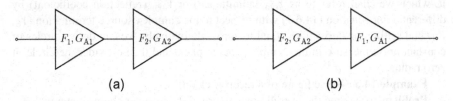

(a)   (b)

FIGURE 14.10

Two ways of cascading two amplifiers.

In general, one of these possibilities will result in a lower overall noise figure than the other. Suppose that putting stage 1 first results in the lowest overall noise figure, that is,

$$F_{12} < F_{21} \qquad (14.6.4)$$

Employing equations (14.6.2) and (14.6.3) results in:

$$F_1 + \frac{F_2 - 1}{G_{A1}} < F_2 + \frac{F_1 - 1}{G_{A2}} \qquad (14.6.5)$$

Equation (14.6.5) can be rearranged to give:

$$\frac{F_1 - 1}{\left(1 - \dfrac{1}{G_{A1}}\right)} < \frac{F_2 - 1}{\left(1 - \dfrac{1}{G_{A2}}\right)} \qquad (14.6.6)$$

Therefore, the lowest overall noise figure results from ensuring that the first stage has the lowest value, not of F, but of the quantity "M" which is defined by:

$$M = \frac{F - 1}{\left(1 - \dfrac{1}{G_A}\right)} \qquad (14.6.7)$$

The quantity M is therefore a more meaningful measure of stage noise performance than noise figure when stages are to be cascaded. It was first identified by Haus and Adler [25] who gave this quantity the name *noise measure*. If several stages with the same noise measure are cascaded then the noise measure of the cascade will be the same as that of each stage. For such a cascade the overall noise figure, assuming large enough stage gains, is given by:

$$F = M + 1 \qquad (14.6.8)$$

We can therefore conclude that, in order to build a multistage amplifier with the minimum overall noise factor, the first stage, and possibly subsequent stages, should be designed for minimum value of noise measure (i.e., M_{min}). We know that noise factor and available gain are functions of the source termination alone, so we deduce that the minimum noise measure can be obtained at a particular value of source termination.

We can determine the value of M_{min} and the source termination required to realize it, which we shall refer to as Y_{om} (admittance) or Γ_{om} (reflection coefficient) by differentiating equation (14.6.7) with respect to the complex source termination (Y_S, Γ_S) and setting the derivatives equal to zero. Alternatively, we can derive circles of constant noise measure in the complex source plane and then consider the circle of zero radius.

Example 14.3 (Noise figure of a receiver chain).

Problem. Calculate the overall noise figure of the receiver chain shown in Figure 14.11, given the following noise figure and gain values for individual elements.

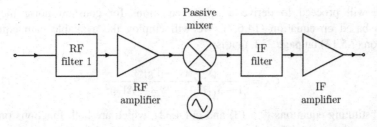

FIGURE 14.11

Overall noise figure of a receiver chain.

Table 14.2 Receiver Chain Parameters

	RF Filter (dB)	RF Amp (dB)	Mixer (dB)	IF Filter (dB)	IF Amp (dB)
F_{dB}	0.5	3	5	1	5
G_{dB}	−0.5	12	−5	−1	18

Solution. First, we convert the dB values in Table 14.2 into numerical ratios, as shown in Table 14.3.

Table 14.3 Receiver Chain Parameters (Numerical)

	RF Filter	RF Amp	Mixer	IF Filter	IF Amp
F	1.122	2.000	3.162	1.259	3.162
G	0.891	15.85	0.316	0.794	63.10

We now apply equation (14.6.1) using the values in Table 14.3:

$$F = 1.122 + \frac{2.000 - 1}{0.891} + \frac{3.162 - 1}{0.891 \times 15.85} +$$
$$\frac{1.259 - 1}{0.891 \times 15.85 \times 0.316} + \frac{3.162 - 1}{0.891 \times 15.85 \times 0.316 \times 0.794}$$
$$F = 3.066$$

which is equal to $F_{dB} = 4.87$ dB.

14.6.1 CIRCLES OF CONSTANT NOISE MEASURE

As is the case when designing to meet a specific noise factor specification, as covered in Section 14.5.1, a graphical representation of the effect of variations in Γ_s on the noise measure of an amplifier is a useful design aid.

We will proceed to derive a set of equations for constant noise measure circles based on equation (14.6.7). We will employ the available gain equation equation (7.5.13) on page 232, that is:

$$G_A = \frac{|S_{21}|^2(1 - |\Gamma_S|^2)}{|1 - S_{11}\Gamma_S|^2 - |S_{22} - \Delta\Gamma_S|^2} \tag{7.5.13}$$

Substituting equations (7.5.13) and (14.4.32), which are both functions only of Γ_S, into equation (14.6.7) and we have:

$$M = \frac{(F_{\min} - 1) + 4r_n\dfrac{|\Gamma_S - \Gamma_{on}|^2}{|1 + \Gamma_{on}|^2(1 - |\Gamma_S|^2)}}{\left(1 - \dfrac{|1 - S_{11}\Gamma_S|^2 - |S_{22} - \Delta\Gamma_S|^2}{|S_{21}|^2(1 - |\Gamma_S|^2)}\right)} \tag{14.6.9}$$

Which can be rearranged as:

$$M = \frac{|S_{21}|^2}{|1 + \Gamma_{on}|^2} \times \frac{|1 + \Gamma_{on}|^2(1 - |\Gamma_S|^2)(F_{\min} - 1) + 4r_n|\Gamma_S - \Gamma_{on}|^2}{|S_{21}|^2(1 - |\Gamma_S|^2) - |1 - S_{11}\Gamma_S|^2 + |S_{22} - \Delta\Gamma_S|^2} \tag{14.6.10}$$

Expanding out equation (14.6.10) and collecting Γ_S terms give:

$$|\Gamma_S|^2[M|1 + \Gamma_{on}|^2(|\Delta|^2 - |S_{21}|^2 - |S_{11}|^2) - |S_{21}|^2(4r_n - |1 + \Gamma_{on}|^2(F_{\min} - 1))]$$
$$+ \Gamma_S(M|1 + \Gamma_{on}|^2 C_1 + 4r_n|S_{21}|^2\Gamma_{on}^*) + \Gamma_S^*(M|1 + \Gamma_{on}|^2 C_1^* + 4r_n|S_{21}|^2\Gamma_{on})$$
$$= |S_{21}|^2[|1 + \Gamma_{on}|^2(F_{\min} - 1) + 4r_n|\Gamma_{on}|^2] - M|1 + \Gamma_{on}|^2(|S_{22}|^2 + |S_{21}|^2 - 1) \tag{14.6.11}$$

Where $C_1 = S_{11} - S_{22}^*\Delta$.
Equation (14.6.11) can be rearranged to give:

$$|\Gamma_S|^2 + \Gamma_S\left[\frac{M|1 + \Gamma_{on}|^2 C_1 + 4r_n|S_{21}|^2\Gamma_{on}^*}{M|1 + \Gamma_{on}|^2(|\Delta|^2 - |S_{21}|^2 - |S_{11}|^2) - |S_{21}|^2(4r_n - |1 + \Gamma_{on}|^2(F_{\min} - 1))}\right]$$
$$+ \Gamma_S^*\left[\frac{M|1 + \Gamma_{on}|^2 C_1^* + 4r_n|S_{21}|^2\Gamma_{on}}{M|1 + \Gamma_{on}|^2(|\Delta|^2 - |S_{21}|^2 - |S_{11}|^2) - |S_{21}|^2(4r_n - |1 + \Gamma_{on}|^2(F_{\min} - 1))}\right]$$
$$= \left[\frac{|S_{21}|^2[|1 + \Gamma_{on}|^2(F_{\min} - 1) + 4r_n|\Gamma_{on}|^2] - M|1 + \Gamma_{on}|^2(|S_{22}|^2 + |S_{21}|^2 - 1)}{M|1 + \Gamma_{on}|^2(|\Delta|^2 - |S_{21}|^2 - |S_{11}|^2) - |S_{21}|^2(4r_n - |1 + \Gamma_{on}|^2(F_{\min} - 1))}\right] \tag{14.6.12}$$

Equation (14.6.12) is in the form:

$$|\Gamma_S|^2 + |C_{Sm}|^2 - \Gamma_S^* C_{Sm} - \Gamma_S C_{Sm}^* = \gamma_m^2 \tag{14.6.13}$$

which describes a circle in the Γ_S plane with center at C_{Sm} and radius γ_{Sm}. From equation (14.6.12) we can see that the center of the constant M circle on the Γ_S plane is located at:

$$C_{S_m} = \frac{M|1 + \Gamma_{on}|^2 C_1^* + 4r_n|S_{21}|^2\Gamma_{on}}{M|1 + \Gamma_{on}|^2(|S_{21}|^2 + |S_{11}|^2 - |\Delta|^2) - |S_{21}|^2(|1 + \Gamma_{on}|^2(F_{min} - 1) - 4r_n)}$$
(14.6.14)

and the radius is given by:

$$\gamma_{S_m} = $$

$$\sqrt{\frac{M|1 + \Gamma_{on}|^2(1 - |S_{22}|^2 - |S_{21}|^2) + |S_{21}|^2[|1 + \Gamma_{on}|^2(F_{min} - 1) + 4r_n|\Gamma_{on}|^2]}{M|1 + \Gamma_{on}|^2(|\Delta|^2 - |S_{21}|^2 - |S_{11}|^2) + |S_{21}|^2(|1 + \Gamma_{on}|^2(F_{min} - 1) - 4r_n)} + |C_{S_m}|^2}$$
(14.6.15)

We can determine the value of the minimum noise measure obtainable with a given device by considering the noise measure circle of zero radius. This means finding a value of M that makes γ_{S_m} in equation (14.6.15) equal to zero. This can be done by trial and error, although closed form solutions have also been proposed [27,28].

The source reflection coefficient which gives rise to M_{min} is the center of the M_{min} noise measure circle. Once the value of M_{min} has been determined, the value of Γ_{om} can therefore be determined from equation (14.6.14) as:

$$\Gamma_{om} = \frac{M_{min}|1 + \Gamma_{on}|^2 C_1^* + 4r_n|S_{21}|^2\Gamma_{on}}{M_{min}|1 + \Gamma_{on}|^2(|S_{21}|^2 + |S_{11}|^2 - |\Delta|^2) - |S_{21}|^2(|1 + \Gamma_{on}|^2(F_{min} - 1) - 4r_n)}$$
(14.6.16)

With the input port of the transistor terminated in Γ_{om}, we can calculate the output reflection coefficient looking into the output port of the transistor by employing equation (6.2.7), that is,

$$\Gamma_{out} = S_{22} + \frac{S_{12}S_{21}\Gamma_{om}}{1 - S_{11}\Gamma_{om}}$$
(14.6.17)

The use of these equations is best illustrated by means of an example.

Example 14.4 (Minimum noise measure design).

Problem. Design a single-stage amplifier for minimum noise measure using an NE71083 GaAs MESFET at a center frequency of 10 GHz and bias conditions $V_{ds} = 3.0$ V, $l_d = 8$ mA. The S-parameters of the transistor in the common source configuration were measured, with a 50 Ω reference impedance, to be as follows:

$$\begin{bmatrix} S_{11} & S_{12} \\ S_{21} & S_{22} \end{bmatrix} = \begin{bmatrix} 0.724\angle 46° & 0.716\angle -47° \\ 1.303\angle -10° & 0.616\angle 64° \end{bmatrix}$$
(14.6.18)

The following noise parameters were supplied by the manufacturer of the FET:

$$F_{min} = 1.7 \text{ dB}$$
$$\Gamma_{on} = 0.620\angle 148°$$
$$R_n = 12 \ \Omega$$

Solution. The stability of the device is first evaluated using the Edwards-Sinsky stability criteria [29] of equations (7.4.36) and (7.4.37), that is,

$$\mu_1 = \frac{1 - |S_{11}|^2}{|S_{22} - \Delta S_{11}^*| + |S_{12}S_{21}|} = \frac{0.4758}{1.3283} = 0.358 \qquad (14.6.19)$$

$$\mu_2 = \frac{1 - |S_{22}|^2}{|S_{11} - \Delta S_{22}^*| + |S_{12}S_{21}|} = \frac{0.6205}{1.1031} = 0.563 \qquad (14.6.20)$$

Since both μ_1 and μ_2 are less than unity we conclude that the device is potentially unstable. We therefore need to draw a source plane stability circle to determine which source terminations we can use. We calculate the center and radius of the source plane stability circle using equations (7.4.14) and (7.4.15) as follows:

$$C_{SS} = \frac{C_1^*}{|S_{11}|^2 - |\Delta|^2} = \frac{0.1702\angle 85°}{-1.3559} = 0.126\angle - 94°$$

$$r_{SS} = \frac{|S_{12}S_{21}|}{||S_{11}|^2 - |\Delta|^2|} = \frac{0.9329}{1.3559} = 0.688$$

where $C_1 = S_{11} - \Delta S_{22}^* = 0.1702\angle - 85°$.

By determining the constant noise measure circle of zero radius the minimum noise measure obtainable with this device was found to be $M_{min} = 0.435$. equation (14.6.16) yielded the value of the associated source reflection coefficient, Γ_{om}, to be $0.729\angle 146.7°$. Figure 14.12 shows the source plane stability circle together with Γ_{om} and circles of constant noise measure on the source reflection coefficient plane, for various values on M, the latter being calculated using equations (14.6.14) and (14.6.15). Since the stability circle encompasses the origin the stable region is represented by the interior of the stability circle.

Figure 14.12 shows that Γ_{om} lies outside the stable region in the source reflection coefficient plane. It is therefore not possible to realize a stable amplifier stage having the theoretical minimum noise measure of $M_{min} = 0.435$. In order to determine the lowest value of M consistent with a stable amplifier design, we draw progressively smaller circles of constant M until we reach the edge of the source plane stability circle. From Figure 14.12 we can see that the $M = 0.5$ circle just overlaps the source plane stability circle, allowing a small range of Γ_S values that will result in a stable amplifier with a value of $M \leq 0.5$. We therefore choose a source termination $\Gamma_S = 0.610\angle 148$, which lies approximately in the center of this overlapping region, as shown in Figure 14.12. We can confirm that the corresponding load termination also lies within the load plane stable region by calculating the output reflection coefficient, Γ_{out}, of the transistor with an input termination $\Gamma_S = 0.610\angle 148°$ as follows:

$$\Gamma_{out} = S_{22} + \frac{S_{12}S_{21}\Gamma_S}{1 - S_{11}\Gamma_S}$$

$$= 0.616\angle 64° + \frac{0.716\angle - 47° \times 1.303\angle - 10° \times 0.610\angle 148°}{1 - 0.724\angle 46° \times 0.610\angle 148°}$$

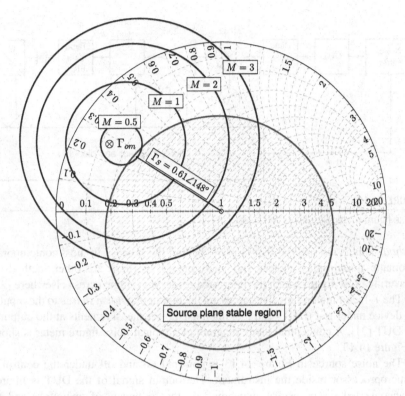

FIGURE 14.12

Source plane constant noise measure circles for NE71083 at 10 GHz ($V_{ds} = 3.0$ V, $I_d = 8$ mA).

$$= 0.616\angle 64° + \frac{0.569\angle 91°}{1.433\angle 4°}$$

$$= 0.994\angle 73°$$

The load required to conjugately match the output port of the device is therefore $\Gamma_L = \Gamma_{out}^* = 0.994\angle -73°$. The fact that $|\Gamma_{out}|$ is close to unity is a reflection of the fact that the chosen value of Γ_S is close to the boundary of the source plane stable region.

14.7 NOISE MEASUREMENTS

14.7.1 NOISE FIGURE MEASUREMENT

The most common method of measuring noise figure of a two-port network, and therefore the only one we will cover in this section, is referred to as the *Y-Factor*

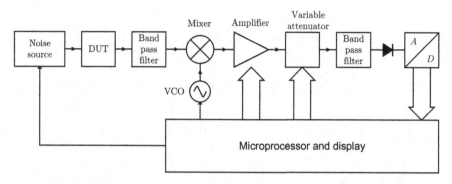

FIGURE 14.13

Noise figure meter simplified block diagram.

method. This is the method that is used "behind the scenes" in most contemporary automatic noise figure meters and analyzers [30]. There are other methods of measuring noise figure which the interested reader may come across elsewhere [23].

The *Y*-factor method involves applying the output of a *noise source* to the input of the device under test (DUT) and making noise power measurements at the output of the DUT [31]. A conceptual block diagram of a typical noise figure meter is shown in Figure 14.13.

The noise source in Figure 14.13 is powered on and off under the control of a microprocessor inside the instrument. The output signal of the DUT is filtered, downconverted (as necessary, depending on the frequency of operation) and the resulting RMS power level measured and digitized. Each time the noise source is turned on or off the noise power at the output of the DUT is thus measured and recorded in the memory of the instrument. The microprocessor carries out the noise figure calculations using the measured data and the equations we will introduce in this section.

The noise source is typically a reverse biased avalanche diode but gas discharge tubes are also used at higher frequencies. A typical commercial solid-state noise source is shown in Figure 14.14.

FIGURE 14.14

Keysight 346B noise source (©Keysight Technologies, Inc. 2014. Reproduced with permission, Courtesy of Keysight Technologies).

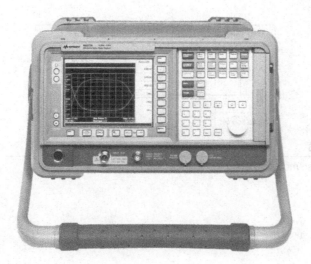

FIGURE 14.15

The Keysight N8975A Noise Figure Analyzer (©Keysight Technologies, Inc. 2014.
Reproduced with permission, Courtesy of Keysight Technologies).

A typical commercial Noise Figure Analyzer, that is capable of measuring noise
figure from 10 MHz to 26.5 GHz is shown in Figure 14.15.

Although we have mostly referred to noise figure throughout this book as
measure of amplifier noise, in the field of microwave noise figure measurements
it is customary to use the *effective noise temperature*, partly because of its greater
precision [13,32].

The noise source in Figure 14.13 can have two levels of noise output power
corresponding to "cold" and "hot" noise temperatures (T_c and T_h), respectively. In
simple terms, these "hot" and "cold" temperatures correspond to the noise source
having its supply switched on and off [33,34].

Assuming the DUT is an amplifier, we can define T_c and T_h in terms of the
corresponding output noise powers, N_1 and N_2, of the amplifier in Figure 14.16,
that is,

$$N_1 = kG\Delta f(T_c + T_a) \tag{14.7.1}$$

and

$$N_2 = kG\Delta f(T_h + T_a) \tag{14.7.2}$$

where G is the numerical power gain of the amplifier and T_a is the effective noise
temperature of the amplifier.

If we measure two noise powers, N_1 and N_2, at noise temperatures T_c and T_h
and plot them on a graph we will get the straight line shown in Figure 14.17. The
slope of the line is the gain bandwidth product of the amplifier scaled by k_B (i.e.,

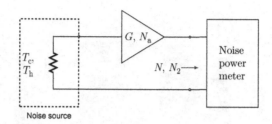

FIGURE 14.16

Amplifier noise model.

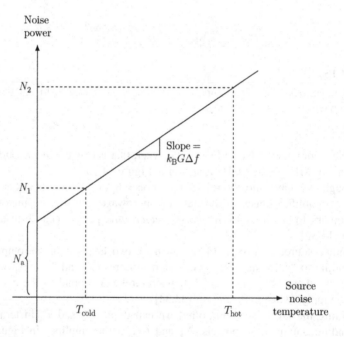

FIGURE 14.17

Effective source temperature versus output noise power.

$k_B G \Delta f$). The line intercepts the noise power axis at a value N_a, which corresponds to the equivalent noise power of the amplifier under test, referred to its input.

The so-called "Y-Factor" is defined as the ratio of "hot" to "cold" measured noise powers, as follows [35]:

$$Y = \frac{N_2}{N_1} \tag{14.7.3}$$

From equations (14.7.1) and (14.7.2) we can write:

$$Y = \frac{T_h + T_a}{T_c + T_a} \qquad (14.7.4)$$

From equation (14.7.4) we can write T_a in terms of the Y-factor as follows [31]:

$$T_a = \frac{T_h - YT_c}{Y - 1} \qquad (14.7.5)$$

The noise factor of the amplifier is related to the effective noise temperature by equation (14.3.3), so we can relate T_a to the system operating temperature as follows:

$$F = \frac{T_a + T_o}{T_o} \qquad (14.7.6)$$

Combining equations (14.7.5) and (14.7.6) we get the noise factor of the amplifier in terms of the Y-factor and the temperatures, T_o, T_c, and T_h as follows [23]:

$$F = \frac{(T_h/T_o - 1) - (T_c/T_o - 1)}{Y - 1} \qquad (14.7.7)$$

Note that equation (14.7.7) is independent of the measurement bandwidth, that has been canceled in the calculation of the Y-factor in equation (14.7.4). This is one of the advantages of the Y-factor technique. The assumption is often made that $T_c = T_o$, in which case (14.7.7) reduces to:

$$F = \frac{(T_h/T_o - 1)}{Y - 1} \qquad (14.7.8)$$

Noise sources are usually specified in terms of the *excess noise ratio* (ENR), which is defined as the power-level difference between hot and cold states, referenced to the thermal equilibrium noise power at the standard operating temperature, T_o. ENR is therefore defined in relation to T_h, T_c, and T_o as:

$$ENR = 10 \ \log_{10}\left(\frac{T_h - T_c}{T_o}\right) \qquad (14.7.9)$$

Again, the assumption is often made that $T_c = T_o$, in which case equation (14.7.9) becomes:

$$ENR \ (dB) = 10 \ \log_{10}\left(\frac{T_h}{T_o} - 1\right) \qquad (14.7.10)$$

Considering equations (14.7.8) and (14.7.10) we can now write the formula for calculating the noise figure of the DUT, in dB, in terms of the measured Y-factor and the ENR of the source, as follows:

$$F \text{ (dB)} = \text{ENR (dB)} - 10 \ \log_{10}(Y - 1) \tag{14.7.11}$$

Example 14.5 (Noise figure measurement: Y-factor method).
Problem. You have been presented with the following measured noise powers:

N_1	−104.5 dBm
N_2	−96.8 dBm

You are using a noise source with an ENR of 13.8 dB.
You can assume the operating temperature of the system is 290 K.
Solution. We start by calculating the Y-factor using equation (14.7.3):

$$Y = \frac{N_2}{N_1} \tag{14.7.12}$$

$$= 10^{\left(\frac{-96.8 + 104.5}{10}\right)} = 10^{0.77} \tag{14.7.13}$$

$$= 5.888 \tag{14.7.14}$$

Assuming $T_c = T_o$, we can apply equation (14.7.11) to get:

$$F\text{(dB)} = \text{ENR (dB)} - 10 \ \log_{10}(Y - 1)$$
$$= 13.8 - 10 \ \log_{10}(5.888 - 1)$$
$$= 6.9 \text{ dB}$$

14.7.2 NOISE CHARACTERIZATION OF MICROWAVE TRANSISTORS

We will now briefly cover the "noise characterization" of microwave transistors, that is, the measurement of the four scalar noise parameters, F_{min}, $|\Gamma_{on}|$, $\angle\Gamma_{on}$, and R_n.

Most noise parameter measurements rely on the noise figure measurement techniques described in Section 14.7.1. In order to measure the four noise parameters of a DUT, we need to measure noise factor, F, of the DUT at different values of source admittance. This can be achieved by inserting a tunable network between the source and input port of the DUT that facilitates the adjustment of the source admittance as seen by the DUT.

The classic approach, recommended since 1959 by the Institution of Radio Engineers (IRE) [36], relies on an application of equation (14.4.24):

$$F = F_{min} + \frac{R_n}{G_S}|Y_S + Y_{on}|^2 \tag{14.4.24}$$

A loss-free admittance tuner (a variable-ratio transformer or a movable stub tuner) is connected to the input of the device so as to present a variable source admittance Y_S. The noise figure of the device is measured with a range of source admittance values. The admittance is adjusted by "trial and error" to find a minimum noise factor, F_{min}. The source admittance associated with F_{min} provides the value of Y_{on}. The value R_n is then extracted from a range of measured (F, Y) data points by linear regression. This technique is referred to as the "source-pull" measurement method [37–39].

A step-by-step procedure for determining the four scalar noise parameters F_{min}, R_n, G_{on}, and B_{on} using a variable tuner was originally proposed by the IRE [36] and can be summarized as follows:

(i) Maintain G_s constant and vary B_s while measuring the noise factor, F. Plot F versus B_s and thereby locate B_{on} in Figure 14.18(a).

(ii) With B_s set at the value determined in (i) above, vary G_s while measuring the noise factor, F. Plot F versus G_s and thereby locate G_{on} (see Figure 14.18(b)).

(iii) Using the data collected in (i) and (ii) above, plot the noise factor, F, against a variable "x," defined by:

$$x = \frac{|Y_S - Y_{on}|^2}{G_S} \qquad (14.7.15)$$

This plot, shown in Figure 14.19, will be a straight line of the form:

$$F = F_{min} + R_n \cdot x \qquad (14.7.16)$$

The slope of this line gives the resistance parameter R_n and its intercept on the F axis gives F_{min}.

The reader will note that the parabolic curves in Figure 14.18 basically represent orthogonal slices through the three-dimensional paraboloid shown in Figure 14.5.

Since the partial derivative of noise figure with respect to source admittance is zero at the noise figure minimum, only small changes in F are observable for large variations in G_s and B_s. This means that values of G_{on} and B_{on} obtained by this simple method are very sensitive to equipment settings and are therefore prone to errors.

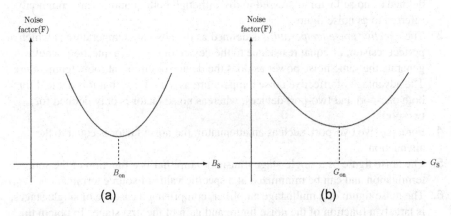

FIGURE 14.18

Determination of $Y_{on} = G_{on} + jB_{on}$. (a) Determination of B_{on}. (b) Determination of G_{on}.

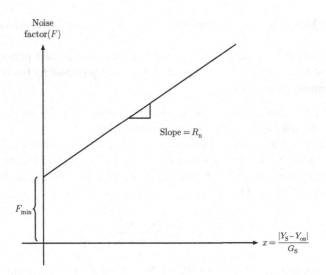

FIGURE 14.19

Determination of R_n.

14.8 TAKEAWAYS

1. Any microwave amplifier design must take into account intrinsic sources of electrical noise, such as thermal noise, shot noise and flicker noise, that will otherwise corrupt and degrade the wanted signal. Although all electrical components exhibit thermal noise other types of noise, such as shot noise and flicker noise, only occur in active devices.

2. Amplifier noise performance is characterized by the *noise factor* which is defined as the ratio of output SNR to input SNR. The *noise figure* is formally defined as noise factor expressed in dB, although both quantities are commonly referred to as noise figure.

3. The *effective noise temperature* is defined as the absolute temperature at which a perfect resistor, of equal resistance to the device or circuit in question, would generate the same noise power as does the device or circuit at room temperature. The advantage of effective noise temperature as a metric is that it is defined for both one-port and two-port devices, whereas noise factor is only defined for two-ports.

4. For a passive two-port, such as an attenuator, the noise factor is equal to the attenuation.

5. The noise figure of a single-stage transistor amplifier is a function of the source termination and can be minimized at a specific value of source termination, Γ_{on}.

6. The noise figure of a multistage amplifier, comprising a cascade of single stages, is largely a function of the noise figure and gain of the first stage. To obtain the minimum overall noise figure for the cascade a compromise must be struck

between the maximum gain and minimum noise figure of each stage. This is achieved by minimising a quantity called the *noise measure*, M.

7. The noise figure of a multistage transistor amplifier is also a function of the source termination and can be minimized at a specific value of source termination Γ_{om}, which is, in general, not the same as Γ_{on} defined above.

REFERENCES

[1] M. Buckingham, Noise in Electronic Devices and Systems, Ellis Horwood Series in Electrical and Electronic Engineering, E. Horwood, 1983, ISBN 9780853122180.

[2] P. Fish, Electronic Noise and Low Noise Design, Macmillan New Electronics Series, Macmillan Press, 1993, ISBN 9780333573105.

[3] I. Darwazeh, L. Moura, Introduction to Linear Circuit Analysis and Modelling, Newnes, 2005, ISBN 0750659327.

[4] J. Johnson, Thermal agitation of electricity in conductors, Phys. Rev. 32 (1928) 97–109, http://dx.doi.org/10.1103/PhysRev.32.97.

[5] M. Keshner, 1/f noise, Proc. IEEE 70 (1982) 212–218.

[6] J. Bernamont, Fluctuations de potential aux bornes d'un conducteur metallique de faible volume parcouru par un courant, Annal. Phys. 7 (1937) 71–140.

[7] A. McWhorter, 1/f noise and related surface effects in germanium, R.L.E. 295 and Lincoln Lab Tech Rep, 80 MIT (1955).

[8] F. Ali, A. Gupta, HEMTs and HBTs: Devices, Fabrication, and Circuits, Artech House Antennas and Propagation Library, Artech House, 1991, ISBN 9780890064016.

[9] H. Rothe, W. Dahlke, Theory of noisy fourpoles, Proc. IRE 44 (6) (1956) 811–818, ISSN 0096-8390, http://dx.doi.org/10.1109/JRPROC.1956.274998.

[10] R. Spence, Linear Active Networks, John Wiley and Sons Inc., New York, 1970.

[11] H. Carlin, A. Giordano, Network Theory: An Introduction to Reciprocal and Non-Reciprocal Circuits, Prentice-Hall Inc., Englewood Cliffs, NJ, 1964.

[12] P. Penfield Jr., Noise in negative-resistance amplifiers, IRE Trans. Circuit Theory 7 (2) (1960) 166–170, ISSN 0096-2007, http://dx.doi.org/10.1109/TCT.1960.1086655.

[13] IRE, IRE standards on electron tubes: definitions of terms, 1957, Proc. IRE 45 (7) (1957) 983–1010, ISSN 0096-8390, http://dx.doi.org/10.1109/JRPROC.1957.278510.

[14] M. Mitama, H. Katoh, An improved computational method for noise parameter measurement, IEEE Trans. Microw. Theory Tech. 27 (6) (1979) 612–615, ISSN 0018-9480, http://dx.doi.org/10.1109/TMTT.1979.1129680.

[15] R. Pengelly, Microwave Field Effect Transistors-Theory, Design and Applications, John Wiley and Sons Ltd, Chichester, 1982.

[16] A. Anastassiou, M. Strutt, Experimental and computed four scattering and four noise parameters of GaAs FET's up to 4 GHz (short papers), IEEE Trans. Microw. Theory Tech. 22 (2) (1974) 138–140, ISSN 0018-9480, http://dx.doi.org/10.1109/TMTT.1974.1128187.

[17] W. Baechtold, M. Strutt, Noise in microwave transistors, IEEE Trans. Microw. Theory Tech. 16 (9) (1968) 578–585, ISSN 0018-9480, http://dx.doi.org/10.1109/TMTT.1968.1126756.

[18] R. Lane, The determination of device noise parameters, Proc. IEEE 57 (8) (1969) 1461–1462, ISSN 0018-9219, http://dx.doi.org/10.1109/PROC.1969.7311.

[19] G. Gonzalez, Microwave Transistor Amplifiers, Analysis and Design, second ed., Prentice Hall Inc., Englewood Cliffs, NJ, 1997.

[20] P. Penfield Jr., Wave representation of amplifier noise, IRE Trans. Circuit Theory 9 (1) (1962) 84–86, ISSN 0096-2007, http://dx.doi.org/10.1109/TCT.1962.1086866.

[21] K. Hartmann, Noise characterization of linear circuits, IEEE Trans. Circuits Syst. 23 (10) (1976) 581–590, ISSN 0098-4094, http://dx.doi.org/10.1109/TCS.1976.1084139.

[22] H. Fukui, Available power gain, noise figure, and noise measure of two-ports and their graphical representations, IEEE Trans. Circuit Theory 13 (2) (1966) 137–142, ISSN 0018-9324, http://dx.doi.org/10.1109/TCT.1966.1082556.

[23] G. Bryant, Institution of Electrical Engineers, Principles of Microwave Measurements, IEE Electrical Measurement Series, P. Peregrinus Limited, 1993, ISBN 9780863412967.

[24] H. Friis, Noise figures of radio receivers, Proc. IRE 32 (7) (1944) 419–422, ISSN 0096-8390, http://dx.doi.org/10.1109/JRPROC.1944.232049.

[25] H. Haus, R. Adler, Optimum noise performance of linear amplifiers, Proc. IRE 46 (8) (1958) 1517–1533, ISSN 0096-8390, http://dx.doi.org/10.1109/JRPROC.1958.286973.

[26] C. Liechti, R. Tillman, Design and performance of microwave amplifiers with GaAs Schottky-gate field-effect transistors, IEEE Trans. Microw. Theory Tech. 22 (5) (1974) 510–517, ISSN 0018-9480, http://dx.doi.org/10.1109/TMTT.1974.1128271.

[27] C. Poole, D. Paul, Optimum noise measure terminations for microwave transistor amplifiers (short paper), IEEE Trans. Microw. Theory Tech. 33 (11) (1985) 1254–1257, ISSN 0018-9480, http://dx.doi.org/10.1109/TMTT.1985.1133207.

[28] J.-C. Liu, S.-S. Bor, P. Lu, D. Paul, P. Gardner, C. Poole, Comments, with reply, on "Optimum noise measure terminations for microwave transistor amplifiers" by C.R. Poole and D.K. Paul, IEEE Trans. Microw. Theory Tech. 41 (2) (1993) 363–364, ISSN 0018-9480, http://dx.doi.org/10.1109/22.216486.

[29] M. Edwards, J. Sinsky, A new criterion for linear 2-port stability using a single geometrically derived parameter, IEEE Trans. Microw. Theory Tech. 40 (12) (1992) 2303–2311, ISSN 0018-9480, http://dx.doi.org/10.1109/22.179894.

[30] Agilent Technologies Inc., Noise figure measurement accuracy–the Y-factor method, Application Note 57-2, 2004.

[31] N. Kuhn, Measure the noise levels of microwave components, Microw. RF (1984) 147–151.

[32] J. Oakes, R. Wickstrom, D. Tremere, T. Heng, A power silicon microwave MOS transistor, IEEE Trans. Microw. Theory Tech. 24 (6) (1976) 305–311, ISSN 0018-9480, http://dx.doi.org/10.1109/TMTT.1976.1128847.

[33] D. Wait, Measurement of amplifier noise, Microw. J. (1973) 25–29.

[34] C. Miller, W. Daywitt, M. Arthur, Noise standards, measurements, and receiver noise definitions, Proc. IEEE 55 (6) (1967) 865–877, ISSN 0018-9219, http://dx.doi.org/10.1109/PROC.1967.5700.

[35] Hewlett Packard, Fundamentals of RF and Microwave Noise Figure Measurements, Hewlett Packard Company, USA, 1983.

[36] IRE, IRE standards on methods of measuring noise in linear two-ports, 1959, Proc. IRE 48 (1) (1960) 60–68, ISSN 0096-8390, http://dx.doi.org/10.1109/JRPROC.1960.287380.

[37] G. Martines, M. Sannino, The determination of the noise, gain and scattering parameters of microwave transistors (HEMT's) using only automatic noise figure test-set,

IEEE Trans. Microw. Theory Tech. 42 (7) (1994) 1105–1113, ISSN 0018-9480, http://dx.doi.org/10.1109/22.299717.

[38] L. Tiemeijer, R. Havens, R. de Kort, A. Scholten, Improved Y-factor method for wideband on-wafer noise-parameter measurements, IEEE Trans. Microw. Theory Tech. 53 (9) (2005) 2917–2925, ISSN 0018-9480, http://dx.doi.org/10.1109/TMTT.2005.854243.

[39] A. Lazaro, L. Pradell, J. O'Callaghan, FET noise-parameter determination using a novel technique based on 50-Ω; noise-figure measurements, IEEE Trans. Microw. Theory Tech. 47 (3) (1999) 315–324, ISSN 0018-9480, http://dx.doi.org/10.1109/22.750233.

Microwave oscillator design

CHAPTER OUTLINE

INTENDED LEARNING OUTCOMES

- *Knowledge*
  - Understand that any electronic oscillator can be understood by employing either a feedback oscillator model or a negative resistance model, and be aware of the equivalence between the two models.
  - Be familiar with the most common RF feedback oscillator topologies, such as Colpitts, Hartley, Clapp/Gouriet, and Pierce.
  - Be familiar with the advantages of the cross-coupled oscillator topology compared with other topologies.
  - Be familiar with the theory and application of various types of high-stability resonator such as crystals, cavities, dielectric resonators, and YIG resonators.

Microwave Active Circuit Analysis and Design. http://dx.doi.org/10.1016/B978-0-12-407823-9.00015-9

- Understand the operation of transistor feedback and negative resistance oscillators.
- Understand how varactors are used to implement an electronically tunable oscillator.
- *Skills*
  - Be able to carry out a simple fixed frequency transistor oscillator design based on standard topologies (Colpitts, Hartley, Clapp/Gouriet).
  - Be able to carry out a simple fixed frequency negative resistance transistor oscillator design.
  - Be able to carry out a simple fixed frequency cross-coupled transistor oscillator design.

15.1 **INTRODUCTION**

Oscillators play a prominent role in all areas of radio frequency (RF) and microwave electronics. Practically all microwave systems will contain at least one oscillator the purpose of which will be to generate a local reference signal at specific frequencies. Even if the device in question is only receiving, not transmitting, radio energy, it will still need to contain a *local oscillator* to facilitate the process of frequency translation via *heterodyning* [1]. For this reason, microwave transistor oscillator design is an extensive and active area of ongoing research, producing a wide variety of integrated (monolithic microwave integrated circuits, MMIC) oscillators operating up to the THz frequency range [2].

The physical world abounds with oscillatory phenomena of all kinds. All readers will be familiar with the pendulum, and some will have a deeper understanding of the fact that the pendulum is basically a system in which mechanical energy is continuously being exchanged between two forms: kinetic and potential. At the top of the pendulum's swing, all the energy is "stored" in its potential form (the pendulum is stationary at that instant, so kinetic energy is zero). As the pendulum passes through the lowest point of its swing, the potential energy is zero and all the energy is "stored" in the form of kinetic energy. These concepts of energy storage and energy exchange are central to an understanding of oscillators.

We now turn our attention to electronic oscillators, and keeping in mind the concept of energy storage and exchange, we note that, in the electrical realm, energy is stored in the form of electrostatic and magnetic fields, in components called capacitors and inductors, respectively. The simplest conceptual oscillator can therefore be modeled as a single capacitor and a single inductor connected together, as shown in Figure 15.1. The circuit of Figure 15.1 is the simplest representation of a "tank" circuit, which is an essential subcircuit in any oscillator.

The switch S is included in Figure 15.1 to define the exact instant t_0 when current begins to flow (i.e., the instant when the switch is closed). We assume that all components in Figure 15.1 are lossless, in other words there is zero resistance around the loop.

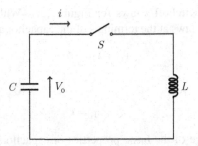

FIGURE 15.1

Simplest possible electronic oscillator.

Let us begin by assuming the capacitor C is charged up to a static voltage V_0 (we could achieve this by connecting the capacitor to a battery, then disconnecting it). At time $t = t_0$, we close the switch, S, thereby connecting the capacitor across the inductor, L. Current starts to flow from the capacitor into the inductor, starting from $i(t) = 0$ at time $t = t_0$ and building up to a maximum some time later. At the instant the inductor current reaches its maximum value, the voltage across the inductor will be zero, since the capacitor has been fully discharged, and the voltage across the capacitor must equal that across the inductor, in order to satisfy Kirchhoff's voltage law.

To satisfy Kirchhoff's current law, however, the same current that is flowing through the inductor at any instant, also has to flow through the capacitor. This means that the current flowing in the inductor will start to "recharge" the capacitor from $v_c(t) = 0$ back up toward $v_c(t) = V_0$, from whence the cycle will repeat. Since the circuit in Figure 15.1 is lossless, the cycle will repeat indefinitely, with energy being exchanged back and forth between the capacitor and inductor, while the total energy in the circuit remains constant, as shown in Figure 15.2.

In Figure 15.2, we have implied that the energy levels in L and C will vary sinusoidally with time. We can prove that the voltage and current variations are

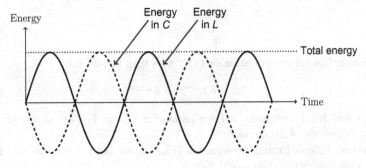

FIGURE 15.2

Energy exchange in an oscillating LC circuit.

sinusoidal by solving Kirchoff's laws for Figure 15.1. With the switch, S, closed we can state Kirchoff's laws at the terminals of the capacitor and inductor as follows:

$$V_C + V_L = 0 \tag{15.1.1}$$

and

$$i_C = i_L = i \tag{15.1.2}$$

From our knowledge of the basic properties of capacitors and inductors we can also write:

$$V_L(t) = L\frac{di(t)}{dt} \tag{15.1.3}$$

and

$$i(t) = C\frac{dV_C(t)}{dt} \tag{15.1.4}$$

Combining equations (15.1.1) and (15.1.3) we can write:

$$V_C + L\frac{di(t)}{dt} = 0 \tag{15.1.5}$$

Differentiating equation (15.1.5) with respect to time we get:

$$\frac{dV_C(t)}{dt} + L\frac{d^2i(t)}{dt^2} = 0 \tag{15.1.6}$$

Substituting equation (15.1.4) into equation (15.1.6) we arrive at a second-order differential equation for the circuit of Figure 15.1:

$$\frac{d^2i(t)}{dt^2} + \left(\frac{1}{LC}\right)i(t) = 0 \tag{15.1.7}$$

If we now define $\omega_0 = 1/\sqrt{LC}$, we can rewrite equation (15.1.7) as:

$$\frac{d^2i(t)}{dt^2} + \omega_0^2 i(t) = 0 \tag{15.1.8}$$

The complete solution to equation (15.1.8) is then in the form:

$$i(t) = A\,e^{-j\omega_0 t} + B\,e^{+j\omega_0 t} \tag{15.1.9}$$

where A and B are constants and the parameter $\omega_0 = 1/\sqrt{LC}$ is known as the *resonant frequency* of the circuit.

Applying Euler's formula to equation (15.1.9), we can see that the variation of current with time, $i(t)$, is sinusoidal, that is,

$$i(t) = [B + A]\cos(\omega_0 t) + j[B - A]\sin(\omega_0 t) \tag{15.1.10}$$

The circuit of Figure 15.1 is an ideal one, wherein the sinusoidal oscillations, once started, will continue indefinitely. In real-world circuits, we have to contend with resistance. This means that the original energy present in the circuit will eventually be dissipated and the oscillatory currents and voltages will decrease in amplitude over time, unless more energy is continuously injected into the circuit. Adding a single resistance, R, to represent all the various resistive elements distributed around the circuit, we now have the circuit of Figure 15.3.

It is left as an exercise for the reader to show that, with the addition of resistance as in Figure 15.3, the generalized version of second-order differential equation of equation (15.1.7) is now:

$$\frac{d^2i(t)}{dt^2} + \left(\frac{R}{L}\right)\frac{di(t)}{dt} + \left(\frac{1}{LC}\right)i(t) = 0 \tag{15.1.11}$$

which can be expressed as:

$$\frac{d^2i(t)}{dt^2} + 2\alpha\frac{di(t)}{dt} + \omega_0^2 i(t) = 0 \tag{15.1.12}$$

The parameter $\alpha = R/2L$ is referred to as *attenuation*, and is a measure of how fast the amplitude of oscillation decays with time. Since attenuation will be present in all real-world circuits, in order to ensure sustained oscillation at constant amplitude, we need a mechanism for injecting fresh energy into the tank circuit at every cycle. Any practical oscillator therefore needs to contain some form of *active* device which can act as an *amplifier*.

We pause at this point to summarize two common constituent parts of all electronic oscillators:

(a) An active device providing positive net signal gain at the frequency of operation.

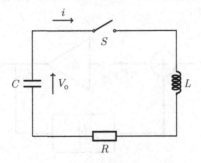

FIGURE 15.3

RLC resonant circuit.

(b) A feedback mechanism providing the correct degree of phase shift at the frequency of operation.

There are two conceptual models for microwave oscillator analysis, namely the feedback oscillator model and the negative resistance oscillator model. Either model may be applied to analyze a particular oscillator, but one model may be more appropriate than the other, depending on the oscillator topology and type of active device used. Both these models will be covered in this chapter.

15.2 RF FEEDBACK OSCILLATORS

15.2.1 THE FEEDBACK OSCILLATOR MODEL

We can model any electronic oscillator as an amplifier with feedback. According to this model, energy from the output of the amplifier is fed back to the input via some kind of feedback mechanism, which may be a distinct electrical circuit or might be some physical process internal to the active device itself, as is the case with an oscillating microwave diode. In any case, we will model the feedback mechanism as a passive electrical feedback network having a particular frequency response, as shown in Figure 15.4.

From Figure 15.4 we can write the output signal voltage, v_o in terms of the input signal voltage, v_i:

$$v_o = A[v_i + \beta v_o] \tag{15.2.1}$$

Which can be written in terms of the transfer function, v_o/v_i, noting that both the amplifier gain, A, and the feedback fraction, β, are functions of frequency:

$$\frac{v_o}{v_i} = \frac{A(j\omega)}{1 - A(j\omega)\beta(j\omega)} \tag{15.2.2}$$

Depending on the sign of the product $A(j\omega)\beta(j\omega)$ in equation (15.2.2), which depends on the combined phases of $A(j\omega)$ and $\beta(j\omega)$, the feedback can be classed

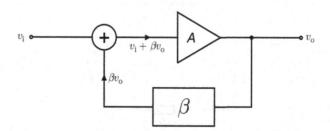

FIGURE 15.4

Conceptual feedback oscillator.

as either *negative* or *positive*. Negative feedback tends to reduce fluctuations in v_o and thereby increase stability of the system, as v_i and βv_o are added destructively. Positive feedback tends to have the opposite effect; variations in v_i and βv_o positively reinforce each other with the result that the amplitude of v_o will increase up to a limit, set by practical circuit constraints such as the power supply voltage or nonlinearities of the active device.

From equation (15.2.2) we can see that the transfer function becomes infinite when:

$$A(j\omega_0)\beta(j\omega_0) = 1 \tag{15.2.3}$$

The condition for oscillation represented by equation (15.2.3) is generally known as the *Barkhausen criterion* [3]. Since both the amplifier gain and the feedback fraction are in general complex, equation (15.2.3) represents both an amplitude and a phase requirement for oscillation. Another way of expressing condition (15.2.3) is in terms of the magnitude and phase of $A(j\omega)\beta(j\omega)$, as follows:

$$|A(j\omega_0)|\,|\beta(j\omega_0)| = 1 \tag{15.2.4}$$

and

$$\angle[A(j\omega_0)\beta(j\omega_0)] = 0 \text{ or } \pm 2n\pi \tag{15.2.5}$$

There are consequences of equation (15.2.4) that apply to all oscillators and should therefore be noted. In order to build an oscillator, one needs a device which is capable of gain ($|A(j\omega_0)| > 1$) at the frequency of interest. Therefore, any criteria which determine the amplification potential of a given active device also potentially determine its oscillation potential. In other words, and put simply, no gain means no oscillation.

We can, for example, implement the passive feedback network $\beta(j\omega)$ as a π or T network of purely reactive elements. Consider a π network shown in Figure 15.5. If we make the reasonable assumption that the amplifier gain, $A(j\omega)$, is more or less flat over the frequency range of interest, then a simple analysis shows that to fulfill equation (15.2.3), the reactances X_1, X_2, and X_3 need to satisfy the following condition:

$$X_3 = -(X_1 + X_2) \tag{15.2.6}$$

Equation (15.2.3) implies that, if X_3 is chosen to be an inductor, then X_1 and X_2 should be capacitors, and vice versa. We can now define two well-known and widely used RF oscillators in Table 15.1.

We will now proceed to analyze the Colpitts and Hartley oscillator in more detail.

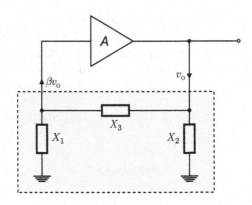

FIGURE 15.5

Feedback oscillator with Π feedback network.

Table 15.1 Classic Feedback Oscillators

Colpitts oscillator	X_3 = Capacitor
	X_2 = Inductor
	X_1 = Inductor
Hartley oscillator	X_3 = Inductor
	X_2 = Capacitor
	X_1 = Capacitor

15.2.2 COLPITTS OSCILLATOR

A feedback oscillator circuit in which the feedback network comprises a low-pass LC π-network is referred to as a Colpitts oscillator, shown conceptually in Figure 15.6. The Colpitts oscillator was invented in 1920 by American engineer Edwin H. Colpitts.

A single bipolar junction transistor (BJT) Colpitts oscillator is shown in Figure 15.7 where the active device is a single BJT and the feedback network consists of the series capacitors C_1 and C_2 in parallel with the inductor, L_1. The transistor is operated in the common base configuration, with the base being AC grounded through the capacitor C_4. In common base configuration, there is 0° phase shift through the transistor amplifier stage (i.e., from emitter to collector). The requirement for oscillation (0° or $n \times 360°$ phase shift around the loop) means that there must be 0° phase shift through the feedback network (C_1 in this case). The capacitor C_3 effectively ensures that the V_{cc} rail is also an AC ground. The DC bias point of the transistor is set by the voltage divider R_1 and R_2.

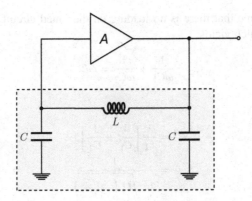

FIGURE 15.6

Conceptual Colpitts oscillator.

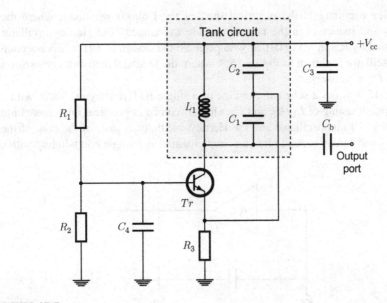

FIGURE 15.7

Common base BJT Colpitts oscillator.

Since we assume that there is no loading on the tuned circuit at resonance, we have $X_L = X_C$, which yields:

$$\frac{1}{\omega C_1} + \frac{1}{\omega C_2} = \omega L \tag{15.2.7}$$

which gives

$$\omega^2 = \frac{1}{L_1}\left[\frac{1}{C_1} + \frac{1}{C_2}\right] \tag{15.2.8}$$

$$f_0 = \frac{1}{2\pi\sqrt{L_1\left(\frac{C_1 C_2}{C_1 + C_2}\right)}} \tag{15.2.9}$$

The Colpitts oscillator is probably the most common oscillator topology and forms the foundation of many oscillator designs well into the millimeter wave frequency range [4].

15.2.3 HARTLEY OSCILLATOR

The Hartley oscillator is the electrical dual of the Colpitts oscillator, where the capacitors and inductors in the tank circuit are exchanged. The Hartley oscillator was invented by Ralph V.L. Hartley who patented the design in 1915. A conceptual Hartley oscillator is shown in Figure 15.8, where the feedback network comprises a high-pass LC π-network.

Figure 15.9 shows a schematic outline of a single BJT Hartley oscillator with a tank circuit consisting of L_1, L_2, and C_1. One difference in practical implementation between the Colpitts oscillator and the Hartley oscillator is that, in the case of the Hartley, L_1 and L_2 can be, and often are, implemented as a single coil winding with a

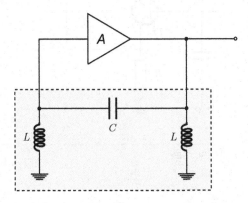

FIGURE 15.8

Conceptual Hartley oscillator.

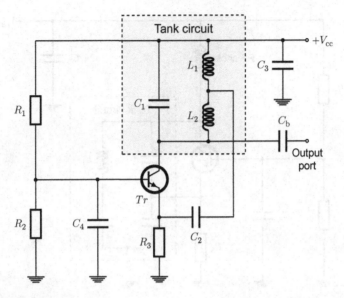

FIGURE 15.9

Common base BJT Hartley oscillator.

center tap. There is no requirement for there to be any mutual coupling between the two coil segments.

Referring to Figure 15.9, once again the transistor is operated in the common base configuration, with the base being AC grounded through the capacitor C_4. C_2 is simply a DC blocking capacitor that prevents L_2 creating a DC short between the emitter and collector of the transistor.

Since we assume that there is no loading on the tuned circuit at resonance, we have $X_L = X_C$, which yields:

$$\omega (L_1 + L_2) = \frac{1}{\omega C_1} \tag{15.2.10}$$

Therefore,

$$f_0 = \frac{1}{2\pi \sqrt{C_1 (L_1 + L_2)}} \tag{15.2.11}$$

The components R_1, R_2, R_3, C_3, and C_4 in Figure 15.9 perform the same functions as the corresponding components in Figure 15.7.

15.2.4 CLAPP/GOURIET OSCILLATOR

The Clapp/Gouriet oscillator topology is basically a Colpitts oscillator with an additional capacitor C_0 added in series with the inductor L_1 in the tank circuit, as

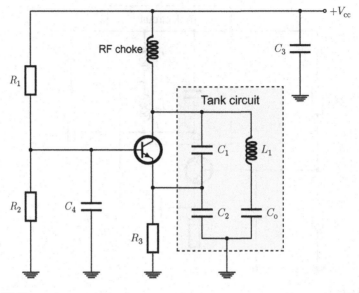

FIGURE 15.10

Single transistor Clapp/Gouriet oscillator.

shown in Figure 15.10. The Clapp/Gouriet topology is preferred over the Colpitts for variable frequency oscillators, where C_o becomes the tuning element. If we were to make a variable frequency oscillator using the Colpitts topology, by varying one of the capacitors in the voltage divider comprising C_1 and C_2, this would cause the feedback voltage to vary, altering the oscillation conditions over a portion of the desired frequency range. This problem is avoided in the Clapp/Gouriet topology by using fixed capacitors in the voltage divider and adding another capacitor, C_o in series with the inductor in the tank circuit in Figure 15.10, which can be varied without changing the feedback conditions around Q_1.

This circuit derives its name from James Kilton Clapp who first published it in 1948. Although Geoffrey G. Gouriet had been reported as employing such a circuit as early as 1938.

Since we assume that there is no loading on the tuned circuit at resonance, we have $X_L = X_C$, which yields:

$$\omega L = \frac{1}{\omega C_1} + \frac{1}{\omega C_2} + \frac{1}{\omega C_o} \tag{15.2.12}$$

Therefore,

$$f_o = \frac{1}{2\pi} \sqrt{\frac{1}{LC_o} \left[1 + \frac{C_o}{C_1} + \frac{C_o}{C_2} \right]} \tag{15.2.13}$$

If we choose component values such that $\left[\frac{C_0}{C_1} + \frac{C_0}{C_2}\right] \ll 1$, equation (15.2.13) simplifies to:

$$f_0 = \frac{1}{2\pi\sqrt{LC_0}} \tag{15.2.14}$$

Clapp/Gouriet oscillators are usually designed with C_1 and C_2 values such that the approximation equation (15.2.14) applies. This allows the oscillator to be tuned by varying only C_0, which can be a varactor, allowing the oscillator to be tuned electronically. This approach has the added advantage of allowing values of C_1 and C_2 to be independently chosen in order to optimize the oscillation conditions of Q_1. It should be noted that in the Clapp/Gouriet circuit (as with the Colpitts), it is the ratio $C_1:C_2$ that is important, not the actual values, within reason.

15.3 CROSS-COUPLED OSCILLATORS

An oscillator topology that has become particularly popular in MMIC implementations, especially where the MMIC is implemented in CMOS technology, is the cross-coupled oscillator. The popularity of the cross-coupled topology is due to the "push-push" operation of the two transistors providing an output amplitude twice that of a typical single transistor oscillator, thereby extending the usable frequency range for a given transistor technology. Since cross-coupled oscillators generally employ LC resonant circuits, the whole circuit, including the resonator, can be fully integrated in MMIC form. The cross-coupled configuration also lends itself readily to electronic tunability, as discussed later in this chapter.

We start by considering the single-stage field-effect transistor (FET) tuned amplifier shown in Figure 15.11. The load of the FET in this case consists of a discrete RLC resonator. The susceptances of the L and C cancel at resonance, resulting in the resonator impedance reaching a maximum, purely real, value equal to R.

We can represent the FET in Figure 15.11 by its absolute simplest small signal model, coupled to the equivalent load impedance of the tank circuit, Z_{tank}, as shown in Figure 15.12.

The small signal gain of the equivalent circuit in Figure 15.12 is given by:

$$v_{out} = -g_m v_{gs} Z_{tank} \tag{15.3.1}$$

With the minus sign representing the 180° phase shift between AC input voltage and output voltage; that is, the circuit behaves as a signal *inverter*. The small signal gain, $A(j\omega)$ of the inverter in Figure 15.11 is, therefore,

$$A_1(j\omega) = \frac{v_{out}}{v_{gs}} = -g_m Z_{tank} \tag{15.3.2}$$

$$= \frac{-g_m}{\frac{1}{R} + j\left(\omega C - \frac{1}{\omega L}\right)} \tag{15.3.3}$$

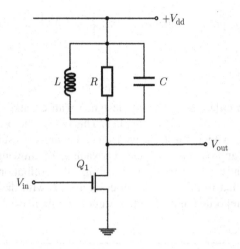

FIGURE 15.11

Single-stage FET tuned amplifier.

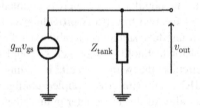

FIGURE 15.12

MOSFET small signal model.

To make an oscillator using the tuned amplifier of Figure 15.11, we need to add one more inversion to achieve an overall phase shift of 360° around the loop. We can do this by adding a second-inverter stage and a feedback path as shown in Figure 15.13.

If we set $R_1 = R_2 = R$, $C_1 = C_2 = C$, and $L_1 = L_2 = L$ then the gain of the second stage in Figure 15.13 will be identical to the first. The overall loop gain of the circuit in Figure 15.13 will therefore be:

$$A_2(j\omega) = \left[\frac{-g_m}{\frac{1}{R} + j\left(\omega C - \frac{1}{\omega L}\right)} \right]^2 \tag{15.3.4}$$

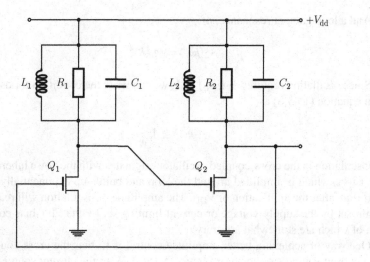

FIGURE 15.13

Two-stage FET tuned amplifier.

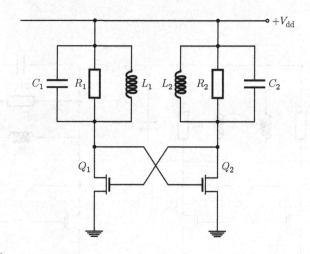

FIGURE 15.14

Cross-coupled oscillator.

The circuit of Figure 15.13 is more commonly drawn in the format shown in Figure 15.14.

Setting the imaginary part of equation (15.3.4) equal to zero gives the oscillation frequency of the circuits of Figures 15.13 and 15.14 as:

$$f_0 = \frac{1}{2\pi\sqrt{LC}} \qquad (15.3.5)$$

And a loop gain, at resonance, of:

$$A_2(j\omega_0) = (g_m R)^2 \tag{15.3.6}$$

Since oscillation requires $A_2(j\omega_0) \geq 1$, we can write the condition for oscillation, from equation (15.3.6) as:

$$g_m R \geq 1 \tag{15.3.7}$$

Oscillation in the cross-coupled oscillator originates with the noise inherent in the transistors, which is amplified around the loop and builds up exponentially within a short time after the application of V_{DD}. The amplitude of oscillation will reach either a limit set by the supply voltage or current limiting set by the FET bias conditions, both of which are somewhat arbitrary.

One way of achieving better amplitude control is to bias the cross-coupled pair via a current source, as shown in Figure 15.15(a). With the current source inserted, the amplitude is controlled by I_{SS}. All of this current is steered between either the left or right side of the differential pair. Thus, the amplitude of the output will be $I_{SS}R$.

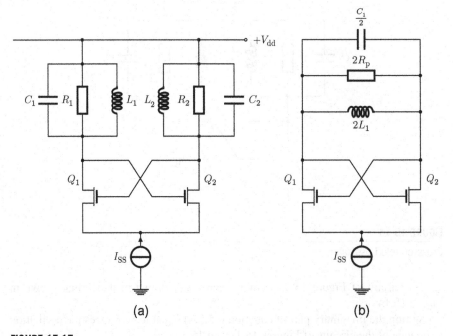

(a)   (b)

FIGURE 15.15

MMIC implementation of the cross-coupled oscillator. (a) Current source bias. (b) Load tanks merged.

The current will vary slightly with changes in V_{dd}, but the output amplitude is much more stable than in the basic circuit of Figure 15.14.

The price to be paid for greater amplitude stability, with the addition of the current source, may be added noise. The channel noise of the device used to implement the current source adds to the total noise of the amplifiers, Q_1 and Q_2, so the phase noise of a current source biased design is somewhat worse than that of the circuit in Figure 15.14.

In many MMIC applications, the two tank circuits of a cross-coupled oscillator are merged, as shown in Figure 15.15(b). In such cases, the equations for oscillation frequency and loop gain, equations (15.3.7) and (15.3.5), still apply.

15.4 NEGATIVE RESISTANCE OSCILLATORS

Although all oscillators are fundamentally feedback oscillators, and conform to the theoretical model shown in Figure 15.4,[1] an alternative model, the *negative resistance model*, is often used to analyze and design microwave oscillators.

The negative resistance model arose in the era of microwave diode oscillators, where the diode behaves as a two-terminal negative resistance device under certain bias conditions, as explained in Chapter 11. The negative resistance model is also frequently used to analyze transistor oscillators, where negative resistance behavior in the transistor has been deliberately engineered by a combination of device configuration and feedback. Since modern microwave oscillators are mostly designed using transistors, we will devote a large part of this section to techniques of generating negative resistance in transistors.

The concept of a "negative resistance" implies that current flow is the reverse of what would be expected when a positive voltage is applied. In other words, signal current flows *out* of the device when a positive signal voltage is applied to it; the opposite of what happens when the same voltage is applied to a "positive" resistor. Whether the negative resistance model is appropriate is largely a question of how we partition the oscillator circuit into active and passive subcircuits, so as to delineate the negative and positive resistance elements of the circuit.

A simple one-port negative resistance circuit is shown in Figure 15.16(a). If we observe the current flowing out of the one-port active device and into the passive load then the resistance of the active device must be:

$$R_d = \frac{V}{-I} = -R_c \qquad (15.4.1)$$

The resistance of the active device is therefore of equal magnitude but of opposite sign to that of an equivalent passive resistor. The condition for sustained current flow in the circuit of Figure 15.16(a) is, therefore,

---

[1] The reader may ask how this statement applies in the case of a one-port active device, such as the microwave diodes discussed in Chapter 11. The answer is that there will always be a feedback mechanism at work somewhere in the circuit, even if this is taking place inside the active device itself.

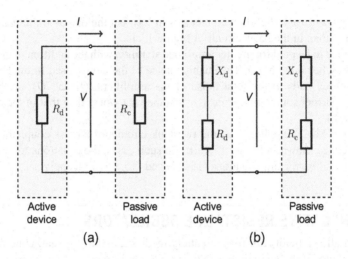

FIGURE 15.16

Conceptual negative resistance. (a) Pure resistances. (b) Complex impedances.

$$R_d + R_c = 0 \qquad (15.4.2)$$

Condition (15.4.2) is independent of frequency, suggesting that the circuit of Figure 15.16(a) is capable of power generation over an infinite bandwidth. In the real world, both source and load will inevitably contain reactive elements which can be represented by X_d and X_c in Figure 15.16(b). We therefore have the oscillation condition for the circuit of Figure 15.16(b) as [5]:

$$Z_d + Z_c = 0 \qquad (15.4.3)$$

In terms of resistive and reactive components, equation (15.4.3) becomes:

$$R_d = -R_c \qquad (15.4.4)$$
$$X_d = -X_c \qquad (15.4.5)$$

Finally, in terms of reflection coefficients, the oscillation condition or the circuit of Figure 15.16(b) can be expressed as:

$$\Gamma_d \Gamma_c = 1 \qquad (15.4.6)$$

The one-port oscillation condition of Equation (15.4.6) is really a special case of a more general oscillation condition for networks with an arbitrary number of ports. S-parameter analysis of n-port negative resistance oscillators is based on the oscillator representation shown in Figure 15.17 [6,7]. The oscillator is considered as being divided into two parts: (a) the n-port active device and (b) a passive n-port

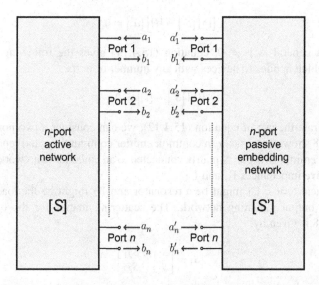

FIGURE 15.17

Generalized *n*-port oscillator model.

network which represents the resonant circuit(s), bias components output matching networks and any other such passive circuitry. In this conception, the passive network is explicitly taken to also include the load.

The terminology in common use refers to the active device as being "embedded" in the active network [8–11]. The object of any design procedure is to synthesize an embedding network which will cause the overall circuit to oscillate at the frequency of interest and deliver useful signal power to the load.

The power wave relationship for the active device of Figure 15.17 is as follows:

$$[b] = [S][a] \tag{15.4.7}$$

where $[S]$ is the $n \times n$ S-matrix of the active device. Similarly, the power wave relationship for the passive embedding network in Figure 15.17 is:

$$[b'] = [S'][a'] \tag{15.4.8}$$

where $[S']$ is the $n \times n$ S-matrix of the embedding network. When the *n*-ports of Figure 15.17 are connected together at each port the following relationships apply:

$$[b'] = [a] \tag{15.4.9}$$

$$[b] = [a'] \tag{15.4.10}$$

From equations (15.4.4) to (15.4.10), the following relationships can be deduced:

$$[a'] = [S][S'][a']$$

$$[[S][S'] - [I]][a'] = 0 \qquad (15.4.11)$$

Since in general $[a'] \neq 0$, equation (15.4.11) gives the following oscillation condition which applies to devices with any number of ports:

$$|[S][S'] - [I]| = 0 \qquad (15.4.12)$$

To illustrate the use of equation (15.4.12), we can consider a two-port example. Figure 15.18 shows a transistor in common emitter configuration and represented by its common emitter two-port S-matrix connected to an embedding network consisting of two passive impedances Γ_1 and Γ_2.

In practical terms, Γ_1 might be a resonator and Γ_2 might be the load, reflected through an output matching network. The scattering matrix for the transistor in Figure 15.18 is given by:

$$[S] = \begin{bmatrix} S_{11} & S_{21} \\ S_{12} & S_{22} \end{bmatrix} \qquad (15.4.13)$$

The S-matrix for the embedding network in Figure 15.18 can be written as:

$$[S'] = \begin{bmatrix} \Gamma_1 & 0 \\ 0 & \Gamma_2 \end{bmatrix} \qquad (15.4.14)$$

Applying equation (15.4.12) to this case gives the following oscillation condition for the circuit of Figure 15.18:

$$\begin{vmatrix} (S_{11}\Gamma_1 - 1) & S_{12}\Gamma_1 \\ S_{21}\Gamma_2 & (S_{22}\Gamma_2 - 1) \end{vmatrix} = 0 \qquad (15.4.15)$$

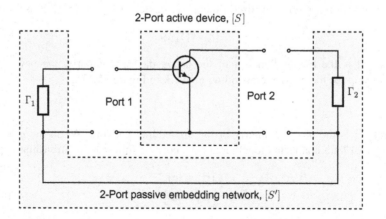

FIGURE 15.18

Two-port oscillator model.

Expanding the determinant in equation (15.4.15) results in the following two simultaneous oscillation conditions:

$$S_{11} + \frac{S_{12}S_{21}\Gamma_2}{1 - S_{22}\Gamma_2} = \frac{1}{\Gamma_1} \tag{15.4.16}$$

$$S_{22} + \frac{S_{12}S_{21}\Gamma_1}{1 - S_{11}\Gamma_1} = \frac{1}{\Gamma_2} \tag{15.4.17}$$

By comparison with equations (6.2.5) and (6.2.7) we can see that the left-hand sides of equations (15.4.16) and (15.4.17) correspond to Γ_{in} with port 2 terminated in Γ_2 and Γ_{out}, and with port 1 terminated in Γ_1, respectively. Equations (15.4.16) and (15.4.17) can therefore be rewritten in simpler form as:

$$\Gamma_{in}\Gamma_1 = 1 \tag{15.4.18}$$
$$\Gamma_{out}\Gamma_2 = 1 \tag{15.4.19}$$

If we arrange for condition (15.4.18) to be satisfied then we will find that equation (15.4.19) is automatically satisfied as well [12]. In other words, if the active two-port exhibits negative resistance at port 1 and Γ_1 is chosen such that condition (15.4.18) is satisfied, then Γ_{out} will assume a value such that:

$$\Gamma_{out} = \frac{1}{\Gamma_2} \tag{15.4.20}$$

If we consider a two-port oscillator with the output at port 2, then Γ_2 would be the matched load, that is $\Gamma_2 = 0$. Then, according to condition (15.4.20):

$$\Gamma_{out} =\rightarrow \infty \tag{15.4.21}$$

Which is consistent with port 2 being a source of signal power (i.e., $b_2 = 0$ when $a_2 = 0$).

The oscillation condition represented by equations (15.4.18) and (15.4.19) implies that, since Γ_1 and Γ_2 are both passive, $|S_{ii}|$ for the transistor should be greater than unity (where $i = 1, 2$). This is rarely the case for a transistor connected in the common emitter configuration without any feedback, as in Figure 15.18.

In Chapter 8, we addressed the question of how to generate negative resistance in transistors using feedback. In the case of a common emitter transistor, for example, capacitive series feedback can be used to generate negative resistance [13].

15.5 FREQUENCY STABILIZATION

All the oscillators we have studied so far have employed some form of LC resonant "tank" circuit as the frequency determining element. In some applications, a higher degree of frequency stability is required than can be provided by a simple LC resonant circuit. In this section, we will introduce some high-stability resonators that are employed at microwave frequencies.

15.5.1 CRYSTAL STABILIZATION

At the lower RFs (below 100 MHz), the most common type of high Q resonator is the crystal.

A crystal is an electro-mechanical resonator that relies on the piezoelectric properties of certain materials. Piezoelectricity, discovered in 1880 by Pierre Curie, is a property whereby electricity is produced from within a material sample when it is subjected to mechanical stress. The piezoelectric effect is reciprocal; that is, a voltage applied to a sample of piezoelectric material will cause it to become mechanically deformed.

Ceramic and silicon dioxide (quartz) are the most commonly used piezoelectric materials for high-frequency crystal resonators. Quartz is a crystalline form of silicon dioxide (SiO_2) which is abundant in nature, forming about 12% of the Earth's crust. The fact that silicon dioxide is abundant does not, however, mean that high-quality natural quartz material is readily available to make electronic components. Most of the quartz used in crystal resonators is cultured; that is, it is grown specifically for this purpose. Large bars of crystal are grown under controlled conditions. These bars are then cut into wafers. The angle at which the wafer is cut is crucial in determining the frequency and temperature stability of the resulting crystal resonator. The most common cut is the AT-cut where the wafer is cut from the bar of crystal at an angle of approximately 35°.

The resonant frequency of a crystal resonator is a function of the thickness of the crystal. The fundamental frequency of AT-cut crystals can be up to around 45 MHz. Higher frequencies, up to several hundred MHz, can be achieved by operating the crystal at odd overtones (3rd, 5th, 7th, 9th, 11th, and so on).

The equivalent circuit for the quartz crystal is usually represented as an *RLC* series circuit, which represents the mechanical characteristics of the crystal sample, in parallel with a capacitance, C_p which represents the package and connection parasitics, as shown in Figure 15.19(b). The inductance, L, represents the electrical equivalent of the crystal mass; the capacitor, C_s, represents the stiffness or elasticity of the crystal sample; and r represents heat losses due to mechanical friction.

Quartz crystal oscillators operate at "parallel resonance," and the equivalent impedance of the crystal has a series resonance where C_s resonates with inductance, L and a parallel resonance where L resonates with the series combination of C_s and C_p.

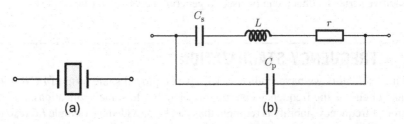

FIGURE 15.19

Crystal equivalent circuit. (a) Crystal circuit symbol. (b) Crystal equivalent circuit.

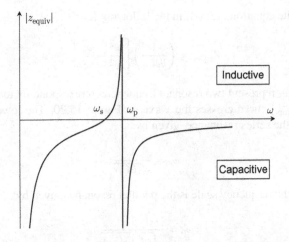

FIGURE 15.20

Quartz crystal reactance characteristic.

The slope of the reactance against frequency curve in Figure 15.20 shows that the series reactance at frequency ω_p is inversely proportional to C_s because below ω_s and above ω_p the crystal appears capacitive.

Between frequencies ω_s and ω_p, the crystal appears inductive as the two parallel capacitances cancel out. The point where the reactance values of the capacitances and inductance cancel each other out $X_c = X_L$ is the fundamental frequency of the crystal.

The Q of a crystal can be related to Figure 15.19(b) as follows:

$$Q = \frac{\omega L}{r} \tag{15.5.1}$$

A typical Q value for a quartz crystal oscillator is between 10^4 and 10^6 (compared with around 10^2 for an LC oscillator). This implies that, with reasonable values of L (i.e., in the Henries range), the series resistance, r, will be very small and can be safely ignored in most cases. We can therefore express the total impedance of the crystal in Figure 15.19(b) as the following pure reactance:

$$Z_{equiv} = \frac{1}{j\omega C_p + \frac{1}{j\omega L + 1/j\omega C_s}} \tag{15.5.2}$$

which can be rearranged as:

$$Z_{equiv} = \left(\frac{1}{j\omega C_p}\right)\left[\frac{\omega^2 - 1/LC_s}{\omega^2 - \left(\frac{C_p + C_s}{LC_p C_s}\right)}\right] \tag{15.5.3}$$

We can write equation (15.5.3) in the following form:

$$Z_{equiv} = \left(\frac{1}{j\omega C_p}\right)\left[\frac{\omega^2 - \omega_s^2}{\omega^2 - \omega_p^2}\right] \tag{15.5.4}$$

where ω_p and ω_s represent two resonant frequencies, corresponding to the two points where the $|Z_{equiv}|$ curve crosses the x-axis in Figure 15.20. The lowest of the two frequencies is the series resonance, given by:

$$\omega_s = \frac{1}{\sqrt{(LC_s)}} \tag{15.5.5}$$

Higher up the frequency scale is the parallel resonance given by:

$$\omega_p = \frac{1}{\sqrt{L\left(\frac{C_p C_s}{C_p + C_s}\right)}} \tag{15.5.6}$$

Comparing equations (15.5.5) and (15.5.6) allows us to express ω_p in terms of ω_s as:

$$\omega_p = \omega_s \sqrt{1 + \frac{C_s}{C_p}} \tag{15.5.7}$$

Since, in general, $C_p \gg C_s$ we can see that the two resonant frequencies, ω_p and ω_s, in equation (15.5.7) will be quite close together, but ω_p will always be higher than ω_s. From Figure 15.20 we can also observe that only in the narrow region between ω_s and ω_p is the crystal impedance inductive. At every other location along the frequency axis the crystal impedance is capacitive. Crystals are usually operated in this inductive region and can be used to replace the tank inductor in some of the common oscillator topologies already described. The stable operating frequency of such an oscillator will be close to, but a little higher than ω_s.

One oscillator topology not so far introduced, but very commonly used with crystals is the Pierce circuit shown in Figure 15.21. This circuit is essentially a common emitter amplifier with a feedback network, from collector to base, consisting of the π-network comprising C_1, C_2, and the crystal, X_1.

The load impedance seen by the transistor in Figure 15.21, at the collector, can be written as:

$$Z_L = jX_2 \parallel \left[Z_{equiv} + jX_1 \parallel h_{ie}\right] \tag{15.5.8}$$

where Z_{equiv} is the equivalent resonant impedance of the crystal, h_{ie} is the input impedance looking into the base of the transistor at that bias point, and the symbol "$\parallel$" indicates a parallel connection. As we are operating the crystal close to ω_s, the resonant impedance will be inductive, that is, $Z_{equiv} = r_e + j\omega L_e$. If $|X_1| \ll h_{ie}$ we can write:

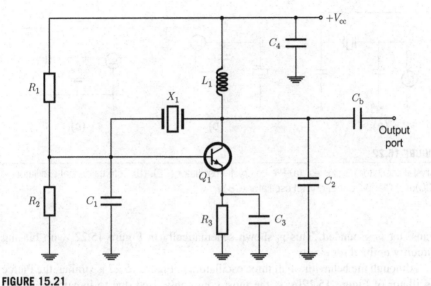

FIGURE 15.21

BJT Pierce crystal oscillator.

$$Z_L = \frac{jX_2\left[r_e + j(X_1 + X_e)\right]}{r_e + j(X_1 + X_2 + X_e)} \tag{15.5.9}$$

If we assume that r_e is very small we can see that the imaginary part of equation (15.5.9), resonance occurs when:

$$(X_1 + X_2 + X_e) = 0 \tag{15.5.10}$$

Solving equation (15.5.10) for ω_o results in:

$$\omega_o = \frac{1}{\sqrt{L_e\left(\dfrac{C_1 C_2}{C_1 + C_2}\right)}} \tag{15.5.11}$$

The gain condition for oscillation of the circuit in Figure 15.21 is given by Gonzalez as [3]:

$$\frac{g_m}{\omega_o^2 R_e C_1 C_2} > 1 \tag{15.5.12}$$

where g_m is the transconductance of Q_1.

As a final word on crystal oscillators, it is worth mentioning here an observation also made by Gonzalez [3] that the Pierce, Colpitts, and Clapp/Gouriet oscillator topologies are very closely related and only distinguished by which terminal of the

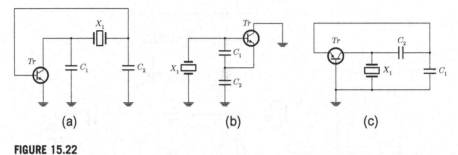

FIGURE 15.22

Crystal oscillator topologies. (a) Pierce crystal oscillator (CE). (b) Colpitts crystal oscillator (CC). (c) Clapp/Gouriet crystal oscillator (CB).

transistor is grounded. This is shown schematically in Figure 15.22, with biasing circuitry omitted for clarity.

Although the behavior of all three oscillators in Figure 15.22 is similar, the Pierce oscillator of Figure 15.22(a) is the most commonly used due to its simplicity and stability when a crystal is used as the main frequency stabilizing element.

15.5.2 **CAVITY STABILIZATION**

A simple method of frequency stabilization at microwave frequencies involves the use of a metal cavity, the physical dimensions of which will define the resonant frequency. Cavity stabilization has the great advantages of low cost and simplicity of manufacture.

The two most popular shapes of microwave cavities are cylindrical and rectangular. These are preferred over more complex shapes because of their relative ease of manufacture. A cavity behaves as a high Q resonator which resonates at frequencies determined by its geometry and the dielectric constant of the material inside. The Q of a well-designed microwave cavity resonator can be as high as 10^6. Although cavities may have more than one resonant frequency, only one resonant frequency is normally used in a given oscillator design. A large body of theoretical work exists to predict the resonant frequencies of metallic cavities.

Figure 15.23 shows how we can consider a cylindrical microwave cavity as being a logical evolution of the familiar discrete component LC resonant circuit. In Figure 15.23(a), we see the conventional LC parallel tank circuit consisting of a single discrete capacitor and inductor. As frequency increases, the required value of both L and C are reduced. Eventually we reach a point where the capacitor is comprised of just two parallel circular plates, of diameter d, separated by a distance x. We also envisage the inductor as being comprised of a single hollow cylindrical conductor, of diameter w, and length l, as shown in Figure 15.23(b). As the frequency increases further, the values of L and C reduce further, meaning that both the spacing between the capacitor plates and the diameter of the hollow cylindrical conductor increase.

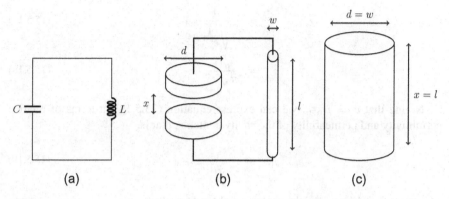

FIGURE 15.23

Evolution of a cylindrical cavity resonator. (a) *LC* circuit. (b) *LC* intermediate stage.
(c) Cylindrical cavity.

Ultimately, as frequency continues to increase, we reach a point where $d = w$ and $x = l$, so we can merge the two components together as shown in Figure 15.23(c).

At resonance, the electric field energy stored in the cavity of Figure 15.23(c) is equal to the stored magnetic energy. This is analogous to energy storage in the resonant *LC* circuit of Figure 15.23(a) where the equivalent values of L and C for the cavity of Figure 15.23(c) are given by [14]:

$$L_{mnl} = \mu k_{mnl}^2 V \tag{15.5.13}$$

and

$$C_{mnl} = \frac{\epsilon}{k_{mnl}^4 V} \tag{15.5.14}$$

where V represents the cavity volume, k_{mnl} is the mode wavenumber, and ϵ and μ are permittivity and permeability, respectively, of the cavity contents. The wavenumber can be considered as the "spatial frequency" of a wave, either in cycles per unit distance or radians per unit distance, and is given by:

$$k = \frac{2\pi}{\lambda} = \frac{\omega}{v_p} \tag{15.5.15}$$

where v_p is the phase velocity.

The resonant frequency for a given "mnl" mode can be derived from the equivalent capacitance and inductance, equations (15.5.13) and (15.5.14), as follows:

$$f_{mnl} = \frac{1}{2\pi \sqrt{L_{mnl} C_{mnl}}} \tag{15.5.16}$$

$$= \frac{1}{2\pi \sqrt{\frac{1}{k_{mnl}^2} \mu\epsilon}} \tag{15.5.17}$$

$$= \frac{k_{mnl}}{2\pi \sqrt{\mu\epsilon}} \tag{15.5.18}$$

Noting that $c = \mu_0\epsilon_0$, we can express equation (15.5.18) in terms of relative permittivity and permeability of the cavity contents, that is,

$$f_{mnl} = \frac{c}{2\pi \sqrt{\mu_r\epsilon_r}} k_{mnl} \tag{15.5.19}$$

where $c = 3 \times 10^8$ m s^{-1}, the speed of light in free space.

Figure 15.24 shows three popular configurations used for microwave cavity stabilized oscillators [15]. The cavity can be coupled to the rest of the circuit via a small aperture, a wire probe, or a loop.

In the reaction stabilized circuit of Figure 15.24(a), the cavity is situated between the active device and the load, and is coupled to a transmission line connecting them. The cavity acts as a band reject filter and at the resonant frequency a small amount of power is reflected back toward the active device. This mode of operation provides high external Q but is susceptible to parasitic oscillation and mode jumping. The reflection stabilized circuit of Figure 15.24(b) is well suited to situations where the active device can be configured as a two-port network. The cavity provides a passive termination at one-port of the active device which satisfies the oscillation condition, equation (15.4.18), at a particular frequency. This circuit has a slightly lower external Q than the reaction stabilized oscillator, but is relatively free from parasitic oscillations. In the transmission stabilized circuit (sometimes called the Kurokawa circuit) of Figure 15.24(c), the cavity acts as a bandpass filter between the active device and the load. This configuration has a very high external Q and is practically free from parasitic oscillation, but the output power may be low due to coupling losses.

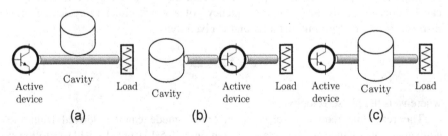

FIGURE 15.24

Cavity stabilization of a transistor oscillator: three possible configurations. (a) Reaction. (b) Reflection. (c) Transmission.

The frequency of a cavity stabilized oscillator can be varied by changing the mechanical dimensions of the cavity or by inserting dielectric material into it. Electronic tuning can be achieved by placing an electronically variable reactance, such as a varactor diode, inside the cavity.

15.5.3 DIELECTRIC RESONATOR STABILIZATION

The physical size of metallic cavity resonators limits their application to higher microwave frequencies. At any given frequency, the physical size of the cavity can be reduced by filling the cavity with a high dielectric constant material. If the dielectric constant of the material is high enough, we can dispense with the metal cavity walls altogether. What we then have is known as a *dielectric resonator*.

The term dielectric resonator was first used by Richtmyer who postulated that if a cylindrical dielectric waveguide were bent round to form a torus, and joined at the ends, as shown in Figure 15.25(a), then electromagnetic radiation would propagate inside the torus at certain distinct frequencies determined by the requirement that the fields be continuous at the junction [16]. The original torus shape proposed by Richtmyer is not widely used in practice as it is difficult to fabricate. The "puck" shape shown in Figure 15.25(b) has very similar electrical properties to the torus and is much easier to fabricate. Most practical dielectric resonators therefore employ the puck format.

There are three types of resonant modes that can be excited in a dielectric resonator, namely the transverse electric (TE), transverse magnetic (TM), or hybrid electromagnetic (HEM) modes. Theoretically, there is an infinite number of modes in each of these three categories. In practice, it is the TE_{01n} mode that is most widely used in oscillator applications. The approximate resonant frequency, in GHz, of the TE_{01n} mode for an isolated cylindrical dielectric resonator, of the type shown in Figure 15.25(b) is given by [17]:

$$f_{GHz} = \frac{34}{a\sqrt{\epsilon_r}} \left(\frac{a}{h} + 3.45 \right)$$ (15.5.20)

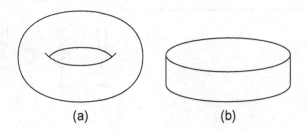

FIGURE 15.25

Dielectric resonators. (a) Dielectric torus. (b) Dielectric puck.

where a is the radius of the resonator puck and h is its height. Both a and h are in millimeters in equation (15.5.20), which is accurate to about $\pm 2\%$ in the range:

$$0.5 < \frac{a}{L} < 2$$

$$30 < \epsilon_r < 50$$

If the material of the dielectric resonator has a large dielectric constant (i.e., the majority of the field is confined to a region close to the resonator) then the Q of the resonator is approximately given by:

$$Q = \frac{1}{\tan(\delta)} \tag{15.5.21}$$

where δ is the dielectric loss of the material.

The linear dimension of a dielectric resonator is in the order of $\lambda_0/\sqrt{\varepsilon_r}$, where λ_0 is the free space wavelength and $\sqrt{\varepsilon_r}$ is the dielectric constant. For $\sqrt{\varepsilon_r} = 100$, the resonator is approximately one-tenth of the size of an equivalent metallic cavity, making such resonators very attractive for use in practical oscillators.

The puck-shaped dielectric resonator is well suited to coupling with microstrip circuits by simply placing the puck at a suitable position next to a microstrip transmission line. The placement of the resonator relative to the line, and hence the degree of coupling, can be easily varied. Figure 15.26(a) shows a plan view of a cylindrical dielectric resonator coupled to a microstrip line. The coupling between the resonator and a microstrip line is adjusted by varying the spacing between the resonator and the line, d.

Figure 15.26(b) shows an equivalent circuit of Figure 15.26(a) in which the dielectric resonator is modeled as an LCR resonant circuit. The normalized input

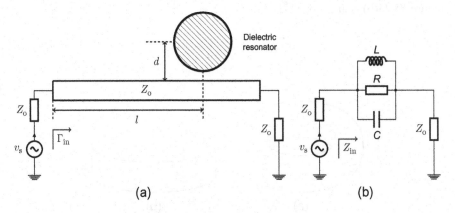

(a)   (b)

FIGURE 15.26

Dielectric resonator coupled to a microstrip line. (a) Physical configuration. (b) Equivalent circuit.

impedance of the transmission line in Figure 15.26(a) at a frequency $\Delta\omega$ away from the resonant frequency, ω_0, is given by [18–20]:

$$Z_{in} = 1 + \frac{\beta}{1 + jQ_0\dfrac{\Delta\omega}{\omega_0}} \tag{15.5.22}$$

where β is the coupling coefficient of the resonator to the line and Q_0 is the unloaded Q of the resonator. β can be varied by moving the resonator relative to the line. In this way, the impedance, Z_{in}, can be continuously varied and a position can be found where the oscillation condition is satisfied.

The magnitude of the source reflection coefficient, Γ_{in}, is a function of lateral spacing, d, between the resonator and the microstrip line in Figure 15.26(a). The phase angle of Γ_{in} is controlled by the length, 1, of the input line up to the point at which the resonator is located.

Figure 15.27 shows a reflection mode FET oscillator using a dielectric resonator as the stabilizing element. The FET is made to present a negative resistance at the gate port by adding a series feedback reactance, jX_{fb} in the common source lead. The value of feedback reactance required can be calculated using equation (8.6.4). With a known negative resistance looking into the gate of the FET, the dielectric resonator is physically positioned with respect to the microstrip line in Figure 15.27 so as to satisfy the oscillation condition:

$$\Gamma_{in}\Gamma_r = 1 \tag{15.5.23}$$

The DC bias voltages, V_g and V_d, are applied through bias networks. The output matching network transforms the 50 Ω load impedance to a value required to

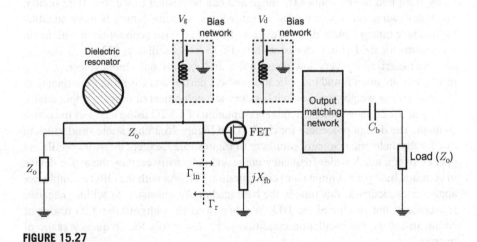

FIGURE 15.27

FET-based dielectric resonator oscillator.

conjugately match the drain port of the FET at the oscillation frequency. The output load is isolated from the oscillator DC voltages by the low-impedance DC blocking capacitor C_b.

15.6 VOLTAGE CONTROLLED OSCILLATORS

15.6.1 YIG TUNED OSCILLATORS

A sphere of Yttrium Iron Garnet (YIG) material, typically less than 1 mm in diameter, can act as a microwave resonator that can be coupled to an active device via a wire loop. YIG resonators exhibit very high Q, similar to that of dielectric resonators, but have the added advantage that the resonant frequency can be tuned over a very wide range by varying an externally applied DC magnetic field. The frequency of resonance is a very linear function of the strength of the applied magnetic field and typically increases at a rate of 2.8 MHz per gauss. Electronic control of the magnetic field must, however, be achieved by means of powerful electromagnets, which tends to make the YIG tuned oscillator (YTO) quite bulky and heavy.

The broadband capability of YTOs can only be obtained with slow sweep rates and high power consumption. The relatively large inductance of the magnetic coils limits the tuning rate to about 1 Hz s^{-1} [21].

A typical main coil of a YTO is made of copper wire and may have 500-1000 turns. As the resistance of this coil will vary over temperature, a constant current source is required to maintain constant control current over the tuning range of the oscillator. A secondary "FM" coil is normally provided to frequency modulate or phase lock the oscillator. This coil is smaller than the main coil, with 10-15 turns being typical. With very low inductance and low resistance, the FM coil can be modulated at a much faster rate than the main coil. FM tuning bandwidths are nominally from mid to very high kHz range and can be pushed to the low MHz region. For faster tuning rates over narrower bandwidths, varactor tuning is more suitable. Multioctave tuning bandwidths (in excess of 10 GHz) have been reported with metal semiconductor field-effect transistor (MESFET) YIG oscillators [22,23]. YTOs are also characterized by very low-phase noise and exceptional tuning linearity. As a result, they are often found in applications where performance is more important than physical size or weight, such as in laboratory test equipment or military applications.

Figure 15.28 shows a schematic representation of a YTO using a BJT as the active element. The design procedure for the YTO in Figure 15.28 has some similarities to that for the dielectric resonator oscillator of Figure 15.27, except that YIG oscillators operate over a much wider frequency range and so usually employ the active device in common base (or common gate) configuration [24]. As with the DRO example, we apply series feedback, this time in the base lead of the transistor, to achieve negative resistance at the emitter of the BJT. We then need to configure the YIG resonator circuit to satisfy the oscillation condition (15.5.23) across the frequency range of interest.

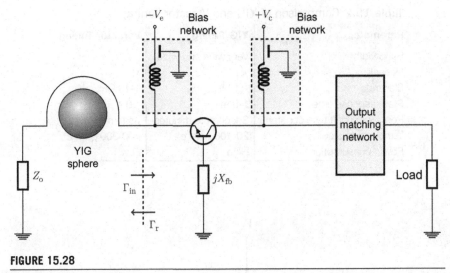

FIGURE 15.28

Transistor YIG tuned oscillator.

The DC bias voltages, V_e and V_c, are applied through bias networks. The output matching network transforms the 50 Ω load impedance to a value required to conjugately match the collector side of the BJT across the frequency range. In contrast with the FET dieletric resonator oscillator (DRO) of Figure 15.27, the output matching network of a YTO needs to be a broadband matching network as the tuning range of the YTO usually covers multiple octaves.

15.6.2 VARACTOR TUNED OSCILLATORS

The advantages of varactor tuning are low power consumption, small size, and high tuning speed. The price to be paid for these advantages are a relatively narrow tuning range of around 3% and a relatively low Q.

Table 15.2 gives a comparison of the current state of the art for YIG and varactor tuning.

A very common voltage controlled oscillator (VCO) configuration, especially in MMIC format, is based on the cross-coupled topology described in Section 15.3. To convert the fixed frequency oscillator shown in Figure 15.15 into a VCO we replace the tuning capacitors by varactor. The basic layout of a typical cross-coupled VCO topology is shown in Figure 15.29. The capacitance of the two varactors VC_1 and VC_2 will be inversely proportional to the voltage, V_c, applied to their common connection point. The output frequency of the VCO will therefore be directly proportional to V_c.

Varactors can also be used to tune a dielectric resonator oscillator, but as expected, there is a fundamental trade-off between electronic tuning range and resonator Q,

Table 15.2 Comparison of YIG and Varactor Tuning

Parameter	YIG Tuning	Varactor Tuning
Bandwidth	Very wide and wide	
Linearity	<+1	12
Slew rate	<1 MHz/ 1S	1-10 GHz/ 4S
Step response time	1-3 ms	<0.1 μs
Post tuning drift time constant	Seconds to minutes	μs to ms
Temperature stability	20-100 ppm c^{-1}	100-300 ppm c^{-1}
Power consumption	High	Low

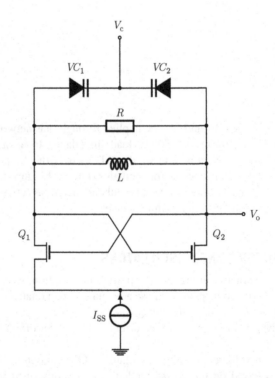

FIGURE 15.29

Cross-coupled VCO using varactors.

cp_currentsource

leading to a compromise between tunability and stability. This problem has been thoroughly investigated by Van Der Heyden [25] who has studied the possibilities of tuning dielectric resonator oscillators over a narrow band by coupling a varactor to the dielectric resonator.

15.7 INJECTION LOCKED AND SYNCHRONOUS OSCILLATORS

An interesting subclass of microwave oscillators are those where a reference signal from one oscillator is "injected" into another oscillator, and "locks" the fundamental output of the injected oscillator. This phenomenon can occur, either intentionally or unintentionally, when the difference in free-running frequency between two oscillators is not that great and they are somehow coupled together. Any kind of oscillator can, in theory, be locked by injection [26].

Since the oscillator being injected is having its free-running frequency "pulled" off-center by the injecting signal, the maximum fractional frequency difference for which injection locking can occur ultimately depends on the loaded Q of the injected oscillator tank circuit, and the amplitude of the injected signal.

When the amplitude of the injected signal is small, the maximum injection locking range is given by [27]:

$$\Delta\omega = \left(\frac{\omega_0}{2Q}\right)\left(\frac{A_i}{A_0}\right) \tag{15.7.1}$$

where

$\Delta\omega$ = frequency locking range;
ω_0 = resonant frequency of the injected oscillator tank circuit;
Q = the injected oscillator tank circuit loaded Q;
A_i = the amplitude of the injected signal at the oscillator tank; and
A_0 = the amplitude of the unperturbed oscillator tank signal.

The injection locking range is reduced by increasing the oscillator tank circuit Q, operating with high tank circuit power and reducing the amplitude of the injected signal. If the frequency difference between the two oscillators is greater than the injection locking range, then the injected oscillator will drop out of lock and will revert to its free-running frequency.

Two highly useful properties of injection-locked oscillators are as follows:

1. Under certain conditions, the injected oscillator will lock to a harmonic or a subharmonic of the injected signal, provided that the harmonic lies in the oscillator synchronization range. This allows a high-frequency oscillator to be stabilized by a lower-frequency reference source.
2. Once locked, the locked oscillator replicates the phase noise of the injected signal. This allows the phase noise of a given source to be improved by locking to a lower phase-noise reference source, which may at a lower frequency.

A particular type of transistorized injection-locked oscillator is known as the *synchronous oscillator* [28]. The main distinguishing feature of the synchronous oscillator is that it contains a separate active device whose role is to modulate the bias point of the primary oscillating device. The most commonly used topology for a synchronous oscillator uses a modified Colpitts oscillator with some additional components.

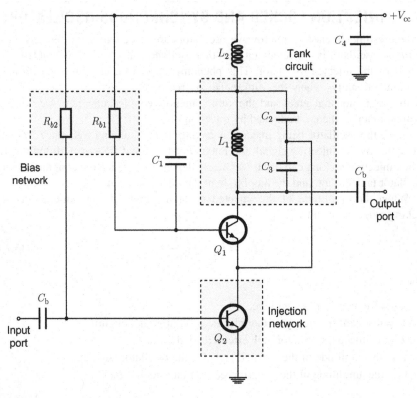

FIGURE 15.30

Synchronous oscillator.

Figure 15.30 shows a schematic representation of an elementary synchronous oscillator according to Uzunoglu [28,29]. The reader will notice the similarity with the Colpitts oscillator of Figure 15.7, where L_1, C_2, and C_3 constitute the main tank circuit. An additional feedback path, C_1 is added to enhance regeneration [30]. Q_1 is the main oscillatory active device and the purpose of Q_2 is to modulate the bias point of Q_1 in accordance with the synchronizing signal injected at the input port. R_{b1} and R_{b2} set the DC bias points of Q_1 and Q_2. The output signal is available from the collector of Q_1 via the DC blocking capacitor C_b.

In the absence of an input signal, the circuit of Figure 15.30 will oscillate at a natural resonant frequency determined by the tank circuit comprised of L_1, C_2, and C_3. When an input signal is applied, the oscillator will be "pulled" to match either the fundamental or one of the harmonics of the input signal, depending on the relative amplitudes of the input signal and the free-running frequency set by the tank circuit.

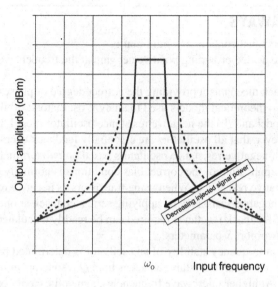

FIGURE 15.31

Gain synchronization curve.

Figure 15.31 shows the relationship between output signal amplitude and input signal frequency. The curves shown in Figure 15.31 are referred to as "gain-synchronization" curves, and have a characteristic flat topped shape, with the width of the flat portion indicating the locking range of the oscillator. With reference to Figure 15.31 we observe that the output power drops sharply at the edges of the bandwidth. This is a result of the signal becoming so small that it cannot compensate for losses in the tank circuit.

The output of the standard synchronous oscillator varies in phase from $+90°$ to $-90°$ with respect to the input, across the bandwidth of the oscillator. The zero phase difference occurs when the input signal corresponds to the resonant frequency ω_0 (see Figure 15.31).

A number of theories have been advanced to explain the operation of synchronous oscillators [27,31,32]. The consensus view seems to be that, provided that the synchronizing process does not modify the oscillation magnitude at the load admittance, then the synchronizing signal can be modeled as a small variation of the admittance load current synchronization [33].

The synchronous oscillator can be driven by a locking signal which is a subharmonic of the free-running frequency. This is a particularly useful characteristic, as it allows a high-frequency RF oscillator to be stabilized by a low-frequency reference source. Subharmonic locking, with frequencies as low as one-tenth of the free-running frequency, has been demonstrated with the synchronous oscillator circuit shown in Figure 15.30 [34].

15.8 TAKEAWAYS

1. All oscillators must contain, at a minimum:
   (i) an active device providing positive net gain at the frequency of operation and
   (ii) a feedback mechanism providing the correct degree of phase shift.
2. There are two theoretical models used to analyze oscillators: (a) the feedback oscillator model and (b) the negative resistance oscillator model. It could be argued, however, that all oscillators are essentially feedback oscillators.
3. In the case of negative resistance oscillators based on two terminal active devices, the application of the correct bias conditions is normally enough to generate negative resistance. When using transistors as negative resistances we can achieve negative resistance by applying series feedback at one of the ports. The value of feedback reactance required can be readily calculated from the transistor's two-port S-parameters.
4. The long-term frequency stability of an oscillator is determined by the Q of the resonator circuit. Quartz crystals can act as high Q resonators at medium frequencies. At higher microwave frequencies, a metallic cavity or dielectric resonator is more suitable.
5. The cross-coupled oscillator has become popular in MMIC implementations due to its stability, flexibility, and a higher output amplitude (than single transistor designs) for a given supply voltage.
6. Oscillators can be tuned electronically by substituting varactor diodes for some of the frequency-determining capacitors.
7. Injection locking is a method by which the output frequency of an oscillator can be "locked" to the frequency of another signal source injected into it. This enables a high-frequency but low-stability source to attain the same frequency stability as a lower frequency, high-stability reference. One type of injection locking topology is the *synchronous oscillator*.

REFERENCES

[1] H. Taub, D. Schilling, Principles of Communication Systems, McGraw-Hill Series in Electrical Engineering, McGraw-Hill, 1986, ISBN 9780070629561.

[2] O. Momeni, E. Afshari, High power terahertz and millimeter-wave oscillator design: a systematic approach, IEEE J. Solid State Circuits 46 (3) (2011) 583–597, ISSN 0018-9200, http://dx.doi.org/10.1109/JSSC.2011.2104553.

[3] G. Gonzalez, Foundations of Oscillator Circuit Design, Artech House, Boston, MA, 2007.

[4] H. Zirath, R. Kozhuharov, M. Ferndahl, A x^2 coupled Colpitt VCO with ultra low phase noise, in: IEEE Compound Semiconductor Integrated Circuit Symposium, ISSN 1550-8781, 2004, pp. 155–158, http://dx.doi.org/10.1109/CSICS.2004.1392520.

[5] K. Kurokawa, Some basic characteristics of broadband negative resistance oscillator circuits, Bell Syst. Tech. J. 48 (1969) 1937–1955.

[6] A. Khanna, J. Obregon, Microwave oscillator analysis, IEEE Trans. Microw. Theory Tech. 29 (11) (1981) 606–607.

[7] J. Cote, Matrix analysis of oscillators and transistor applications, IRE Trans. Circuit Theory 5 (1958) 181–188.

[8] K. Johnson, Large signal GaAs MESFET oscillator design, IEEE Trans. Microw. Theory Tech. 27 (3) (1979) 217–227.

[9] K. Kotzebue, W. Parrish, The use of large-signal S-parameters in microwave oscillator design, in: Proceedings of IEEE International Microwave Symposium on Circuits and Systems, 1975.

[10] J.K. Plourde, C.L. Ren, Application of dielectric resonators in microwave components, IEEE Trans. Microw. Theory Tech. 29 (8) (1981) 754-770.

[11] G. Vendelin, Design of Amplifiers and Oscillators by the S-Parameter Method, John Wiley and Sons, New York, 1982.

[12] G. Basawapatna, R. Stancliff, A unified approach to the design of wideband microwave solid state oscillators, IEEE Trans. Microw. Theory Tech. 27 (5) (1979) 379–385.

[13] C. Poole, I. Darwazeh, A simplified approach to predicting negative resistance in a microwave transistor with series feedback, in: IEEE MTT-S International Microwave Symposium, Seattle, USA, 2013.

[14] C. Montgomery, R. Dicke, E. Purcell, Principles of Microwave Circuits, IEE Electromagnetic Waves Series, Institution of Engineering and Technology, 1948, ISBN 9780863411007.

[15] R. Pengelly, Microwave Field Effect Transistors: Theory, Design and Applications, John Wiley and Sons Ltd, Chichester, 1982.

[16] R.D. Richtmyer, Dielectric resonators, J. Appl. Phys. 10 (1939) 391-398.

[17] D. Kajfez Jr., W. Wheless, Invariant definitions of the unloaded Q factor (short paper), IEEE Trans. Microw. Theory Tech. 34 (7) (1986) 840–841, ISSN 0018-9480, http://dx.doi.org/10.1109/TMTT.1986.1133452.

[18] A. Podcameni, et al., Unloaded quality factor measurement for MIC dielectric resonator applications, Electron. Lett. 17 (18) (1981) 656–657.

[19] C. Tsironis, V. Pauker, Temperature stabilisation of GaAs MESFET oscillators using dielectric resonators, IEEE Trans. Microw. Theory Tech. 31 (11) (1988) 312–314.

[20] A. Khanna, Y. Garault, Determination of loaded, unloaded, and external quality factors of a dielectric resonator coupled to a microstrip line, IEEE Trans. Microw. Theory Tech. 31 (3) (1983) 261–264, ISSN 0018-9480, http://dx.doi.org/10.1109/TMTT.1983.1131473.

[21] H. Howes, R. Tebbenham, Design of microwave oscillators, in: Microwave Circuit and System Design, Leeds University Summer School, 1983.

[22] R. Trew, Octave-band GaAs FET YIG-tuned oscillators, Electron. Lett. 13 (21) (1977) 529–530.

[23] R. Trew, Design theory for broadband YIG-tuned oscillators, IEEE Trans. Microw. Theory Tech. 27 (1) (1979) 8–14.

[24] N. Osbrink, et al., YIG-tuned oscillator fundamentals, Microw. Syst. Design. Handbook Microw. Syst. News 13 (12) (1983) 207–225.

[25] B. Van Der Heijden, Dielectric resonator oscillators: a new microwave signal source, Philips Electron. Compon. Appl. 4 (4) (1982) 227–232.

[26] R. Adler, A study of locking phenomena in oscillators, Proc. IEEE 61 (10) (1973) 1380–1385, ISSN 0018-9219, http://dx.doi.org/10.1109/PROC.1973.9292.

[27] B. Razavi, A study of injection locking and pulling in oscillators, IEEE J. Solid State Circuits 39 (9) (2004) 1415–1424, ISSN 0018-9200, http://dx.doi.org/10.1109/JSSC. 2004.831608.

[28] V. Uzunoglu, Coherent phase-locked synchronous oscillator, Electron. Lett. 22 (20) (1986) 1060–1061, ISSN 0013-5194, http://dx.doi.org/10.1049/el:19860726.

[29] V. Uzunoglu, Synchronous oscillator outperforms the PLL, EDN, 1999, www.ednmag. com.

[30] V. Uzunoglu, M. White, Synchronous and the coherent phase-locked synchronous oscillators: new techniques in synchronization and tracking, IEEE Trans. Circuits Syst. 36 (7) (1989) 997–1004, ISSN 0098-4094, http://dx.doi.org/10.1109/31.31335.

[31] Y. Deval, J.-B. Begueret, H. Lapuyade, P. Fouillat, E. Kerherve, The synchronous oscillator in frequency generation: an overview, in: The 14th International Conference on Microelectronics, ICM, 2002, pp. 148–151, http://dx.doi.org/10.1109/ICM-02.2002. 1161517.

[32] M. Tam, M. White, Z. Ma, Theoretical analysis of a coherent phase synchronous oscillator, IEEE Trans. Circuits Syst. I Fundam. Theory Appl. 39 (1) (1992) 11–18, ISSN 1057-7122, http://dx.doi.org/10.1109/81.109238.

[33] R. Huntoon, A. Weiss, Synchronization of oscillators, Proc. IRE 35 (12) (1947) 1415–1423, ISSN 0096-8390, http://dx.doi.org/10.1109/JRPROC.1947.226202.

[34] C. Poole, Subharmonic injection locking phenomena in synchronous oscillators, Electron. Lett. 26 (21) (1990) 1748–1750.

CHAPTER OUTLINE

INTENDED LEARNING OUTCOMES

- *Knowledge*
  - Understand the nature and impact of phase noise in oscillators.
  - Be familiar with the origin, application, and limitations of Leeson's equation.
  - Be familiar with analytical phase-noise models for both feedback and negative resistance oscillators and be able to compare the two models.

- Understand how a simple phase-noise model implies certain design guidelines for low-noise oscillators and be able to apply these guidelines.
- Be familiar with various phase-noise measurement techniques.
- *Skills*
  - Be able to select suitable transistors and resonators for use in low-noise oscillator design.
  - Be able to design an oscillator with low phase noise based on the tools and techniques presented.
  - Be able to carry out basic phase-noise measurements.

16.1 INTRODUCTION

When we covered oscillators in Chapter 15 we postulated that, at power on, oscillations originate from noise perturbations inherent in the various components, that then get amplified by the gain element, resulting in a signal that builds up in amplitude and is pulled toward the oscillator design frequency by a frequency selective element (the *resonator* or *tank* circuit). Aside from the consequent, profound observation that noise is essential to initiating oscillation, this description also implies that the gain element is inherently nonlinear. The reasoning is as follows: Although the Barkhausen amplitude criterion of equation (15.2.4) implies unity loop gain in steady-state conditions, if the oscillator did have unity loop gain for all signal levels then the initial noise perturbations would never get amplified sufficiently to build up to the final signal amplitude. Notwithstanding the Barkhausen criterion, the effective loop gain must be significantly greater than unity at very low signal levels, reducing to exactly unity as the signal amplitude approaches the steady-state level. Likewise, under steady-state conditions, if the signal amplitude should drop for any reason, the higher loop gain at lower amplitudes, due to this loop gain nonlinearity, will drive the amplitude back up to a steady-state level, which is limited, at its upper bound, by the power supply voltage. In short, it is the inherent nonlinearity of the gain element (i.e., high gain at low signal levels and low gain at high signal levels) that primarily acts to maintain the oscillator output signal at a constant amplitude. This fundamental amplitude stabilization mechanism in oscillators was first described by Lord Rayleigh as early as 1896 [1].

A further implication of the nonlinear loop gain characteristic is that, as the instantaneous voltage of the steady-state signal transitions through a single oscillation cycle, it will experience different levels of amplification. Such gain nonlinearity will result in the generation of complex harmonics and intermodulation components in the output signal. In summary, therefore, there is a fundamental paradox in the design of an oscillator. On the one hand, we require the output to be as "clean" as possible, with all the energy concentrated at a single design frequency. On the other hand, the very imperfections that make oscillation possible in the first place will inevitably produce unwanted frequency components, that will manifest themselves as sidebands around the nominal center frequency.

We generally refer to oscillator noise as *phase noise*. This is because, although the noise generated within the oscillator will be random in both amplitude and phase, the amplitude randomness tends to be suppressed by the gain compression mechanism just described. In fact, it is relatively easy to ensure a stable output amplitude by good design. What remains of the noise components at the output of a well-designed oscillator therefore appears as random phase (and hence frequency) fluctuations, which are much harder to remove and must be "designed out."

An empirical equation for oscillator phase noise was first defined by Professor David B. Leeson [2], who produced a descriptive equation that formed the basis of much of the subsequent work in this field [3]. A number of very good explanations of phase noise are provided in books by Robins [4], Gonzalez [5], Odyniec [6], Mukherjee [7], and Rubiola [8].

One of the clearest explanations of phase-noise mechanisms in oscillators is provided by Everard, whose 2001 book explains phase noise in terms of the feedback oscillator model [9]. Everard later extended this model to cover the negative resistance oscillator model [10]. Due to the clarity of Everard's analysis, it is this approach that we have adopted for the following explanations.

16.2 DEFINITION OF PHASE NOISE

An ideal oscillator will produce only one sinusoidal frequency component at the nominal center frequency. Real-world oscillators produce a signal which is more complex, containing other, unintended, frequency components. We need a figure of merit by which to measure the purity of a given oscillator output.

The signal output from an ideal oscillator can be described as:

$$V(t) = \cos(\omega_0 t) \qquad (16.2.1)$$

where ω_0 is the center frequency and the amplitude is defined, in this case, to be unity. The output of a real oscillator output will contain both amplitude and phase-noise components. We can represent the output of a real oscillator as:

$$V(t) = [1 + e(t)] \cos(\omega_0 t + \Phi(t)) \qquad (16.2.2)$$

where $\Phi(t)$ is the time varying phase component and $e(t)$ is the amplitude noise component added to the nominal unit signal amplitude. Due to the inherent amplitude limiting mechanism already described, $e(t)$ will be significantly attenuated, so that $[1 + e(t)] \approx 1$. Nevertheless, the presence of both amplitude and phase fluctuations cause sidebands in the output voltage spectrum of the oscillator. Figure 16.1 shows the difference in the output spectrum of an ideal oscillator with that of a real oscillator.

The figure of merit used to describe the quality of an oscillator output signal is called *phase noise*, being a ratio quantity similar to the signal-to-noise ratio (SNR). Instead of measuring the ratio of the noise to the signal power in the same

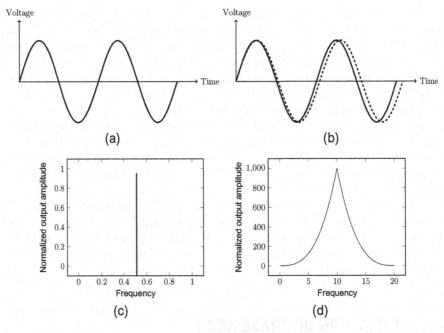

FIGURE 16.1

Ideal and real oscillator output signals and spectra. (a) Ideal oscillator signal. (b) Real oscillator signal. (c) Ideal oscillator spectrum. (d) Real oscillator spectrum.

frequency range, however, phase noise is defined as a ratio of noise power in a 1 Hz bandwidth at a specified frequency offset, Δf, from the center frequency to the nominal carrier power. The US National Institute of Standards and Technology defines single sideband (SSB) phase noise as the ratio of the spectral power density in a 1 Hz bandwidth, measured at an offset frequency from the carrier to the total power of the carrier signal, as shown in Figure 16.2. The symbol used to represent this quantity is $\mathscr{L}\{\Delta f\}$, which is pronounced "script L of delta f," or sometimes simply $\mathscr{L}\{f_m\}$ ("script L of f m"), in cases where f_m is taken to represent the frequency offset. The basic definition of $\mathscr{L}\{\Delta f\}$, which is usually expressed in decibels, is as follows:

$$\mathscr{L}\{\Delta f\}_{dB} = 10 \, \log \left[\frac{\text{Power spectral density in 1 Hz bandwidth at } (f_0 + \Delta f)}{P_{carrier}} \right] \quad (16.2.3)$$

The unit of phase noise is dBc Hz^{-1} (dB relative to carrier frequency, per 1 Hz offset). A typical phase noise plot is shown in Figure 16.3, which shows four distinct regions, in order of distance from the carrier:

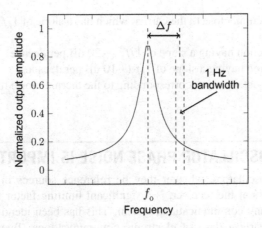

FIGURE 16.2

Noise at offset from a carrier.

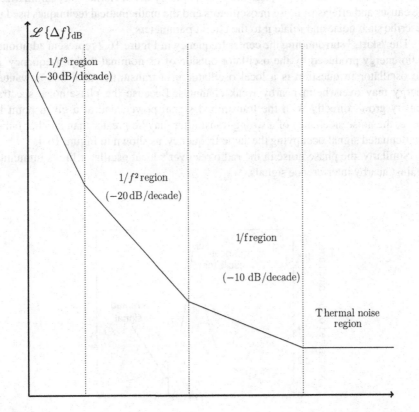

FIGURE 16.3

Typical oscillator phase-noise profile.

(i) An initial section, close to the carrier, which has a slope of $1/f^3$ (-30 dB per decade).

(ii) A second section having a slope of $1/f^2$ (-20 dB per decade).

(iii) A third section having a slope of $1/f$ (-10 dB per decade).

(iv) A final horizontal section corresponding to the thermal noise floor.

16.3 WHY OSCILLATOR PHASE NOISE IS IMPORTANT

Phase noise in oscillators, whether they be reference sources in the transmitter, or local oscillators at the receiver, is a significant limiting factor affecting overall performance of any communications system. This has been identified as a critical issue since the earliest days of electronic communications. Two seminal papers appeared in the 1960s specifically addressing the problem of noise in oscillators. The 1960 paper by Edson [11] and the 1966 paper by Erich Hafner [12] summarized the causes and effects of noise in oscillators and the mathematical techniques used to describe such noise and relate it to the circuit parameters.

The "skirts" surrounding the center frequency in Figure 16.2 represent additional radio energy produced by the oscillator outside of its nominal design frequency. If the oscillator in question is a local oscillator in a transmitter then this unwanted energy may overwhelm nearby weak channels. Because the phase-noise spectral density grows directly with the transmitted signal power, and at a given point in space, the noise sidebands of a strong transmitter may be greater than another faded or attenuated signal occupying the same frequency, as shown in Figure 16.4.

Similarly, the phase noise in the radio receiver's local oscillator limits immunity against nearby interference signals.

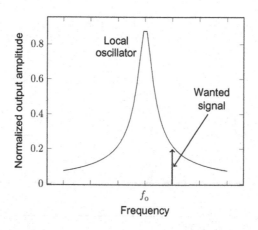

FIGURE 16.4

Interference caused by phase noise.

16.4 ROOT CAUSES OF PHASE NOISE

So far we have established that the fundamental cause of phase noise in oscillators is the inherent noise in the various oscillator circuit components. Thermal noise has already been covered in some detail in Chapter 14 and this type of noise will be present in the passive oscillator components, such as the *LC* tank circuit (which will always contain resistance, whether intentional or parasitic). In addition to thermal noise, the presence of active devices also introduces other specific types of noise, most notably shot and flicker noise.

16.4.1 THE THERMAL NOISE COMPONENT OF PHASE NOISE

As explained in Section 14.2.1, thermal noise is ubiquitous in any electronic system. It can therefore be expected to form an important component of any phase noise in an oscillator.

In any perfectly resistively matched system at room temperature, which is normally taken as being 290 K which is equivalent to 17 °C, the background thermal noise spectral density level is approximated by:

$$N = k_B T = 1.38 \times 10^{-23} \times 290 = -174 \text{ dBm Hz}^{-1} \qquad (16.4.1)$$

This is the minimum level of noise that exists in any 1 Hz bandwidth at room temperature, and is commonly referred to as the *thermal noise floor*.

This thermal noise is composed of both amplitude noise and phase-noise components which are assumed to be approximately equal; a concept embodied in the *equipartition theorem* [9,13]. While the treatment of noise in Chapter 14 focused on amplitude noise, the discussion in this chapter focuses on the phase component of the noise, as it is normally assumed that, when talking about oscillators, the amplitude noise portion of the thermal noise floor is suppressed by the amplitude limiting mechanism discussed previously. This means that the thermal phase-noise floor is 3 dB below the noise floor given by equation (16.4.5), that is,

$$N_{\text{phase}} = \frac{k_B T}{2} = -177 \text{ dBm Hz}^{-1} \qquad (16.4.2)$$

The active element in an oscillator will also inevitably contribute an additional quantity of noise to the thermal noise floor. The amount of noise added by the amplifier is measured as the ratio of output noise to input noise power, adjusted for the gain of the amplifier. This is quantified in terms of the *noise factor* (*F*), which, when expressed in dB, is referred to as the *noise figure* (NF). We can thus write the equivalent total thermal phase noise at the input of an amplifier of noise factor *F* as:

$$N_{\text{phase}} = \frac{F k_B T}{2} \qquad (16.4.3)$$

In terms of dB, equation (16.4.3) can be expressed as:

$$N_{\text{phase(dB)}} = 10 \, \log \left[\frac{Fk_BT}{2} \right] = -177 \text{ dBm Hz}^{-1} + (\text{NF})_{\text{dB}} \qquad (16.4.4)$$

If we now apply a signal of power, P_{in}, and a specific frequency, f_0, to the input of the amplifier in question, we can compute the thermal phase-noise power in a 1 Hz bandwidth at some frequency offset Δf away from f_0 by the ratio:

$$\mathscr{L}\{\Delta f\} = \left(\frac{k_BTF}{2P_{\text{in}}} \right) \qquad (16.4.5)$$

The reader will notice that the right-hand side of equation (16.4.5) does not contain term Δf. This is because this equation describes a simplistic case for an oscillator where all the noise is thermal noise having a response which is flat irrespective of offset frequency. Equation (16.4.5) is adequate to describe the noise of any oscillator if we look far enough away from the nominal oscillation frequency, f_0. As Δf increases, the phase-noise profile asymptotically approaches the thermal noise floor defined by equation (16.4.5). As we get closer to f_0, on the other hand, additional sources of noise attributable to active devices become increasingly influential, eventually becoming dominant. We will now proceed to add terms to equation (16.4.5) to take these additional noise effects into account.

16.4.2 FLICKER NOISE IN OSCILLATORS

While thermal or Johnson noise is primarily responsible for the phase-noise spectrum at large offsets from the carrier, *flicker noise*, due to its $1/f$ characteristic, plays an increasingly important role at smaller frequency offsets. Flicker noise was introduced in Section 14.2.3 in the context of amplifier design, although it has a more important influence on oscillator behavior than it does on amplifiers.

The baseband $1/f$ characteristic of a typical semiconductor amplification device is up-converted in a microwave oscillator through nonlinearities in the device and forms a symmetrical "skirt" around f_0, sloping away at -10 dB per decade either side, as shown in Figure 16.5 (one sideband shown).

The flicker corner frequency, f_c, is defined at the point at which the flicker noise power equals the underlying thermal noise power. In other words, this is the point at which the total noise power increases by 3 dB as we approach f_0 from far away. At this point, to avoid confusion, we will standardize on the use of f_m to represent the offset frequency in Hz distant from the carrier, that is, the same quantity that we have referred to as Δf up until this point. The reader should be aware that other texts will use either of these symbols to represent the frequency offset, and sometimes simply f, which is quite confusing. From Figure 16.5 we can write an empirical expression for $\mathscr{L}\{f_m\}$ that extends equation (16.4.5) to include the flicker noise component as follows:

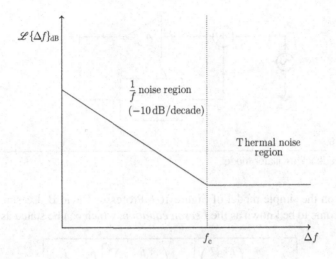

FIGURE 16.5

Phase-noise profile with $1/f$ (flicker) noise component.

$$\mathscr{L}\{f_m\} = \frac{k_B T F}{2 P_{in}} \left(1 + \frac{f_c}{f_m} \right) \tag{16.4.6}$$

By inspection we can see that, when $f_m \gg f_c$, equation (16.4.6) approximates equation (16.4.5).

16.5 MODELING OSCILLATOR PHASE NOISE

There have been numerous different approaches to analyzing and modeling the phase noise in electronic oscillators. Most of these models are based on a feedback oscillator model which has then been extended to cover negative resistance oscillators. We will describe these two approaches here.

The most well-known model of oscillator phase noise is that originally proposed by Professor D.B. Leeson in 1966 [2]. Leeson's model assumes a simple feedback oscillator model shown in Figure 16.6, and is based on the following assumptions:

  (i) The amplifier is noiseless, has a high gain and limits at a level corresponding to the nominal output power.

 (ii) The resonator is a bandpass type centered at the frequency of oscillation and has a loaded Q of Q_L.

(iii) The noise source represents all noise sources in the oscillator, including those introduced by the amplifier and the resonator.

(iv) The limiting action of the amplifier removes the AM component of the noise, leaving only the phase component.

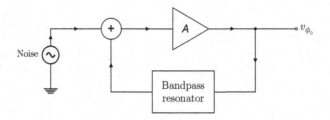

FIGURE 16.6

Leeson's feedback oscillator model.

Based on the simple model of Figure 16.6 Professor David B. Leeson proposed what has come to be known as the *Leeson equation*, which can be stated as follows:

$$\mathcal{L}\{f_m\} = \left(\frac{Fk_B T}{2P_S}\right)\left[1 + \left(\frac{f_0}{2Q_L f_m}\right)^2\right]\left(1 + \frac{f_c}{f_m}\right) \qquad (16.5.1)$$

where

F = the "effective noise factor" of the amplifier.
P_S = the oscillation signal power.
f_0 = the oscillator center frequency.
Q_L = the loaded Q of the resonator.
f_m = the offset frequency from the carrier at which phase noise is measured (in a 1 Hz bandwidth).
f_c = the corner frequency between the $1/f^2$ and $1/f^3$ slope region.

The "effective noise factor" parameter F in equation (16.5.1) is numerically not the same the "noise factor" of the amplifier as conventionally understood. This is because the operating conditions of the active devices are different when the amplifier is used as an oscillator, and amplifiers operating in oscillator mode are rarely matched for optimum noise at the input port. In practice, this F parameter is a fitting factor that is used to account for the fact that the measured phase noise is usually higher than that predicted by Leeson's model.

The reader will notice the similarity between Leeson's equation of equations (16.5.1) and (16.4.6). The difference being the addition of the following term:

$$\left[1 + \left(\frac{f_0}{2Q_L f_m}\right)^2\right] \qquad (16.5.2)$$

The new term equation (16.5.2) accounts for the $1/f^2$ phenomenon observed in an oscillator phase-noise spectrum, while also taking into account the effect of resonator loaded Q. The combined term $f_0/2Q_L$ is actually the half-power bandwidth of the loaded resonator, in Hz. The term $f_0/2Q_L f_m$ therefore approximates the fractional bandwidth of the resonator. With this in mind we could, as an aid to understanding, break down Leeson's equation into the elements shown in Figure 16.7.

Leeson's equation is essentially an empirical description of the observed phase-noise profile shown in Figure 16.8. One key message of Leeson's equation is that the spectrum increases as $1/f^2$ when f_m is less than $f_o/2Q$ and as $1/f^3$ when f_m is also less than f_c, which is almost always less than the measured device $1/f$ corner frequency because the modulation conversion does not raise the noise above the thermal floor.

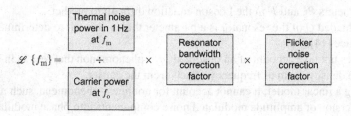

FIGURE 16.7

Diagrammatic representation of Leeson's equation.

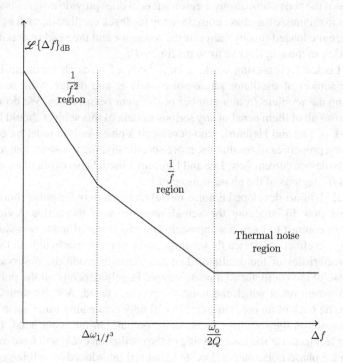

FIGURE 16.8

Phase-noise profile according to Leeson's model.

Although Leeson's model formed the basis of much subsequent work on phase noise, and has been widely cited, it has some significant limitations, as follows:

1. It assumes that the oscillator operates at the center frequency of the resonator. The fact this condition is rarely met in practice, especially at high frequencies, means that real resonators will be less effective at suppressing phase noise than the Leeson equation predicts.
2. The amplifier will not be matched for maximum power transfer in the input circuit, neither will it be matched for minimum noise factor, making the parameters P_S and F in the Leeson equation difficult to predict.
3. The loaded Q of the resonator is a parameter that is difficult to determine in practice [14].
4. It does not take account of an observed $1/f^3$ phenomenon that results in higher noise density at small frequency offsets from the carrier.
5. Being a linear model, it cannot account for nonlinear phenomena, such as the conversion of amplitude modulated noise components into phase modulated noise components, referred to as "AM-PM conversion" [15].
6. Being a linear time invariant (LTI) model, it is not suitable for modeling some common classes of oscillators, such as relaxation and ring oscillators.

In spite of the above limitations, Leeson's model does provide insights into design techniques to minimize the phase noise in many feedback oscillators, such as the need to maximize the loaded quality factor of the resonator and the need to maximize the voltage swing of the amplifier so as to maximize P_S.

Since Leeson's pioneering work, a large body of research literature has built up on the subject of oscillator phase-noise analysis and design, with researchers approaching the problem from a number of different perspectives. We do not have scope to cover all of them here but any serious student of this subject should be aware of the work of Lee and Hajimiri, who developed a phase-noise model based on the time-varying properties of oscillators, more specifically, the transient characteristics of the active device current flow. Lee and Hajimiri's model also explicitly address the $1/f^2$ and $1/f^3$ regions of the phase-noise profile.

Lee and Hajimiri developed a noise model that accounts for noise sources from this current flow by studying the actual waveforms of the active device in an oscillator. In contrast to Leeson's approach, Lee and Hajimiri made no assumptions concerning the effects of device flicker noise and resonator bandwidth on the phase-noise characteristics of the oscillator. Lee and Hajimiri made the observation that the response of the oscillator to a noise sample largely depends on the point in the oscillation waveform at which the noise sample is injected. A noise sample that is injected at the peak of an oscillation cycle will only create amplitude noise and will have no impact on the oscillator phase-noise output. On the other hand, the same noise sample injected at the zero crossing of the oscillation cycle will have maximum impact on the phase-noise output. Lee and Hajimiri introduced something called the *impulse sensitivity function* (ISF), which encodes information about the sensitivity of the oscillator to an impulse injected at a certain phase. The ISF is a periodic function having the same period as the oscillation waveform and attaining a maximum value near the zero crossing of the oscillation waveform, falling to zero at the waveform peaks. The calculation of the ISF is quite laborious and depends upon the oscillator topology.

As with Leeson's model, Lee and Hajimiri's model predicts that minimizing phase noise requires maximizing voltage swing and resonator Q. An additional interesting prediction arising from Lee and Hajimiri's model is that, since the noise is injected into the resonator when the transistor is conducting, a narrower collector current pulse tends to give better phase-noise performance. The Colpitts oscillator topology and its variants with high drive level tend to have this property.

Interested readers are referred to Lee and Hajimiri's tutorial paper on phase noise [13], as well as Ulrich L. Rohde's book [16] and an interesting review article on the history of phase noise by Riddle [17].

16.5.1 ANALYTICAL PHASE-NOISE MODELS: FEEDBACK OSCILLATOR

Of all the numerous feedback oscillator models of phase noise that have been proposed, the one set out by Everard [9] is probably the easiest for the first time reader to follow in detail. For this reason we have chosen to present Everard's analysis and resulting equations in Sections 16.5.1 and 16.5.2 with the kind permission of Professor Everard himself. Everard's model also has the advantage that it can be readily extended to cover negative resistance oscillators as well as feedback oscillators. Negative resistance oscillator noise analysis is covered in Section 16.5.2.

Everard proceeds to analyze phase noise in a feedback oscillator using the model shown in Figure 16.9, in which a single input amplifier is modeled as having two inputs with equal input impedance. One input is used only for noise injection and the other is used for the feedback signal injection. The signal voltages at these two inputs are simply added together. A transfer function for the amplifier, $V_{OUT(node2)}/V_{IN2}$, can be calculated, this can then be converted to a power ratio and used to derive a phase-noise equation. A simple *LCR* circuit is used to model the resonator having an unloaded Q of $Q_o = \omega L/R_{loss}$. The value of any coupling transformers and other components are incorporated into the model values of L, C, and R_{loss}.

Everard uses two definitions of power, that were originally defined by Parker [18] and will be used in the following derivations:

(i) P_{RF} = the power dissipated in the source, load, and resonator loss resistance.
(ii) P_{AVO} = the maximum power available from the output of the amplifier, delivered into a matched load.

The voltage transfer function of the amplifier in Figure 16.9 is given by:

$$v_{out} = G\left(v_{in1} + v_{in2}\right) = G\left(\beta v_{out} + v_{in2}\right) \tag{16.5.3}$$

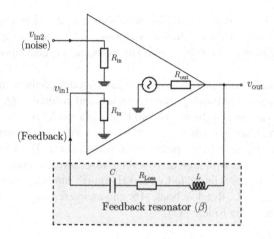

FIGURE 16.9

Oscillator feedback model [9].

where G is the voltage gain of the amplifier between either of the inputs separately and the output, and β is the voltage feedback factor defined by the resonator. In this case, we are primarily interested in the effect of the injected noise voltage, V_{in2} on the output. From equation (16.5.3), therefore, we can write:

$$\frac{v_{out}}{v_{in2}} = \frac{G}{1 - (\beta G)} \tag{16.5.4}$$

We now turn our attention to the resonator. The voltage feedback factor, from output to input 1 of Figure 16.9 can be derived by inspection:

$$\beta = \frac{R_{in}}{R_{out} + R_{loss} + R_{in} + j\left(\omega L - 1/\omega C\right)} \tag{16.5.5}$$

We are interested in the behavior of the resonator at frequencies very close to the carrier, that is, $\omega = \omega_o \pm \Delta\omega$, where ω_o is the carrier frequency and $\Delta\omega$ is a small frequency offset. Let us consider just the imaginary term, $(\omega L - 1/\omega C)$, in the denominator of equation (16.5.5), which can be rewritten as follows:

$$(\omega L - 1/\omega C) = \frac{\left(\omega^2 LC - 1\right)}{\omega C} \tag{16.5.6}$$

Substituting $\omega = \omega_o \pm \Delta\omega$ for ω in equation (16.5.6) gives:

$$\frac{(\omega_o \pm \Delta\omega)^2 LC - 1}{(\omega_o \pm \Delta\omega)C} \tag{16.5.7}$$

As we are only interested in very small values of $\Delta\omega$, we can apply the binomial approximation in this specific case, which allows us to write:

$$(\omega_0 \pm \Delta\omega)^2 \approx \omega_0 (\omega_0 \mp 2\Delta\omega) \qquad (16.5.8)$$

Applying equation (16.5.8) to equation (16.5.7) gives:

$$\frac{(\omega_0 \pm \Delta\omega)^2 LC - 1}{(\omega_0 \pm \Delta\omega)C} \approx \frac{\omega_0(\omega_0 \mp 2\Delta\omega)LC - 1}{(\omega_0 \pm \Delta\omega)C} \qquad (16.5.9)$$

Noting that $\omega_0^2 = 1/LC$, we can reduce equation (16.5.9) to:

$$\frac{\mp 2\omega_0 \Delta\omega L}{(\omega_0 \pm \Delta\omega)} \qquad (16.5.10)$$

Since $\omega_0 \gg \Delta\omega$ we can further approximate equation (16.5.10) to simply:

$$\mp 2\Delta\omega L \qquad (16.5.11)$$

Now, we note that the loaded Q of the resonator, Q_{L}, is defined by:

$$Q_{\mathrm{L}} = \frac{\omega_0 L}{(R_{\mathrm{out}} + R_{\mathrm{loss}} + R_{\mathrm{in}})} \qquad (16.5.12)$$

We can now combine equation (16.5.5) with equations (16.5.11) and (16.5.12) to give us the feedback factor in terms of $\Delta\omega$ and Q_{L}, for small values of $\Delta\omega$:

$$\beta = \frac{R_{\mathrm{in}}}{(R_{\mathrm{out}} + R_{\mathrm{loss}} + R_{\mathrm{in}})\left[1 \pm 2jQ_{\mathrm{L}}\dfrac{\Delta\omega}{\omega_0}\right]} \qquad (16.5.13)$$

Let us now consider the feedback factor at the center frequency (i.e., at resonance), β_0. If we set $\Delta\omega = 0$ in equation (16.5.13) we then have:

$$\beta_0 = \frac{R_{\mathrm{in}}}{R_{\mathrm{out}} + R_{\mathrm{loss}} + R_{\mathrm{in}}} \qquad (16.5.14)$$

Which accords with our understanding that the net reactance of the resonator ($\omega L - 1/\omega C$) will be zero at resonance, resulting in a purely real value of β_0.

Since the unloaded Q of the resonator, Q_0, is simply:

$$Q_0 = \frac{\omega_0 L}{R_{\mathrm{loss}}} \qquad (16.5.15)$$

We can define the ratio:

$$\frac{Q_{\mathrm{L}}}{Q_0} = \frac{R_{\mathrm{loss}}}{R_{\mathrm{out}} + R_{\mathrm{loss}} + R_{\mathrm{in}}} \qquad (16.5.16)$$

which implies:

$$\left[1 - \frac{Q_L}{Q_o}\right] = \frac{R_{out} + R_{in}}{R_{out} + R_{loss} + R_{in}} \tag{16.5.17}$$

Now, by employing equation (16.5.17) in equation (16.5.14) we can write:

$$\beta_o = \frac{R_{in}}{R_{out} + R_{in}} \left[1 - \frac{Q_L}{Q_o}\right] \tag{16.5.18}$$

With reference to equation (16.5.13), and replacing angular frequency by $\omega = 2\pi f$, we can write the overall resonator response in terms of β_o as:

$$\beta = \beta_o \left[\frac{1}{1 \pm 2jQ_L \dfrac{\Delta f}{f_o}}\right] \tag{16.5.19}$$

or in its fuller form as:

$$\beta = \frac{R_{in}}{R_{out} + R_{in}} \left[1 - \frac{Q_L}{Q_o}\right] \left[\frac{1}{1 \pm 2jQ_L \dfrac{\Delta f}{f_o}}\right] \tag{16.5.20}$$

We are now in a position to define the voltage transfer characteristic of the amplifier in Figure 16.9 by incorporating equation (16.5.20) into equation (16.5.4), thus:

$$\frac{V_{out}}{V_{in2}} = \frac{G}{1 - G\left[\dfrac{R_{in}}{R_{out} + R_{in}}\right]\left[1 - \dfrac{Q_L}{Q_o}\right]\left[\dfrac{1}{1 \pm 2jQ_L \dfrac{\Delta f}{f_o}}\right]} \tag{16.5.21}$$

Under steady-state oscillation conditions the voltage gain of the amplifier, at $f = f_o$, according to the Barkhausen criterion, is defined by:

$$G = \frac{1}{\beta_o} = \frac{1}{\dfrac{R_{in}}{R_{out} + R_{in}}\left[1 - \dfrac{Q_L}{Q_o}\right]} \tag{16.5.22}$$

Replacing the term G in equation (16.5.21) by equation (16.5.22) gives:

$$\frac{V_{out}}{V_{in2}} = \frac{G}{1 - \dfrac{1}{1 \pm 2jQ_L \dfrac{\Delta f}{f_o}}} \tag{16.5.23}$$

$$= \cfrac{1}{\left[\cfrac{R_{\text{in}}}{R_{\text{out}} + R_{\text{in}}}\right]\left[1 - \cfrac{Q_{\text{L}}}{Q_{\text{o}}}\right]\left[1 - \cfrac{1}{1 \pm 2jQ_{\text{L}}\cfrac{\Delta f}{f_{\text{o}}}}\right]} \qquad (16.5.24)$$

The Q multiplication process causes the noise to fall to the thermal noise floor within the 3 dB bandwidth of the resonator [9]. The noise of interest therefore occurs within the boundaries of $Q_{\text{l}}(\Delta f / f_{\text{o}}) \ll 1$, so we can apply the binomial approximation to equation (16.5.23) resulting in a further simplification:

$$\frac{V_{\text{out}}}{V_{\text{in2}}} = \frac{G}{\pm 2jQ_{\text{L}}\dfrac{\Delta f}{f_{\text{o}}}} \qquad (16.5.25)$$

$$= \cfrac{1}{\left[\cfrac{R_{\text{in}}}{R_{\text{out}} + R_{\text{in}}}\right]\left[1 - \cfrac{Q_{\text{L}}}{Q_{\text{o}}}\right]\left[\pm 2jQ_{\text{L}}\cfrac{\Delta f}{f_{\text{o}}}\right]} \qquad (16.5.26)$$

The gain of the feedback system has been incorporated into equation (16.5.25) in terms of Q/Q_{o}, since the gain is set by the insertion loss of the resonator.

We now turn our attention to the thermal noise input voltage at input 2 of the amplifier in Figure 16.9, which we can write as:

$$V_{\text{in2}} = \sqrt{4k_{\text{B}}TBR_{\text{in}}} \qquad (16.5.27)$$

where k_{B} is Boltzmann's constant, T is the absolute temperature, and B is the bandwidth of interest. Since we are interested in the ratio of noise to signal power, we will deal in terms of squared voltages. If the source resistance is equal to R_{in}, the total noise power available at the input of the amplifier is equal to $k_{\text{B}}TB$. Since we are specifically interested in noise power in a 1 Hz bandwidth at an offset Δf from the carrier, we set $B = 1$. We relate the noise power at the amplifier input to that at the output via the amplifier noise factor, F.

If we apply the above reasoning to equation (16.5.25) we can write the square of the output voltage in a 1 Hz bandwidth at an offset, f_{m}, from the carrier as:

$$(V_{\text{out}}(f_{\text{m}}))^2 = \frac{Fk_{\text{B}}TR_{\text{in}}}{4Q_{\text{L}}^2\left[\dfrac{R_{\text{in}}}{R_{\text{out}} + R_{\text{in}}}\right]^2\left[1 - \dfrac{Q_{\text{L}}}{Q_{\text{o}}}\right]^2}\left(\frac{f_{\text{o}}}{f_{\text{m}}}\right)^2 \qquad (16.5.28)$$

Since Q_{o} is fixed by the type of resonator, but the ratio $(Q_{\text{L}}/Q_{\text{o}})$ can be varied by adjusting the resonator coupling, we can rewrite equation (16.5.28) in a more useful form that separates constants and variables, thus:

$$(V_{\text{out}}(f_{\text{m}}))^2 = \frac{Fk_{\text{B}}TR_{\text{in}}}{4Q_{\text{o}}^2\left(\dfrac{Q_{\text{L}}}{Q_{\text{o}}}\right)^2\left[\dfrac{R_{\text{in}}}{R_{\text{out}} + R_{\text{in}}}\right]^2\left[1 - \dfrac{Q_{\text{L}}}{Q_{\text{o}}}\right]^2}\left(\frac{f_{\text{o}}}{f_{\text{m}}}\right)^2 \qquad (16.5.29)$$

Equation (16.5.29) incorporates both amplitude and phase components of the input noise signal. In practice, the amplitude fluctuations are largely suppressed by the amplitude limiting function of the amplifier. This has the effect of halving the total input noise power defined in equation (16.5.28). A more correct equation for the square of the output voltage is therefore:

$$(V_{out}(f_m))^2 = \frac{F k_B T B R_{in}}{8 Q_0^2 \left(\frac{Q_L}{Q_0}\right)^2 \left[\frac{R_{in}}{R_{out} + R_{in}}\right]^2 \left[1 - \frac{Q_L}{Q_0}\right]^2} \left(\frac{f_o}{f_m}\right)^2 \tag{16.5.30}$$

The generally accepted definition of phase noise is the ratio of output noise power in a 1 Hz bandwidth at a frequency offset f_m to the total output power. If the total output power is $V_{out\ max\ rms}$, then we can write:

$$\mathscr{L}\{f_m\} = \frac{(V_{out}(f_m))^2}{(V_{out\ max\ rms})^2} \tag{16.5.31}$$

Applying equation (16.5.30) gives:

$$\mathscr{L}\{f_m\} = \frac{F k_B T R_{in}}{8 Q_0^2 \left(\frac{Q_L}{Q_0}\right)^2 \left[\frac{R_{in}}{R_{out} + R_{in}}\right]^2 \left[1 - \frac{Q_L}{Q_0}\right]^2 (V_{out\ max\ rms})^2} \left(\frac{f_o}{f_m}\right)^2 \tag{16.5.32}$$

P_{RF} is limited by the maximum voltage swing at the output of the amplifier and the value of $R_{out} + R_{loss} + R_{in}$, that is,

$$P_{RF} = \frac{(V_{out\ max\ rms})^2}{R_{out} + R_{loss} + R_{in}} \tag{16.5.33}$$

Equation (16.5.32) now becomes:

$$\mathscr{L}\{f_m\} = \frac{F k_B T (R_{out} + R_{in})^2}{8 Q_0^2 \left(\frac{Q_L}{Q_0}\right)^2 R_{in} \left[1 - \frac{Q_L}{Q_0}\right]^2 P_{RF}(R_{out} + R_{loss} + R_{in})} \left(\frac{f_o}{f_m}\right)^2 \tag{16.5.34}$$

We note that:

$$\frac{R_{out} + R_{in}}{R_{out} + R_{loss} + R_{in}} = \left[1 - \frac{Q_L}{Q_0}\right] \tag{16.5.35}$$

So, the ratio of sideband noise in a 1 Hz bandwidth at an offset Δf to the total power given in equation (16.5.34) therefore becomes:

$$\mathscr{L}\{f_m\} = \frac{F k_B T}{8 Q_0^2 \left(\frac{Q_L}{Q_0}\right)^2 \left[1 - \frac{Q_L}{Q_0}\right] P_{RF}} \left[\frac{R_{out} + R_{in}}{R_{in}}\right] \left(\frac{f_o}{f_m}\right)^2 \tag{16.5.36}$$

Everard outlines three possible cases, as follows.

Case 1: High efficiency oscillator

If $R_{out} \approx 0$, as would be the case for any high efficiency oscillator then equation (16.5.36) simplifies to:

$$\mathscr{L}\{f_m\} = \frac{Fk_B T}{8Q_o^2 \left(\dfrac{Q_L}{Q_o}\right)^2 \left[1 - \dfrac{Q_L}{Q_o}\right] P_{RF}} \left(\frac{f_o}{f_m}\right)^2 \qquad (16.5.37)$$

Case 2: Microwave amplifier

If $R_{out} = R_{in}$, as would be the case for most microwave amplifiers then equation (16.5.36) simplifies to:

$$\mathscr{L}\{f_m\} = \frac{Fk_B T}{4Q_o^2 \left(\dfrac{Q_L}{Q_o}\right)^2 \left[1 - \dfrac{Q_L}{Q_o}\right] P_{RF}} \left(\frac{f_o}{f_m}\right)^2 \qquad (16.5.38)$$

$$P_{AVO} = \frac{(V_{out\ max\ rms})^2}{4R_{out}} \qquad (16.5.39)$$

Equation (16.5.36) then becomes:

$$\mathscr{L}\{f_m\} = \frac{Fk_B T R_{in}}{8Q_o^2 \left(\dfrac{Q_L}{Q_o}\right)^2 \left[\dfrac{R_{in}}{R_{out} + R_{in}}\right]^2 \left[1 - \dfrac{Q_L}{Q_o}\right]^2 P_{AVO}(4R_{out})} \left(\frac{f_o}{f_m}\right)^2 \qquad (16.5.40)$$

which can be rearranged as:

$$\mathscr{L}\{f_m\} = \frac{Fk_B T R_{in}}{32Q_o^2 \left(\dfrac{Q_L}{Q_o}\right)^2 \left[1 - \dfrac{Q_L}{Q_o}\right]^2 P_{AVO}} \left[\frac{(R_{out} + R_{in})^2}{R_{out}R_{in}}\right]^2 \left(\frac{f_o}{f_m}\right)^2 \qquad (16.5.41)$$

Case 3: Matched output

The term:

$$\frac{(R_{out} + R_{in})^2}{R_{out}R_{in}} \qquad (16.5.42)$$

attains a minimum value of 4 when $R_{out} = R_{in}$. In this case, equation (16.5.36) becomes:

$$\mathscr{L}\{f_m\} = \frac{Fk_B T}{8(Q_o)^2 \left(\dfrac{Q_L}{Q_o}\right)^2 \left(1 - \dfrac{Q_L}{Q_o}\right)^2 P_{AVO}} \left(\frac{f_o}{f_m}\right)^2 \qquad (16.5.43)$$

A general equation can be written which describes all three cases:

$$\mathcal{L}\{f_m\} = A \frac{Fk_B T}{8(Q_o)^2 \left(\dfrac{Q_L}{Q_o}\right)^2 \left(1 - \dfrac{Q_L}{Q_o}\right)^N P} \left(\frac{f_o}{f_m}\right)^2 \qquad (16.5.44)$$

where the parameters A and N are determined as follows:

(1) $N = 1$ and $A = 1$ if P is defined as P_{RF} and $R_{OUT} = 0$ (16.5.37).
(2) $N = 1$ and $A = 2$ if P is defined as P_{RF} and $R_{OUT} = R_{IN}$ (16.5.41).
(3) $N = 2$ and $A = 1$ if P is defined as P_{AVO} and $R_{OUT} = R_{IN}$ (16.5.43).

If we take assumption 3, equation (16.5.44) becomes:

$$\mathcal{L}\{f_m\} = 1 \frac{Fk_B T}{8(Q_o)^2 \left(\dfrac{Q_L}{Q_o}\right)^2 \left(1 - \dfrac{Q_L}{Q_o}\right)^2 P_{AVO}} \left(\frac{f_o}{f_m}\right)^2 \qquad (16.5.45)$$

It should be noted that the thermal noise source resistance for the amplifier in this analysis is assumed to be the same as the input impedance of the amplifier, so the noise power available is, therefore, $k_B TB$. This means that the noise factor, F, takes into account the excess noise generated by the amplifier and the change in noise resulting from a change in source impedance, which varies with Q_L/Q_o. The variation of F with Q_L/Q_o is often a second-order effect, but it can sometimes be important. The noise factor may also show a power dependence that can be affected by both the device type and the amplifier topology.

By differentiating equation (16.5.45) with respect to Q_L/Q_o, it can be seen that the power in the resonator (P_{RLOSS}) is maximum and hence phase noise is minimum when $Q_L/Q_o = 1/2$. At this point, half the power is dissipated in the resonator, one quarter is transmitted, and one quarter is reflected.

This, therefore, means that the phase noise is *minimum when maximum power is dissipated in the resonator*. The phase-noise equation for minimum noise is, therefore,

$$\mathcal{L}\{f_m\} = \frac{2k_B TF}{P_{AVO}} \left(\frac{f_o}{f_m}\right)^2 \qquad (16.5.46)$$

It should be noted that this equation is half the value of Leeson's equation when Q_L/Q_o is optimum (i.e., when we set $Q_L/Q_o = 1/2$). It has been reported [10] that the factor of 2 was omitted by mistake from the original Leeson equation [2].

16.5.2 ANALYTICAL PHASE-NOISE MODELS: NEGATIVE RESISTANCE OSCILLATOR

The analytical model of phase noise for a feedback oscillator can be extended to cover the negative resistance oscillators also. Once again, we will follow Professor Everard's analysis [10] here, with reference to Figure 16.10.

The simple model shown in Figure 16.10 consists of the parallel addition of a noise current source, $\bar{i}_n$, a negative admittance/resistance, $-R$, a parallel LCR resonant tank circuit (L, C), and a load. The noise current is generated by R_{loss} and, for this analysis, the load resistance and NF are ignored. This is similar to Lee and Hajimiri's model [13], using a slightly different approach and terminology.

In this case, the transfer function is calculated as a forward transresistance (V_{OUT}/I_{IN}) and then converted to a power ratio. The effect of limiting is incorporated by assuming that the phase-noise element of the noise is $k_B T/2$, that is, -177 dBm Hz^{-1} at room temperature.

The admittance of the resonator in Figure 16.10 is simply:

$$Y_{RES} = \frac{1}{j\omega L} + j\omega LC + \frac{1}{R_{LOSS}} \tag{16.5.47}$$

If we consider a given frequency, ω, as being the sum of a center angular frequency and offset frequency, that is $(\omega_0 + \Delta\omega)$, then:

$$Y_{RES} = \frac{-j}{(\omega_0 \pm \Delta\omega) L} + j(\omega_0 \pm \Delta\omega) C + \frac{1}{R_{LOSS}} \tag{16.5.48}$$

Which can be rewritten as:

$$Y_{RES} = \frac{-j}{\omega_0 \left(1 \pm \dfrac{\Delta\omega}{\omega_0}\right) L} + j\omega_0 \left(1 \pm \frac{\Delta\omega}{\omega_0}\right) C + \frac{1}{R_{LOSS}} \tag{16.5.49}$$

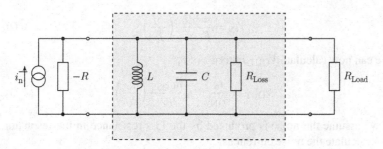

FIGURE 16.10

Negative resistance oscillator model [10].

At resonance, $\omega_0 L = 1/\omega_0 C$, therefore,

$$Y_{RES} = \frac{-j\omega_0 C}{\left(1 \pm \dfrac{\Delta\omega}{\omega_0}\right)} + j\omega_0 \left(1 \pm \frac{\Delta\omega}{\omega_0}\right) C + \frac{1}{R_{LOSS}} \qquad (16.5.50)$$

where ω is the angular frequency, which can be replaced by $\omega = 2\pi f$. Assuming that $\Delta f \ll f_0$, and applying the binomial approximation we get:

$$Y_{RES} = j\omega_0 C\left[-\left(1 \mp \frac{\Delta f}{f_0}\right) + \left(1 \pm \frac{\Delta f}{f_0}\right)\right] + \frac{1}{R_{LOSS}} \qquad (16.5.51)$$

$$Y_{RES} = 2j\omega_0 C\left(\frac{\Delta f}{f_0}\right) + \frac{1}{R_{LOSS}} \qquad (16.5.52)$$

We note that:

$$Q_0 = R_{LOSS}\omega_0 C \qquad (16.5.53)$$

Therefore,

$$\omega_0 C = \frac{Q_0}{R_{LOSS}} \qquad (16.5.54)$$

and so, replacing equation (16.5.54) in equation (16.5.52) we get:

$$Y_{RES} = \frac{1}{R_{LOSS}}\left[1 \pm 2jQ_0\left(\frac{\Delta f}{f_0}\right)\right] \qquad (16.5.55)$$

Equation (16.5.55) describes the resonant response around the carrier for most other forms of parallel resonant circuits. If we now complete the oscillator circuit of Figure 16.10 by adding the negative resistance, the input admittance becomes:

$$Y_{IN} = \frac{1}{R_{LOSS}}\left[1 \pm 2jQ_0\left(\frac{\Delta f}{f_0}\right)\right] - \frac{1}{R} \qquad (16.5.56)$$

Under steady-state oscillating conditions, the negative resistance of the active device exactly cancels the positive loss resistance of the resonator. We therefore have:

$$Y_{IN} = \frac{\pm 2jQ_0}{R_{LOSS}}\left(\frac{\Delta f}{f_0}\right) \qquad (16.5.57)$$

We can now calculate V_{OUT} in terms of i_n:

$$V_{OUT} = \frac{i_n}{Y_{IN}} = i_n \frac{R_{LOSS}}{2jQ_0}\left(\frac{f_0}{\Delta f}\right) \qquad (16.5.58)$$

If we assume the noise is produced by the loss resistance in the resonator, then we can calculate the noise current as:

$$\left(\frac{i_n}{2}\right)^2 R_{LOSS} = k_B TB \qquad (16.5.59)$$

Therefore,

$$i_n = \sqrt{\frac{4k_B TB}{R_{LOSS}}} \tag{16.5.60}$$

The output voltage at offset Δf according to equation (16.5.58) is, therefore,

$$V_{OUT}(\Delta f) = \sqrt{\frac{4k_B TB}{R_{LOSS}}} \frac{R_{LOSS}}{2jQ_0} \left(\frac{f_0}{\Delta f}\right) \tag{16.5.61}$$

We are more accustomed to talking in terms of noise power. The output power at a finite frequency offset (i.e., $\Delta f \neq 0$) is, therefore,

$$\frac{(V_{OUT}(\Delta f))^2}{R_{LOSS}} = \frac{k_B TB}{Q_0^2} \left(\frac{f_0}{\Delta f}\right)^2 \tag{16.5.62}$$

Because this theory is a linear theory, the sideband noise is effectively amplified narrowband noise that has been processed by a Q multiplication filter.

Once again, we introduce the factor of $1/2$ to account for the fact that the noise power had been halved by the removal of the amplitude component due to the limiting mechanism. Equation (16.5.62) is therefore halved to give:

$$\frac{(V_{OUT}(\Delta f))^2}{R_{LOSS}} = \frac{4k_B TB}{2Q_0^2} \left(\frac{f_0}{\Delta f}\right)^2 \tag{16.5.63}$$

To obtain $\mathscr{L}\{f_m\}$, we now just need to divide equation (16.5.63) by the output power of the oscillator (P), and set $B = 1$, since we are talking about the noise in a 1 Hz bandwidth. We will also now switch notation, replacing Δf with f_m, the frequency offset (as discussed previously). With these changes we can now write:

$$\boxed{\mathscr{L}\{f\} = \frac{k_B T}{2Q_0^2 P} \left(\frac{f_0}{f}\right)^2} \tag{16.5.64}$$

This equation is very similar to that shown by Edson [11], Grebennikov [19], and Lee and Hajimiri [13]. Note that a noise factor can be added to take into account the added noise from the actual negative resistance. One feature is that the source impedance is known and set by the resonator/load combination.

16.5.3 COMPARISON OF FEEDBACK AND NEGATIVE RESISTANCE OSCILLATOR PHASE-NOISE EQUATIONS

It is interesting to compare the phase-noise equation for a negative-resistance oscillator equation (16.5.64) with the phase-noise equation for a feedback oscillator equation (16.5.46), as follows:

Feedback Oscillator	Negative Resistance Oscillator
$\mathscr{L}\{f_m\} = \dfrac{2k_BT}{Q_0^2 P_{AVO}} \left(\dfrac{f_0}{f_m}\right)^2$	$\mathscr{L}\{f_m\} = \dfrac{k_BT}{2Q_0^2 P} \left(\dfrac{f_0}{f_m}\right)^2$

We observe that the equation for a feedback oscillator is four times larger than that for the negative resistance oscillator. A difference of a factor of 2 can be explained by the fact that P in the negative resistance is the power dissipated in the resonator, whereas P_{AVO} is twice the value of the power dissipated in the resonator under optimum operating conditions. The other factor of 2 is due to the fact that the SNR is set by the power at the input of the amplifier, which is $1/4$ in the case of the feedback oscillator.

It may still, therefore, be important in some circumstances (crystal oscillators, for example) in which the power in the resonator must be kept low, that the negative-resistance oscillator improves the phase noise by a factor of 2 for the same power dissipated in the resonator [10].

16.6 LOW-NOISE OSCILLATOR DESIGN

Based on the analysis of oscillator phase noise in the previous sections, and the resulting equations, we can now contemplate some general rules pertaining to low-noise oscillator design. The two key areas of attention in low phase-noise oscillator design are (a) the resonator and (b) the active device.

16.6.1 LOW-NOISE DESIGN: RESONATOR

We can take steps to reduce oscillator noise by maximizing the reactive energy in the resonator. This can be done by maintaining a high radio frequency (RF) voltage across the resonator, which implies a low L/C ratio. To construct a resonant structure with a high Q-factor, low losses are required in all of the constituent parts, such as:

1. Q of the resonator device itself.
2. Series resistance of capacitors and series resistance of any tuning diodes.
3. Losses in the surrounding circuitry (printed circuit board or integrated circuit).

16.6.2 LOW-NOISE DESIGN: TRANSISTOR

Since flicker noise is a dominant factor close to the carrier, we should concentrate on ways of reducing flicker noise in the transistor. The basic rules to select the right transistor are:

1. The best oscillator transistor is a device with the lowest possible f_T commensurate with the required frequency of operation. A commonly used criteria is $f_T \leq 2f_c$, where f_c is the design center frequency of the oscillator.

2. Flicker noise is directly related to the current density in the transistor, so transistors with high $I_{C_{max}}$ used at low currents have best $1/f$ performance.

3. On the other hand, the f_T of a transistor drops as collector current decreases. Additionally, the parasitic capacitances of high current transistors are higher due to the larger transistor geometry. There is therefore a trade-off between (1) and (2) above.

It has been generally well established in the literature [20–22] that the above general design rules help reduce the phase noise of practical microwave oscillators.

16.7 PHASE-NOISE MEASUREMENTS

Phase-noise measurement is a complex and challenging task that has occupied researchers and equipment manufacturers for many decades. The main challenge associated with phase-noise measurement is limited dynamic range of the equipment used [23], as the noise power levels being measured are usually very small and in very close frequency proximity to carrier signals many orders of magnitude larger. Any reference source used in the measuring equipment will also be required to have a higher spectral purity than the device under test (DUT). There are several methods to measure phase noise, three of which we will briefly outline in this section.

16.7.1 THE DIRECT METHOD

The simplest and most intuitive method of measuring phase noise is to connect the oscillator to a high-quality spectrum analyzer, as shown in Figure 16.11. The power of the carrier is measured and a measurement of the power spectral density of the oscillator noise, at a specified offset frequency, is also made and referenced to the carrier power level.

Although the direct method of measurement is simple and uses a readily available piece of test equipment (the spectrum analyzer), it has some serious limitations. The biggest limiting factor is the quality of the spectrum analyzer being used, but all spectrum analyzers share some common limitations, as follows:

1. The 3 dB bandwidth and the noise bandwidth of the analyzer's resolution bandwidth filters are not identical and correction factors must be used.

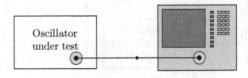

FIGURE 16.11

Phase-noise measurement: direct method.

2. There are errors associated with the way the peak detector inside most spectrum analyzers responds to noise, meaning that root mean square (RMS) noise power will be under-reported by a factor of 1.05 dB. In addition, the logging process in spectrum analyzers tend to amplify noise peaks less than the rest of the noise signal resulting in a reported power that is less than the actual noise power. Combining these two effects results in a noise power measurement that is 2.5 dB below the actual noise power.

3. The noise floor of the spectrum analyzer and the residual FM of the analyzer's own local oscillator will limit the accuracy of the measurement.

4. Lastly, spectrum analyzers generally only measure the scalar magnitude of noise sidebands of the signal and are not able to differentiate between amplitude noise and phase noise. Finally, the measurement process using a spectrum analyzer involves having to make a noise measurement at each frequency offset of interest. This will be very time consuming if done manually, although it can be automated if the spectrum analyzer is programmable.

16.7.2 **THE PHASE DETECTOR METHOD**

The basic phase detector method is shown in Figure 16.12. At the heart of this method is the phase detector. Two sources, at the same frequency and in phase quadrature, are presented to a double-balanced mixer which, together with a low-pass filter, acts as a phase detector. The difference frequency emerging from the low-pass filter has an average voltage level of 0 V. Riding on this DC signal are AC voltage fluctuations proportional to the combined phase noise of the two sources. The baseband signal is amplified and then fed into a baseband spectrum analyzer.

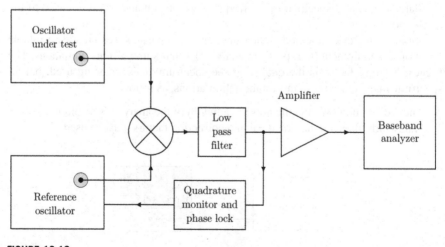

FIGURE 16.12

Phase-noise measurement: phase detector method.

In order for this method to work, phase quadrature between the oscillator under test and the reference oscillator must be strictly maintained. This is achieved by making the frequency of the reference oscillator electronically tunable and driving this from a quadrature detector connected to the output of the phase detector.

Any phase deviation from quadrature between the two sources will result in a corresponding voltage fluctuation at the output and a value other than zero. This will interfere with the measurement.

The phase detector method requires a reference source that has a lower phase noise than the oscillator being measured, and measurements made inside the loop bandwidth of the system require special correction. This means that the phase detector method is complex and requires very high-quality equipment.

16.7.3 THE DELAY LINE/FREQUENCY DISCRIMINATOR METHOD

In contrast to the phase detector method, the frequency discriminator method of phase-noise measurement has the advantage of not requiring a reference source phase locked to the oscillator under test, and is therefore somewhat simpler and lower cost to implement. The key element in the measurement set-up is an analog delay line, as shown in Figure 16.13. Short-term frequency fluctuations in the oscillator under test are converted into voltage fluctuations that can be measured by a baseband analyzer. This process is accomplished in two stages: first, frequency fluctuations are converted into phase fluctuations and then these phase fluctuations are converted into voltage fluctuations.

With reference to Figure 16.13, we can see that the output signal of the oscillator under test is split into two channels. One path is delayed relative to the other by the adjustable delay τ_d. The delay line has the effect of converting frequency variations into phase fluctuations. The other channel is passed through an adjustable phase shifter, the purpose of which is to maintain quadrature between the signals applied to the subsequent mixer.

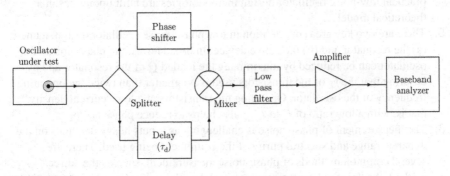

FIGURE 16.13

Phase-noise measurement: frequency discriminator method.

The double balanced mixer, followed by a low-pass filter acts as a phase detector which converts instantaneous phase fluctuations into voltage fluctuations that are then measured by the baseband analyzer.

For this method to work properly it is important that both inputs to the mixer be maintained in phase quadrature to ensure maximum phase sensitivity. This is achieved by adjusting the delay time, τ_d.

This concludes our brief introduction to the measurement of phase noise. Any readers interested in delving deeper into the field of phase-noise measurement are referred to some of the numerous books and papers on this subject [24–26]. Manufacturer's application notes are another very useful source of information on this topic [27,28]. Finally, readers are also referred to a useful good practice guide on phase-noise measurement published by the National Physical Laboratory [29].

16.8 TAKEAWAYS

1. Phase noise is a measure of the spectral purity of an oscillator and is a significant limiting factor affecting overall performance of microwave systems.
2. Oscillator noise is referred to as *phase noise* because, although the noise generated within the oscillator will be random in both amplitude and phase, the amplitude randomness tends to be suppressed by *gain compression*, which is a mechanism inherent to all oscillators.
3. Phase noise in oscillators originates from fundamental noise processes within the components being used. In addition to thermal noise, which is present in all electronic components, the use of active devices in oscillators also introduces other specific types of noise, most notably flicker noise.
4. A number of phase-noise models have been proposed, most of which have been based on the classical feedback oscillator topology. The most famous of such models is that first proposed by Leeson [2]. An alternative model proposed by Lee and Hajimiri is based on the time-varying properties of oscillators. Most practical low-noise oscillator design methodologies are built upon Leeson's theoretical model.
5. There are two key areas of attention in low phase-noise oscillator design, namely (a) the resonator and (b) the active device. In broad terms, the phase noise of an oscillator can be reduced by maximizing the loaded Q of the resonator and by ensuring that the f_T of the transistors used is no greater than twice the operating frequency of the oscillator. Operating the transistors at a low "current density," that is, with a low ratio of I_c to $I_{c_{max}}$ also helps to reduce phase noise.
6. The measurement of phase noise is challenging as it puts heavy demands on the dynamic range and spectral purity of the equipment being used. There are several common methods of phase-noise measurement, such as the "direct" method, the "phase-detector" method, and the "delay-line/frequency discriminator" method.

REFERENCES

[1] J. Rayleigh, The Theory of Sound, second ed., Macmillan and Co., London, 1896.

[2] D. Leeson, A simple model of feedback oscillator noise spectrum, Proc. IEEE 54 (2) (1966) 329-330, ISSN 0018-9219, http://dx.doi.org/10.1109/PROC.1966.4682.

[3] D. Scherer, Design principles and test methods for low phase noise RF and microwave sources, in: HP RF and Microwave Measurement Symposium, Hewlett Packard, 1978.

[4] W. Robins, Phase Noise in Signal Sources, Peter Peregrinus Ltd, London, 1982.

[5] G. Gonzalez, Foundations of Oscillator Circuit Design, Artech House, Boston, MA, 2007.

[6] M. Odyniec, RF and Microwave Oscillator Design, Artech House, Inc, Norwood, MA, 2002, ISBN 9781580533201.

[7] J. Mukherjee, P. Roblin, S. Akhtar, An analytic circuit-based model for white and flicker phase noise in LC oscillators, IEEE Trans. Circuits Syst. Regul. Pap. 54 (7) (2007) 1584-1598, ISSN 1549-8328, http://dx.doi.org/10.1109/TCSI.2007.898673.

[8] E. Rubiola, Phase Noise and Frequency Stability in Oscillators, Cambridge Books Online, Cambridge University Press, 2009, ISBN 9780521886772.

[9] J. Everard, Fundamentals of RF Circuit Design: With Low Noise Oscillators, John Wiley and Sons Ltd, Chichester, 2001, ISBN 9780471497936.

[10] J. Everard, M. Xu, S. Bale, Simplified phase noise model for negative-resistance oscillators and a comparison with feedback oscillator models, IEEE Trans. Ultrason. Ferroelectr. Freq. Control 59 (3) (2012) 382-390, ISSN 0885-3010, http://dx.doi.org/10.1109/TUFFC.2012.2207.

[11] W. A. Edson, Noise in oscillators, Proc. IRE 48 (8) (1960) 1454-1466, ISSN 0096-8390, http://dx.doi.org/10.1109/JRPROC.1960.287573.

[12] E. Hafner, The effects of noise in oscillators, Proc. IEEE 54 (2) (1966) 179-198, ISSN 0018-9219, http://dx.doi.org/10.1109/PROC.1966.4631.

[13] T. Lee, A. Hajimiri, Oscillator phase noise a tutorial, IEEE J. Solid State Circuits 35 (3) (2000) 326-336, ISSN 0018-9200, http://dx.doi.org/10.1109/4.826814.

[14] J.-C. Nallatamby, M. Prigent, M. Camiade, J. Obregon, Extension of the Leeson formula to phase noise calculation in transistor oscillators with complex tanks, IEEE Trans. Microw. Theory Tech. 51 (3) (2003) 690-696, ISSN 0018-9480, http://dx.doi.org/10.1109/TMTT.2003.808670.

[15] A. Takaoka, K. Ura, Noise analysis of nonlinear feedback oscillator with AM-PM conversion coefficient, IEEE Trans. Microw. Theory Tech. 28 (6) (1980) 654-662, ISSN 0018-9480, http://dx.doi.org/10.1109/TMTT.1980.1130134.

[16] U. Rohde, The Design of Modern Microwave Oscillators for Wireless Applications: Theory and Optimization, first ed., Wiley-Interscience, Hoboken, NJ, 2005, ISBN 13: 978-0471723424.

[17] A. Riddle, A long, winding road, IEEE Microw. Mag. 11 (6) (2010) 70-81, ISSN 1527-3342, http://dx.doi.org/10.1109/MMM.2010.937734.

[18] T. Parker, Current developments in SAW oscillator stability, in: 31st Annual Symposium on Frequency Control, 1977, pp. 359-364.

[19] A. Grebennikov, RF and Microwave Transistor Oscillator Design, Wiley, 2007, ISBN 9780470512081.

[20] V. Lam, P. Yip, C. Poole, Microwave oscillator design with phase noise optimisation, Int. J. Circuit Theory Appl. 21 (1993) 287-292.

[21] V. Lam, P. Yip, Microwave oscillator phase noise reduction using negative resistance compensation, Electron. Lett. 29 (4) (1993) 379-381, ISSN 0013-5194, http://dx.doi.org/10.1049/el:19930255.

[22] H.-C. Chang, Phase noise in self-injection-locked oscillators: theory and experiment, IEEE Trans. Microw. Theory Tech. 51 (9) (2003) 1994-1999, ISSN 0018-9480, http://dx.doi.org/10.1109/TMTT.2003.815872.

[23] A. Poddar, U. Rohde, E. Rubiola, Phase noise measurement: challenges and uncertainty, in: IEEE International Microwave and RF Conference (IMaRC), 2014, pp. 342-345, http://dx.doi.org/10.1109/IMaRC.2014.7039043.

[24] G. Bryant, Institution of Electrical Engineers, Principles of Microwave Measurements, IEE Electrical Measurement Series, Peter Peregrinus Limited, 1993, ISBN 9780863412967.

[25] R. Collier, A. Skinner, Microwave Measurements, third ed., IET Electrical Measurement Series, Institution of Engineering and Technology, 2007, ISBN 9780863417351.

[26] C. Nelson, Tutorial session 1A: phase noise measurements, in: IEEE International Frequency Control Symposium and Exposition, 2006, p. nil17, http://dx.doi.org/10.1109/FREQ.2006.275326.

[27] D. Scherer, The "art" of phase noise measurement, HP Application Note 34, 1985, www.hparchive.com.

[28] J. Wolf, Phase noise measurements with spectrum analyzers of the FSE family, Rohde and Schwarz: Application Note 1EPAN 16E, 1995, ISSN 1368-6550.

[29] D. Owen, Good Practice Guide to Phase Noise Measurement, National Physical Laboratory, 2004, ISSN 1368-6550.

Microwave mixers

17

INTENDED LEARNING OUTCOMES

- *Knowledge*
  - Understand the role and function of mixers in microwave systems.
  - Understand various figures of merit used to characterize mixers.
  - Understand the strengths and weaknesses of various mixer topologies.
- *Skills*
  - Be able to design a single balanced diode passive mixer.
  - Be able to design a double balanced diode passive mixer.
  - Be able to design a Gilbert cell mixer, based on BJT or FETs.

Microwave Active Circuit Analysis and Design. http://dx.doi.org/10.1016/B978-0-12-407823-9.00017-2

17.1 INTRODUCTION

The ability to translate a band of frequencies from one part of the frequency spectrum to another is an essential function within almost all radio frequency (RF) and microwave systems, whether it be radio transmitters, receivers, or test equipment. This frequency translation function, also known as "mixing," or "heterodyning," is carried out by a circuit called a *mixer*.

A mixer is a three-port circuit (two inputs and one output) that generates output frequencies equal to the sum and difference of the two input frequencies and their harmonics.

Downconversion refers to the case when a radio frequency (RF) input signal is mixed with a high-frequency local oscillator (LO) signal, both in a similar frequency range, to produce a lower intermediate frequency (IF) output signal. The process of *upconversion* on the other hand refers to the case where a lower-frequency IF signal is one of the inputs and is mixed with a higher-frequency LO signal to produce a higher-frequency RF output signal. The respective output signals are the product of the LO signal and the respective input signals. Downconversion is normally applied in receivers whereas upconversion is typically applied in transmitters. The mixer symbol, as applied in downconversion and upconversion modes, is shown in Figure 17.1.

Mixers are sometimes described as "nonlinear" circuits because they generate frequency components at the output that occupy a different part of the frequency spectrum from the input signals, but also because we often use nonlinear devices to accomplish the mixing operation, as will be described shortly. In fact, mixers can be also considered linear circuits in the sense that the relationship between input signal amplitude and output signal amplitude is linear over a specified operating range. In a well-designed mixer, the properties of the signal (phase and amplitude) will be maintained and there should be a linear relationship between these quantities at the input and those at the output, albeit shifted in frequency.

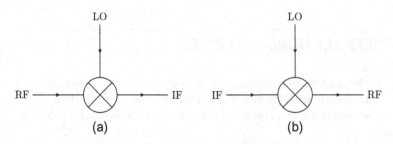

FIGURE 17.1

Mixer circuit symbols. (a) Downconversion mixer. (b) Upconversion mixer.

The ideal mixer is a pure analog signal multiplier in the time domain. In other words, the instantaneous time domain signal voltage relationships in, for example, Figure 17.1(a) are:

$$v_{IF} = v_{RF} \times v_{LO} \tag{17.1.1}$$

The effect of this time domain multiplication is to translate the entire RF signal spectrum from one region to another in the frequency domain. This can be demonstrated by representing v_{RF} and v_{LO} as sinusoids of amplitude V and angular frequency ω as follows:

$$v_{IF} = V_{RF} \cos(\omega_{RF} t) \times V_{LO} \cos(\omega_{LO} t) \tag{17.1.2}$$

Which is equal to:

$$v_{IF} = \frac{V_{RF} V_{LO}}{2} [\cos(\omega_{RF} + \omega_{LO})t + \cos(\omega_{RF} - \omega_{LO})t] \tag{17.1.3}$$

The resultant signal in equation (17.1.3) contains two frequency components, at $(\omega_{RF} + \omega_{LO})$ and $(\omega_{RF} - \omega_{LO})$, which are referred to as the sum and difference frequencies. For a *downconversion* mixer we are interested in the difference frequency, $(\omega_{RF} - \omega_{LO})$, which is lower than either ω_{RF} or ω_{LO}. For an *upconversion* mixer we are interested in the sum frequency, $(\omega_{RF} + \omega_{LO})$, which is higher than either ω_{RF} or ω_{LO}. In either case, the unwanted frequency component is removed at the output by filtering.

In a downconversion mixer, the RF input signal is usually small in amplitude when compared with the LO. When the mixer is used for upconversion the input signal is the IF signal, which is of a similar amplitude to the LO. These differences have implications for mixer topology and performance, as will be discussed later in this chapter.

Mixers may be classified into *passive* and *active* types. The essential difference is that active mixers employ transistors as the primary active elements and can therefore provide positive signal gain, in addition to their primary function of frequency translation. Passive mixers, on the other hand, usually employ only diodes and are therefore characterized by conversion loss, rather than gain.

Aside from frequency translation (upconversion and downconversion), mixers are also used as phase detectors in phase locked loops and frequency synthesizers and in frequency multipliers. It is hard to imagine any piece of microwave equipment that does not contain at least one mixer.

17.2 MIXER CHARACTERIZATION

As mixers are frequency translation devices, there are some specific metrics involved in their characterization that do not arise in the case of amplifiers and oscillators. In particular, the degree of nonlinearity is usually explicitly specified for mixers, in the form of such parameters as the "Third-Order Intercept Point (IP3)." In this section, we will briefly cover some of the key parameters that are used to characterize microwave mixers.

17.2.1 CONVERSION GAIN

The conversion gain (or loss) of a mixer is defined as the ratio of the desired output signal power to the input signal power. In the case of a downconversion mixer, the desired output signal is the IF, and the input signal is the RF. In the case of an upconversion mixer, the desired output signal is the RF and the input signal is the IF. Note that the LO power does not feature in conversion gain calculations. If the input impedance and the load impedance of the mixer are both equal to the source impedance, then the voltage conversion gain and the power conversion gain of the mixer in dB terms will be the same.

Passive and active mixers are distinguished by the fact that active mixers provide a positive conversion gain, whereas passive mixers do not. Conversion gain is also dependent on impedance matching at the input and output ports (i.e., it depends on power transfer at these ports). Although IF signal power does not feature in the conversion gain calculation, the level of the LO will affect the conversion gain so this needs to be specified. Conversion gain of a typical active mixer is approximately +10 dB, while the conversion loss of a typical passive (diode) mixer is approximately −6 dB.

17.2.2 ISOLATION

Isolation is a measure of the amount of "leakage" or "feed through" between the mixer ports, especially leakage of the LO signal, as this tends to be the largest of the three. Typically, mixer isolation tends to deteriorate with frequency due to circuit imperfections, such as imbalances in the transformers, capacitive unbalance between diodes and other parasitic effects.

17.2.3 DYNAMIC RANGE

Dynamic range (DR) is defined as the range of input power levels for which the output power is linearly proportional to the input power. The proportionality constant is the conversion loss or gain of the mixer.

The lower bound of DR is set by the noise floor which defines the *minimum detectable signal* (MDS).

As the input power is increased we eventually reach a point where the output power can no longer keep increasing linearly with input power. At some point the gain of the system deviates from the linear relationship by −1 dB and this point is called the *1 dB compression point*. The 1 dB compression point is commonly used to define the upper bound of the linear region. Therefore, the system's DR can be defined as the power difference from the MDS to the 1 dB compression point.

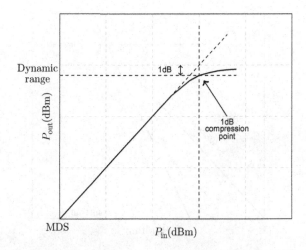

FIGURE 17.2

1 dB compression point.

This can be defined either in terms of input power or output power. Figure 17.2 shows that how DR is defined in terms of the relationship between input power and output power.

17.2.4 THIRD-ORDER INTERCEPT POINT: IP3

The third-order intercept point (or "IP3") is the input (or output) power level at which the nonlinear intermodulation products caused by the third-order nonlinearities are equal to the desired signal. The intercept point is a theoretical concept which cannot be directly measured in practice. This is because it lies well beyond the saturation level of the active devices and, in most cases, beyond the damage threshold of these devices. Although not practically measurable, the value of defining the IP3 point lies in it being a figure of merit for distortion at lower power levels. The IP3 point can be defined at either the input or output ports, as shown in Figure 17.3.

Two different definitions for intercept points are in use:

1. *Based on harmonics*: The device is tested using a single input tone. The nonlinear products caused by nth-order nonlinearity appear at n times the frequency of the input tone.
2. *Based on intermodulation products*: The device is fed with two sine tones with a small frequency difference. The nth-order intermodulation products then appear at n times the frequency spacing of the input tones. This is the more commonly used approach.

Note: the above two definitions differ by $10 \log_{10}(3) = 4.8$ dB.

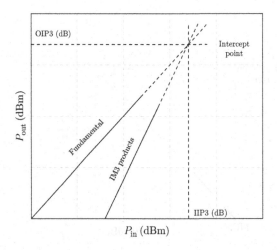

FIGURE 17.3

Third-order intercept point.

The intercept point is obtained graphically by plotting the output power versus the input power on log-log axes (i.e., both input and output power scales are in dB). On a logarithmic scale, the function $y = x^n$ is a straight line with slope of n. Therefore, the linearly amplified signal will exhibit a slope of 1. A third-order nonlinear product will have a slope of 3, and so on. We can draw both curves as shown in Figure 17.3. The two straight lines are extrapolated beyond their normal linear ranges (i.e., beyond the 1 dB compression point) to a point where they intersect. This is the third-order intercept point, which can be read off from the input or output power axis, leading to the input intercept point (IIP3) or output intercept point (OIP3), respectively. These input and output intercept points differ by the small signal gain of the mixer.

17.2.5 NOISE FIGURE

In Chapter 14, we looked at how amplifiers could be characterized in terms of noise figure. The same definitions of noise figure may be applied to mixers provided we take account of the fact that the noise generated by the mixer will be upconverted or downconverted to a different frequency range. For passive mixers, where there is no gain, only loss, the noise figure is approximately equal to the insertion loss (as per Section 14.5.2). Active mixers typically exhibit higher noise figure than passive mixers at comparable linearity, which is as one would expect from when comparing any active circuit with its passive equivalent [1].

17.3 BASIC MIXER OPERATION

The simplest single balanced mixer (SBM) can be modeled as an electronic switch which is driven by the LO waveform, as shown in Figure 17.4.

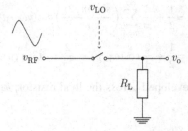

FIGURE 17.4

Single balanced mixer equivalent circuit.

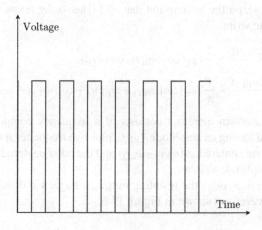

FIGURE 17.5

Single balanced mixer LO switching waveform.

The switch is closed during positive half cycles of the LO signal and open during the negative half cycles, resulting in the LO switching waveform shown in Figure 17.5. The LO switching waveform can be represented by a square wave switching function, $S(v_{LO})$, which is defined by:

$$S(v_{LO}) = \begin{cases} 1 & \text{if } v_{LO} \geq 0 \\ 0 & \text{if } v_{LO} < 0 \end{cases} \qquad (17.3.1)$$

The Fourier series representation of the square wave switching function, $S(v_{LO})$, is given in equation (17.3.2) and consists of a DC term plus all the odd harmonics of the LO frequency, decreasing as $1/n$. The positive DC term in equation (17.3.2) is the average value of the $S(v_{LO})$ switching waveform and is always a positive value because the square waveform in Figure 17.5 never goes below zero as it simply represents the opening and closing of a virtual switch, as per Figure 17.4.

$$S(v_{LO}) = \frac{1}{2} + \frac{2}{\pi} \sum_{n=1}^{\infty} \left(\frac{\sin(n\pi/2)}{n} \right) \cos(n\omega_{LO}t) \tag{17.3.2}$$

$$= \frac{1}{2} + \frac{2}{\pi} \cos(\omega_{LO}t) - \frac{2}{3\pi} \cos(3\omega_{LO}t) + \cdots \tag{17.3.3}$$

The output voltage developed across the load resistor, R_L, is the product of v_{RF} and $S(v_{LO})$, as follows:

$$v_0 = v_{RF} \cdot S(v_{LO}) \tag{17.3.4}$$

$$= \frac{v_{RF}}{2} + \frac{2}{\pi} v_{RF} \cos(\omega_{LO}t) - \frac{2}{3\pi} v_{RF} \cos(3\omega_{LO}t) + \cdots \tag{17.3.5}$$

If we set $v_{RF} = V_{RF} \cos(\omega_{RF}t)$ and we consider only the first and second terms of equation (17.3.4), on the assumption that all higher-order terms will be removed by filtering, we can write:

$$v_0 = \frac{V_{RF} \cos(\omega_{RF}t)}{2} + \frac{2}{\pi} V_{RF} \cos(\omega_{RF}t)\cos(\omega_{LO}t) \tag{17.3.6}$$

$$= \frac{V_{RF} \cos(\omega_{RF}t)}{2} + \frac{1}{\pi} V_{RF} [\cos[(\omega_{LO} + \omega_{RF})t] + \cos[(\omega_{LO} - \omega_{RF})t]] \tag{17.3.7}$$

The output waveform therefore consists of a frequency component at angular frequency ω_{RF} and having an amplitude $V_{RF}/2$ plus two frequency translated versions of the RF signal, one centered at $(\omega_{RF} - \omega_{LO})$ and the other centered at $(\omega_{RF} + \omega_{LO})$, both having an amplitude V_{RF}/π.

If we select $\omega_{LO} > \omega_{RF}$, the resulting output voltage, V_o, developed across the load R_L, has the waveform shown in Figure 17.6.

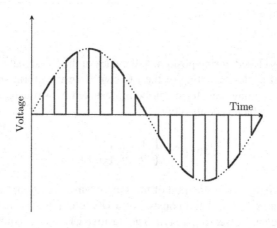

FIGURE 17.6

Single balanced mixer output waveform.

By contrast, we can model a double balanced mixer by the switching circuit in Figure 17.7.

In this case, alternate half-cycles of the LO signal are used to select between two positions of a double pole switch. When the switch is in the upper position, the switch passes the RF signal to the output. When the switch is in the lower position, an inverted version of the RF signal is passed to the output.

In this case, the LO switching waveform is shown in Figure 17.8, and can be represented by a new square wave switching function $S(v_{LO})$:

$$S(v_{LO}) = \begin{cases} 1 & \text{if } v_{LO} \geq 0 \\ -1 & \text{if } v_{LO} < 0 \end{cases} \tag{17.3.8}$$

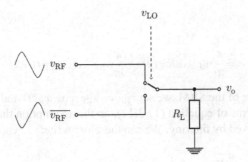

FIGURE 17.7

Double balanced mixer equivalent circuit.

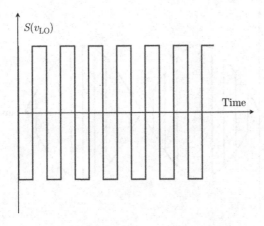

FIGURE 17.8

Double balanced mixer LO switching waveform.

The Fourier series representation of the square wave switching waveform in Figure 17.8 is given in equation (17.3.9) and consists of all the odd harmonics of the LO frequency, decreasing by $1/n$. In contrast to the SBM drive waveform in equation (17.3.2), the waveform of Figure 17.8 is symmetrical about zero and, consequently, there is no DC term in equation (17.3.9).

$$S(v_{LO}) = \frac{4}{\pi} \sum_{n=1}^{\infty} \left(\frac{\sin(n\pi/2)}{n} \right) \cos(n\omega_{LO}t) \tag{17.3.9}$$

$$= \frac{4}{\pi} \cos(\omega_{LO}t) - \frac{4}{3\pi} \cos(3\omega_{LO}t) + \cdots \tag{17.3.10}$$

The resulting output voltage, V_o, is developed across the load R_L and has the waveform shown in Figure 17.9. This waveform represents the product of v_{RF} and $S(v_{LO})$, as follows:

$$v_o = v_{RF} \cdot S(v_{LO}) \tag{17.3.11}$$

$$= \frac{4}{\pi} v_{RF} \cos(\omega_{LO}t) - \frac{4}{3\pi} v_{RF} \cos(3\omega_{LO}t) + \cdots \tag{17.3.12}$$

As with the case of the SBM, we set $v_{RF} = V_{RF} \cos(\omega_{RF}t)$ and consider only the first and second terms of equation (17.3.11), on the assumption that all higher-order terms will be removed by filtering. We can therefore write:

$$v_o = \frac{4}{\pi} V_{RF} \cos(\omega_{RF}t) \cos(\omega_{LO}t) \tag{17.3.13}$$

$$= \frac{2}{\pi} V_{RF} [\cos[(\omega_{RF} + \omega_{LO})t] + \cos[(\omega_{RF} - \omega_{LO})t]] \tag{17.3.14}$$

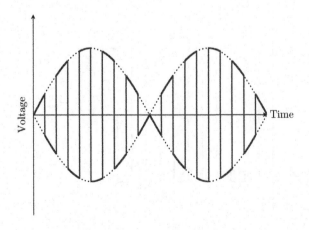

FIGURE 17.9

Double balanced mixer output waveform.

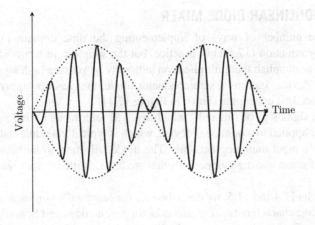

FIGURE 17.10

Double balanced mixer output waveform after filtering.

The output waveform therefore consists of two frequency translated versions of the RF signal, one centered at $(\omega_{LO} - \omega_{RF})$ and the other centered at $(\omega_{LO} + \omega_{RF})$, both having an amplitude $2V_{RF}/\pi$. This time there is no RF signal "feedthrough" component, as we had seen in the case of the SBM.

After low-pass filtering, the output waveform looks like that shown in Figure 17.10. The reader will notice that the waveform in Figure 17.10 resembles the waveform of an AM modulated signal where the carrier has been suppressed, that is a "double sideband suppressed carrier" signal [2].

Aside from the absence of RF signal feedthrough, the other advantage of the double balanced mixer topology over the single balanced topology is twice the voltage conversion gain, which is directly attributable to the use of a differential LO drive signal. We can confirm this by comparing equation (17.3.6) with equation (17.3.13). The conversion gain of a "theoretical" double balanced mixer is $2/\pi$ (≈ -4 dB) compared with $1/\pi$ (≈ -10 dB) in the case of an SBM.

17.4 PASSIVE MIXER CIRCUITS

The definition of a passive mixer is one that does not provide positive conversion gain, only loss. The majority of passive mixers involve the use of diodes as switches or as nonlinear devices. Diodes have been used in this way since the very early days of electronics and are still very widely used in mixers all across the microwave frequency range, and into the THz frequency range [3,4]. In this section, we will describe the most common circuit topologies used to implement both single balanced and double balanced diode mixers.

17.4.1 NONLINEAR DIODE MIXER

There are a number of ways of implementing the time domain multiplication described by equation (17.1.1) in practice, but the simplest, in terms of implementation, is to accomplish the multiplication indirectly by passing both signals through a nonlinear device, such as a semiconductor diode, that has a roughly square law characteristic. This concept is shown in Figure 17.11.

The two signals to be "mixed," which we will refer to as v_1 and v_2, are added together and applied to a nonlinear device, which is typically a semiconductor diode, via a suitable input matching network. The diode will have to be biased correctly by means of some biasing components that we have omitted from Figure 17.11 for simplicity.

In Sections 11.4 and 11.5, we described the fundamentally nonlinear nature of the current/voltage characteristic of semiconductor *pn*-junctions and Schottky junctions. This nonlinearity may be put to good effect to convert a signal containing the sum of two components into a signal containing their products, as we will now explain. Recall the well-known current/voltage characteristic for a semiconductor diode given by equation (11.4.5) on page 369:

$$I = I_S \left(e^{qV_D/k_B T} - 1 \right) \tag{17.4.1}$$

where V_D is the voltage across the diode, I is the forward current, and I_S is the reverse saturation current. We can replace the exponential function e^x by the following equivalent Taylor series [5]:

$$e^x = \sum_{n=0}^{\infty} \frac{x^n}{n!} \tag{17.4.2}$$

For small values of x, equation (17.4.2) can be approximated by the first few terms of that series, so we can write:

$$e^x - 1 \approx x + \frac{x^2}{2} \tag{17.4.3}$$

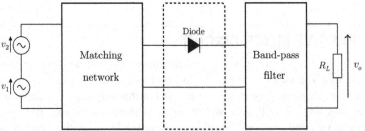

FIGURE 17.11

Mixing using a nonlinear device.

Now suppose that the sum of the two input voltage signals $v_1 + v_2$ is applied to a diode, and that the output voltage, v_O, across the load in Figure 17.11 is proportional to the current through the diode. Based on equations (17.4.1)–(17.4.3), therefore, and ignoring DC terms, the output signal voltage will be of the form:

$$v_o = (v_1 + v_2) + \frac{1}{2}(v_1 + v_2)^2 + \cdots \quad (17.4.4)$$

which can be rewritten as:

$$v_o = (v_1 + v_2) + \frac{1}{2}v_1^2 + v_1 v_2 + \frac{1}{2}v_2^2 \quad (17.4.5)$$

Equation (17.4.5) contains the product term $v_1 v_2$, which is the only component we are interested in. In addition, the signal described by equation (17.4.5) contains the two original input signals, v_1 and v_2, which are undesirable leakage components, plus frequency components at double that of the two original input signals, which are also undesirable. There will also be higher-frequency components generated by the higher-order terms we have neglected in the assumption equation (17.4.3). All of these undesirable components are removed by the filter in Figure 17.11. This is, in principle, how we obtain the product of two signals by passing their sum through a nonlinear device.

17.4.2 PASSIVE DOUBLE BALANCED DIODE MIXER

The most common form of double balanced mixer is the diode double balanced mixer. In its simplest form, it consists of two "Baluns" (unbalanced to balanced transformers) and a diode ring consisting of four diodes, as shown in Figure 17.12. Schottky barrier diodes are most commonly used in this type of mixer due to their low "on" resistance and good high-frequency response. In this circuit, the diodes behave as switches that are either "on" or "off." In other words, this topology does not make use of the diode nonlinearity discussed in the previous section.

A well-designed passive double balanced diode mixer will have good isolation (LO to IF, LO to RF, and RF to IF) and good linearity and DR. Being passive, however, the conversion loss (RF to IF) will typically be around −6 dB.

The forward voltage drop of the diodes determines the optimum LO drive level. RF mixers designed to handle a high RF input level will need a correspondingly high LO input level. As a rule of thumb, the LO signal level should have a voltage amplitude 100 times greater (i.e., a power level 20 dB higher) than either the RF or IF signal [6]. This ensures that it is the LO signal, rather than the RF or IF signal, that switches the diodes on and off. This is a key factor in reducing intermodulation distortion, IMD, and also maximizing the DR of a passive mixer. In the event that we need to use a higher drive signal level we can accommodate this by placing multiple diodes in series in each leg of the diode ring.

The Balun transformers are key elements within the overall mixer design and achieving the required bandwidth and performance from the Baluns can be

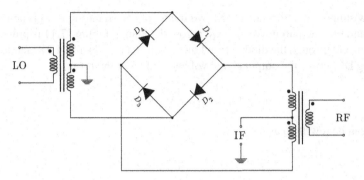

FIGURE 17.12

Double balanced diode mixer.

challenging. The matching between the input and output Balun transformers, and the individual transformer legs are important factors in determining the balance of the RF mixer. The Balun transformers also play an important role in determining the conversion loss and drive level of the mixer. As discrete Balun transformers are generally wound on ferrite cores using copper wire, the core loss, copper loss, and impedance mismatch all contribute to overall conversion losses.

The operation of the circuit in Figure 17.12 can be understood by considering the diodes as switches, which are turned "on" and "off" during alternative half-cycles of the LO signal [7]. This approach assumes that the LO drive signal is of sufficiently large amplitude so that the diode conductance waveform is a square wave. With this assumption we can see that the LO signal turns on one arm of the diode ring ($D3$, $D4$), and then the other ($D1$, $D2$) alternately within each cycle. Since the points where the LO signal enters the diode ring at the junction of $D1$ and $D4$ appear as a virtual earth to the RF signal, the points where the RF signal enters are alternatively connected to ground as the diodes turn on and off.

During positive half cycles of the LO signal we therefore have the situation shown in Figure 17.13, where $D3$ and $D4$ have been replaced by their ON resistances, R_{on} and $D1$ and $D2$ are OFF (open circuit). The junction of $D3$ and $D4$ effectively becomes a virtual earth for the lower winding of the RF transformer in Figure 17.13. This means that the RF signal is connected to the IF port via the lower winding, while the upper winding is unconnected.

During negative half cycles of the LO signal we have the reverse situation, shown in Figure 17.14, where $D1$ and $D2$ have been replaced by their ON resistances, R_{on}, and $D3$ and $D4$ are OFF (open circuit). This time the junction of $D1$ and $D2$ becomes a virtual earth for the upper winding of the RF transformer in Figure 17.14. The RF signal is thus connected to the IF port via this winding, while the lower winding is unconnected.

The crucial point to note is that, because the two secondary windings of the RF transformer are wound in opposite directions away from the center tap (note the

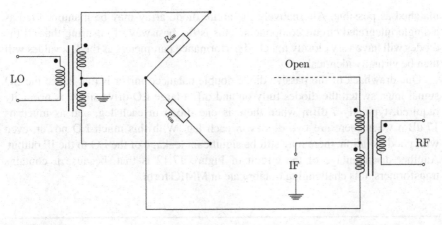

FIGURE 17.13

Double balanced diode mixer model: positive LO cycle.

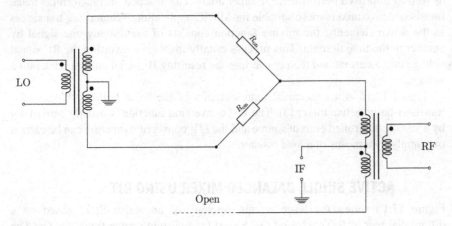

FIGURE 17.14

Double balanced diode mixer model: negative LO cycle.

position of the dots in Figure 17.12), the signal appearing at the IF port alternates between the RF signal and an inverted version of the RF signal every half cycle of the LO waveform. In this way, the signal at the IF port is multiplied by a square wave at the LO frequency. The output waveform will thus resemble that shown in Figure 17.9.

The currents flowing in the R_{on} resistors in Figures 17.13 and 17.14 should be made equal (or as close as possible) in order to ensure good isolation. If individual discrete components are used, care needs to be taken to make sure they are as closely matched as possible. Alternatively the entire diode array may be manufactured as a single integrated circuit component. This is the best way of ensuring that all the diodes will have very closely matched performance parameters as the R_{on} values will then be virtually identical.

One drawback of the passive diode double balanced mixer is that, since the LO signal must switch the diodes fully on and off, a large LO drive signal is normally required, typically 7 dBm when there is one diode in each leg, and as much as 17 dBm when there are two diodes in each leg. With this much LO power, even with good isolation, there may still be significant leakage of the LO to the IF output. Another disadvantage of the circuit of Figure 17.12 is that, because it contains transformers, it is challenging to fabricate in MMIC form.

17.5 ACTIVE MIXER CIRCUITS

Active mixers can be constructed using transistors (bipolar junction transistor, BJT or field-effect transistor, FET) that have the advantage of providing conversion gain as well as improved performance in other areas. The absence of transformers make transistor active mixers more suitable for MMIC application. When using transistors as the active elements, the mixing function consists of multiplying one signal by another in the time domain. This process usually involves converting the RF signal voltage into a current and then converting the resulting IF signal current back into a voltage.

Figure 17.15 is a conceptual representation of the basic building blocks of a transistor-based active mixer [8]. The V/I conversion function is usually carried out by a voltage controlled current source and the I/V conversion function can be carried out simply by a means of a load resistor.

17.5.1 ACTIVE SINGLE BALANCED MIXER USING BJT

Figure 17.16 shows the basic circuit topology of an active SBM based on a differential pair of BJTs, Q_1 and Q_2, biased by a current source transistor Q_3. The functional blocks of Figure 17.15 are overlayed on Figure 17.16 for illustration. The current source transistor, Q_3, sets the total current in the upper two transistors, which then compete for a fraction of this current. As the total current is fixed by Q_3, more current in Q_1 means less current in Q_2, and vice versa, hence the differential operation of this circuit. The load resistors, R_L, convert the collector currents into voltages. The gain of a differential pair depends on the transconductance, g_m, of the two transistors which, in turn, depends on their collector current, I_C, according to $g_m = I_C/V_T$. The circuit of Figure 17.16 can therefore be used as a mixer when the current in Q_3 is modulated by the RF signal voltage. We apply the LO signal as a differential voltage to the bases of Q_1 and Q_2, and take the IF signal as a differential voltage across the collectors of Q_1 and Q_2. The degenerative feedback resistance, R_e, is added to the current source to improve its linearity.

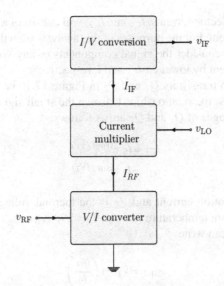

FIGURE 17.15

Active mixer building blocks.

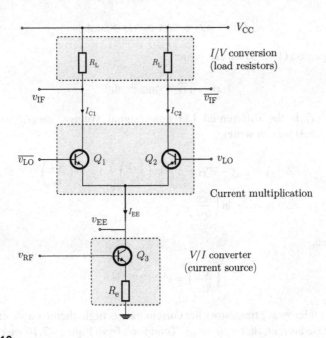

FIGURE 17.16

Single-balanced active mixer using BJTs.

The transistor collector currents, I_{C1} and I_{C2}, will consist of a DC bias component plus a signal component. For the purposes of this analysis we will ignore the DC bias components and just consider the signal components of any voltages and currents, which we will represent by lower case v and i, respectively.

Assuming the two transistors Q_1 and Q_2 in Figure 17.16 have identical voltage-current characteristics, the relationships between the small signal collector currents and base-emitter voltages of Q_1 and Q_2 are as follows:

$$i_{C1} = I_S \, e^{(v_{BE1}/V_T)} \tag{17.5.1}$$

$$i_{C2} = I_S \, e^{(v_{BE2}/V_T)} \tag{17.5.2}$$

where I_S is the saturation current and V_T is the thermal voltage defined by $V_T = kT/q \approx 26$ mV at room temperature (around $T = 290$ K).

Alternatively we can write:

$$v_{BE1} = V_T \, \ln\left(\frac{i_{C1}}{I_S}\right) \tag{17.5.3}$$

$$v_{BE2} = V_T \, \ln\left(\frac{i_{C2}}{I_S}\right) \tag{17.5.4}$$

The voltage at the common emitter point, v_{EE}, in Figure 17.16 is given by:

$$v_{EE} = \overline{v_{LO}} - v_{BE1} = v_{LO} - v_{BE2} \tag{17.5.5}$$

From equation (17.5.5) we can write:

$$v_{LO} - \overline{v_{LO}} = v_{BE2} - v_{BE1} \tag{17.5.6}$$

$v_{LO} - \overline{v_{LO}}$ is the differential LO input signal voltage, Δv_{LO}, so applying equation (17.5.3) we can write:

$$\Delta v_{LO} = v_{LO} - \overline{v_{LO}} = V_T \, \ln\left(\frac{i_{C2}}{I_S}\right) - V_T \, \ln\left(\frac{i_{C1}}{i_S}\right) \tag{17.5.7}$$

$$\Delta v_{LO} = V_T \, \ln\left(\frac{i_{C2}}{i_{C1}}\right) \tag{17.5.8}$$

Therefore,

$$\left(\frac{i_{C1}}{i_{C2}}\right) = e^{(-\Delta v_{LO}/V_T)} \tag{17.5.9}$$

For most microwave transistors the current gain is high, therefore we can usually ignore the base current, that is, $i_E \approx i_C$. Therefore, from Figure 17.16 we can write:

$$i_{EE} = i_{C1} + i_{C2} \tag{17.5.10}$$

Combining equations (17.5.9) and (17.5.10) we can now write the collector signal currents of Q_1 and Q_2 in Figure 17.16 in terms of Δv_{LO} and I_{EE} as:

$$i_{C1} = \frac{i_{EE}}{1 + e^{(\Delta v_{LO}/V_T)}} \tag{17.5.11}$$

$$i_{C2} = \frac{i_{EE}}{1 + e^{(-\Delta v_{LO}/V_T)}} \tag{17.5.12}$$

The difference between the two collector currents can now be written as follows:

$$\Delta i_{IF12} = i_{C1} - i_{C2} \tag{17.5.13}$$

$$= I_{EE}\left(\frac{1}{1 + e^{(\Delta v_{LO}/V_T)}} - \frac{1}{1 + e^{(-\Delta v_{LO}/V_T)}}\right) \tag{17.5.14}$$

Which can be rewritten as:

$$\Delta i_{IF12} = i_{EE}\left(\frac{e^{(-\Delta v_{LO}/2V_T)} - e^{(\Delta v_{LO}/2V_T)}}{e^{(-\Delta v_{LO}/2V_T)} + e^{(\Delta v_{LO}/2V_T)}}\right) \tag{17.5.15}$$

Equation (17.5.15) can be more neatly expressed by using the definition of the hyperbolic tangent (tanh) function, which is defined as:

$$\tanh(x) = \frac{e^x - e^{-x}}{e^x + e^{-x}} \tag{17.5.16}$$

We can now rewrite equation (17.5.13) as:

$$\Delta i_{IF12} = i_{EE} \tanh\left(\frac{\Delta v_{LO}}{2V_T}\right) \tag{17.5.17}$$

Now, if we assume that the bias current I_{EE} is modulated by v_{RF}, as shown in Figure 17.16, we can replace the simple term I_{EE} in equation (17.5.17) by $(I_{EE_0} + g_{m3}v_{RF})$ where I_{EE_0} is the quiescent DC bias current and g_{m3} is the transconductance of the current source transistor, Q_3. We can now write:

$$\Delta i_{IF12} = (I_{EE_0} + g_{m3}v_{RF}) \tanh\left(\frac{\Delta v_{LO}}{2V_T}\right) \tag{17.5.18}$$

Which can be expanded to:

$$\Delta i_{IF12} = I_{EE_0} \tanh\left(\frac{\Delta v_{LO}}{2V_T}\right) + g_{m3}v_{RF} \tanh\left(\frac{\Delta v_{LO}}{2V_T}\right) \tag{17.5.19}$$

Consider the Maclaurin series expansion of $\tanh(x)$, as follows [5]:

$$\tanh(x) = x - \frac{1}{3}x^3 + \frac{2}{15}x^5 - \frac{17}{315}x^7 + \frac{62}{2835}x^9 + \cdots \tag{17.5.20}$$

From equation (17.5.20) we can show that $\tanh(x) \approx x$ for small values of x (i.e., for values of x up to $x = 0.5$ we have $\tanh(x)/x > 0.92$). So we can approximate the tanh in equation (17.5.19) on the assumption that v_{RF} and v_{LO} are small (i.e., less than $V_T = 26$ mV at room temperature). We therefore have:

$$\Delta i_{IF_{12}} \approx I_{EE_0} \left(\frac{\Delta v_{LO}}{2V_T} \right) + g_{m3} \left(\frac{\Delta v_{LO} \; v_{RF}}{2V_T} \right) \tag{17.5.21}$$

The first term in equation (17.5.21) represents the LO leakage component, which is proportional to the DC bias current, i_{EE_0}. The second term contains the product term we are interested in, namely ($v_{RF} \, \Delta v_{LO}$).

The SBM of Figure 17.16 has the advantage of simplicity and relatively low-noise figure, as there are fewer noise generating components than in a double balanced design. On the other hand, the direct feed-through of LO frequency components can be a serious problem in some applications and is a significant disadvantage of the SBM topology. Although these unwanted frequency components are typically at a much higher frequency than the IF, and can therefore be filtered out, the additional unwanted signal energy can affect the linearity of the circuit and cause saturation of subsequent circuit stages. A further drawback of the SBM of Figure 17.16 is that the RF input is single ended, not balanced. This makes it susceptible to common mode noise and interference [9].

17.5.2 ACTIVE DOUBLE BALANCED MIXER: THE GILBERT CELL

The deficiencies of the single balanced mixer described in the previous section can be overcome by adopting a double balanced design, and using a balanced RF signal feed. This can be achieved by combining two single balanced circuits, one being driven by v_{RF} and the other driven by its inverse, $\overline{v_{RF}}$. The balanced LO inputs and IF outputs are combined by connecting the respective nodes together. The resultant circuit is known as a *Gilbert cell* [10], the basic topology of which is shown in Figure 17.17. The two current source transistors in Figure 17.17 are driven by a balanced RF signal; that is, Q_3 is driven by v_{RF} and the Q_6 is driven by its inversion, $\overline{v_{RF}}$.

We can analyze the Gilbert cell by employing the SBM analysis in the previous section as a starting point. The second transistor pair in Figure 17.17, Q_4, Q_5, has a signal response similar to the original pair, Q_1, Q_2, but $180°$ out of phase, as this pair is being driven by an inverted version of the LO signal (relative to that driving Q_1 and Q_2). By analogy with equation (17.5.21) we can therefore write:

$$\overline{\Delta i_{IF_{45}}} \approx I_{EE} \left(\frac{\Delta v_{LO}}{2V_T} \right) - g_{m6} v_{RF} \left(\frac{\Delta v_{LO}}{2V_T} \right) \tag{17.5.22}$$

where, in this case, $\Delta i_{IF_{45}} = i_{C4} - i_{C5}$.

If we ensure that all the transistors being used have identical characteristics, so the g_m values are the same in all cases and specifically $g_{m3} = g_{m6}$ (we can get close to this in practice if the circuit is implemented in MMIC form), the differential IF output current of the circuit of Figure 17.17 is given by:

$$\Delta i_{IF} = \Delta i_{IF_{12}} - \overline{\Delta i_{IF_{45}}} \tag{17.5.23}$$

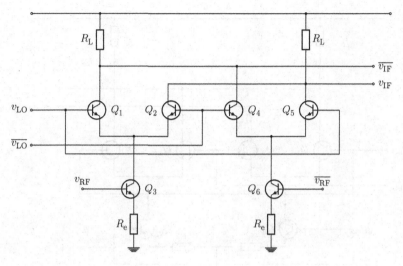

FIGURE 17.17

Gilbert cell mixer topology.

$$\approx 2 g_m v_{RF} \left(\frac{\Delta v_{LO}}{2 V_T} \right) = \left(\frac{g_m}{V_T} \right) v_{LO} \cdot v_{RF} \qquad (17.5.24)$$

The above subtraction of $\overline{i_{IF_{45}}}$ from $i_{IF_{12}}$ has the effect, to a first-order approximation, of canceling the common term containing the unmodulated v_{LO} signal, leaving only the desired product term $v_{LO} \cdot v_{RF}$. Thus, the IF feed-through component that was present at the output of the SBM described by equation (17.5.19) has been removed. This is the primary benefit of using the double balanced topology in Figure 17.17.

Figure 17.18 shows a practical Gilbert cell implementation such as would be implemented in MMIC form. The common bias current, I_{EE}, is set by the fixed current source Q_7. The gain of the two differential amplifiers, formed of Q_1 and Q_2, and Q_4 and Q_5, is controlled by modulating the emitter bias current via the transistors Q_3 and Q_6. In normal operation, the LO signal is applied differentially to the bases of Q_1 and Q_5 (positive phase) and Q_2 and Q_4 (negative phase), while the RF signal is applied differentially to the bases of Q_3 (positive phase) and Q_6 (negative phase). The IF signal is taken differentially from the collectors of the two upper transistor pairs.

Another enhancement applied to the Gilbert cell in Figure 17.18 is the addition of the emitter degeneration resistors, R_E. These are used in practical circuits to improve linearity at the expense of some conversion gain [11]. The effect of R_E is to reduce the transconductance of the lower transistors, Q_3 and Q_6 by a factor of $1/(1 + g_m R_E)$, due to the action of local feedback.

The collector signal currents of the upper pair transistors shown in Figure 17.18, once again ignoring DC bias and base currents, are as follows:

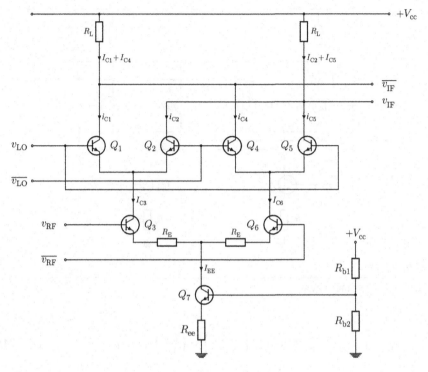

FIGURE 17.18

Gilbert cell mixer implementation using BJT.

$$i_{C1} = \frac{i_{C3}}{1 + e^{-v_{LO}/V_T}} \tag{17.5.25}$$

$$i_{C2} = \frac{i_{C3}}{1 + e^{v_{LO}/V_T}} \tag{17.5.26}$$

$$i_{C4} = \frac{i_{C6}}{1 + e^{v_{LO}/V_T}} \tag{17.5.27}$$

$$i_{C5} = \frac{i_{C6}}{1 + e^{-v_{LO}/V_T}} \tag{17.5.28}$$

For the lower pair transistors, Q_3 and Q_6, we can write:

$$i_{C3} = \frac{I_{EE}}{1 + e^{-v_{RF}/V_T}} \tag{17.5.29}$$

$$i_{C6} = \frac{I_{EE}}{1 + e^{v_{RF}/V_T}} \tag{17.5.30}$$

Noting that, in the case of the lower transistors, I_{EE} is the fixed bias current provided by Q_7. Combining equation (17.5.25) through equation (17.5.30), we obtain expressions for the collector currents I_{C1}, I_{C2}, I_{C4}, and I_{C5} in terms of input signal voltages v_{RF} and v_{LO}.

$$i_{C1} = \frac{I_{EE}}{\left[1 + e^{-v_{LO}/V_T}\right]\left[1 + e^{-v_{RF}/V_T}\right]} \tag{17.5.31}$$

$$i_{C2} = \frac{I_{EE}}{\left[1 + e^{-v_{RF}/V_T}\right]\left[1 + e^{v_{LO}/V_T}\right]} \tag{17.5.32}$$

$$i_{C4} = \frac{I_{EE}}{\left[1 + e^{v_{LO}/V_T}\right]\left[1 + e^{v_{RF}/V_T}\right]} \tag{17.5.33}$$

$$i_{C5} = \frac{I_{EE}}{\left[1 + e^{v_{RF}/V_T}\right]\left[1 + e^{-v_{LO}/V_T}\right]} \tag{17.5.34}$$

With reference to Figure 17.18 the differential output current is given by:

$$\Delta i_{IF} = (i_{C1} + i_{C4}) - (i_{C2} + i_{C5}) \tag{17.5.35}$$

Which can be rearranged as:

$$\Delta i_{IF} = (i_{C1} - i_{C5}) - (i_{C2} - i_{C4}) \tag{17.5.36}$$

Applying equation (17.5.31) to equation (17.5.36) and using the definition of tanh, we can rewrite equation (17.5.36) as:

$$\Delta i_{IF} = I_{EE} \, \tanh\left(\frac{v_{LO}}{2V_T}\right) \cdot \tanh\left(\frac{v_{RF}}{2V_T}\right) \tag{17.5.37}$$

The differential output current is converted into a differential voltage by the load resistors, R_L, so we can write:

$$v_{IF} = I_{EE}R_L \, \tanh\left(\frac{v_{LO}}{2V_T}\right) \cdot \tanh\left(\frac{v_{RF}}{2V_T}\right) \tag{17.5.38}$$

If we again assume that both v_{RF} and v_{LO} are small (i.e., less than V_T), we can apply the approximation $\tanh(x) \approx x$ to equation (17.5.38) and thereby obtain:

$$v_{IF} \approx \left(\frac{I_{EE}R_L}{4V_T^2}\right) v_{LO} \, v_{RF} \tag{17.5.39}$$

Equation (17.5.39) means that, for small enough signals, the differential output IF voltage is directly proportional to the product of the RF and LO input signal voltages. Equation (17.5.39) is true irrespective of the polarities of v_{RF} and v_{LO}. The output of a Gilbert cell is therefore a true four quadrant multiplication of the differential base voltages of the LO and RF inputs. For this reason the circuit, although applied as a mixer in this context, is often simply referred to as a *Gilbert cell multiplier*.

In practice, there are three distinct operating modes for the Gilbert cell, according to the magnitudes of v_{RF} and v_{LO} relative to V_T, as follows:

1. If both v_{RF} and v_{LO} are much less than V_T, then the hyperbolic tangent function is approximately linear and the circuit behaves as a true four quadrant analog voltage multiplier, as per equation (17.5.39). The input voltage range can be extended by adding "predistortion" circuits at the inputs, which have an approximately $\tanh^{-1}$ characteristic. This technique is sometimes used at lower frequencies but is not common at microwave frequencies [10].

2. If one of the input voltages significantly exceeds V_T then one of the transistor pairs will be driven into saturation and will behave like on/off switches. This is effectively equivalent to multiplying the other input signal by a square wave. This mode of operation is quite common in downconversion mixers where the IF signal is a lot larger than the RF signal. The "squaring" of the LO signal in this mode is not a problem as the information content of the RF signal is preserved.

3. If both of the input voltages significantly exceed V_T then all of the transistors are operating as switches. This mode is sometimes employed when the mixer is being used as a phase detector, as the phase relationship between the two input signals is preserved, even though any information contained in the signal amplitudes will have been lost.

The Gilbert cell mixer topology was first used as a mixer by Barrie Gilbert in 1968 [12]. Although Gilbert himself did not suggest the use of the name, it is widely used to describe this mixer circuit topology. The same idea had previously been used and patented as a synchronous detector, although not a mixer by Jones [13]. The Gilbert cell is by far the most popular topology for active mixers based on transistors and can be widely found, applied in MMIC form, right across the microwave frequency range, up to THz frequencies [14,15].

17.5.3 CONVERSION GAIN OF THE GILBERT CELL

Consider equation (17.5.38) for the case of a downconversion mixer, where the RF signal is of a small amplitude. Using the approximation $\tanh(x) \approx x$ for $x \ll 1$, we can write:

$$v_{IF} = \frac{I_{EE}R_L v_{RF}}{2V_T} \tanh\left(\frac{v_{LO}}{2V_T}\right) \tag{17.5.40}$$

where I_{EE} is the DC bias current through Q_7. We will simplify the analysis by firstly considering the absence of emitter degeneration resistors; that is, we set $R_E = 0$. We can then define the transconductance of the bias transistor, Q_7, in Figure 17.18 as:

$$g_{m7} = \frac{I_{EE}}{V_T} \tag{17.5.41}$$

We now substitute equation (17.5.41) into equation (17.5.40) to give:

$$v_{IF} = \tanh\left(\frac{v_{LO}}{2V_T}\right) \frac{g_{m7}R_L v_{RF}}{2} \tag{17.5.42}$$

which gives the conversion gain of the circuit of Figure 17.18 as:

$$\frac{v_{IF}}{v_{RF}} = \tanh\left(\frac{v_{LO}}{2V_T}\right)\frac{g_{m7}R_L}{2} \tag{17.5.43}$$

For a downconversion mixer we expect the RF input to be of small amplitude and we need its treatment to be as linear as possible, so as to preserve the information content [16]. The handling of the LO input, on the other hand, need not be linear, since the LO is of known amplitude and frequency and therefore has no information content. Distortion of the LO signal is of no consequence, and so we may choose the LO signal amplitude so as to maximize conversion efficiency. In fact, the LO input is usually designed to switch the upper transistor quad so that for half the cycle Q_1 and Q_5 are "on" and taking all of the current i_{C3} and i_{C6}. For the other half of the LO cycle, Q_1 and Q_5 are "off" and Q_2 and Q_4 are on, so all of i_{C3} and i_{C6} flows through these respective transistors. In other words, for a switching Gilbert cell where $v_{LO} \gg 2V_T$, then equation (17.5.43) can be approximated as:

$$\frac{v_{IF}}{v_{RF}} = u(v_{LO})\left(\frac{g_m R_L}{2}\right) \tag{17.5.44}$$

where

$$u(v_{LO}) = \begin{cases} 1 & \text{if } v_{LO} \geq 0 \\ -1 & \text{if } v_{LO} < 0 \end{cases} \tag{17.5.45}$$

We can represent the square wave function $u(v_{LO})$ by its Fourier series expansion:

$$u(v_{LO}) = \frac{4}{\pi}\left(\sin(\omega_{LO}t) + \frac{1}{3}\sin(3\omega_{LO}t) + \frac{1}{5}\sin(5\omega_{LO}t) + \cdots\right) \tag{17.5.46}$$

The conversion gain of a downconversion mixer is defined as the ratio of IF signal amplitude to RF signal amplitude. We are therefore only interested in the first term in equation (17.5.46), which has an amplitude of $\pi/4$. The conversion gain given by equation (17.5.44) now becomes:

$$\frac{v_{RF}}{v_{IF}} \approx \frac{2}{\pi}g_m R_L \tag{17.5.47}$$

We now consider the effect of finite values of R_E, which has the effect of reducing the transconductance of the lower transistors, Q_3 and Q_6 by a factor of $(1 + g_m R_E)$. With finite R_E the conversion gain given by equation (17.5.43) now becomes:

$$\frac{v_{IF}}{v_{RF}} = -\tanh\left(\frac{v_{LO}}{2V_T}\right)\left(\frac{g_m R_L}{1 + g_m R_E}\right) \tag{17.5.48}$$

Employing the function $u(v_{LO})$ defined above yields:

$$\frac{v_{IF}}{v_{RF}} = u(v_{LO})\left(\frac{g_m R_L}{1 + g_m R_E}\right) \tag{17.5.49}$$

The voltage gain of the downconversion mixer with the emitter degeneration resistors added is given by:

$$\frac{V_{RF}}{V_{IF}} \approx \frac{2}{\pi}\left(\frac{g_m R_L}{1 + g_m R_E}\right) \tag{17.5.50}$$

17.5.4 THE FET GILBERT CELL

Modern Gilbert cell mixers, that are usually implemented in MMIC form, tend to use FETs rather than bipolar transistors [17–20]. A typical NMOS Gilbert cell circuit is shown in Figure 17.19. Readers will note the similarity with Figure 17.18, but we have replaced the source degeneration resistors with inductors L_S, which is more often the case in practical MMIC implementation. This has the advantage of lower DC voltage drop, meaning that the circuit can operate at lower supply voltages, and better noise performance due to the low parasitic ohmic resistance of the inductors [21].

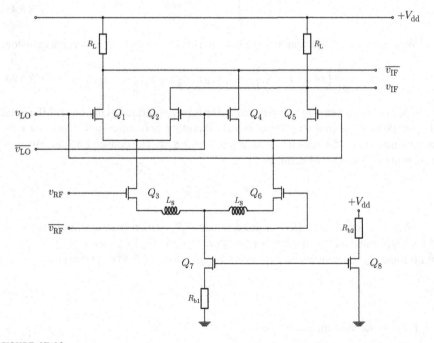

FIGURE 17.19

Integrated FET Gilbert cell mixer.

17.6 TAKEAWAYS

1. Mixers are a fundamental building block in any microwave system that requires frequency translation. Consequently, they are widely used in communications, radar, and instrumentation systems.

2. A mixer is a three-port circuit (two inputs and one output) that generates output frequencies equal to the sum and difference of the two input frequencies and their harmonics. Whether the sum or difference frequencies are selected depends on how the output of the mixer is postfiltered.

3. When a high-frequency RF input signal is mixed with a high-frequency LO signal, both being in a similar frequency range, to produce a lower IF output signal, the process is referred to as "downconversion." When a lower-frequency IF signal is one of the inputs and is mixed with a higher-frequency LO signal to produce a high-frequency RF output signal, the process is referred to as "upconversion."

4. Passive mixers are those that do not have conversion gain, only conversion loss. Such mixers are mostly constructed using diodes.

5. Passive mixers can be classified as either single balanced or double balanced. Double balanced mixers have the advantage of better linearity and suppression of spurious products (e.g., even order products of the LO and RF inputs), at the expense of greater circuit complexity. The use of *Balun* transformers in the double balanced mixer topology makes them difficult to implement in MMIC form.

6. Active mixers are able to provide positive conversion gain, in addition to the mixing function. The most common form of active mixer topology uses the *Gilbert cell* analog multiplier. Gilbert cell mixers are widely used, especially in MMIC form, and can employ either BJT or FETs.

REFERENCES

[1] S. Maas, Microwave Mixers, The Artech House Microwave Library, Artech House Inc., Norwood, MA, 1993, ISBN 9780890066058.

[2] A. Carlson, P. Crilly, Communication Systems: An Introduction to Signals and Noise in Electrical Communication, McGraw-Hill, 2010, ISBN 9780071263320.

[3] P. Siegel, R. Smith, M. Graidis, S. Martin, 2.5-THz GaAs monolithic membrane-diode mixer, IEEE Trans. Microw. Theory Tech. 47 (5) (1999) 596-604, ISSN 0018-9480, http://dx.doi.org/10.1109/22.763161.

[4] M. Morschbach, A. Muller, C. Schollhorn, M. Oehme, T. Buck, E. Kasper, Integrated silicon Schottky mixer diodes with cutoff frequencies above 1 THz, IEEE Trans. Microw. Theory Tech. 53 (6) (2005) 2013-2018, ISSN 0018-9480, http://dx.doi.org/10.1109/TMTT.2005.848831.

[5] K. Stroud, D. Booth, Advanced Engineering Mathematics, Palgrave Macmillan Limited, 2011, ISBN 9780230275485.

[6] I. Poole, Basic Radio: Principles and Technology, Newnes, 1998, ISBN 9780750626323.

[7] J. Everard, Fundamentals of RF Circuit Design: With Low Noise Oscillators, John Wiley and Sons Ltd, Chichester, 2001, ISBN 9780471497936.

[8] B. Razavi, RF Microelectronics, second ed., Prentice Hall, Englewood Cliffs, 2011.

[9] C. Toumazou, G. Moschytz, B. Gilbert, Trade-Offs in Analog Circuit Design: The Designer's Companion (pt. 1), Springer, 2004, ISBN 9781402080463.

[10] B. Gilbert, The multi-tanh principle: a tutorial overview, IEEE J. Solid State Circuits 33 (1) (1998) 2-17, ISSN 0018-9200, http://dx.doi.org/10.1109/4.654932.

[11] R.-H. Lee, J.-Y. Lee, S.-H. Lee, B. Shrestha, S.-J. Kim, G. Kennedy, C. Park, N.-Y. Kim, S.-H. Cheon, Circuit techniques to improve the linearity of an up-conversion double balanced mixer with an active Balun using InGaP/GaAs HBT technology, in: Microwave Conference Proceedings, APMC 2005, Asia-Pacific Conference Proceedings, vol. 2, 2005, p. 4, http://dx.doi.org/10.1109/APMC.2005.1606499.

[12] B. Gilbert, A precise four-quadrant multiplier with subnanosecond response, IEEE J. Solid State Circuits 3 (4) (1968) 365-373, ISSN 0018-9200, http://dx.doi.org/10.1109/JSSC.1968.1049925.

[13] H. Jones, Dual output synchronous detector utilizing transistorized differential amplifiers, 1966, US Patent 3,241,078, http://www.google.co.uk/patents/US3241078.

[14] Y. Yan, Y. Karandikar, S. Gunnarsson, M. Urteaga, R. Pierson, H. Zirath, 340 GHz integrated receiver in 250 nm InP DHBT technology, IEEE Trans. Terahertz Sci. Technol. 2 (3) (2012) 306-314, ISSN 2156-342X, http://dx.doi.org/10.1109/TTHZ.2012.2189912.

[15] J.-H. Tsai, H.-Y. Yang, T.-W. Huang, H. Wang, A 30-100 GHz wideband sub-harmonic active mixer in 90 nm CMOS technology, IEEE Microw. Wireless Compon. Lett. 18 (8) (2008) 554-556, ISSN 1531-1309, http://dx.doi.org/10.1109/LMWC.2008.2001021.

[16] J. Rogers, C. Plett, Radio Frequency Integrated Circuit Design, Artech House Microwave Library, Artech House Inc., 2014, ISBN 9781607839804.

[17] D. Parveg, M. Varonen, M. Karkkainen, K. Halonen, Design of mixers for a 130-GHz transceiver in 28-nm CMOS, in: 9th Conference on Ph.D. Research in Microelectronics and Electronics (PRIME), 2013, pp. 77-80, http://dx.doi.org/10.1109/PRIME.2013.6603100.

[18] S. Maas, A GaAs MESFET mixer with very low intermodulation, IEEE Trans. Microw. Theory Tech. 35 (4) (1987) 425-429, ISSN 0018-9480, http://dx.doi.org/10.1109/TMTT.1987.1133665.

[19] H. Qi, C. Junning, P. Hao, M. Jian, A 1.2 V High Linearity Mixer Design, in: 8th International Conference on Electronic Measurement and Instruments, ICEMI '07, 2007, pp. 2-74-2-77, http://dx.doi.org/10.1109/ICEMI.2007.4350787.

[20] P. Sullivan, B. Xavier, D. Costa, W. Ku, A Low Voltage Evaluation of a 1.9 GHz Silicon MOSFET Gilbert Cell Downconversion Mixer, in: Proceedings of the 22nd European Solid-State Circuits Conference, ESSCIRC '96, 1996, pp. 212-215.

[21] C. Bredendiek, N. Pohl, T. Jaeschkel, K. Aufinger, A. Bilgic, A highly-linear low-power down-conversion mixer for monostatic broadband 80 GHz FMCW-radar transceivers, in: Progress in Electromagnetics Research Symposium Proceedings, 2012.

Parameter conversion tables

A.1 TWO-PORT IMMITTANCE PARAMETER CONVERSIONS

	[Y]	**[Z]**	**[h]**	**[ABCD]**
[Y]	$\begin{bmatrix} Y_{11} & Y_{12} \\ Y_{21} & Y_{22} \end{bmatrix}$	$\begin{bmatrix} \dfrac{Y_{22}}{\Delta Y} & -\dfrac{Y_{12}}{\Delta Y} \\ -\dfrac{Y_{21}}{\Delta Y} & \dfrac{Y_{11}}{\Delta Y} \end{bmatrix}$	$\begin{bmatrix} \dfrac{1}{Y_{11}} & -\dfrac{Y_{12}}{Y_{11}} \\ \dfrac{Y_{21}}{Y_{11}} & \dfrac{\Delta Y}{Y_{11}} \end{bmatrix}$	$\begin{bmatrix} -\dfrac{Y_{22}}{Y_{21}} & \dfrac{1}{Y_{21}} \\ -\dfrac{\Delta Y}{Y_{21}} & \dfrac{Y_{11}}{Y_{21}} \end{bmatrix}$
[Z]	$\begin{bmatrix} \dfrac{Z_{22}}{\Delta Z} & -\dfrac{Z_{12}}{\Delta Z} \\ -\dfrac{Z_{21}}{\Delta Z} & \dfrac{Z_{11}}{\Delta Z} \end{bmatrix}$	$\begin{bmatrix} Z_{11} & Z_{12} \\ Z_{21} & Z_{22} \end{bmatrix}$	$\begin{bmatrix} \dfrac{\Delta Z}{Z_{22}} & \dfrac{Z_{12}}{Z_{22}} \\ -\dfrac{Z_{21}}{Z_{22}} & \dfrac{1}{Z_{22}} \end{bmatrix}$	$\begin{bmatrix} \dfrac{Z_{11}}{Z_{21}} & \dfrac{\Delta Z}{Z_{21}} \\ \dfrac{1}{Z_{21}} & \dfrac{Z_{22}}{Z_{21}} \end{bmatrix}$
[h]	$\begin{bmatrix} \dfrac{1}{h_{11}} & \dfrac{h_{12}}{h_{11}} \\ \dfrac{h_{21}}{h_{11}} & \dfrac{\Delta h}{h_{11}} \end{bmatrix}$	$\begin{bmatrix} \dfrac{\Delta h}{h_{22}} & \dfrac{h_{12}}{h_{22}} \\ -\dfrac{h_{21}}{h_{22}} & \dfrac{1}{h_{22}} \end{bmatrix}$	$\begin{bmatrix} h_{11} & h_{12} \\ h_{21} & h_{22} \end{bmatrix}$	$\begin{bmatrix} -\dfrac{\Delta h}{h_{21}} & -\dfrac{h_{11}}{h_{21}} \\ -\dfrac{h_{22}}{h_{21}} & -\dfrac{1}{h_{21}} \end{bmatrix}$
[ABCD]	$\begin{bmatrix} \dfrac{D}{B} & -\dfrac{\Delta(ABCD)}{B} \\ -\dfrac{1}{B} & \dfrac{A}{B} \end{bmatrix}$	$\begin{bmatrix} \dfrac{A}{C} & \dfrac{\Delta(ABCD)}{C} \\ \dfrac{1}{C} & \dfrac{D}{C} \end{bmatrix}$	$\begin{bmatrix} \dfrac{B}{D} & \dfrac{\Delta(ABCD)}{D} \\ -\dfrac{1}{D} & \dfrac{C}{D} \end{bmatrix}$	$\begin{bmatrix} A & B \\ C & D \end{bmatrix}$

Note: In all above cases, ΔX represents the determinant of the two-port matrix [X], for example, $\Delta h = h_{11}h_{22} - h_{12}h_{21}$, etc.

A.2 TWO-PORT *S*-PARAMETERS TO IMMITTANCE PARAMETER CONVERSIONS

[Z]

$$Z_{11} = Z_0 \frac{(1 + S_{11})(1 - S_{22}) - S_{12}S_{21}}{(1 - S_{11})(1 - S_{22}) - S_{12}S_{21}}$$

$$Z_{12} = Z_0 \frac{2S_{12}}{(1 - S_{11})(1 - S_{22}) - S_{12}S_{21}}$$

$$Z_{21} = Z_0 \frac{2S_{21}}{(1 - S_{11})(1 - S_{22}) - S_{12}S_{21}}$$

$$Z_{22} = Z_0 \frac{(1 - S_{11})(1 + S_{22}) - S_{12}S_{21}}{(1 - S_{11})(1 - S_{22}) - S_{12}S_{21}}$$

[Y]

$$Y_{11} = \frac{(1 - S_{11})(1 + S_{22}) + S_{12}S_{21}}{Z_0(1 + S_{11})(1 + S_{22}) - S_{12}S_{21}}$$

$$Y_{12} = \frac{-2S_{12}}{Z_0(1 + S_{11})(1 + S_{22}) - S_{12}S_{21}}$$

$$Y_{21} = \frac{-2S_{21}}{Z_0(1 + S_{11})(1 + S_{22}) - S_{12}S_{21}}$$

$$Y_{22} = \frac{(1 + S_{11})(1 - S_{22}) + S_{12}S_{21}}{Z_0(1 + S_{11})(1 + S_{22}) - S_{12}S_{21}}$$

[h]

$$h_{11} = Z_0 \frac{(1 + S_{11})(1 + S_{22}) - S_{12}S_{21}}{(1 - S_{11})(1 + S_{22}) + S_{12}S_{21}}$$

$$h_{12} = \frac{2S_{12}}{(1 - S_{11})(1 + S_{22}) + S_{12}S_{21}}$$

$$h_{21} = \frac{-2S_{21}}{(1 - S_{11})(1 + S_{22}) + S_{12}S_{21}}$$

$$h_{22} = \frac{(1 - S_{11})(1 - S_{22}) - S_{12}S_{21}}{Z_0(1 - S_{11})(1 + S_{22}) + S_{12}S_{21}}$$

[ABCD]

$$A = \frac{(1 + S_{11})(1 - S_{22}) + S_{12}S_{21}}{2S_{21}}$$

$$B = \frac{(1 + S_{11})(1 + S_{22}) - S_{12}S_{21}}{2S_{21}}$$

$$C = \frac{(1 - S_{11})(1 - S_{22}) - S_{12}S_{21}}{2S_{21}}$$

$$D = \frac{(1 - S_{11})(1 + S_{22}) + S_{12}S_{21}}{2S_{21}}$$

Note: Normalized immittance parameters are obtained by setting $Z_0 = 1$ in the above.

A.3 TWO-PORT IMMITTANCE PARAMETER TO *S*-PARAMETER CONVERSIONS

[Z]

$$S_{11} = \frac{(Z_{11} - Z_0)(Z_{22} + Z_0) - Z_{12}Z_{21}}{(Z_{11} + Z_0)(Z_{22} + Z_0) - Z_{12}Z_{21}}$$

$$S_{12} = \frac{2Z_{12}Z_0}{(Z_{11} + Z_0)(Z_{22} + Z_0) - Z_{12}Z_{21}}$$

$$S_{21} = \frac{2Z_{21}Z_0}{(Z_{11} + Z_0)(Z_{22} + Z_0) - Z_{12}Z_{21}}$$

$$S_{22} = \frac{(Z_{11} + Z_0)(Z_{22} - Z_0) - Z_{12}Z_{21}}{(Z_{11} + Z_0)(Z_{22} + Z_0) - Z_{12}Z_{21}}$$

[Y]

$$S_{11} = \frac{(1 - Z_0Y_{11})(1 + Z_0Y_{22}) + Y_{12}Y_{21}Z_0^2}{(1 + Z_0Y_{11})(1 + Z_0Y_{22}) + Y_{12}Y_{21}Z_0^2}$$

$$S_{12} = \frac{-2Y_{12}Z_0}{(1 + Z_0Y_{11})(1 + Z_0Y_{22}) + Y_{12}Y_{21}Z_0^2}$$

$$S_{21} = \frac{-2Y_{21}Z_0}{(1 + Z_0Y_{11})(1 + Z_0Y_{22}) + Y_{12}Y_{21}Z_0^2}$$

$$S_{22} = \frac{(1 + Z_0Y_{11})(1 - Z_0Y_{22}) + Y_{12}Y_{21}Z_0^2}{(1 + Z_0Y_{11})(1 + Z_0Y_{22}) + Y_{12}Y_{21}Z_0^2}$$

[h]

$$S_{11} = \frac{(h_{11}/Z_0 - 1)(h_{22}Z_0 + 1) - h_{12}h_{21}}{(h_{11}/Z_0 + 1)(h_{22}Z_0 + 1) - h_{12}h_{21}}$$

$$S_{12} = \frac{2h_{12}}{(h_{11}/Z_0 + 1)(h_{22}Z_0 + 1) - h_{12}h_{21}}$$

$$S_{21} = \frac{-2h_{21}}{(h_{11}/Z_0 + 1)(h_{22}Z_0 + 1) - h_{12}h_{21}}$$

$$S_{22} = \frac{(h_{11}/Z_0 + 1)(1 - h_{22}Z_0) + h_{12}h_{21}}{(h_{11}/Z_0 + 1)(h_{22}Z_0 + 1) - h_{12}h_{21}}$$

[ABCD]

$$S_{11} = \frac{A + B/Z_0 + CZ_0 - D}{A + B/Z_0 + CZ_0 + D}$$

$$S_{12} = \frac{2(AD - BC)}{A + B/Z_0 + CZ_0 + D}$$

$$S_{21} = \frac{2}{A + B/Z_0 + CZ_0 + D}$$

$$S_{22} = \frac{-A + B/Z_0 + CZ_0 + D}{A + B/Z_0 + +CZ_0 + D}$$

Note: When using normalized immittance parameters set $Z_0 = 1$ in the above.

A.4 TWO-PORT *T*-PARAMETER AND *S*-PARAMETER CONVERSIONS

[S] to [T]	$T_{11} = \left(\dfrac{S_{12}S_{21} - S_{11}S_{22}}{S_{21}} \right)$	$T_{12} = \dfrac{S_{11}}{S_{21}}$
	$T_{21} = - \left(\dfrac{S_{22}}{S_{21}} \right)$	$T_{22} = \dfrac{1}{S_{21}}$
[T] to [S]	$S_{11} = \dfrac{T_{12}}{T_{22}}$	$S_{12} = \left(\dfrac{T_{11}T_{22} - T_{12}T_{21}}{T_{22}} \right)$
	$S_{21} = \dfrac{1}{T_{22}}$	$S_{22} = - \left(\dfrac{T_{21}}{T_{22}} \right)$

Physical constants

Quantity	Symbol	Value	Units
Speed of light in a vacuum	c	2.99792×10^{8}	meter/second
Permittivity of free space	ϵ_0	8.85418×10^{-12}	farad/meter
Permeability of free space	μ_0	$4\pi \times 10^{-7}$	henry/meter
Boltzmann's constant	k_B	1.38066×10^{-23}	joule/kelvin
Electron charge	q	1.60218×10^{-19}	coulomb
Free electron mass	m_0	9.1×10^{-31}	kg
Impedance of free space	η_0	376.7	ohm

Forbidden regions for *L*-sections

C.1 FORBIDDEN REGIONS FOR *LC L*-SECTIONS

The type 1 *LC L*-section has the shunt element across the load and the series elements on the source side, as shown in Figure C.1.

- *Type 1a* can only match capacitive loads lying outside unit admittance circle or inductive loads lying inside unit resistance circle.
- *Type 1b* can only match inductive loads lying outside unit admittance circle or capacitive loads lying inside unit resistance circle.

The type 2 *LC L*-section has the shunt element across the source and the series elements on the load side, as shown in Figure C.2.

- *Type 2a* can only match capacitive loads lying outside unit resistance circle or inductive loads lying inside unit admittance circle.
- *Type 2b* can only match inductive loads lying outside unit resistance circle or capacitive loads lying inside unit admittance circle.

C.2 FORBIDDEN REGIONS FOR *LL* AND *CC L*-SECTIONS

The type 1 *LL* or *CC L*-section has the shunt element across the load and the series elements on the source side, as shown in Figure C.3.

- *Type 1c (LL)* can only match capacitive loads lying outside both unit circles.
- *Type 1d (CC)* can only match inductive loads lying outside both unit circles.

The type 2 *LL* or *CC L*-section has the shunt element across the source and the series elements on the load side, as shown in Figure C.4.

- *Type 2c* can only match capacitive loads lying outside both unit circles.
- *Type 2d* can only match inductive loads lying outside both unit circles.

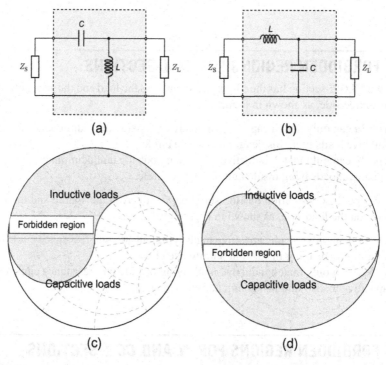

FIGURE C.1

L-section matching: forbidden region for type 1 (*LC*). (a) *L*-section type 1a (*LC*).
(b) *L*-section type 1b (*LC*). (c) Forbidden region for *LC* type 1a. (d) Forbidden region
for *LC* type 1b.

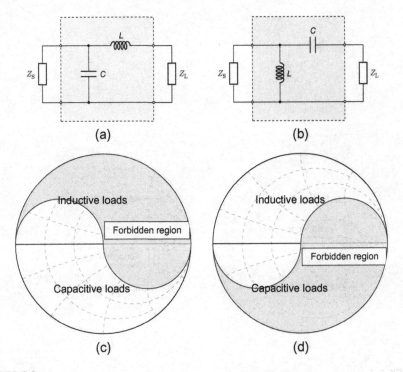

FIGURE C.2

L-section matching: forbidden region for type 2 (*LC*). (a) *L*-section type 2a (*LC*).
(b) *L*-section type 2b (*LC*). (c) Forbidden region for *LC* type 2a. (d) Forbidden region
for *LC* type 2b.

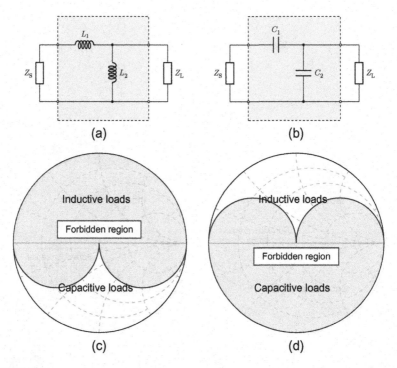

(a)   (b)

(c)   (d)

FIGURE C.3

L-section matching: forbidden region for type 1 (*LL* or *CC*). (a) *L*-section type 1c (*LL*).
(b) *L*-section type 1d (*CC*). (c) Forbidden region for *LL* type 1c. (d) Forbidden region for
CC type 1d.

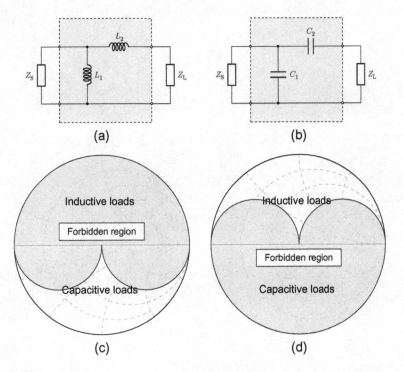

FIGURE C.4

L-section matching: forbidden region for type 2 (*LL/CC*). (a) *L*-section type 2c (*LL*).
(b) *L*-section type 2d (*CC*). (c) Forbidden region for *LL* type 2c. (d) Forbidden region for
CC type 2d.

Index

Note: Page numbers followed by *f* indicate figures and *t* indicate tables.

Printed in the United States
by Baker & Taylor Publisher Services